Transparent Electronics: From Synthesis to Applications

Transparent Electronics

From Synthesis to Applications

Edited by

ANTONIO FACCHETTI

Department of Chemistry and the Materials Research Center, Northwestern University, Evanston, IL, USA

TOBIN J. MARKS

Department of Chemistry, Northwestern University, Evanston, IL, USA

WILEY

A John Wiley and Sons, Ltd, Publication

This edition first published 2010

Registered office
John Wiley & Sons Ltd, The Atrium, Southern Gate, Chichester, West Sussex, PO19 8SQ, United Kingdom

For details of our global editorial offices, for customer services and for information about how to apply for permission to reuse the copyright material in this book please see our website at www.wiley.com.

Library of Congress Cataloging-in-Publication Data

Facchetti, Antonio.
Transparent electronics : from synthesis to applications / Antonio Facchetti, Tobin J. Marks.
p. cm.
Includes bibliographical references and index.
ISBN 978-0-470-99077-3 (cloth)
1. Transparent electronics. 2. Transparent semiconductors. I. Marks, Tobin J. II. Title.
TK7835.F25 2010
621.381–dc22 2009049248

A catalogue record for this book is available from the British Library.

ISBN: 978-0-470-99077-3 (H/B)

Set in 10/12pt Times Roman by Thomson Digital, Noida, India
Printed and bound in Great Britain by CPI Antony Rowe, Chippenham, Wiltshire

AF would like to dedicate this book to Rinaldo and Celestina, his parents.

TJM would like to dedicate this book to Miriam and Indrani
for all their patience and encouragement.

Contents

Preface

Transparent electronics is an emerging science and technology field focused on producing 'invisible' electronic circuitry and opto-electronic devices. Applications include consumer electronics, new energy sources, and transportation; for example, automobile windshields could transmit visual information to the driver. Glass in almost any setting could also double as an electronic device, possibly improving security systems or offering transparent displays. In a similar vein, windows could be used to produce electrical power. Other civilian and military applications in this research field include real-time wearable displays. As for conventional Si/III–V-based electronics, the basic device structure is based on semiconductor junctions and transistors. However, the device building block materials, the semiconductor, the electric contacts, and the dielectric/passivation layers, must now be transparent in the visible –a true challenge! Therefore, the first scientific goal of this technology must be to discover, understand, and implement transparent high-performance electronic materials. The second goal is their implementation and evaluation in transistor and circuit structures. The third goal relates to achieving application-specific properties since transistor performance and materials property requirements vary, depending on the final product device specifications. Consequently, to enable this revolutionary technology requires bringing together expertise from various pure and applied sciences, including materials science, chemistry, physics, electrical/electronic/circuit engineering, and display science.

During the past 10 years, the classes of materials available for transparent electronics applications have grown dramatically. Historically, this area was dominated by transparent conducting oxides (oxide materials that are both electrically conductive and optically transparent) because of their wide use in antistatic coatings, touch display panels, solar cells, flat panel displays, heaters, defrosters, 'smart windows' and optical coatings. All these applications use transparent conductive oxides as *passive* electrical or optical coatings. The field of transparent conducting oxide (TCO) materials has been reviewed and many treatises on the topic are available. However, more recently there have been tremendous efforts to develop new *active* materials for functional transparent electronics. These new technologies will require new materials sets, in addition to the TCO component, including conducting, dielectric and semiconducting materials, as well as passive components for full device fabrication.

In this book we intend to describe fundamental scientific information and recent breakthroughs concerning both the basic science and real-world applications of transparent electronic materials, circuits and devices. We bring together renowned experts from both academia and industry working in this field from all around the world, including the USA, Germany, Japan, Sweden, Italy and Portugal. This book is structured to strike a balance between introductory and advanced topics, fundamental scientific versus technological/application issues, and materials versus device structure/applications. Considering the fundamental device structures and diverse possible application fields, the first section of this book is devoted to fundamental materials issues and properties. The second section deals with transparent electronic devices, including thin-film transistors, photovoltaic cells, electronic circuits, displays, sensors, solar cells, and electro-optic devices. We hope that this book will attract the attention of young scientists, as well as more senior industrial and academic researchers interested in electronic materials and devices. We also believe that this book will provide stimulating ideas for curious chemists, physicists, materials scientists, and electrical engineers seeking new opportunities in this exciting area.

We conclude by thanking all the authors for contributing their very hard work, expertise, and insightful suggestions. This work would not have been possible without their knowledge, dedication, and enthusiasm. Furthermore, we express our gratitude to Alexandra Carrick and Richard Davies at John Wiley & Sons, Ltd for their help and guidance through the editorial process.

Antonio Facchetti and Tobin J. Marks

List of Contributors

Camilla Baratto University of Brescia and CNR-INFM SENSOR Laboratory, Brescia, Italy

Teresa M. Barnes National Renewable Energy Laboratory, Golden, CO, USA

Pedro Barquinha CENIMAT/I3N, Materials Science Department, Faculdade de Ciências e Tecnologia, Universidade Nova de Lisboa, Caparica, Portugal

Jeffrey L. Blackburn National Renewable Energy Laboratory, Golden, CO, USA

Clark I. Bright Condor Group Technical Leader, 3M Corporate Research Process Laboratory, Tucson, AZ, USA

Peter F. Carcia DuPont CR&D Experimnetal Station, Wilmington, DE, USA

Li-Min Chen Department of Materials Science and Engineering, UCLA, Los Angeles, CA, USA

Elisabetta Comini University of Brescia and CNR-INFM SENSOR Laboratory, Brescia, Italy

Antonio Facchetti Department of Chemistry and the Materials Research Center, Northwestern University, Evanston, IL, USA

Guido Faglia University of Brescia and CNR-INFM SENSOR Laboratory, Brescia, Italy

Matteo Ferroni University of Brescia and CNR-INFM SENSOR Laboratory, Brescia, Italy

Elvira Fortunato CENIMAT/I3N, Materials Science Department, Faculdade de Ciências e Tecnologia, Universidade Nova de Lisboa, Caparica, Portugal

Gonçalo Gonçalves CENIMAT/I3N, Materials Science Department, Faculdade de Ciências e Tecnologia, Universidade Nova de Lisboa, Caparica, Portugal

Claes G. Granqvist Department of Engineering Sciences, The Angstrom Laboratory, Uppsala University, Sweden

Sang Ho MCC Electrical Engineering and Computer Science, Northwestern University, Evanston, IL, USA

Seng-Tiong Ho Department of Electrical Engineering and Computer Science, Northwestern University, Evanston, IL, USA

Hideo Hosono Frontier Research Center & Materials and Structures Laboratory, Tokyo Institute of Technology, Nagatsuta, Japan

Jinsong Huang Department of Materials Science and Engineering, UCLA, Los Angeles, CA, USA

Manabu Ito Technical Research Institute, Toppan Printing Co., Ltd., Sugito, Japan

Howard E. Katz Department of Materials Science and Engineering, Johns Hopkins University, Baltimore, MD, USA

Douglas A. Keszler Department of Chemistry, Oregon State University, Corvallis, OR, USA

Gang Li Department of Materials Science and Engineering, UCLA, Los Angeles, CA, USA

Juo-Hao Li Department of Materials Science and Engineering, UCLA, Los Angeles, CA, USA

Tobin J. Marks Department of Chemistry and the Materials Research Center, Northwestern University, Evanston, IL, USA

Rodrigo Martins CENIMAT/I3N, Materials Science Department, Faculdade de Ciências e Tecnologia, Universidade Nova de Lisboa, Caparica, Portugal

Julia E. Medvedeva Department of Physics, Missouri University of Science and Technology, Rolla, MO, USA

Rocio Ponce Ortiz Department of Chemistry and the Materials Research Center, Northwestern University, Evanston, IL, USA

Luis Pereira CENIMAT/I3N, Materials Science Department, Faculdade de Ciências e Tecnologia, Universidade Nova de Lisboa, Caparica, Portugal

Dale L. Perry Lawrence Berkeley National Laboratory, Berkeley, CA, USA

Andrea Ponzoni University of Brescia and CNR-INFM SENSOR Laboratory, Brescia, Italy

Thomas Riedl Advanced Semiconductors Group, Institut für Hochfrequenztechnik, Technische Universität Braunschweig, Germany

Giorgio Sberveglieri University of Brescia and CNR-INFM SENSOR Laboratory, Brescia, Italy

Kevin Cua See Department of Materials Science and Engineering, Johns Hopkins University, Baltimore, MD, USA

Vivek Subramanian EECS, University of California, Berkeley, CA, USA

Janet Tate Department of Physics, Oregon State University, Corvallis, OR , USA

Alberto Vomiero University of Brescia and CNR-INFM SENSOR Laboratory, Brescia, Italy

Yang Yang Department of Materials Science and Engineering, UCLA, Los Angeles, CA, USA

Fei Yi Department of Electrical Engineering and Computer Science, Northwestern University, Evanston, IL, USA

1

Combining Optical Transparency with Electrical Conductivity: Challenges and Prospects

Julia E. Medvedeva

Department of Physics, Missouri University of Science and Technology, USA

1.1 Introduction

Transparent conductors are neither 100% optically transparent nor metallically conductive. From the band structure point of view, the combination of the two properties in the same material is contradictory: a transparent material is an insulator which possesses completely filled valence and empty conduction bands; whereas metallic conductivity appears when the Fermi level lies within a band with a large density of states to provide high carrier concentration.

Efficient transparent conductors find their niche in a compromise between a sufficient transmission within the visible spectral range and a moderate but useful in practice electrical conductivity [1–6]. This combination is achieved in several commonly used oxides – In_2O_3, SnO_2, ZnO and CdO. In the undoped stoichiometric state, these materials are insulators with optical band gap of about 3 eV. To become a transparent conducting oxide (TCO), these TCO hosts must be degenerately doped to displace the Fermi level up into the conduction band.

The key attribute of any conventional n-type TCO host is a highly dispersed single free-electron-like conduction band [7–13] (Figure 1.1). Degenerate doping then provides both (i) the high mobility of extra carriers (electrons) due to their small effective mass and (ii) low optical absorption due to the low density of states in the conduction band. The high energy dispersion of the conduction band also ensures a pronounced Fermi energy

Transparent Electronics: From Synthesis to Applications Edited by Antonio Facchetti and Tobin J. Marks

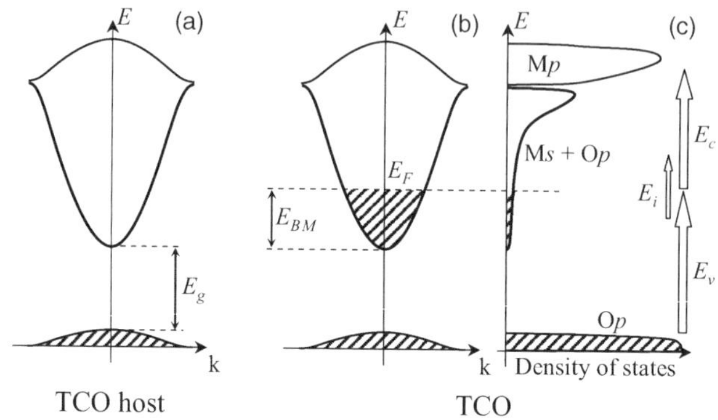

Figure 1.1 (a) Schematic electronic band structure of a TCO host – an insulator with a band gap E_g and a dispersed parabolic conduction band which originates from interactions between metal s and oxygen p states. (b) and (c) Schematic band structure and density of states of a TCO, where a degenerate doping displaces the Fermi level (E_F) via a Burstein-Moss shift, E_{BM}, making the system conducting. The shift gives rise to inter-band optical transitions from the valence band, E_v, and from the partially filled conduction band up into the next empty band, E_c, as well as to intra-band transitions within the conduction band, E_i

displacement up above the conduction band minimum, the Burstein–Moss (BM) shift [14, 15]. The shift helps to broaden the optical transparency window and to keep the intense optical transitions from the valence band out of the visible range. This is critical in oxides which are not transparent throughout the entire visible spectrum, for example, in CdO where the optical (direct) band gap is 2.3 eV.

Achieving the optimal performance in a TCO is a challenging because of the complex interplay between the electronic and optical properties [16, 17]. The large carrier concentrations desired for a good conductivity may result in an increase of the optical absorption [18] (i) at short wavelengths, due to inter-band transitions from the partially filled conduction band and (ii) at long wavelengths, due to intra-band transitions within this band. In addition, plasma oscillations may affect the optical properties by reflecting the electromagnetic waves of frequency below that of the plasmon. Furthermore, ionized impurity scattering on the electron donors (native point defects or substitutional dopants) have a detrimental effect on the charge transport, while the structural relaxation around the impurities may alter the electronic and optical properties of the host, leading to a nonrigid-band shift of the Fermi level.

This chapter is devoted to *ab initio* electronic band structure investigations of common TCOs and related oxide materials. We demonstrate here that a thorough understanding of the microscopic properties of metal oxides provides an insight into the underlying phenomena and also suggests that the range of efficient TCO materials can be significantly broadened.

1.2 Electronic Properties of Conventional TCO Hosts

Conventional n-type TCO hosts (In_2O_3, SnO_2, CdO and ZnO) share similar chemical, structural and electronic properties. Exclusively oxides of the post-transition metals with

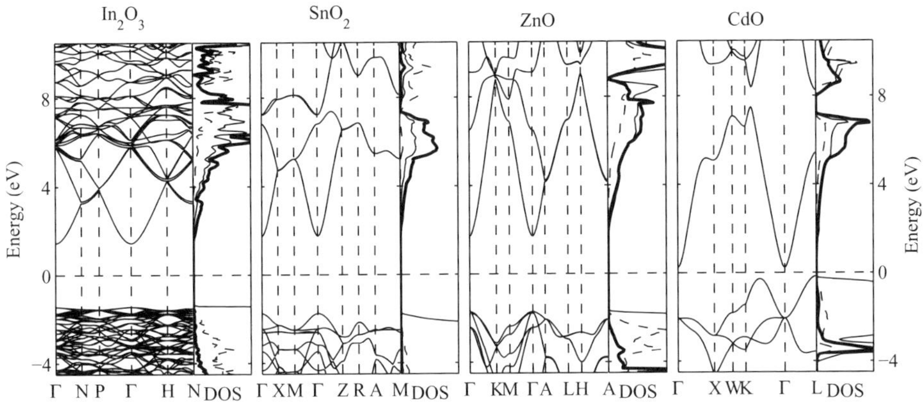

Figure 1.2 *Electronic band structure and partial density of states of TCO hosts, In_2O_3, SnO_2, ZnO and CdO, as obtained within the screened-exchange local-density approximation [19]. In the density of states plots, the thick, dashed and thin lines represent metal s, metal p and oxygen p states, respectively. The plots should be compared with the schematic band structure shown in Figure 1.1(a)*

$(n-1)d^{10}ns^2$ electronic configurations, they have densely packed structures with four- or six-coordinate metal ions. Strong interactions between the oxygen 2*p* and metal *ns* orbitals give rise to electronic band structures qualitatively similar for all these oxides (cf. Figures 1.1 and 1.2): the bonding and nonbonding O 2*p* states form the valence band while the conduction band arises from the antibonding M*s*–O*p* interactions. The empty *p* states of the metal ion form the following band at a higher energy. The partial density of states plots (Figure 1.2), reveal that the oxygen 2*p* and metal *ns* states make similar contributions to the conduction band. This provides a three-dimensional M*s*–O*p* network for charge transport once extra carriers fill the band.

M*s*–O*p* interactions result in a gap between the valence and the conduction bands. In ZnO, the gap is direct whereas in CdO, In_2O_3 or SnO_2 the valence band maximum is at the L point ([111]), H point ($[1\bar{1}1]$) or R point ([011]), respectively, giving rise to an indirect band gap of 0.4 eV, 2.6 eV or 2.7 eV, respectively. Table 1.1 lists the direct optical

Table 1.1 *Basic properties of conventional TCO hosts. The optical band gaps and the electron effective masses are determined within screened-exchange local-density approximation (sX-LDA) [19]. Anisotropy of the electron effective mass is defined as $\delta = (m^{[100]} + m^{[010]})/2m^{[001]}$*

Oxide	Lattice	Coordination of		Optical (direct) band gap (eV)	Electron effective mass, m_e			Effective mass anisotropy
		cation	anion		$m^{[100]}$	$m^{[010]}$	$m^{[001]}$	
ZnO	Wurtzite	4	4	3.41	0.35	0.35	0.35	1.008
CdO	Rocksalt	6	6	2.28	0.23	0.23	0.23	1.000
In_2O_3	Byxbyite	6	4	3.38	0.28	0.28	0.28	1.000
SnO_2	Rutile	6	3	3.50	0.33	0.33	0.28	1.179

band gaps which are of primary importance for TCO applications. These values are obtained from the electronic band structure calculations within screened-exchange local density approximation (sX-LDA) [19], which gives good agreement with the reported experimental values (3.5–3.7 eV for In_2O_3, 2.3 eV for CdO, 3.1–3.6 eV for ZnO and 3.6–4.0 eV for SnO_2) [25–29].

The M*s*–O*p* overlap also determines the energy dispersion of the conduction band in these materials. Within the framework of **k·p** theory [30], the electron effective mass can be found within the second-order perturbation:

$$\frac{m_e}{m_{ii}^{(c)}} = 1 + \frac{2}{m_e} \sum_{l \neq c} \frac{|\langle u^{(c)} | \hat{p}_i | u^{(l)} \rangle|^2}{E^{(c)} - E^{(l)}}, \tag{1.1}$$

where p is the momentum operator, $|u^{(l)}\rangle$ is the Bloch wave function of the l's band at the Γ point (wave vector $\mathbf{k} = 0$) and $E^{(l)}$ is its energy. Band label c represents the conduction band, while the sum runs over all other bands. In the oxides under consideration here, the electron effective mass is less than the mass of the electron, m_e. As it follows from Equation (1.1), it is determined primarily by the valence band contributions ($E^{(l)} < E^{(c)}$), i.e. by the oxygen 2p states.

From the orbital symmetry considerations (Figure 1.3) coordination of cations *by the oxygen atoms* have little effect on the M*s*–O*p* overlap owing to the spherical symmetry of the *s* orbitals. The largest M*s*–O*p* overlap is attained when the oxygen atom is coordinated octahedrally *by the cations*, i.e. when each of the oxygen p_x, p_y and p_z orbitals connects two *s* orbitals (Figure 1.3). Accordingly, the octahedral coordination of the oxygen atoms in rocksalt CdO gives rise to the largest dispersion and, hence, the smallest electron effective mass among the TCO materials (Table 1.1). However, it was found [31] that variations in the oxygen coordination and strong distortions in the polyhedra have little effect on the electron effective mass which varies insignificantly when the symmetry of the same-cation oxide

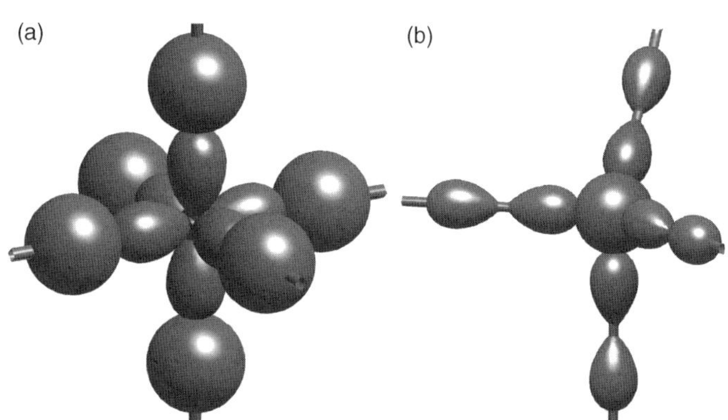

Figure 1.3 *Octahedral coordination of oxygen atoms by cations (a) provides the largest overlap between the oxygen p_x, p_y and p_z orbitals and the s orbitals of the metal ions. Coordination of cations by oxygen atoms as well as local distortions (b) have little effect on the Ms–Op overlap owing to the spherical symmetry of the metal s orbitals*

is changed. For example, for ZnO in rocksalt (octahedral coordination) or wurtzite (tetrahedral coordination) structures, and for In_2O_3 in $Ia\bar{3}$ (byxbyite), $R\bar{3}c$ (corundum) or $I2_13$ structures, the effective masses vary by about 15%. Moreover, the effective mass remains nearly isotropic in all phases of the oxides – including those with irregular atomic arrangements or large structural voids [31, 32].

Little sensitivity of the M*s*–O*p* overlap and, hence, of the electron effective mass to structural variations may explain the success of amorphous TCOs whose optical and electrical properties remain similar to those in the crystalline state [6, 33–37]. This is in marked contrast to, for example, amorphous Si where the directional interactions between the conduction *p* orbitals lead to strong anisotropy of the transport properties which are sensitive to the orbital overlap and, hence, to the distortions in the atomic chains [36].

Thus, the network of alternating metal and oxygen atoms ensures the small electron effective mass in the TCO hosts. A direct overlap between metal *s* orbitals is not possible in these materials except for SnO_2 where Sn atoms may bond along the edge-sharing rutile chain (along the [001] crystallographic direction). However, the fact that the calculated [11, 31] (Table 1.1) and the observed [38] electron effective mass in this oxide is nearly isotropic suggests that the *s*–*s* interactions do not govern the transport properties of TCOs.

In the next section, where we will consider the conversion of the TCO hosts from insulators to conductors, the M*s*–O*p* origin of the conduction band will play a critical role.

1.3 Carrier Generation in Conventional TCO Hosts

The optical and transport properties of a conventional TCO are governed by the efficiency and the specifics of the carrier generation mechanism employed. Even in the most favorable situation, i.e. when the effects of dopant solubility, clustering, secondary phase formation and charge compensation can be avoided, large concentrations of electron donors (substitutional dopants and/or native point defects) not only promote the charge scattering but also may significantly alter the electronic band structure of the host oxide, leading to a nonrigid-band shift of the Fermi level. A detailed band structure analysis of the doped oxides helps to elucidate the role of different factors involved.

1.3.1 Substitutional Doping

Substitutional doping with aliovalent ions is the most widely used approach to generate free carriers in TCO hosts. Compared with native point defects, it allows a better control over the resulting optical and transport properties as well as better environmental stability of the TCO films. Traditionally, same-period, next-row elements, e.g, Sn^{4+} for In^{3+} and In^{3+} for Cd^{2+}, are thought to provide better compatibility and, thus, less disturbance in the host crystal and electronic structure. However, other dopants may prove beneficial for optimizing the properties for a specific application. For example, transparent conducting ZnO films have been prepared by doping with Group III (Al, Ga, In and B), Group IV (Si, Ge, Ti, Zr and Hf) and a Group VII element (F substituted at an oxygen site), giving rise to a wide range of electrical conductivities [39].

Here we will give a detailed consideration to rocksalt CdO, where the high crystal symmetry and the densely packed structure ensures the most uniform charge density distribution via the isotropic M*s*–O*p* network. Compared with more complex In_2O_3 or SnO_2, one can expect fewer ionized and neutral scattering centers and, hence, longer relaxation times. At the same time, introduction of dopants into the densely packed structure may significantly influence the Cd*s*–O2*p* hybridization and, therefore, alter the structural, electronic and optical properties of the host. A systematic comparison of CdO doped with In, Ga, Sc or Y, whose ionic radius and electronic configuration differ from those of the host cation, has revealed [40–42] that:

(i) Substitutional dopants with smaller ionic radii compared with that of Cd shrink the lattice. The shrinkage, however, is not as large as expected from the Vegard's law weighted average [43] of the six-coordinated X^{3+} and Cd^{2+} ionic radii. Moreover, in the case of X = In or Y, the lattice parameter is similar or even slightly greater than that of CdO (cf. Table 1.2). One of the possible explanations is that the doping-induced shrinkage is compensated by an expansion mechanism which originates from the antibonding character of the conduction band formed from Cd 5*s* and O 2*p* states [44, 45]. The antibonding mechanism is dominant in In or Y doped CdO, while Sc or Ga have sufficiently smaller ionic radii to weaken M*s*–O*p* hybridization and, thus, to compress the lattice.
(ii) Weaker Cd5*s*–O2*p* hybridization associated with strong structural relaxation around dopant with a smaller ionic radius results in a smaller optical band gap (cf. Table 1.2). Doping with Ga whose ionic radius is significantly smaller than that of Cd, reduces the optical band gap (to 2.53 eV) so that it becomes smaller than the one in undoped CdO (2.82 eV) – despite the doping-introduced BM shift of 2.3 eV. The smallest optical

Table 1.2 *Properties of 12.5 at% doped CdO as obtained from electronic band structure calculations within the screened-exchange local-density approximation [40–42]. Dopant X^{3+}, where X = Y, In, Sc or Ga, substitutes Cd^{2+} (ionic radius 1.09 Å), while F^- substitutes O^{2-} atoms. The electron velocity is calculated in the [100](Δ), [110](Σ) and [111](Λ) directions. Values for CdO with one extra electron (CdO + e^-) are found from a rigid band shift. Reprinted with permission from [41]. Copyright 2005 American Chemical Society*

	e^-	Y	In	Sc	Ga	F
Dopant ionic radius (Å)	–	1.04	0.94	0.89	0.76	1.19
Lattice parameter (Å)	4.66	4.67	4.66	4.63	4.62	4.65
Distance X-O or Cd-F (Å)	–	2.28	2.24	2.18	2.08	2.38
Distance Cd-O (Å)	2.33	2.39	2.42	2.45	2.54	2.27
Optical band gap (transitions E_v) (eV)	4.56	3.38	3.03	3.02	2.53	2.73
Optical transitions E_c (eV)	8.30	0.70	0.51	0.83	0.94	0.73
Fundamental band gap (eV)	2.82	2.99	2.54	3.13	2.42	2.64
Hybridization gap (eV)	0.95	–	–	0.55	0.65	–
Band width (eV)	7.29	3.36	3.91	2.57	3.01	4.17
Electron velocity, v^{Δ} ($\times 10^5$ m s^{-1})	10.39	9.12	10.54	4.65	8.25	9.45
v^{Σ} ($\times 10^5$ m s^{-1})	10.43	9.61	10.29	7.66	7.46	10.24
v^{Λ} ($\times 10^5$ m s^{-1})	9.43	9.17	9.23	7.95	2.94	9.36
Density of states at the Fermi level	0.96	1.34	1.16	2.00	1.74	1.21

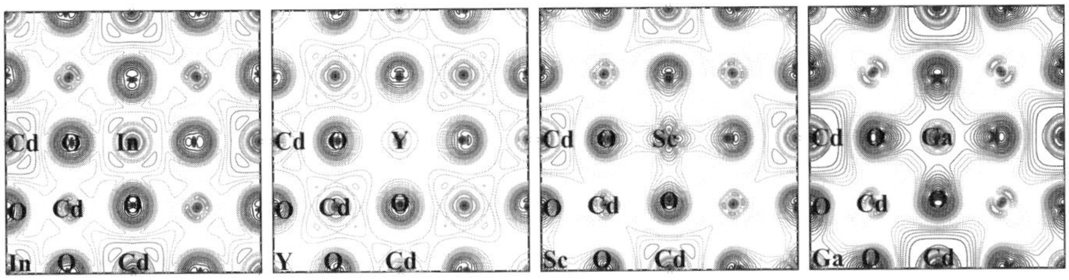

Figure 1.4 *Contour plots of the charge density distribution in In, Y, Sc and Ga-doped CdO illustrate considerable electron localization around Sc and Ga ions as compared with In and Y cases where the charge density is more uniform. The plots are calculated in the xy plane within the 2kT energy window near the Fermi level. The grey scale increases with charge; the same scale is used for all plots. Atoms within one unit cell are labeled. Reprinted with permission from [41]. Copyright 2005 American Chemical Society*

band gap in Ga-doped CdO as compared with In, Y and Sc cases was observed experimentally [40–42].

(iii) In and Y dopants preserve the uniform charge density distribution while Sc and Ga lead to significant electron localization around the dopant (Figure 1.4). The difference originates from the mismatch of the electronic configuration of the dopants and the energy location of the dopant empty *p* or *d* states with respect to the Fermi level. The Sc 3*d* states and Ga 4*p* states are energetically compatible with the conduction 5*s* states of Cd, while the Y 4*d* and In 5*p* are located higher in energy. As a result, the contributions from the Sc *d* or Ga *p* orbitals become significant near the Fermi level: the Sc *d* orbital contribution is dominant (85% of the Sc total) and the Ga *p* and *s* orbitals give comparable contributions (60% and 40%, respectively). The anisotropic Sc *d* or Ga *p* orbitals form strong directional bonds with the orbitals of the nearest oxygen atoms resulting in significant charge localization which is clearly seen from the charge density distribution plots (Figure 1.4).

(iv) The electron localization in Sc and Ga doped CdO results in a narrower conduction band and, hence, a reduction of the electron velocity as compared with In or Y (Table 1.2). Moreover, due to the high anisotropy of the Sc *d* or Ga *p* orbitals, a significantly reduced velocity is found in the Δ (Sc *d* orbitals) or Λ (Ga *p* orbitals) directions so that anisotropic transport properties are expected.

(v) The electron binding in Sc and Ga-doped CdO also leads to larger (in energy) optical transitions from the Fermi level (E_c in Figure 1.1), in contrast to the In and Y cases where the charge delocalization deminishes the second (hybridization) gap.

(vi) Finally, we note that even in the In, Y and F cases where the dopant ionic radius and electronic configuration are similar to that of Cd or O, the optical properties are worse than expected from the rigid band shift ($CdO+e^-$) (Table 1.2). However, the calculated electron velocity and the density of states for In, Y and F-doped CdO are similar to those obtained from the rigid-band model (Table 1.2). Both factors contribute to the conductivity σ, given by the expression:

$$\sigma = \frac{2e^2}{\Omega}\sum_{\mathbf{k}\lambda}|v_{\mathbf{k}\lambda}|^2\tau_{\mathbf{k}\lambda}\delta(E_{\mathbf{k}\lambda}-E_F), \tag{1.2}$$

so that the relaxation time τ will play the dominat role in determining the final carrier transport. [In Equation (1.2) e is the electron charge, Ω is the volume of the Brillouin zone, **k** is the wave vector, λ is the band index, v is the electron group velocity and E_F is the Fermi energy.] Assuming that τ is similar for all X^{3+}-doped systems, estimates of the Fermi electron velocity and the density of states at the Fermi level result in the trend In > Y > Sc > Ga, which is in agreement with experimental observations of the conductivity [40–42].

1.3.2 Oxygen Reduction

Removal of an oxygen atom from a metal oxide leaves two extra electrons in the crystal. Whether one or both of these electrons become free carriers or remain localized at the vacancy site correlates with the oxide free energy of formation. In light metal oxides, such as CaO or Al_2O_3, where the formation energy is high, oxygen vacancies create deep charge-localized states within the electronic band gap known as color or F centers. A relatively low formation energy of the conventional TCOs [46] favors large oxygen deficiencies even under equilibrium growth conditions, giving rise to the free-carrier densities of 10^{17}–10^{19} cm^{-3} for In_2O_3 and ZnO [47–49].

Electronic band structure investigations of oxygen deficient oxides [49–51] showed that the oxygen defect $V_O^{\bullet\bullet}$ (in Kröger–Vink notation the superscript $^\bullet$ stands for effective positive charge) corresponds to a nonconducting state associated with the filling of the lowest single conduction band by the two vacancy-induced electrons. Only if the vacancy is excited, e.g. via a photoexcitation [49], or partially compensated to $V_O^{\bullet}$, does the single conduction band become half-occupied and conducting behavior may occur.

In oxygen deficient TCOs, the conduction band wave function resembles the one in the corresponding hosts [50, 52], i.e. it is derived from the M s and O p states (Figure 1.1). A relatively uniform charge density distribution suggests that the vacancy-induced electrons are delocalized [52]. However, a more thorough analysis of reduced In_2O_3 reveals [50] that the metal atoms nearest to the oxygen defect give about two times larger contributions than the rest of the In atoms in the cell. As a result, there is a notable build-up of the charge density near the vacancy site. Importantly, the In atoms nearest the vacancy exhibit a reduction of the s-orbital contribution: the relative orbital contributions from the In s, p and d states are 81%, 8% and 11%, respectivly, in contrast to 97% s-orbital contributions from other In atoms in the cell. The high anisotropy of the p and d orbitals favors stronger covalent (directional) bonds between the In atoms which surround the defect and their oxygen neighbors. These In–O pairs trap about 31% of the total charge density at the bottom of the conduction band. Similar behavior is found for other TCOs: in oxygen deficient CdO and ZnO, 18% and 39%, respectively, of the total charge density belong to the nearest (cation) and next nearest (oxygen) neighbors of the oxygen vacancy [50].

The presence of oxygen vacancies leads to significant changes in the electronic band structure of a TCO host. To illustrate the typical behavior, we compare the results obtained for oxygen deficient and Sn-doped In_2O_3 (cf. Table 1.3 and Figure 1.5):

(i) Strong structural relaxation around the vacancy reduces the distance between the In and O atoms nearest to the defect to 2.12 Å (on average). This leads to an increased In–O distances for the atoms located further from the defect and, hence, to a weaker

Table 1.3 *Properties of oxygen-deficient and Sn-doped In_2O_3 as obtained from electronic band structure calculations within local density approximation. Values for undoped stoichiometric In_2O_3 found from a rigid band shift are given for comparison. The electron concentration is 1.95×10^{21} cm^{-3} for all systems. The plasma frequency is calculated from Equation (1.3)*

	Optical transitions (eV)		Fundamental band gap (eV)	Plasma frequency (eV)	Electron velocity ($\times 10^5$ m s^{-1})			$N(E_F)$
	E_v	E_c			$v^{[001]}$	$v^{[111]}$	$v^{[1\bar{1}\bar{1}]}$	
$In_2O_3 + e^-$	3.01	0.54	1.16	2.35	9.42	9.45	8.60	1.51
$In_2O_3 + Sn^{\bullet}_{In}$	2.72	0.71	0.98	2.25	8.93	9.17	8.66	1.73
$In_2O_3 + V^{\bullet}_O$	2.07	1.11	0.71	1.32	5.58	6.42	4.81	3.36

In*s*–O*p* hybridization. As a result, the fundamental band gap and the optical transitions from the valence band (E_v) are significantly reduced in oxygen-deficient In_2O_3 as compared with Sn-doped oxide.

(ii) Owing to the stronger binding between the In and O atoms nearest to the defect, the lowest single conduction state occupied by the vacancy-induced electrons is split from the rest of the conduction band by a second gap. In marked contrast, the second gap is absent in the substitutionally doped oxide. This is a manifestation of a more uniform spatial charge density distribution, i.e. the charge delocalization. Note, the second gap previously reported for Sn-doped In_2O_3 [9] vanishes upon structural relaxation around Sn ions [cf. Figure 1.5(b)].

(iii) The increased charge density in the vicinity of the oxygen vacancy and the related narrowing of the conduction band give rise to the reduced electron velocity (Table 1.3). At the same time, the density of states near the Fermi level increases. Since both factors contribute to the conductivity [cf. Equation (1.2)], the difference in the charge

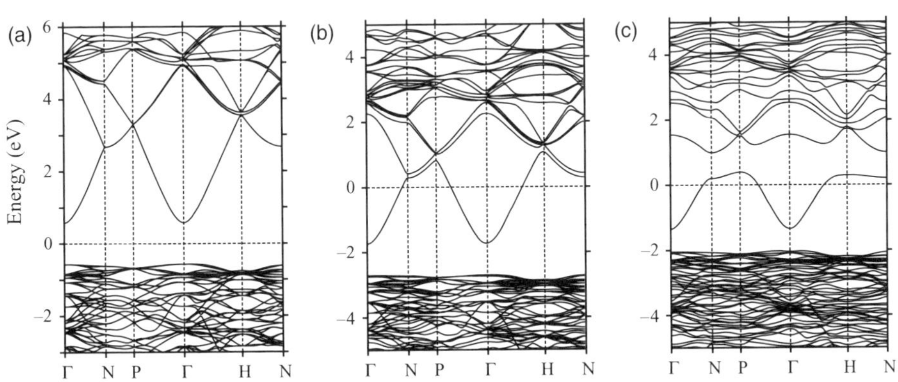

Figure 1.5 *Electronic band structure of (a) undoped stoichiometric In_2O_3. Reprinted with permission from [13]. Copyright (2007) Springer Science + Business Media (b) 6.25 at% Sn-doped In_2O_3 and (c) oxygen-deficient In_2O_3 as obtained within the local density approximation [19]. Reprinted with permission from [54]. Copyright (2006) American Physical Society*

transport of the oxygen-deficient and Sn-doped In_2O_3 will be determined primarily by the relaxation time τ in the same equation. Qualitatively, the stronger structural relaxation with the atomic displacements around the oxygen vacancy being twice as large as those around Sn ions [50], implies a stronger charge scattering in oxygen-deficient oxide. In addition, a shorter electron relaxation time in this case should be expected due to the Coulomb attraction of the free carriers to $V_O^{\bullet}$ associated with its higher formation energy compared with that of $V_O^{\bullet\bullet}$, which is the ground-state defect [49]. Moreover, due to the strong preference of the extra electrons to bind with $V_O^{\bullet}$ to form $V_O^{\bullet\bullet}$, the charge transport will be adversely affected since the latter defect corresponds to a nonconducting state [49, 50] (a completely filled single conduction band).

(iv) Due to the narrower conduction band in the oxygen-deficient oxide, the plasma frequency is expected to be significantly smaller than that in Sn-doped material. The plasma oscillations affect the optical properties: the electromagnetic waves of frequency below (and wavelength above) ω_p are reflected due to the electron screening. The plasma frequency is given by the expression:

$$\omega_p^2 = \frac{8\pi e^2}{3\Omega} \sum_{\mathbf{k}\lambda} |v_{\mathbf{k}\lambda}|^2 \delta(E_{\mathbf{k}\lambda} - E_F), \tag{1.3}$$

where e is the electron charge, Ω is the volume of the Brillouin zone, $\mathbf{k}$ is the wave vector, λ is the band index, v is the electron group velocity and E_F is the Fermi energy. Our estimates for ω_p in the oxygen-reduced and Sn-doped In_2O_3 as well as the one obtained from the rigid band model are given in Table 1.3.

In summary, compared with substitutional doping, oxygen reduction of a TCO host may result in higher carrier densities but would limit the electron mobility due to shorter relaxation times and considerable charge trapping near the vacancy site. Also, a weaker M*s*–O*p* hybridization due to stronger structural relaxation around the vacancy significantly reduces the optical transparency window.

There may be other native point defects that give rise to a conducting behavior in a TCO. For example, it was shown that interstitial Sn ions in SnO_2 have low formation energies and produce donor levels inside the conduction band of this material [53]. In this case, significant structural rearrangement associated with the formation of Sn(II)O bonds as in SnO [53] is expected to have an even stronger effect on the properties of the oxide host and to increase electron scattering.

The above considerations demonstrate the advantages of employing substitutional doping as a primary carrier generation mechanism in conventional TCO hosts. However, notwithstanding the above limitations, we believe that varying the degree of nonstoichiometry may serve as a versatile tool for optimizing a TCO's overall performance.

1.4 Magnetically Mediated TCO

Transition metal dopants opened up an avenue for alternative carrier generation in conventional TCO hosts [54]. Initially, carrier mobility with more than twice the value

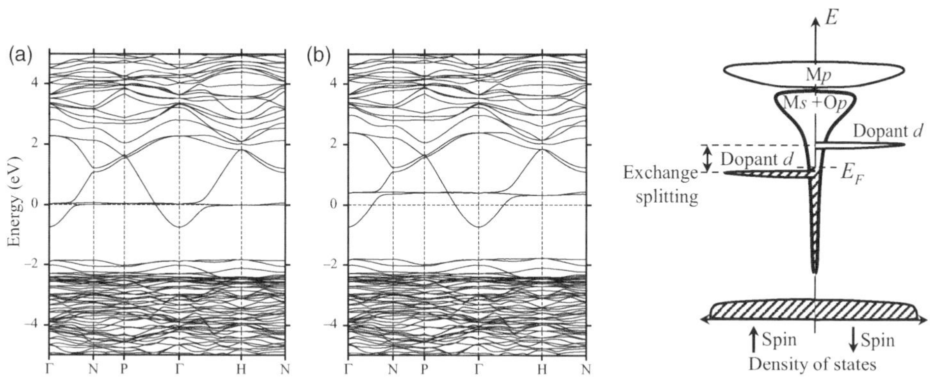

Figure 1.6 *Electronic band structure for (a) the majority and (b) the minority spin channels of 6.25% Mo-doped In_2O_3 compensated with an oxygen vacancy. In (c), the results of (a) and (b) are shown schematically. Reprinted with permission from [13, 54]. Copyright (2007) Springer Science + Business Media; Copyright (2006) by the American Physical Society*

of the commercial Sn-doped indium oxide (ITO) was observed in Mo-doped In_2O_3 (IMO), and the resulting enhanced conductivity appeared with no changes in the spectral transmittance upon doping with Mo [55–57]. Surprisingly, introduction of the transition metal Mo^{6+} which is expected to donate two more carriers per substitution compared with Sn^{4+}, does not lead to the expected increase of the optical absorption or a decrease of the mobility due to the scattering on the localized Mo *d* states.

Electronic band structure investigations of IMO revealed [54] that both high carrier mobility and low optical absorption originate from the Mo-induced magnetic interactions. Strong exchange interactions split the Mo *d* states so that the occupied *d* states with spin up lie just below the Fermi level while the empty spin down *d* states are well above it (Figure 1.6). The partial density of states suggests that in the majority spin channel about 58% of the total density near the Fermi level comes from the Mo *d* states while the rest is spread uniformly throughout the cell. In the minority spin channel, contributions from the Mo *d* states at the Fermi level are negligible and the conduction charge density distribution resembles that in In_2O_3. Therefore, the spin-up *d* states are resonant states, while the charge transport occurs through the In*s*–O*p* network. In other words, the free carriers in the system flow in a background of the Mo defects which serve as scattering centers. As a result of the exchange splitting of the Mo *d* states, the carriers of one spin are affected by only a half of the scattering centers, i.e. only by the Mo *d* states of the same spin. Therefore, the concentration of the Mo scattering centers is effectively lowered by half compared with the Mo doping level. The lack of long-range magnetic order leads to the formation of two interpenetrating networks transporting efficiently the carriers of opposite spin.

Significantly, the BM shift is less pronounced in the IMO case – despite the fact that Mo^{6+} donates two extra carriers as compared with Sn^{4+} at the same doping level. Such a low sensitivity to doping appears from the resonant Mo *d* states located at the Fermi level that facilitates the *d*-band filling (pinning) and thus hinders further displacement of the Fermi level deep into the conduction band. The smaller BM shift in IMO leads to the following advantageous features to be compared with those of ITO:

(i) Smaller increase in the electron effective mass with respect to the value in undoped stoichiometric In_2O_3 is expected upon Mo doping. This is borne out in experimental observations [56] showing that the effective mass does not vary with doping (up to 12% of Mo) and carrier concentration.
(ii) Larger (in energy) optical transitions from the partially occupied band (E_c in Figure 1.1) ensure lower short-wavelength optical absorption.
(iii) The calculated plasma frequency, ω_p, in IMO (1.3 for $[Mo_{In}^{\bullet\bullet\bullet}O''_i]^{\bullet}$ and 1.6 eV for $Mo_{In}^{\bullet\bullet\bullet}$ defects) is significantly smaller than that of ITO (2.3 eV). This finding suggests the possibility to introduce larger carrier concentrations without sacrificing the optical transmittance in the long wavelength range.

It should be pointed out that smaller BM shift in IMO does not lead to the appearance of the intense inter-band transitions from the valence band, E_v, in the visible range due to the large optical band gap in pure indium oxide (3.4 eV). Furthermore, in contrast to ITO where the band gap narrowing has been demonstrated both experimentally [26] and theoretically [9], the fundamental band gap of IMO was found to increase upon introduction of Mo [54].

The properties of IMO can be further optimized by varying ambient oxygen pressure [57]. An increased oxygen content in IMO facilitates the formation of the oxygen compensated complexes, e.g. $[Mo_{In}^{\bullet\bullet\bullet} + 2In_{In}O''_i]^{\bullet}$ or $[Mo_{In}^{\bullet\bullet\bullet}O''_i]^{\bullet}$, where the subscript stands for the site position and the superscript stands for effective negative (′) or positive (•) charge. These complexes reduce the number of free carriers – from three to one per Mo substitution – but, at the same time, enhance the carrier mobility due to smaller ionized impurity scattering and, hence, longer relaxation times. However, the interstitial oxygen significantly supresses the magnetic interactions which should be strong enough to split the transition metal *d* states in order to provide good conductivity in one (or both) spin channels [54].

In summary, we have shown that transition metal dopants offer the possibility to enhance conductivity via an increased mobility (due to smaller BM shift) of the free carriers and not their concentration (since half of the carriers are trapped on the *d* states of Mo). The latter usually leads to reduction of the optical transparency.

The advantages of carrier generation via *d*-element doping will be also discussed in Section 1.7.

1.5 Multicomponent TCO Hosts

Multicomponent TCOs, complex oxides which contain a combination of In, Zn, Cd and Sn metal ions, have been developed to broaden the range of the TCO materials required for a variety of specialized applications. Binary and ternary compounds and solid solutions with electrical, optical and mechanical properties controlled via chemical composition, have been the subject of numerous investigations [1, 2, 4, 58–60]. Since the 1990s, multi-cation TCOs which include metal ions beyond the traditional Sn, Cd, In and Zn have emerged. For example, $MgIn_2O_4$ [61], $GaInO_3$ [62] and Ga_2O_3-containing 2-3-3 or 3-3-4 systems, where

Table 1.4 *Net contributions to the conduction band at the Γ point from the states of the atoms that belong to the $InO_{1.5}$ layer, N_1, or Ga(Al)Zn(Cd,Mg)$O_{2.5}$ layers, N_2, in per cent; the electron effective masses m, in m_e, along the specified crystallographic directions; and the components of the electron effective-mass tensor, m_{ab} and m_z, calculated via simple averaging of those of the corresponding single-cation oxides [Equations (1.6) and (1.7)]. Reprinted with permission from [13, 31]. Copyright (2007) Springer Science + Business Media and (2007) Institute of Physics Publishing respectively*

Compound	N_1 (%)	N_2 (%)	$m_{[100]}$	$m_{[010]}$	$m_{[001]}$	m_{ab}	m_z
$InGaZnO_4$	48	52	0.23	0.22	0.20	0.23	0.23
$InAlCdO_4$	54	46	0.26	0.25	0.20	0.27	0.27
$InGaMgO_4$	58	42	0.27	0.27	0.24	0.28	0.29
$InAlMgO_4$	72	28	0.32	0.31	0.35	0.31	0.34

the numbers correspond to divalent, trivalent and tetravalent cations [8], have attracted wide attention.

Electronic band structure investigations [31, 50] of multi-cation oxides with layered structures [63–69], $InGaZnO_4$, $InAlCdO_4$, $InGaMgO_4$ and $InAlMgO_4$, identified the key electronic feature of complex oxides – the hybrid nature of the conduction band associated with the strong hybridization between the states of *every* cation in the cell with the states of its neighboring oxygen atoms. Strikingly, despite the substantially different values of the band gaps in the constituent single-cation oxides (e.g. the band gap of Al_2O_3 and MgO is about two times larger than the one in In_2O_3, CdO and ZnO), the states of all cations were found to have comparable contributions to the bottom of the conduction band of the multicomponent oxides (Table 1.4 and Figure 1.7). This results in a uniform three-dimensional charge density distribution within and across the structurally and chemically distinct layers (Figure 1.7) as well as in isotropic electron effective mass (Table 1.4).

Moreover, because the states of all cations contribute to the bottom of the conduction band, the electron effective mass of the complex oxides is an 'effective' average over the effective masses of the single-cation constituents. This averaging can be shown analytically within the tight-binding approximation. A one-dimensional chain consisting of two types of metal atoms which alternate with oxygen atoms (Figure 1.8) captures the key features of complex oxides. The Hamiltonian for such model system where nearest-neighbor interactions are given by the hopping integrals β_1 and β_2, is:

$$H = \sum_{n,l} |n,l\rangle \varepsilon_l \langle n,l| + \sum_{n,n',l,l'} |n',l'\rangle \beta_l \langle n,l|. \tag{1.4}$$

Here l is the atom index in the unit cell, n enumerates the cells and n', l' is the second sum run over the nearest neighbors. For the bottom of the conduction band, the dispersion relation can be simplified to:

$$\varepsilon(\mathbf{k}) = \frac{\varepsilon_1 + \varepsilon_2}{2} + \frac{1}{\frac{1}{2}\left(\frac{\Delta}{\beta_1^2} + \frac{\Delta}{\beta_2^2}\right)} (\mathbf{k}a)^2 \tag{1.5}$$

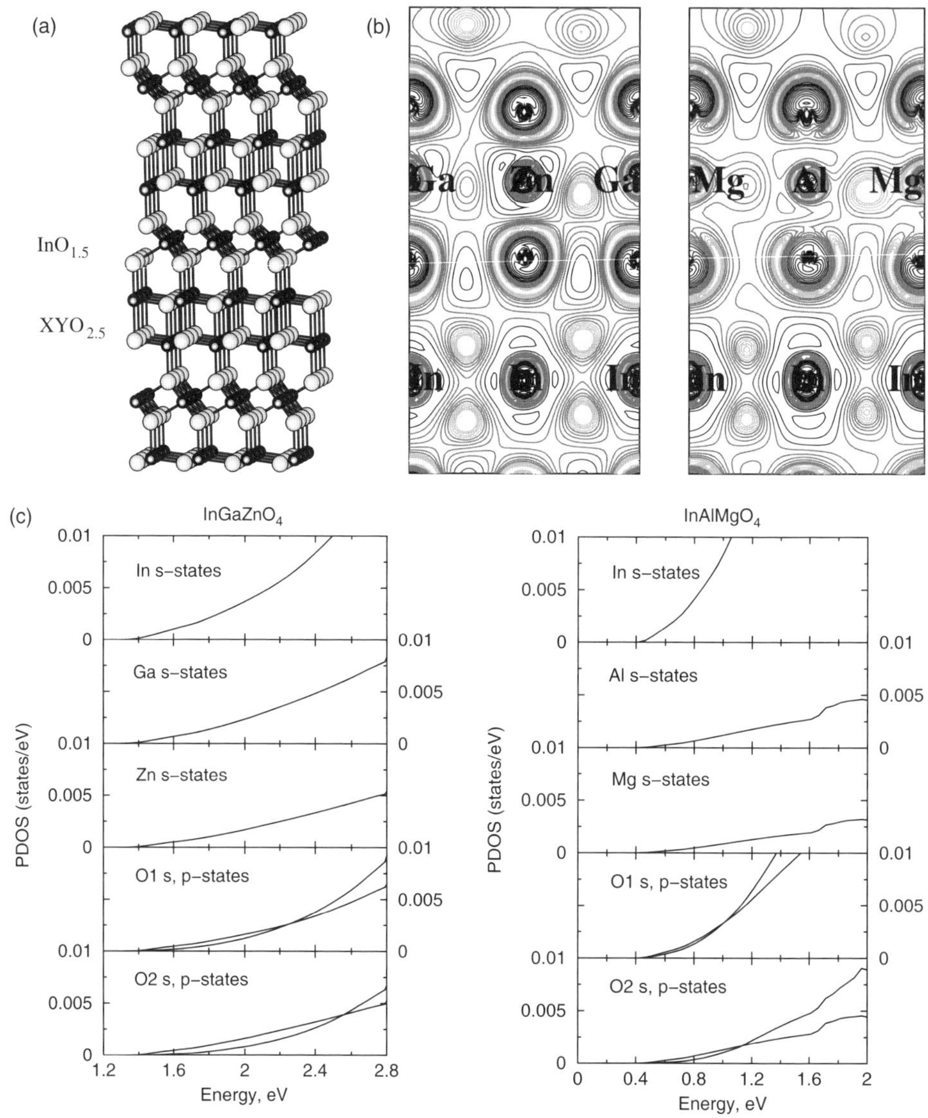

Figure 1.7 *(a) The unit cell of $InXYO_4$ (X = Ga, Al and Y = Zn, Cd, Mg) has three similar blocks, each consisting of one $InO_{1.5}$ layer with octahedral oxygen coordination of In atoms and a double layer $XYO_{2.5}$ with tetrahedral oxygen coordination of the cations. The layers alternate along the [0001] direction. The X^{3+} and Y^{2+} atoms are distributed randomly. (b) Contour plots of the charge density distribution calculated in the (011) plane for the conduction band in $InGaZnO_4$ and $InAlMgO_4$ with extra electron concentration of $\sim 1 \times 10^{18}$ cm^{-3}. The uniform interatomic charge density distribution within and across the chemically and structurally distinct layers implies isotropic electron transport. (c) Partial density of states at the bottom of the conduction band of $InGaZnO_4$ and $InAlMgO_4$. Although the contributions from Al s and Mg s states are notably reduced, these states will participate in charge transport once the electrons fill the band. Reproduced from [13, 31] by permission of Europhysics Letters Asiciation and of Springer Science + Business Media*

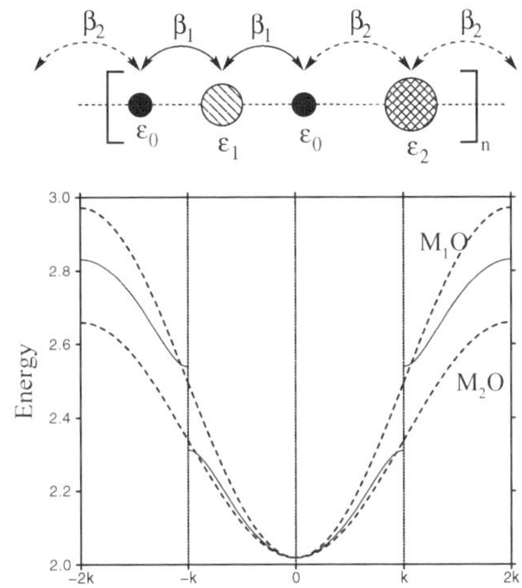

Figure 1.8 *Tight-binding conduction band (solid line) calculated for one-dimensional atomic chain depicted above the plot. Two types of metal atoms (\ and × filled cirles) alternate with oxygen atoms (filled circles) and only the nearest-neighbor hopping β is assumed. To illustrate the effective mass averaging, the conduction bands for the corresponding single-metal oxide chains (dashed lines) are aligned with* $(\varepsilon_1+\varepsilon_2)/2$. *The following parameters were used:* $\varepsilon_0 = 1.00$, $\varepsilon_1 = 2.00$, $\varepsilon_2 = 2.05$, $\beta_1 = 0.4$ *and* $\beta_2 = 0.5$. *Reproduced with permission from [31]. Copyright (2007) Institute of Physics*

if $|\varepsilon_1-\varepsilon_2| < 2\frac{\beta_1^2-\beta_2^2}{\Delta}$. Here ε_0, ε_1 and ε_2 are the atomic level energies of the oxygen and two types of metal atoms, respectively, and it is assumed that $\varepsilon_0 < \varepsilon_{1,2}$ and $\varepsilon_1 \sim \varepsilon_2$; $\Delta = \frac{1}{2}(\varepsilon_1+\varepsilon_2)-\varepsilon_0$ and a is half of the lattice parameter. Similar considerations for the chain consisting of only one type of metal atom alternating with oxygen atoms show that the quantity $\frac{\Delta}{\beta^2}$ represents the effective mass of that system. Therefore, Equation (1.5) represents the effective mass averaging over those in the corresponding single-metal 'oxide' chains (Figure 1.8).

First-principle calculations confirm the results found in the tight-binding model. For the layered oxides, the effective mass can be estimated as follows. As the resistivity along the z direction, i.e. across the layers, is a sum of the resistivities of each layer, the z component of the average effective-mass tensor can be found as:

$$m_z = (m_1+m_2+m_3)/3, \tag{1.6}$$

where $m_{1,2,3}$ are the effective masses of the corresponding single metal oxides, e.g. In_2O_3, Ga_2O_3 and ZnO in the case of $InGaZnO_4$. For the in-plane charge transport, the effective-mass tensor components can be found in a parallel manner:

$$\frac{1}{m_{a,b}} = \frac{1}{3}\left[\frac{1}{m_1}+\frac{2}{\frac{1}{2}(m_2+m_3)}\right]. \tag{1.7}$$

Here, one needs to average the effective mass for the mixed $GaZnO_{2.5}$ layers (Figure 1.7). The obtained $m_{a,b}$ and m_z values nearly coincide with the corresponding calculated effective masses of the multi-cation oxides.

The above effective-mass averaging procedure can be generalized for materials consisting of any number of layers, e.g. for $InGaO_3(ZnO)_m$, where m is an integer. Moreover, because the intrinsic transport properties are determined by the Ms–Op interactions which show little sensitivity to the oxygen coordination and the distortions in the metal-oxygen chains (as discussed earlier), the effective mass averaging should also apply to the oxides in amorphous state. In this case, one needs to average the components of the effective-mass tensor, $m_{\text{amorph}} = (m_a + m_b + m_z)/3$.

Thus, similar to the single-cation TCO hosts, multicomponent oxides have the conduction band formed from M s and O p states. Nonetheless, no multicomponent oxide has outperformed the conventional single-cation TCOs, in part due to challenges of carrier generation. Targeted doping via aliovalent substitution becomes difficult as the number of multivalent cations increases owing to a possibility of same-valence substitution or anti-site defects [8] which can neutralize the donors. The effects of clustering and second phase formation narrow the range of dopants efficient for a particular multicomponent oxide even further. In addition, the isotropic electronic properties may not be maintained due to a nonuniform distribution of carrier donors in the complex TCO hosts with structural anisotropy, e.g. atomic layers or chains of edge-sharing polyhedron in spinels. For example, preferential distribution of oxygen vacancies as well as Sn, Ti and Zr dopants in $InGaZnO_4$ [50] explains the observed strong anisotropy of the electrical conductivity in $InGaO_3(ZnO)_m$ compounds where the number of ZnO layers is increased [63, 64].

We note that many of these carrier generation bottlenecks may be overcome in the amorphous state of these complex oxides which represents a more uniform mixture of the constituent oxides while preserving the short range structure (alternating metal and oxygen atoms) and, thus, an even stronger hybridization between the states of the different cations mediated by their interactions with the states of shared oxygen atoms. Experimental observations that the mobility and conductivity are independent of the large variations in the composition in amorphous [65] but not in crystalline [8] $InGaO_3(ZnO)_n$ with $n \leq 4$, support the above idea.

Studies [31] of multi-cation TCOs which include light metal ions, such as Al and Mg, motivate an intruguing question: how do these ions influence carrier generation and the resulting transport properties of the multicomponent oxides? Furthermore, we also would like to understand why transparent conducting behavior is unique to SnO_2, In_2O_3, CdO and ZnO but has not been attained in SiO_2, Al_2O_3, CaO or MgO. To address these questions, we will first look at the electronic structure of classic insulators, CaO and Al_2O_3, to determine the origin of the strong electron localization in these oxides.

1.6 Electronic Properties of Light Metal Oxides

Oxides of light metals, such as CaO, MgO, Al_2O_3 or SiO_2, have the same s^2 valence electron configuration of the cations as the conventional TCOs and, therefore, their electronic band structure is similar to the one in a TCO host (cf. Figures 1.1, 1.2 and

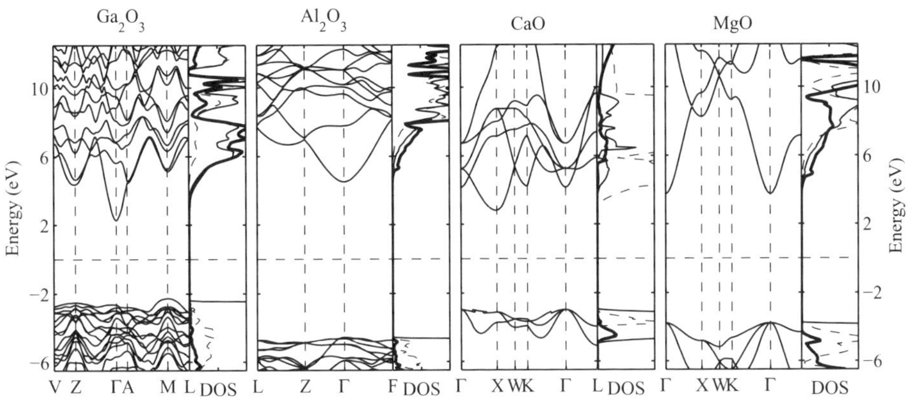

Figure 1.9 *Electronic band structure and partial density of states of several main group metal oxides, Ga_2O_3, Al_2O_3, CaO and MgO, as obtained within the screened-exchange local-density approximation [19]. In the density of states plots, the thick, dashed and thin lines represent metal s, metal p and oxygen p states, respectively*

Figure 1.9). It features a dispersed conduction band resulting in a relatively small electron effective mass of 0.3–0.5 m_e [31] (Table 1.5). However, degenerate doping of these refractory oxides has been a challenge [70–72]. Natural defects, e.g. oxygen vacancy, create deep charge-localized states within the electronic band gap, known as color or F centers.

Electronic band structure investigations of oxygen-deficient MgO, CaO and Al_2O_3 [51] demonstrate the strong localization of the vacancy-induced electrons near the oxygen vacancy – even despite the large concentration of defects used in the calculations, $\sim 0.8\text{–}1.0 \times 10^{21}\ \text{cm}^{-3}$. For these oxides, about 80–87% of the total charge density at the bottom of the conduction band resides at the metal and oxygen atoms which are nearest to the defect. Further analysis reveals large contributions from the Mg *p* states (47%), Ca *d* states (56%) or Al *p* states (47%) while their *s* states contribute only 20–30% to the total

Table 1.5 *Properties of several main group metal oxides. The optical band gaps and the electron effective masses are determined within the screened-exchange local-density approximation (sX-LDA) [24]. The anisotropy of the electron effective mass is defined as $\delta = (m^{[100]} + m^{[010]})/2m^{[001]}$*

Oxide	Lattice	Coordination of		Optical (direct) band gap (eV)	Electron effective mass, m_e			Effective mass anisotropy
		cation	anion		$m^{[100]}$	$m^{[010]}$	$m^{[001]}$	
β-Ga_2O_3	Monoclinic	6,4	4,3	4.86	0.35	0.35	0.32	1.097
Al_2O_3	Corundum	6	4	9.08	0.45	0.45	0.45	1.000
CaO	Rocksalt	6	6	7.15	0.42	0.42	0.42	1.000
MgO	Rocksalt	6	6	7.55	0.46	0.46	0.46	1.000

Table 1.6 *Origin of the electron localization in oxides of main group metals. The larger the p and d orbital contributions on the cations nearest the oxygen vacancy, the more charge is localized near the defect. The conduction band width reflects the degree of the charge localization*

Oxide	Lattice	Charge localized on defect neighboring atoms (%)	Relative orbital contributions for cations nearest to defect (%)			Conduction band width (eV)
			s	*p*	*d*	
CdO	Rocksalt	18	79	3	18	1.55
In_2O_3	Byxbyite	31	81	8	11	1.41
h-ZnO	Wurtzite	39	45	16	39	0.96
c-ZnO	Rocksalt	27	64	8	28	1.16
β-Ga_2O_3	Monoclinic	61	56	33	11	0.45
α-Ga_2O_3	Corundum	53	66	21	13	0.65
CaO	Rocksalt	80	22	22	56	0.06
MgO	Rocksalt	85	29	47	24	0.16
Al_2O_3	Corundum	87	30	47	23	0.14

(Table 1.6). Due to the high anisotropy of the *p* and *d* orbitals, strong covalent M-O bonds are formed near the oxygen defect. The resulting charge confinement is clearly seen from the charge density plots for oxygen-deficient MgO and CaO (Figure 1.10). This is not the case for rocksalt ZnO where the *s* state contributions are more than two times larger (Table 1.6).

A comparison of the electronic band structure of stoichiometric undoped MgO, CaO and Al_2O_3 (Figure 1.9) with the one in the conventional TCOs (Figure 1.2), reveals that the fundamental differences in the electronic properties of these oxides originate from the different energy location of the metal's empty *p* or *d* states with respect to the conduction band bottom. In In_2O_3, SnO_2, CdO or ZnO, the metal *p* band is well above its *s* band (Figures 1.2 and 1.9). As a result, the charge transport occurs via the M*s*–O*p* network, even

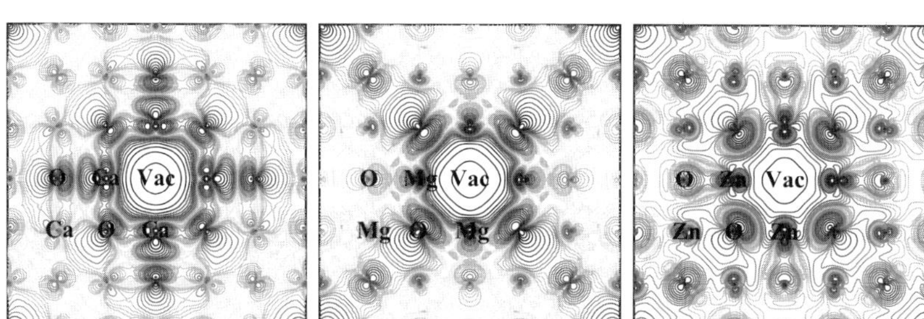

Figure 1.10 *Contour plots of the charge density distribution near the Fermi level in oxygen-deficient rocksalt CaO, MgO and ZnO. The charge confinement near the oxygen vacancy (an F-center defect) is clearly seen in CaO and MgO but not in ZnO where the charge distribution is more uniform*

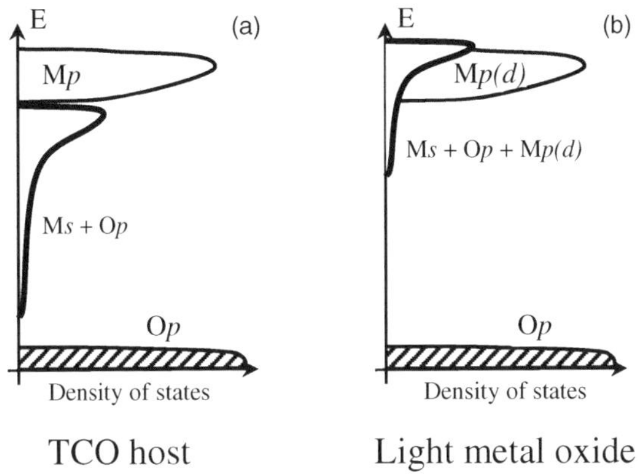

Figure 1.11 *Schematic electronic structure of main group metal oxides. (a) In conventional TCOs (post-transition metal oxides) Ms-type conduction band bottom ensures a uniform Ms–Op network for good carrier transport. (b) In oxides of lighter metals, e.g. Ga, Al, Mg or Ca, a substantial contribution from the metal anisotropic p or d states leads to a strong localization of doping-induced electrons*

for a large carrier concentration, i.e. when the BM shift is large. The spherical symmetry of the metal *s* orbitals and their strong hybridization with the *p* orbitals of the oxygen neighbors provides the most uniform charge distribution throughout the cell and, thus, facilitates good carrier transport.

In striking contrast to the conventional TCO hosts, in oxides of light metals the metal *p* or *d* bands almost coincide with its *s* band (Figures 1.9 and 1.11). When an oxygen vacancy is created, the Mg *p*, Al *p* or Ca *d* orbitals are energetically available for the induced electrons. Strong binding of these highly anisotropic orbitals with the states of the nearest oxygen atoms lowers the total energy of the system. The charge confinement explains the large formation energy of the oxygen vacancy in these refractory oxides as compared with the conventional TCOs where extra electrons are delocalized. Likewise, charge trapping on the anisotropic *p* or *d* states is expected for other carrier generation mechanisms, e.g. substitutional doping.

Naturally, the transition between the conventional TCO hosts (CdO, ZnO, In_2O_3 and SnO_2) and classic insulators (e.g. CaO and Al_2O_3) is not abrupt. The proximity of the metal *p* or *d* states to the *s*-type conduction band bottom in oxide of a main group metal (with ns^2 electronic configuration) will determine the orbital composition of the conduction band wavefunction. As an example, in oxygen-deficient β-Ga_2O_3, about 50% of the total cation contributions to the conduction band wavefunction comes from the Ga atoms nearest to the oxygen defect – to be compared with the same result for In_2O_3 (21%) on one side and Al_2O_3 (81%) or CaO (85%) on the other. Consistent with the degree of electron localization near the vacancy site, the conduction band width is smallest in CaO, MgO and Al_2O_3 followed gradually by Ga_2O_3, ZnO, In_2O_3 and CdO (Table 1.6). The second gap which splits this lowest conduction band from the higher band is found to be 3.80 eV in Al_2O_3, 1.44 eV in Ga_2O_3, 0.79 eV in ZnO, 0.64 eV in In_2O_3 and 0.20 eV in CdO, which also correlates with the degree of the electron binding.

The above band structure considerations generalize the fundamental physical properties in the main group metal oxides and also suggest ways to overcome the electron localization in these materials which we consider below.

1.7 Carrier Delocalization in Complex Oxides

The analysis in the previous section suggests a way to facilitate electrical charge transport in an oxide of main group metal(s) – via reduction of the contributions from the cation(s) *p* or *d* states at the energies near the Fermi level.

In Ga_2O_3, the Ga *p* band is located relatively close to the metal *s* band but does not coincide with it exactly as, for example, in Al_2O_3 (Figure 1.9). This leads to considerable but not dominant contributions from the Ga *p* states near the bottom of the conduction band. Hence, in Ga_2O_3, alternative carrier generation may give satisfactory results. For example, doping with a transition metal ion is expected to lead to a smaller BM shift associated with filling of the localized *d* states of the *d*-metal impurity (see Section 1.4). This would help to keep the detrimental Ga *p* states at energies high enough to make them unavailable for the induced carriers.

The above approach of circumventing carrier localization in Ga_2O_3 will not be successful in CaO or Al_2O_3 where the metal *s* and *p* or *d* bands overlap significantly. In the latter case, the electronic structure of the hosts should be altered via band engineering to attain the desired M*s*–O*p* hybridized conduction state. This approach is considered below.

1.7.1 Multicomponent Oxides with Layered Structures

In Section 1.5, it was shown that layered multicomponent oxides have a hybrid conduction band which consists of the *s* states of all consituent cations including light metal ions (if any). Due to the interaction between the alternating metal and oxygen atoms, the band gap of the complex material is an effective average over those in the corresponding single-cation oxides. For example, the band gap of $InAlZnO_4$ (3.5 eV) or $InAlCaO_4$ (4.6 eV) is smaller than the one in CaO (7.2 eV) or Al_2O_3 (9.1 eV) and larger than the one in In_2O_3 or ZnO ($\sim$3.4 eV) (Tables 1.1 and 1.5). This suggests that the hybrid conduction band of complex oxides can be driven away from the Al *p* and Ca *d* states via proper material composition to reduce the contributions from these states detrimental for carrier transport near the Fermi level (Figure 1.12).

Electronic band structure calculations of undoped stoichiometric $InAlZnO_4$ and $InAlCaO_4$ [51] indeed confirm that the Al *p* and Ca *d* states are at least 3 eV above the bottom of the conduction band as compared with 1.5 eV in Al_2O_3 and –1.2 eV in CaO where the Ca *d* band is *below* its *s* band (Figure 1.9). Nonetheless, Ca and Al will participate in the charge transport – *by providing their states* for extra electrons once the latter fill the conduction band. This is confirmed by isotropic electronic properties found in oxygen-deficient $InAlZnO_4$ and $InAlCaO_4$ [51]. Despite the preferential distribution of the oxygen vacancies which tend to concentrate in the $InO_{1.5}$ layer, the contributions to the conduction band wave function from different layers are comparable (Table 1.7) and the electron velocities calculated within and across the layers have similar values (Table 1.7). We note here that the electron velocity values for these oxides, although reduced, are comparable with

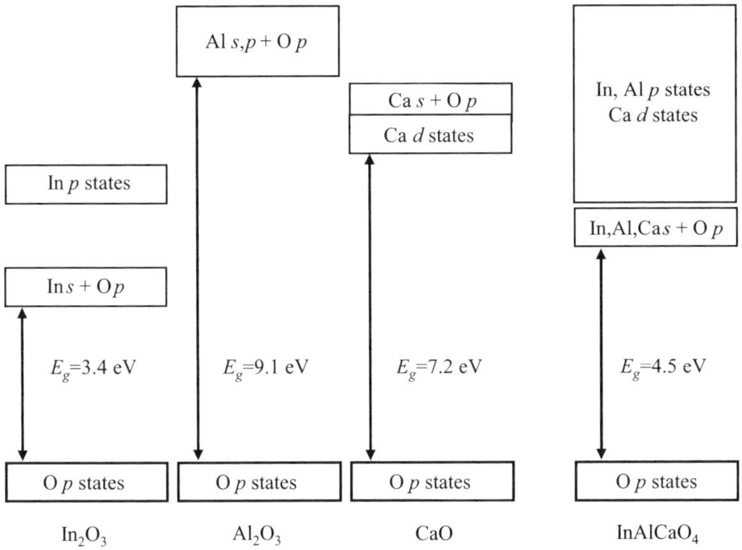

Figure 1.12 *Hybrid nature of the conduction band in single and multi-cation TCO hosts which consists primarily of the cation s states and the p states of oxygen atoms (see also Figures 1.1, 1.2, 1.9 and 1.11). Due to the band gap averaging, the conduction band in* $InAlCaO_4$ *is displaced away from the empty Al p and Ca d states to form a three-dimensional Ms–Op network for transport of extra electrons which fill the band upon doping*

those calculated for oxygen-deficient $InGaZnO_4$ which has been successfully employed as a TCO [36, 37, 69].

To summarize, multicomponent oxides offer a possibility to overcome the electron localization effects by tuning their electronic properties via proper composition. We stress, that both the relative content of the constituent oxides as well as their band gaps are crucial parameters that control the resulting optical and transport properties. For example, larger Al_2O_3 content in a multicomponent oxide where other constituents possess smaller band gaps, e.g. ZnO and In_2O_3, will result in an increase of the band

Table 1.7 *Net contributions from the states of the atoms that belong to the InO1 or* Y_2O2 *layers (Y = Zn, Al and/or Ca) to the conduction band wavefunction near* Γ *point, in per cent; and the Fermi electron group velocity v, in* $\times 10^{-5}$ $m\,s^{-1}$*, calculated along the specified crystallographic directions for oxygen-deficient layered oxides*

InY_2O_4	Contributions				Electron velocity		
	N_{In}	N_{O1}	N_{Y2}	N_{O2}	$v_{[100]}$	$v_{[010]}$	$v_{[001]}$
$InAlZnO_4$	27	37	13	23	3.1	3.3	3.4
$InAlCaO_4$	35	45	9	11	3.0	3.2	3.1
$InGaZnO_4$	24	32	21	23	3.8	4.0	4.6

gap. This may be appealing from the optical properties standpoint; it would also allow optimization of the oxide work function desired for a specific application. However, larger Al_2O_3 content with respect to the other constituents will also effectively move the hybrid conduction band closer to the Al *p* states, increasing their contributions and limiting the charge transport via an increased electron localization. This, along with the challenges of carrier generation in multicomponent oxides outlined in Section 1.5, suggests that careful microscopic analysis is required to produce a viable complex TCO.

1.7.2 Nanoporous Calcium Aluminate

In 2002, a persistent conductivity with a 10-order of magnitude change (from 10^{-10} to $\sim 1\,S\,cm^{-1}$) has been observed in $12CaO{\cdot}7Al_2O_3$ upon H doping followed by UV irradiation [73, 74]. Currently, conductivities as high as $1700\,S\,cm^{-1}$ have been achieved and various conversion approaches – in addition to photoactivation – have been developed [75–82].

$12CaO \cdot 7Al_2O_3$ or mayenite, a member of Portland cements, has a unique zeolite-like crystal structure with spacious cages of about 5.6 Å in diameter [83]. The cage framework includes 32 of the oxygen atoms in the unit cell while the remaining O^{2-} ion, which provides charge neutrality, is located inside one of the six cages. These encaged oxygen ions are loosely bound to the cage walls and can be easily substituted [84, 85] or reduced [76, 80, 81]. Indeed, our calculated formation energy of the oxygen vacancy in $12CaO \cdot 7Al_2O_3$ is 8–10 eV lower than those in the oxygen-deficient CaO or Al_2O_3 [50].

It has been shown that the nanoporous structure of $12CaO{\cdot}7Al_2O_3$ results in the formation of the so-called cage (or cavity) conduction band (CCB) [86, 87]. It consists of five bands (Figure 1.13) associated with the five empty cages in the unit cell. The sixth

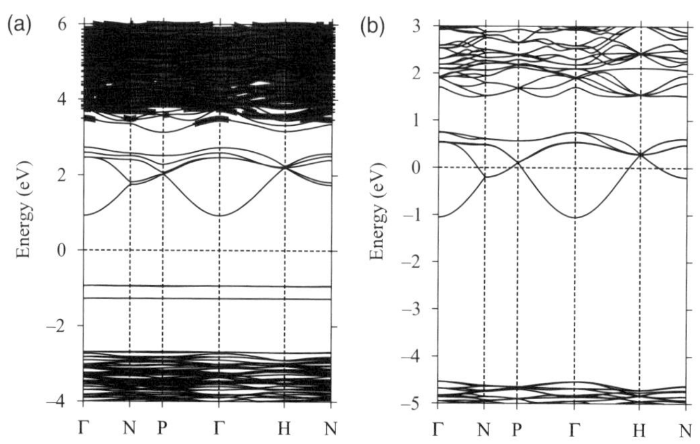

Figure 1.13 *Electronic band structure of (a) stoichiometric undoped and (b) oxygen-deficient* $12CaO \cdot 7Al_2O_3$. *In (a) the Ca d and Al p states are highlighted with (+) symbols. The encaged loosely bound* O^{2-} *ions give rise to the occupied flatbands below the Fermi level, cf. (a). When these* O^{2-} *are removed, cf. (b), the Fermi level shifts up into the cage conduction band formed by the Ca s and O p states giving rise to high electrical conductivity*

cage is filled with O^{2-} giving rise to the fully occupied flatbands (p_x, p_y, p_z) below the Fermi level. Due to the presence of the encaged oxygen ions, the CCB is shifted into the lower energy region and is located well below the framework conduction band (Figure 1.13). Most importantly, it was found that the latter is composed of the Ca *d* and Al *p* states. Hence, these orbitals will not be available to the vacancy-induced electrons, even if all the encaged oxygen ions are removed (which corresponds to the extra electron concentration of 2.33×10^{21} cm^{-3} and to the Fermi level shift of ~1.0 eV counting from the bottom of the CCB).

The analysis of the nature of the CCB in oxygen reduced $12CaO{\cdot}7Al_2O_3$ suggests that the conduction wave function is composed primarily of the Ca *s* and O *p* states (46% and 48% of the total charge in the cell, respectively). Since all Ca atoms in the cell give identical contributions to the conduction band, the resulting charge density distribution is uniform throughout the cage framework. The delocalization of the extra electrons in the reduced $12CaO{\cdot}7Al_2O_3$ [88] manifests itself in a large electron velocity (5.57×10^{-5} m s^{-1} in the [111] direction) to be compared with those calculated for oxygen deficient In_2O_3 (5.88×10^{-5} m s^{-1}) and ZnO (3.90×10^{-5} m s^{-1}) with similar electron concentration.

Another advantage of the unique crystal structure of $12CaO{\cdot}7Al_2O_3$ is that the CCB consists of six bands (Figure 1.13). They appear due to the six cages in a single unit cell of this material. Due to the uniform charge distribution throughout all cages [88], the lowest conduction state does not split off when oxygen vacancy is created. Therefore, oxygen reduction in $12CaO{\cdot}7Al_2O_3$ leads to two carriers per vacancy – in marked contrast to the conventional TCOs where the $V_O^{\bullet\bullet}$ defect leads to the completely occupied single conduction state being split from the rest of the band and only an excited or partially compensated vacancy can lead to conducting behavior (see Section 1.3.2).

Thus, the unusual nanoporous structure of $12CaO{\cdot}7Al_2O_3$ and the presence of the encaged O^{2-} ions result in the formation of the M*s*–O*p* hybridized conduction band located well below the detrimental Ca *d* and Al *p* orbitals. This explains the observed insulator-to-metal transition and the high electrical conductivity in oxygen-reduced $12CaO{\cdot}7Al_2O_3$ [76, 80–82].

1.8 An Outlook: Toward an Ideal TCO

Despite the success of converting the wide band gap $12CaO{\cdot}7Al_2O_3$ into a conductor via oxygen reduction, the conversion process resulted in a greatly increased absorption [76] making it inferior in relation to the conventional TCOs. The absorption arises due to optical transitions E_i within the CCB, which has width of 1.8 eV, as well as the transitions E_c from the Fermi level into the empty framework conduction band which have energies throughout the entire visible range (Figure 1.13). Importantly, band structure analysis suggests that nanoporous $12CaO{\cdot}7Al_2O_3$ belongs to a conceptually new class of transparent conductors [12]. In striking contrast to the conventional TCOs, where there is an unavoidable trade-off between optical absorption and conductivity, nanoporous materials offer a possibility to combine a complete, i.e. 100%, optical transparency with high electrical conductivity.

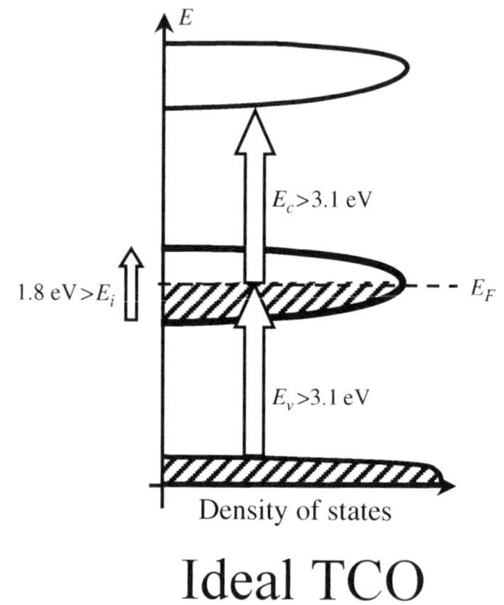

Figure 1.14 *Schematic electronic band structure of an ideal TCO*

The schematic band structure of the proposed [12] 'ideal' TCO is depicted in Figure 1.14. Introduction of a deep 'impurity' band with a high density of states (crossed by the Fermi level to make the system conducting) in the band gap of an insulating host material would help to keep intense inter-band transitions (from the valence band to the impurity band, E_v, and from the impurity band to the conduction band, E_c) above the visible range. This requires the band gap of a host material to be more than 6.2 eV. In addition, the impurity band should be narrow enough (<1.8 eV) to keep intra-band transitions E_i as well as the plasma frequency below the visible range [12]. To achieve high conductivity, the concentration of impurities should be high enough so that their electronic wave functions overlap well to form a *band*. The formation of the band would lead to a high carrier mobility due to the extended nature of these states and a relatively low scattering.

To fabricate such an ideal TCO, a material with a close-packed structure may not be suitable, because the required large concentration of impurities would result in: (i) an increase of ionized impurity scattering which limits electron transport; and (ii) a large structural relaxation in the host material, affecting its electronic structure and, most likely, decreasing the desired optical transparency. These effects are, indeed, observed in conventional TCOs (cf. Section 1.3).

Alternatively, materials with a nanoporous structure may offer a way to incorporate a large concentration of impurities into the pores without any significant changes in the band structure of the host material. Zeolites have been proposed [12] as potential candidates for such ideal TCOs. This class of materials possesses the desired structural and optical features, namely, spacious interconnected pores and large band gaps, as well as exhibit the ability to trap functional 'guest' atoms inside the nanometer-sized cavities which would govern the transport properties of the material.

In conclusion, understanding the principles of the conventional transparent conductors provides a solid foundation for further search of novel TCO host materials as well as efficient carrier generation mechanisms to make them good conductors. *Ab initio* density-functional band structure investigations are instrumental not only in providing a thorough insight into the TCO basics but also in predicting hidden capabilities of the materials beyond those traditionally employed.

Acknowledgements

The author acknowledges the support by the National Science Foundation and by the Petroleum Research Fund of the American Chemical Society. Computational resources for this work were provided by the San Diego Supercomputer Center (SDSC) supported by the National Science Foundation and National Energy Research Scientific Computing Center (NERSC) supported by the Department of Energy.

References

[1] K. L. Chopra, S. Major and D. K. Pandya, Transparent conductors – a status review, *Thin Solid Films*, **102**, 1–46 (1983).
[2] A. L. Dawar and J. C. Joshi, Semiconducting transparent thin films: their properties and applications, *J. Mater. Sci.*, **19**, 1–23 (1984).
[3] H. L. Hartnagel, A. L. Dawar, A. K. Jain and C. Jagadish, *Semiconducting Transparent Thin Films*, Institute of Physics Publishing, London, 1995.
[4] D. S. Ginley and C. Bright (Editors), Special issue on Transparent Conducting Oxides, *MRS Bull.*, **25**, (2000).
[5] E. Fortunato, D. Ginley, H. Hosono and D. C. Paine, Transparent conducting oxides for photovoltaics, *MRS Bull.*, **32**, 242–247 (2007).
[6] G. J. Exarhos and X.-D. Zhou, Discovery-based design of transparent conducting oxide films, *Thin Solid Films*, **515**, 7025–7052 (2007).
[7] J. C. C. Fan and J. B. Goodenough, X-ray photoemission spectroscopy studies of Sn-doped indium-oxide films, *J. Appl. Phys.*, **48**, 3524–3531 (1977).
[8] A. J. Freeman, K. R. Poeppelmeier, T. O. Mason, R. P. H. Chang and T. J. Marks, Chemical and thin-film strategies for new transparent conducting oxides, *MRS Bull.*, **25**, 45–51 (2000).
[9] O. N. Mryasov and A. J. Freeman, Electronic band structure of indium tin oxide and criteria for transparent conducting behavior, *Phys. Rev. B*, **64**, 233 111 (2001).
[10] R. Asahi, A. Wang, J. R. Babcock, N. L. Edleman, A. W. Metz, M. A. Lane, V. P. Dravid, C. R. Kannewurf, A. J. Freeman and T. J. Marks, First-principles calculations for understanding high conductivity and optical transparency in $In_xCd_{1-x}O$ films, *Thin Solid Films*, **411**, 101–105 (2002).
[11] H. Mizoguchi and P. M. Woodward, Electronic structure studies of main group oxides possessing edge-sharing octahedra: implications for the design of transparent conducting oxides, *Chem. Mater.*, **16**, 5233–5248 (2004).
[12] J. E. Medvedeva and A. J. Freeman, Combining high conductivity with complete optical transparency: a band-structure approach, *Europhys. Lett.*, **69**, 583–587 (2005).
[13] J. E. Medvedeva, Unconventional approaches to combine optical transparency with electrical conductivity, *Appl. Phys. A*, **89**, 43–47 (2007).
[14] E. Burstein, Anomalous optical absorption limit in InSb, *Phys. Rev.*, **93**, 632–633 (1954).
[15] T. S. Moss, The interpretation of the properties of indium antimonide, *Proc. Phys. Soc. B*, **67**, 775–782 (1954).

[16] J. R. Bellingham, W. A. Phillips and C. J. Adkins, Intrinsic performance limits in transparent conducting oxides, *J. Mater. Sci. Lett.*, **11**, 263–265 (1992).
[17] T. J. Coutts, D. L. Young and X. Li, Characterization of transparent conducting oxides, *MRS Bull.*, **25**, 58–65 (2000).
[18] G. Frank and H. Köstlin, Electrical properties and defect model of tin-doped indium oxide layers, *Appl. Phys. A*, **27**, 197–206 (1982).
[19] Highly precise all-electron full-potential linearized augmented plane wave (FLAPW) method [20, 21] within the local density approximation (LDA) and the screened-exchange LDA [22–24] was employed for the electronic band structure investigations.
[20] E. Wimmer, H. Krakauer, M. Weinert and A. J. Freeman, Full-potential self-consistent linearized-augmented-plane-wave method for calculating the electronic structure of molecules and surfaces – O_2 molecule, *Phys. Rev. B*, **24**, 864–875 (1981).
[21] M. Weinert, E. Wimmer and A. J. Freeman, Total-energy all-electron density functional method for bulk solids and surfaces, *Phys. Rev. B*, **26**, 4571–4578 (1982).
[22] D. M. Bylander and L. Kleinman, Good semiconductor band gaps with a modified local-density approximation, *Phys. Rev. B*, **41**, 7868–7871 (1990).
[23] A. Seidl, A. Görling, P. Vogl, J. A. Majewski and M. Levy, Generalized Kohn-Sham schemes and the band-gap problem, *Phys. Rev. B*, **53**, 3764–3774 (1996).
[24] R. Asahi, W. Mannstadt and A. J. Freeman, Optical properties and electronic structures of semiconductors with screened-exchange LDA, *Phys. Rev. B*, **59**, 7486–7492 (1999).
[25] R. L. Weiher and R. P. Ley, Optical properties of indium oxide, *J. Appl. Phys.*, **37**, 299–302 (1966).
[26] I. Hamberg, C. G. Granqvist, K. F. Berggren, B. E. Sernelius and L. Engström, Band-gap widening in heavily Sn-doped In_2O_3, *Phys. Rev. B*, **30**, 3240–3249 (1984).
[27] F. P. Koffyberg, Thermoreflectance spectra of CdO: band gaps and band-population effects, *Phys. Rev. B*, **13**, 4470–4476 (1976).
[28] C. Kligshirn, The luminescence of ZnO under high one- and two-quantum excitation, *Phys. Status Solidi B*, **71**, 547 (1975).
[29] D. Fröhlich, R. Kenklies and R. Helbig, Band-gap assignment in SnO_2 by two-photon spectroscopy, *Phys. Rev. Lett.*, **41**, 1750–1751 (1978).
[30] N. W. Ashcroft and N. D. Mermin, *Solid State Physics*, W.B. Saunders, Philadelphia, 1976.
[31] J. E. Medvedeva, Averaging of the electron effective mass in multicomponent transparent conducting oxides, *Europhys. Lett.*, **78**, 57 004 (2007).
[32] J. E. Medvedeva, E. N. Teasley and M. D. Hoffman, Electronic band structure and carrier effective mass in calcium aluminates, *Phys. Rev. B*, **76**, 155 107 (2007).
[33] H. Hosono, Ionic amorphous oxide semiconductors: material design, carrier transport, and device application, *J. Non-Cryst. Solids*, **352**, 851 (2006).
[34] A. J. Leenheer, J. D. Perkins, M. F. A. M. van Hest, J. J. Berry, R. P. O'Hayre and D. S. Ginley, General mobility and carrier concentration relationship in transparent amorphous indium zinc oxide films, *Phys. Rev. B*, **77**, 115 215 (2008).
[35] K. Nomura, T. Kamiya, H. Ohta, T. Uruga, M. Hirano and H. Hosono, Local coordination structure and electronic structure of the large electron mobility amorphous oxide semiconductor In-Ga-Zn-O: experiment and ab initio calculations, *Phys. Rev. B*, **75**, 035 212 (2007).
[36] K. Nomura, H. Ohta, A. Takagi, T. Kamiya, M. Hirano and H. Hosono, Room temperature fabrication of transparent flexible thin-film transistors using amorphous oxide semiconductors, *Nature*, **432**, 488–492 (2004).
[37] K. Nomura, H. Ohta, K. Ueda, T. Kamiya, M. Hirano and H. Hosono, Thin-film transistor fabricated in single-crystalline transparent oxide semiconductor, *Science*, **300**, 1269–1272 (2003).
[38] K. J. Button, C. G. Fonstad and W. Dreybrodt, Determination of the electron masses in stannic oxide by submillimeter cyclotron resonance, *Phys. Rev. B*, **4**, 4539–4542 (1971).
[39] T. Minami, New n-type transparent conducting oxides, *MRS Bull.*, **25**, 38–43 (2000).
[40] S. Jin, Y. Yang, J. E. Medvedeva, J. R. Ireland, A. W. Metz, J. Ni, C. R. Kannewurf, A. J. Freeman and T. J. Marks, Dopant ion size and electronic structure effects on transparent

conducting oxides. Sc-doped CdO thin films grown by MOCVD, *J. Am. Chem. Soc.*, **126**, 13 787–13 793 (2004).

[41] Y. Yang, S. Jin, J. E. Medvedeva, J. R. Ireland, A. W. Metz, J. Ni, M. C. Hersam, A. J. Freeman and T. J. Marks, CdO as the archetypical transparent conducting oxide. Systematics of dopant ionic radius and electronic structure effects on charge transport and band structure, *J. Am. Chem. Soc.*, **127**, 8796–8804 (2005).

[42] S. Jin, Y. Yang, J. E. Medvedeva, L. Wang, S. Li, N. Cortes, J. R. Ireland, A. W. Metz, J. Ni, M. C. Hersam, A. J. Freeman and T. J. Marks, Tuning the properties of transparent oxide conductors. Dopant ion size and electronic structure effects on CdO-based transparent conducting oxides. Ga- and In-doped CdO thin films grown by MOCVD, *Chem. Mater.*, **20**, 220–230 (2008).

[43] L. Vegard, Die Konstitution der Mischkristalle und die Raumfullung der Atome, *Z. Phys.*, **5**, 17–26 (1921).

[44] L. V. Morozova and A. V. Komarov, Solid solutions based on cadmium oxide in the CdO-In_2O_3 system, *Russ. J. Appl. Chem.*, **68**, 1240–1242 (1995).

[45] Y. Dou, R. G. Egdell, T. Walker, D. S. L. Law and G. Beamson, N-type doping in CdO ceramics: a study by EELS and photoemission spectroscopy, *Surf. Sci.*, **398**, 241–258 (1998).

[46] T. B. Reed, *Free Energy of Formation of Binary Compounds*, MIT Press, Cambridge, MA, 1971.

[47] J. H. W. de Wit, G. van Unen and M. Lahey, Electron concentration and mobility in In_2O_3, *J. Phys. Chem. Solids*, **38**, 819–824 (1977).

[48] F. A. Kröger, *The Chemistry of Imperfect Crystals*, North-Holland, Amsterdam, 1974.

[49] S. Lany and A. Zunger, Dopability, intrinsic conductivity and nonstoichiometry of transparent conducting oxides, *Phys. Rev. Lett.*, **98**, 045 501 (2007).

[50] J. E. Medvedeva and C. L. Hettiarachchi, Tuning the properties of complex transparent conducting oxides: role of crystal symmetry, chemical composition and carrier generation, to be published in, *Phys. Rev. B.*

[51] J. E. Medvedeva,Toward conductive main group metal oxides: overcoming electron localization, to be published.

[52] I. Tanaka, K. Tatsumi, M. Nakano and H. Adachi, First-principles calculations of anion vacancies in oxides and nitrides, *J. Am. Ceram. Soc.*, **85**, 68–74 (2002).

[53] Ç. Kiliç and A. Zunger, Origins of coexistence of conductivity and transparency in SnO_2, *Phys. Rev. Lett.*, **88**, 95 501 (2002).

[54] J. E. Medvedeva, Magnetically mediated transparent conductors: In_2O_3 doped with Mo, *Phys. Rev. Lett.*, **97**, 086 401 (2006).

[55] Y. Meng, X. Yang, H. Chen, J. Shen, Y. Jiang, Z. Zhang and Z. Hua, A new transparent conductive thin film In_2O_3:Mo, *Thin Solid Films*, **394**, 219–223 (2001).

[56] Y. Yoshida, D. M. Wood, T. A. Gessert and T. J. Coutts, High-mobility sputtered films of indium oxide doped with molybdenum, *Appl. Phys. Lett.*, **84**, 2097–2099 (2004).

[57] S. Sun, J. Huang and D. Lii, Effects of oxygen contents on the electrical and optical properties of indium molybdenum oxide films fabricated by high density plasma evaporation, *J. Vac. Sci. Technol. A*, **22**, 1235–1239 (2004).

[58] R. D. Shannon, J. L. Gillson and R. J. Bouchard, Single crystal synthesis and electrical properties of $CdSnO_3$, Cd_2SnO_4, In_2TeO_6, $CdIn_2O_4$, *J. Phys. Chem. Solids*, **38**, 877–881 (1977).

[59] H. Kawazoe and K. Ueda, Transparent conducting oxides based on the spinel structure, *J. Am. Ceram. Soc.*, **82**, 3330–3336 (1999).

[60] B. J. Ingram, G. B. Gonzalez, D. R. Kammler, M. I. Bertoni and T. O. Mason, Chemical and structural factors governing transparent conductivity in oxides, *J. Electroceram.*, **13**, 167–175 (2004).

[61] H. Un'no, N. Hikuma, T. Omata, N. Ueda, T. Hashimoto and H. Kawazoe, *Jpn. J. Appl. Phys.*, **32**, L1260–L1262 (1993).

[62] J. M. Phillips, J. Kwo and G. A. Thomas, Transparent conducting thin films of $GaInO_3$, *Appl. Phys. Lett.*, **65**, 115–117 (1994).

[63] H. Hiramatsu, H. Ohta, W. S. Seo and K. J. Koumoto, Thermoelectric properties of $(ZnO)_5In_2O_3$ thin films prepared by r.f. sputtering method, *J. Jpn. Soc. Powder Powder Metall.*, **44**, 44–49 (1997).

[64] H. Kaga, R. Asahi and T. Tani, Thermoelectric properties of highly textured Ca-doped (ZnO)(m) In_2O_3 ceramics, *Jpn. J. Appl. Phys.*, **43**, 7133–7136 (2004).

[65] M. Orita, H. Ohta, M. Hirano, S. Narushima and H. Hosono, Amorphous transparent conductive oxide $InGaO_3(ZnO)_m$ ($m \leq 4$): a Zn 4s conductor, *Phil. Mag. B*, **81**, 501–515 (2001).

[66] M. Orita, M. Takeuchi, H. Sakai and H. Tanji, New transparent conductive oxides with $YbFe_2O_4$ structure, *Jpn. J. Appl. Phys.*, **34**, L1550–L1552 (1995).

[67] K. Von Kato, I. Kawada, N. Kimizuka and T. Katsura Die Kristallstructur von $YbFe_2O_4$, *Z. Krist.*, **141**, 314–320 (1975).

[68] N. Kimizuka and T. Mohri, Spinel, $YbFe_2O_4$, and $Yb_2Fe_3O_7$ types of structures for compounds in the In_2O_3 and Sc_2O_3-A_2O_3-BO systems [A: Fe, Ga or Al; B: Mg, Mn, Fe, Ni, Cu or Zn] at temperatures over 1000 °C, *J. Solid State Chem.*, **60**, 382–384 (1985).

[69] N. Kimizuka and T. Mohri, Structural classification of $RAO_3(MO)_n$ compounds (R = Sc,In,Y or lanthanides; A = Fe(III), Ga, Cr, or Al; M = Divalent Cation; $n = 1$–11), *J. Solid State Chem.*, **78**, 98–107 (1989).

[70] G. F. Neumark, Defects in wide band gap II–VI crystals, *Mater. Sci. Eng. R*, **21**, 1–46 (1997).

[71] C. G. Van de Walle, Strategies for controlling the conductivity of wide-band-gap semiconductors, *Phys. Status Solidi B*, **229**, 221–228 (2002).

[72] A. Zunger, Practical doping principles, *Appl. Phys. Lett.*, **83**, 57–59 (2003).

[73] K. Hayashi, S. Matsuishi, T. Kamiya, M. Hirano and H. Hosono, Light-induced conversion of an insulating refractory oxide into a persistent conductor, *Nature*, **419**, 462–465 (2002).

[74] J. E. Medvedeva, A. J. Freeman, M. I. Bertoni and T. O. Mason, Electronic structure and light-induced conductivity in a transparent refractory oxide, *Phys. Rev. Lett.*, **93**, 16 408 (2004).

[75] Y. Toda, S. Matsuishi, K. Hayashi, K. Ueda, T. Kamiya, M. Hirano and H. Hosono, Field emission of electron anions clathrated in subnanometer-sized cages in $[Ca_{24}Al_{28}O_{64}]^{4+}$ $(4e^-)$, *Adv. Mater.*, **16**, 685–689 (2004).

[76] S. Matsuishi, Y. Toda, M. Miyakawa, K. Hayashi, T. Kamiya, M. Hirano, I. Tanaka and H. Hosono, High-density electron anions in a nanoporous single crystal: $[Ca_{24}Al_{28}O_{64}]^{4+}$ $(4e^-)$, *Science*, **301**, 626–629 (2003).

[77] S. W. Kim, M. Miyakawa, K. Hayashi, T. Sakai, M. Hirano and H. Hosono, Simple and efficient fabrication of room temperature stable electride: melt-solidification and glass ceramics, *J. Am. Chem. Soc.*, **127**, 1370–1371 (2005).

[78] S. W. Kim, Y. Toda, K. Hayashi, M. Hirano and H. Hosono, Synthesis of a room temperature stable 12CaO·7Al(2)O(3) electride from the melt and its application as an electron field emitter, *Chem. Mater.*, **18**, 1938–1944 (2006).

[79] S. W. Kim, K. Hayashi, M. Hirano, H. Hosono and I. Tanaka, Electron carrier generation in a refractory oxide 12CaO·7Al(2)O(3) by heating in reducing atmosphere: conversion from an insulator to a persistent conductor, *J. Am. Ceram. Soc.*, **89**, 3294–3298 (2006).

[80] S. W. Kim, S. Matsuishi, T. Nomura, Y. Kubota, M. Takata, K. Hayashi, T. Kamiya, M. Hirano and H. Hosono, Metallic state in a lime-alumina compound with nanoporous structure, *Nano Lett.*, **7**, 1138–1143 (2007).

[81] M. Bertoni, PhD Thesis, Northwestern University, Evanston, IL, 2006.

[82] M. Bertoni, J. Medvedeva, Y. Q. Wang, A. Freeman, K. R. Poeppelmeier and T. Mason, Enhanced electronic conductivity in Si-substituted calcium aluminate, *J. Appl. Phys.*, **102**, 113 704 (2007).

[83] A. N. Christensen, Neutron powder diffraction profile refinement studies on Ca11.3Al14O32.3 and CaClO(D0.88H0.12), *Acta Chem. Scandinavica*, **A41**, 110–112 (1987).

[84] J. Jeevaratnam, F. P. Glasser and L. S. Dent Glasser, Anion substitution and structure of $12CaO.7Al_2O_3$, *J. Am. Ceram. Soc.*, **47**, 105–106 (1964).

[85] G. I. Zhmoidin and G. S. Smirnov, Characteristics of the crystals of derivatives of $12CaO{\cdot}7Al_2O_3$, *Inorganic Mater.*, **18**, 1595–1601 (1982).

[86] P. V. Sushko, A. L. Shluger, K. Hayashi, M. Hirano and H. Hosono, Hopping and optical absorption of electrons in nano-porous crystal $12CaO{\cdot}7Al_2O_3$, *Thin Solid Films*, **445**, 161–167 (2003).
[87] Z. Y. Li, J. L. Yang, J. G. Hou and Q. S. Zhu, Is mayenite without clathrated oxygen in inorganic electride?, *Angew. Chem., Int. Ed.*, **43**, 6479–6482 (2004).
[88] J. E. Medvedeva and A. J. Freeman, Hopping versus bulk conductivity in transparent oxides: $12CaO{\cdot}7Al_2O_3$, *Appl. Phys. Lett.*, **85**, 955–957 (2004).

2

Transparent Oxide Semiconductors: Fundamentals and Recent Progress

Hideo Hosono
Frontier Research Center & Materials and Structures Laboratory, Tokyo Institute of Technology, Japan

2.1 Introduction

It is generally believed that high optical transparency is incompatible with high electronic conduction, since optical transparency requires band gaps larger than 3.3 eV and such a large gap makes carrier doping very difficult. In this sense, transparent conductive oxides (TCOs) are exceptional materials. Figure 2.1 summarizes the progress made in materials for TCOs and transparent oxide semiconductors (TOSs). The first TCO, In_2O_3:Sn (ITO), was reported by Rupperecht [1] in 1954, followed by other TCOs (SnO_2 and ZnO). Although TCOs have been commercialized intensively as transparent window electrodes and inter connections, there is almost no application as transparent semiconductors because of the absence of p-type TCOs and the uncontrollability of conductance by applied voltage. No active electronic devices such as bipolar transistors and diodes can be fabricated without the above properties. Although bipolarity is not needed for application to an active layer in thin film transistors (TFTs), conductance control by voltage is indispensable.

There have been three distinct advances in this area in recent years. One was the discovery of a p-type TCO, $CuAlO_2$, with delafossite structure in 1997 [2], which triggered the development of a series of p-type TCOs and transparent pn junction devices such as UV light emitting diodes (LEDs) [3]. So far, the conduction type of TCO materials had been limited to n-type. This achievement has significantly changed our understanding of TCOs and has opened a new frontier, that of TOSs [4]. Therefore, we now consider that TOSs have the

Transparent Electronics: From Synthesis to Applications Edited by Antonio Facchetti and Tobin J. Marks

Year	Material (Ref.)	Year	Device (Ref.)
		1935	FET proposed [13]
1954	In_2O_3 TCO [1]	1954	Solar Cell [14]
1975	a-Si:H [11]		
		1979	a-Si:H TFT [12]
		1987	OLED [15]
1996	TAOS [5]		
1997	p-TCO, $CuAlO_2$ [2]		
		2000	All oxide pn-junction UV-LED [3]
2002	C12A7 transparent conductor [9]	2002	TOS homo-junction diode $CuInO_2$ [16]
		2004	TAOS-TFT [6]
2005	TiO_2:Nb TCO [8]	2005	ZnO homo-junction blue LED [17]
		2006	Front Drive Structure [18]
			ITO / organic gate TFT [19]
		2007	4.5" OLED using a-IGZO TFT [20]
		2008	12" OLED using a-IGZO TFT [7]
			15" LED using a-IGZO TFT [21]
			p-channel oxide TFT [22]

Figure 2.1 *Progress made in transparent oxide semiconductors and relevant devices*

potential to develop new functionalities useful for novel optoelectronic devices that are hard to realize by current Si-based semiconductor technology. The second advance was the discovery of transparent amorphous oxide semiconductors (TAOSs), which have unique electron transport properties compared with conventional amorphous semiconductors such as a-Si:H and a-chalcogenides [5]. The Fermi level can be controllable from the mid gap state to above the electron mobility gap by chemical doping. As a result, the Hall mobility exceeds $>10\,cm^2\,(V\,s)^{-1}$, which is comparable with those in the corresponding crystals and is larger by an order of magnitude than that in a-Si:H, notwithstanding that room temperature (RT) fabrication is easily possible by a conventional sputtering method on any type of substrate including plastics. These characteristics led to the realization of high performance TFTs fabricated on plastic substrates in 2004 [6], and serious studies are being performed by industry for the application of these TFT arrays as the driver backplane of organic LED displays and next generation liquid crystal displays [7].

The third advance was the extension of TCO candidate materials. TCO materials are restricted to oxides of p-block metal cations such as In_2O_3, SnO_2, ZnO and CdO. However, two TCO materials which do not belong to p-block metal oxides were reported, i.e. TiO_2:Nb [8] belonging to d-block metal oxides and electron-doped $12CaO{\cdot}7Al_2O_3$ (C12A7) belonging to s-block metal oxides. One feature of both materials is that they are composed of elements that are abundant. In particular, C12A7 is made from the abundant and typical insulator CaO and Al_2O_3. This representative transparent insulating material, known as a constituent of commercial alumina cement, has been transformed to a transparent semiconductor [9], metal and eventually a superconductor by doping electrons to the sub-nanometer-sized cages constituting the unit cell of the crystal structure [10]. Such a series of discoveries has opened a route to realize transparent semiconductors using naturally abundant oxides by successfully utilizing built-in nanostructures embedded in crystal structures.

In this chapter, I outline TOSs based on their electronic structure and describe recent advances along with some device applications.

2.2 Electronic Structure in Oxides: Carrier Transport Paths in Semiconductors

To design carrier transport properties in new semiconductors, we should take two factors, carrier doping and carrier mobility (μ), into account since electronic conductivity (σ) is expressed as $\sigma = en\mu$, where e is the elementary electric charge and n the carrier density. First, we concentrate mainly on the mobility issue. Basic carrier transport properties in crystals are often discussed based on band theory. In this case, electronic structures are represented in reciprocal space, and curvatures at the conduction band minimum (CBM) and valence band maximum (VBM) determine the effective masses of the electron and hole, respectively. Smaller effective masses correspond to larger hybridization in the CBM and VBM bands, which can result in larger carrier mobilities (difference in carrier scattering time is neglected in this discussion). From this, we can easily find that carrier transport characteristics in oxides are different from those in covalent semiconductors such as Si. Figure 2.2 shows the schematic electronic structure of oxides. In many oxides, especially oxides of main group metals, band gaps are made primarily of occupied O 2p antibonding bands and unoccupied

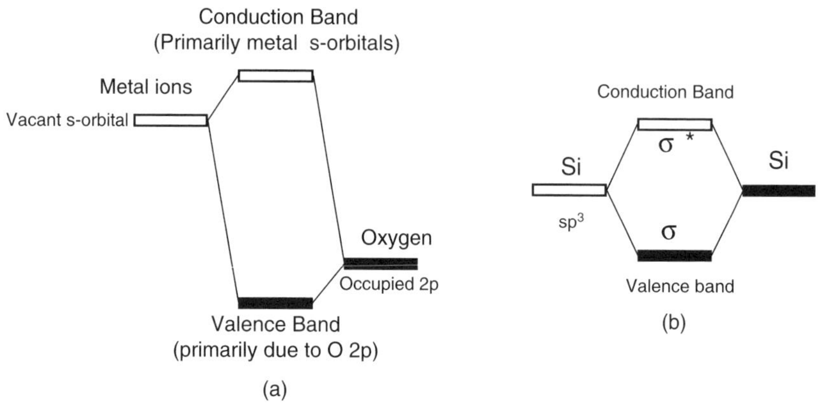

Figure 2.2 Schematic energy bands. (a) Ionic representative metal oxides (non-transition metal). (b) Covalent compounds (Si)

cations bonding bands for VBM and CBM, respectively. Such electronic structure forms favorable transport paths for electrons because the electron transport paths (i.e. CBM) are formed by spatially extended metal s orbitals with a spherical shape. It causes larger overlaps between the neighboring metal orbitals and increases band dispersion at CBM, resulting in small electron effective masses [23, 24]. Therefore, it is not difficult to attain good electronic conduction in oxides if high-density electron doping is possible. This is the reason why there are not a few good n-type TCOs such as SnO_2, indium–tin oxide (ITO) and ZnO.

Next, we discuss carrier doping. The thermodynamic stability of band holes and electrons is determined by the magnitude of the ionization potential (IP) and electron affinity (EA). The hole becomes stable as the IP decreases, whereas the electron becomes stable as the EA increases. Figure 2.3 shows the band line-up of various oxide semiconductors along with

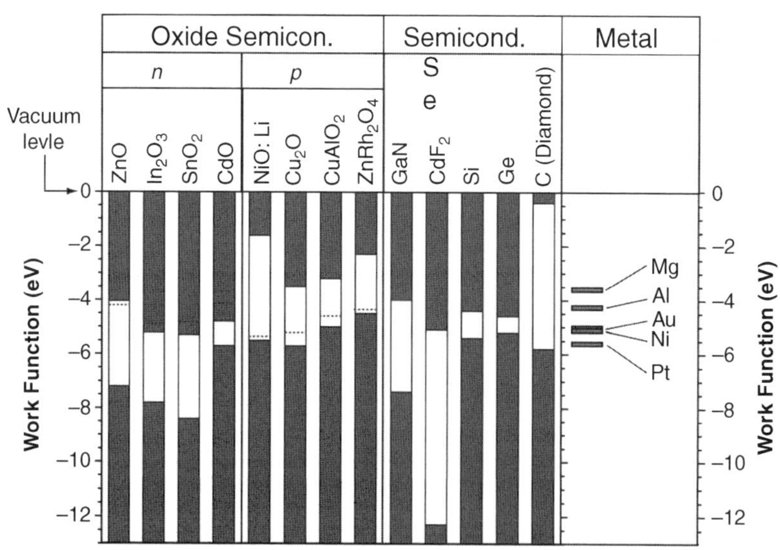

Figure 2.3 Band line-up of oxide semiconductors and relevant materials

group VI compound semiconductors [4]. One may note that the CBM for n-type TCOs (TOSs) is below –4 eV from the vacuum level, whereas the VBM of p-type oxides is located above –6 eV. These experimental observations agree with the thermodynamic argument. However, there is an exception, namely CdO. Although according to the band line-up CdO should be bipolar, no p-type conduction has been reported to date. This means that killer defect formation for holes is easy in CdO. Band line-up tells us the p/n orientation in thermodynamics but "the ease of killer defect" formation must be taken into account. This is discussed in the work by Zunger [25].

2.3 Materials Design of p-Type TOSs [23, 26]

In contrast to n-type TOSs, it is very difficult to attain good hole conduction because O 2p orbitals, which form hole transport paths in many oxides, are rather localized, therefore hole effective masses are rather large and VBM levels are deep. The former results in small hole mobilities and the latter causes difficulty in hole doping. These are the reasons why only n-type oxides were known for TCOs before 1997. Considering the electronic structure discussed above, we expected that dispersion of VBM could be modified by (i) decreasing the nearest neighboring oxygen–oxygen distance, (ii) using hybridization of metal orbitals whose energy levels are close to those of O 2p or (iii) employing more extended orbitals for anions. We first adopted approach (ii) and selected Cu^+-based oxides because the energy levels of Cu 3d are close to those of O 2p and the closed-shell configuration of Cu^+ $3d^{10}$ was expected not to give optical absorption in the visible region due to d–d or O 2p–Cu 3d transitions, which therefore met the requirements to keep optical transparency in the visible region. Following this strategy (Figure 2.4), we found a delafossite-type crystal, $CuAlO_2$, as the first p-type TOS in 1997 [2], followed by the discovery of a series of p-type TOSs, $CuGaO_2$ and $SrCu_2O_2$.

The above consideration also gives an idea of how to obtain a bipolar TOS in which either hole or electronic conduction is selected by intentional impurity doping. It suggested that n-type conduction can be imparted to p-type TOSs if heavy metals having largely extended s orbitals are incorporated into a delafossite oxide, which led to the finding of the first bipolar TOS, $CuInO_2$ [27] and fabrication of pn homojunctions made of TOSs [16]. We also applied the p-type TOS $SrCu_2O_2$ to pn heterojunctions using ZnO for the n-layer to fabricate near UV LEDs [3]. These achievements have demonstrated the

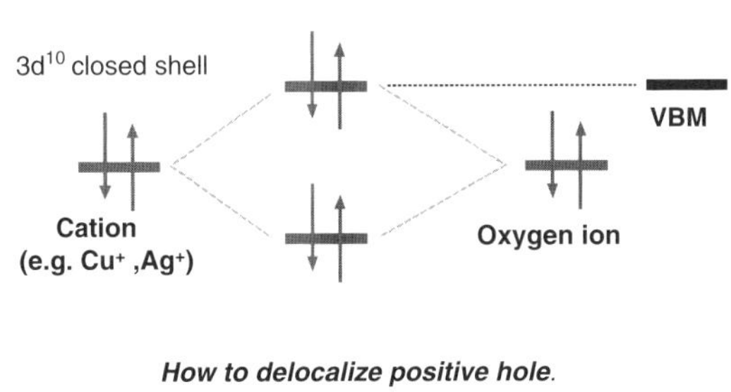

Figure 2.4 *A strategy to realize p-type transparent oxide semiconductors. It is the key to making isolated O 2p levels delocalize by forming covalent bonds. with metal cations. Transition metal cations with $3d^{10}$ have 3d levels comparable with O 2p levels*

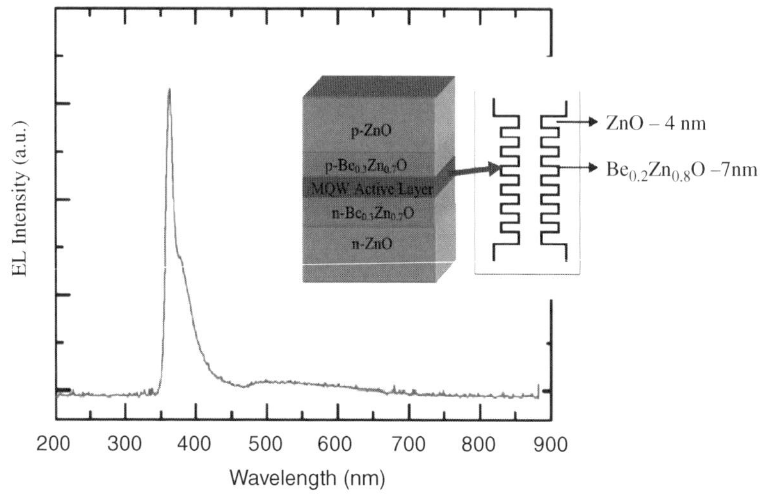

Figure 2.5 *Blue LED from ZnO pn-homojunction*

capability of TOSs for optoelectronic device applications, which launched 'transparent oxide electronics' (see [28, 29] for reviews). The realization of p-type ZnO is a long standing issue. Although many papers have been reported to date, papers reporting reliable data are few. A pn homojunction LED was fabricated by Tsukazaki *et al.* [17] and Rye *et al.* [30]. The former adopted the temperature modulation technique and nitrogen as the dopant to make the p-type, whereas the latter used arsenic as the dopant. Figure 2.5 shows the excitonic luminescence from a ZnO homojunction by current injection along with the device's structure [30]. The progress made in ZnO studies is summarized in the review by Özgür *et al.* [31].

2.4 Layered Oxychalcogenides: Improved p-Type Conduction and Room-Temperature Stable Excitons [26, 32, 33]

2.4.1 Improved Hole Transport in p-Type TOSs

First, we realized transparent p-type oxides using Cu^+-based TOSs. However, the hole mobilities and concentrations were far from practical levels. Thus, we then extended the guiding principle to approach (iii) given above. Figure 2.6 shows the crystal structure and energy band structure of LaCuOCh. Chalcogen ions were employed to form largely hybridized VBM with Cu^+ $3d^{10}$ orbitals. To satisfy the condition to maintain large band gaps, we chose layered oxychalcogenides, LnCuOCh (Ln = La, Ce, Pr, Nd; Ch = S, Se, Te), which have a layered crystal structure composed of alternately stacked $(Ln_2O_2)^{2+}$ and $(Cu_2Ch_2)^{2-}$ layers along the c-axis. We found that LnCuOCh exhibits intriguing optoelectronic properties, and it has been revealed that these properties are associated with their two-dimensional electronic structures. A large hole mobility ($\sim$8 $cm^2\,V^{-1}\,s^{-1}$) was obtained in nondoped LaCuOSe, and degenerate p-type conduction with moderately large hole mobility of $\sim$4 $cm^2\,V^{-1}\,s^{-1}$ was attained in Mg-doped LaCuOSe (LaCuOSe:Mg). In

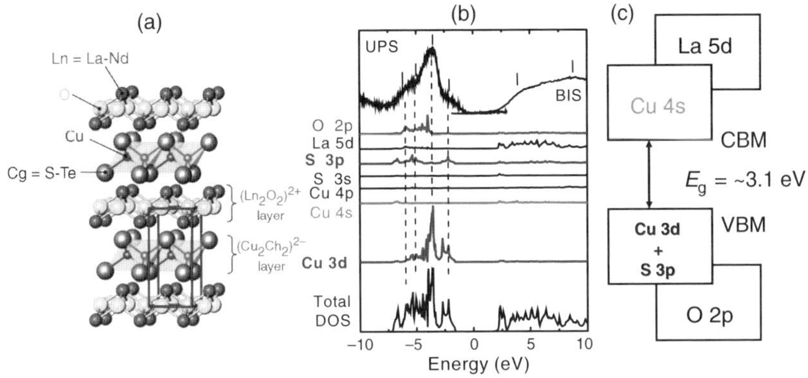

Figure 2.6 *LuCuOCh. (a) Crystal structure, (b) the calculated electronic structure and DOS probed by photoemission and (c) a simplified energy band*

addition, it was found that excitons were stable even at room temperature and sharp excitonic photoluminescence (PL) in the blue-to-UV region was observed. Here, we review these properties of the layered oxychalcogenides in relation to their electronic structures.

2.4.2 Epitaxial Film Fabrication: Reactive Solid-Phase Epitaxy (R-SPE) [34]

The epitaxial films were not obtained for LnCuOCh by simple vapor phase deposition in a vacuum chamber at high temperatures. Epitaxial films used for optical and electrical characterizations in our study were fabricated by the R-SPE method. Figure 2.7(a) shows the procedure for R-SPE. In R-SPE, a very thin (~5 nm), discontinuous Cu layer was first deposited on the (001) MgO surface for an epitaxial template layer. Then a thick amorphous LnCuOCh layer was subsequently deposited at room temperature. The samples were taken out from the chamber and sealed in an evacuated silica tube with LnCuOCh powder, followed by post-thermal annealing at ~1000 °C. This or similar procedures produced

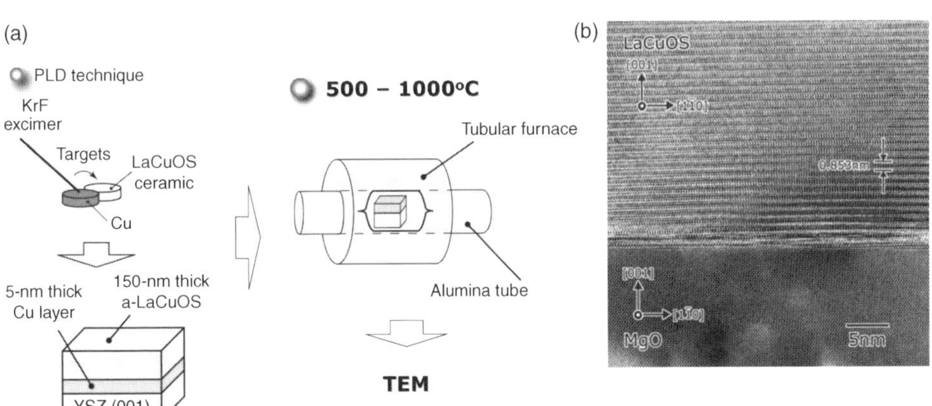

Figure 2.7 *(a) Procedure for R-SPE for fabrication of $LaCuS_{1-x}Se_xO$ epitaxial thin films and (b) TEM image of near the interface region of the resulting thin films*

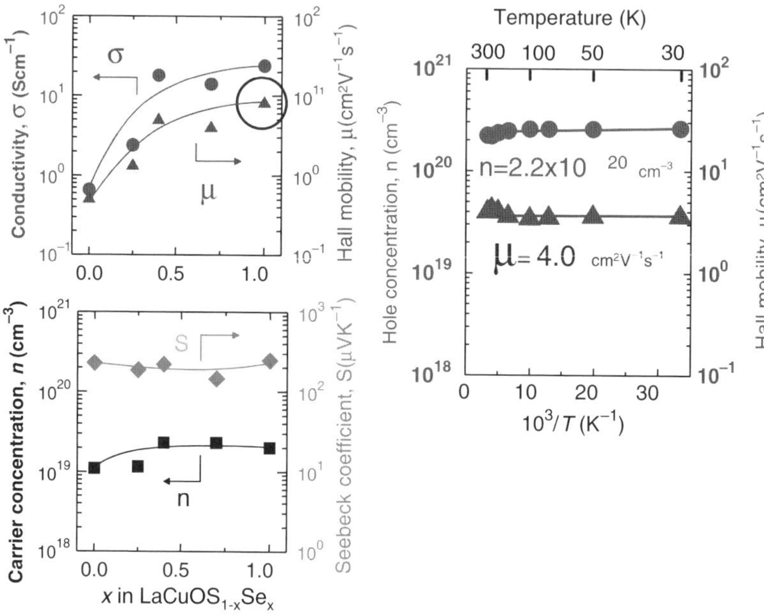

Figure 2.8 *Hole transport properties in LaCuS$_{1-x}$Se$_x$O epitaxial thin films at RT and temperature dependence with x = 1*

epitaxial films of a variety of LnCuOCh materials [20]. [Figure 2.7(b) shows a high-resolution TEM image of a LaCuOS film].

2.4.3 Carrier Transport, Light Emission and Excitonic Properties

Optical and electronic properties were examined using the epitaxial films prepared by R-SPE. All the LnCuOCh films exhibited p-type electrical conduction. Figure 2.8 shows the temperature dependence of conductivity and hole concentration and mobility estimated by Hall effect measurements. Mobility becomes larger with an increase in the Se content in LaCuO($S_{1-x}Se_x$), which agrees with our expectation described above. Since Se 4p orbitals have larger spatial spread than S 3p orbitals, hybridization with Cu 3d becomes larger and thereby hole mobility increases as the Se content increases. The largest hole mobility $\sim$8.0 cm^2 V^{-1} s^{-1}, which is comparable with that of p-type GaN:Mg, was obtained in LaCuOSe. By contrast, LaCuOSe has distinct differences from GaN:Mg. Mg doping to LaCuOSe increased the hole concentration up to 2×10^{20} cm^{-3} and carrier transport changed to degenerate conduction. (Note that degenerate p-type conduction has not been attained in any type.)

2.4.4 Two-Dimensional Electronic Structure in LnCuOCh [35]

Ab initio band calculations showed that the band structure is highly anisotropic and hole effective mass is smaller in the Γ–X direction than in the Γ–Z direction [Figure 2.9(a)], which implies that the electronic structure has a highly two-dimensional nature. Optical

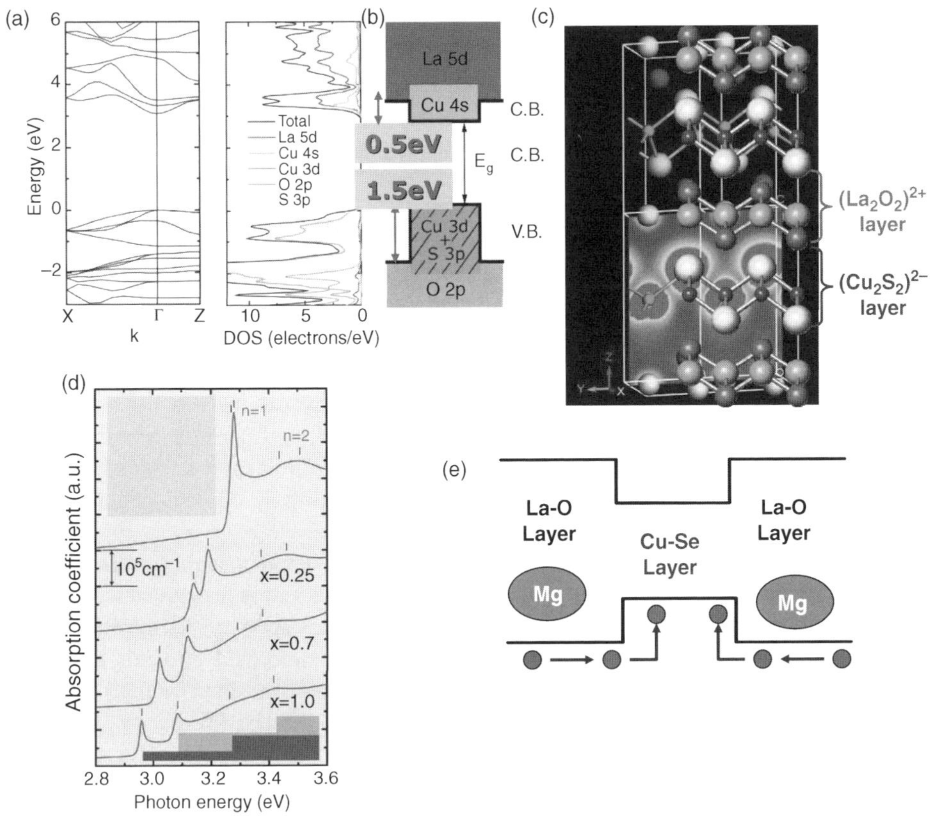

Figure 2.9 *Two-dimensional electronic structure in LnCuOCh. (a) Band structure of LaCuOS, (b) schematic illustration of electronic structure near the band gap, (c) hole density map, (d) two-dimensional optical absorption spectra obtained from LaCuO($S_{1-x}Se_x$) and (e) schematic illustration of local electronic structure to explain natural modulation doping of wide band gap semiconductors. See color plate section*

absorption spectra measured at 10 K [Figure 2.9(d)] showed a step-wise structure with sharp peaks just on the edges of the steps. Such a step-wise structure is similar to those observed in semiconductor artificial superlattices, and we speculated that it reflects two-dimensional density of states (DOS). A density plot of electrons in the vicinity of the VBM [Figure 2.9(c)] shows that hole transport paths spread only in the $(Cu_2Ch_2)^{2-}$ layers and holes are confined two dimensionally, explaining the step-wise structure of the optical absorption spectra. It was also confirmed that the split of the sharp peaks comes from multi-level excitons split due to spin–orbit interaction in Ch ions. Projected density of states (PDOS) showed that the VBM is mainly composed of hybridized orbitals of Cu^+ 3d and Ch p orbitals and the CBM of Cu 4s. Thus, the band gap is formed almost solely of the $(Cu_2Ch_2)^{2-}$ layer in LnCuOCh. In contrast, the $(Ln_2O_2)^{2+}$ layer (except for Ln = Ce) has a larger energy gap than the $(Cu_2Ch_2)^{2-}$ layer [Figure 2.9(b)]. This characteristic structure explains the moderately large hole mobility and intense photoluminescence in the heavily doped LaCuOSe:Mg. In LaCuOSe:Mg, Mg ions are thought to be doped in the$(Ln_2O_2)^{2+}$

layer. Holes are generated from acceptor levels made from the Mg dopants, which are then transferred to the $(Cu_2Ch_2)^{2-}$ layer [Figure 2.9(e)] because it forms hole transport paths [Figure 2.9(c)]. This electronic structure spatially separates the conducting holes from the ionized acceptors (Mg^{2+} ions at Ln^{3+} sites). As a consequence, impurity-carrier scattering is effectively reduced. Modulation doping, which is realized artificially in semiconductor superlattices and high electron mobility transistors (HEMTs), occurs in these layered oxychalcogenides. Third-order nonlinear optical susceptibility $\chi(3)$ is used to examine the nature of an optical band. $\chi(3)$ spectra showed peaks just at the excitonic absorption peaks [36]. The maximum $\chi(3)$ values are 4×10^{-9} esu for LaCuOS and $\sim 2 \times 10^{-9}$ esu for LaCuOSe, which are larger than that for ZnO films (1×10^{-9} esu). These results suggest that the large $\chi(3)$ values are enhanced due to excitons in LaCuOCh. As exciton binding energy is larger for ZnO (~60 meV) than for LaCuOCh (~40–50 meV), these large $\chi(3)$ values in LaCuOCh are attributable to the confinement of excitons in the $(Cu_2Ch_2)^{2-}$ layer. It was also confirmed that the exciton levels split by the spin–orbit interaction in Ch ions quantum mechanically interfered with each other if the split energy was small and the excitions came close, which was confirmed as quantum beating signals in degenerate four-wave mixing (DFWM) measurements on LaCuOS.

Finally, we would like to add that iron-based superconductors were discovered in the course of extending p-type TCOs to magnetic semiconductors within the crystal structure of LaCuOCh, i.e. when the Ch^{2-} ion is replaced with the pnictide ion Pn^{3-}, a 3d transition metal cation, TM^{2+}, with a 3d open shell structure, can be incorporated into this structure. Since the magnetic interaction between the $(TM_2Pn_2)^{2-}$ layer and intervening insulating $(La_2O_2)^{2-}$ layer is weak, we expected that magnetic ordering could be destroyed by injecting charge carriers to the (TMPn) layers. As a result, $LaFeAsO_{1-x}F_x$ with a $T_c = 26$ K [37, 38] (onset $T_c = 32$ K) was discovered as a by-product.

2.5 Nanoporous Crystal, C12A7: New Functions Created by Subnanometer Cages and Clathrated Anions [39, 40]

2.5.1 Crystal Structure of C12A7

Diversity in the crystal structure is a characteristic of oxides. Successful utilization of built-in nanostructure embedded in the characteristic crystal structure is an approach for the realization of new active electronic functions in oxides. In this section, we introduce the other exotic ionic crystal, the so-called C12A7 ($12CaO{\cdot}7Al_2O_3$). C12A7 is a typical electrical insulator and well known as a good refractory oxide and a constituent of alumina cement. However, it can be converted to a persistent electronic conductor and exhibits active chemical/optoelectronic functions by utilizing its intrinsic nanostructure.

The crystal lattice of C12A7 belongs to the space group $I\bar{4}3d$ with a lattice constant 1.199 nm, and the unit cell [Figure 2.7(a)] includes 12 cages (~0.4 nm in inner diameter) (Figure 2.10). The chemical formula for the unit cell is represented as $[Ca_{24}Al_{28}O_{64}]^{4+} + 2O^{2-}$: the former denotes the cage framework and the latter are called 'free oxygen ions'. Therefore, each cage is formally charged $+1/3e$ on average. Two free oxygen ions are clathrated in the unit cell to compensate the positive charge of the cage framework. A free oxygen ion is octahedrally coordinated with six Ca^{2+} ions that constitute part of the

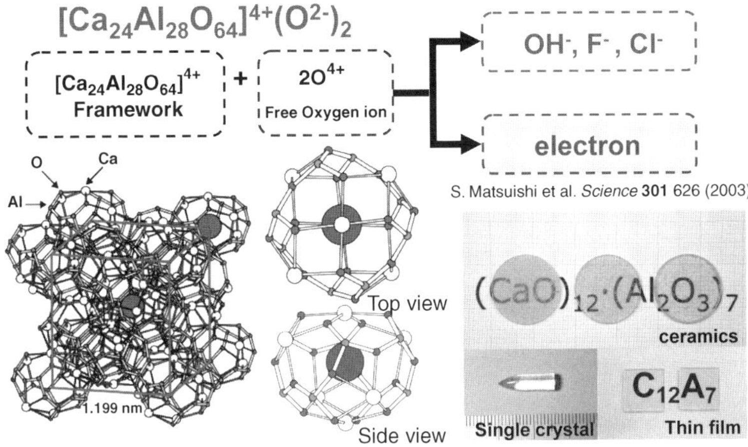

Figure 2.10 *Crystal structure of 12CaO·7Al$_2$O$_3$. Unit cell has Z = 2 and is composed of 12 cages with an inner diameter of ~0.4 nm. Two free O^{2-} are trapped as counter anions. Also shown are photos of various forms of samples fabricated*

cage wall. The separation between Ca^{2+} and free O^{2-} ion is longer by ~50% than the sum (0.24 nm) of the respective ionic radii, indicating that the free oxygen ion is loosely bound in the cage.

Although the crystal structure looks rather complex, the synthesis is quite easy (by heating $CaCO_3$ and Al_2O_3 at temperatures >1000 °C in an ambient atmosphere).

2.5.2 Electronic Structure of Clathrated Ions

As noted in Section 2.2, band gaps in typical transparent oxides are made of O 2p bands for VBM and metal s orbitals for CBM. This is because the Madelung potential raises energy levels of electrons in cations and lowers those in anions, which thereby stabilizes both the cation and anion states and causes the formation of the large band gaps. Common oxides do not contain an O^- ion in a bulk crystal because the lowest unoccupied atomic orbital (LUAO) level of O^- is lower than the highest occupied atomic orbital (HOAO) level of O^{2-}, and therefore the O^- state is not stabilized unless a hole is introduced.

This consideration implies that more electronegative anionic sites are required to stabilize the O^- state. Such electronegative sites and the presence of O^- ions are found at surfaces and in the vicinity of the defects with specific structures. The crystal structure of C12A7 offers such electronegative sites in the sub-nanometer-sized cages. Since the cage framework is charged positive, the Madelung potential in the cage is positive and provides anionic sites (here the Madelung potential V_M is defined so as to express the ion energy as qeV_M, where q is the ionic charge and e is the elementary electric charge, therefore a positive Madelung potential stabilizes anionic states). The distance between the cage center (i.e. the free O^{2-} ion site) and the nearest neighboring Ca^{2+} is longer by ~50% than the usual distance, therefore the Madelung potential of the free O^{2-} ion is shallower ($V_M = 8.4$ eV) than those of the cage wall O^{2-} ions ($V_M = 24.1$–25.4 eV). The electronegative potential at the cage centre raises the HOAO level of the free oxygen ion to above the VBM level, which consequently

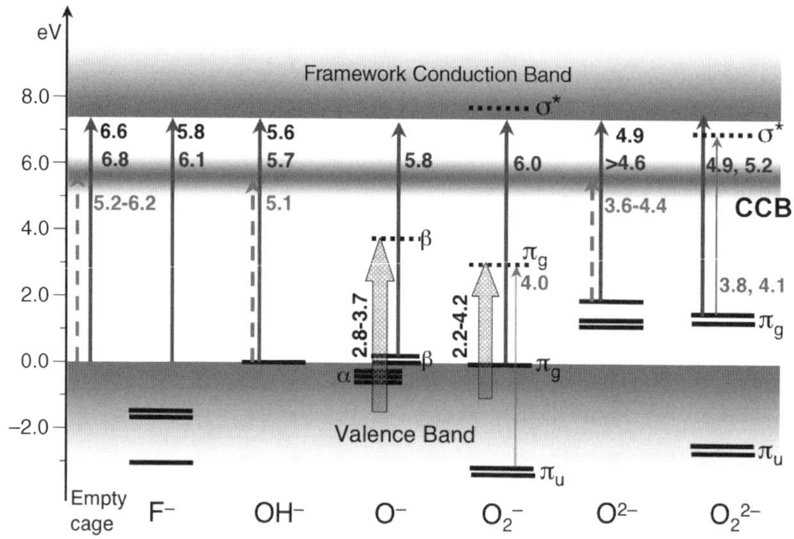

Figure 2.11 *Calculated energy levels of various anions in the cages of C12A7*

stabilizes the O^- state. This is the case for other anions such as O^{2-}, H^-, Cl^-, F^- and OH^-, and probably provides a preferable site also for forming other anions such as Au^- and Na^- if these atoms are introduced in the cages (electronic levels of some of these clathrated anions were calculated in [41]). Multivalent anionic states such as O_2^- and O_2^{2-} can also be stable as long as their HOAO levels are in the fundamental band gap of the cage framework of C12A7 and do not exceed the CBM. It should be noted that we need to take effects of local structural relaxation into account because the cage structure is flexible and easily deforms to stabilize the clathrated anions, which lowers the energy levels of the electrons in the clathrated anions significantly. Figure 2.11 summarizes the energy levels of various anion species entrapped in the cages [41].

It should be noted that the cages in C12A7 have a moderate size to give a Madelung potential suitable for stabilizing many anionic states. If the cage size was much larger, the Madelung potential becomes shallower and HOAO levels of multivalent anions would exceed the CBM of the cage framework. If the cage size was much smaller, the Madelung potential becomes deeper and approaches the VBM level, therefore O^- and other monovalent anionic states are not stabilized anymore.

2.5.3 C12A7:H$^-$: Reversible Insulator–Conductor Conversion by UV Irradiation and Thermal Heating [9, 42]

Substitution of the free oxygen ions with hydrogen creates high-density H^- ions in the cages (C12A7:H^-), which leads to the first demonstration of electronic conduction in oxides composed only of light ions. C12A7:O^{2-} and C12A7:O^- have large band gaps (>5 eV; the fundamental band gap of the cage framework would be ~6 eV), therefore they are good electrical insulators with conductivities below the experimental limit of our set-up ($<10^{-10}$ S cm^{-1}). C12A7:H^- is also an electrical insulator, while it exhibits conductivities as

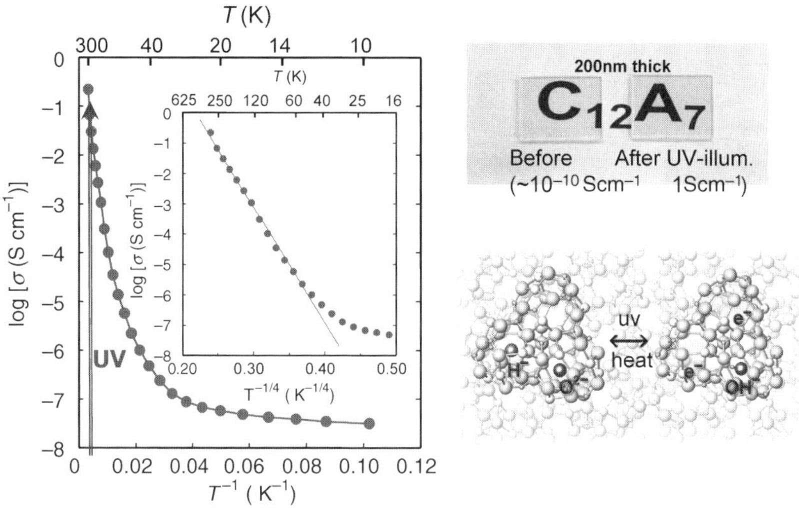

Figure 2.12 *Change in conductivity of C12A7:H⁻ with UV illumination. Transparent conductive thin films are obtained after UV illumination. Carrier electrons are released from H⁻ ions by photoionization*

large as ~1 S cm^{-1} at RT after irradiation with UV light (Figure 2.12). New optical absorption bands appear at the same time at ~0.4 eV and 2.8 eV, changing the sample color from colorless transparent to green. The insulating and colorless transparent properties are recovered by heating at ~350 °C. It is thought that H^- ions clathrated in the cages are ionized by UV light to a pair of $H^0 + e^-$. The electrons released from H^- are trapped in other cages. The electrons then migrate through cages and contribute to the electronic conduction. The detailed mechanism of electronic conduction and coloration were examined by *ab initio* embedded cluster calculations. They showed that the optical absorption at ~0.4 eV came from an inter-cage transition of the electrons trapped in the cages and that at ~2.8 eV from an intra-cage s–p transition. The electrons hop from a cage to a neighboring cage with the assistance of cage deformation (i.e. phonon), which thereby forms a polaron and contributes to electronic conduction.

2.5.4 C12A7:e⁻: Room-Temperature Stable Inorganic Electrode [43, 44]

The role of H^- in C12A7:H^- is to provide an electron to the cage through photoionization. This suggests that intrinsic C12A7 would have an electroconductive nature if electrons were successfully injected to the empty cages in C12A7. So, we tried to replace the free O^{2-} in the cages with electrons by chemical reduction processes and found it was possible to replace all the free oxygen ions with electrons (C12A7:e^-), forming a structure that may be regarded as an 'electride' [45]. An electride is a crystalline salt in which stoichiometric amounts of electrons serve as anions. A typical example of an electride is an ionic compound in which a cesium cation is complexed by cryptand or crown ether to form a giant cation. Potential applications such as an electron emitter are expected. However, they decompose below −40 °C or upon exposure to ambient atmosphere. This is why electrides still remain as exotic materials and there have been no applications and a limited amount of detailed fundamental

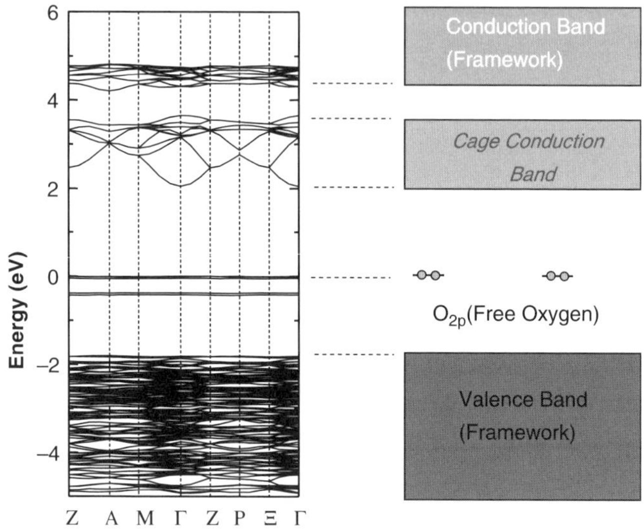

Figure 2.13 *Calculated band structure of C12A7*

research. C12A7:e^- exhibits electronic conduction with conductivities greater than 1000 S cm^{-1} without UV irradiation [46], which is different from the case of C12A7:H^-. The free oxygen ions may be replaced with electrons by reducing processes such as Ca or Ti vapor treatment. The complete replacement leads to the formation of $[Ca_{24}Al_{28}O_{64}]^{4+}(4e^-)$, which is regarded as a new type of inorganic electrode. Theoretical analyses and experimental approaches have revealed that the cages in C12A7 form an additional conduction band called the 'cage conduction band' (CCB*)*. Figure 2.13 shows the band structure of C12A7 (parent state, i.e. free O^{2-} ions are accommodated in the cages) and C12A7:e^-. The CCB is located 1–2 eV below the bottom of the 'framework conduction band', which is primarily composed of Ca 5s orbitals. At low electron concentrations, the electrons induce a large lattice deformation due to the Coulomb attractive force between the entrapped electron and the two Ca^{2+} ions on the cage wall, causing electron localization. Consequently, the conduction occurs via the hopping of the electron from a localized deformed cage (an isolated quantum dot state) to the CCB (i.e. by *polaron*) as in the UV-illuminated. This is the reason why the drift mobility is much smaller than 1 $cm^2\,V^{-1}\,s^{-1}$ and the conductivity shows a thermally activated behavior in the electron-doped C12A7 reported previously. Here we show that the complete replacement of the free oxygen ions by electrons in C12A7 results in a metallic conduction with a sharp enhancement of the mobility from ~ 0.1 to 4 $cm^2\,V^{-1}\,s^{-1}$ [47]. Figure 2.14 shows the electron transport properties in C12A7:e^- as a function of electron concentrations in the cages.

When the metallic conductive C12A7:e^- is cooled down to lower temperatures, it exhibits a metal-superconductor transition at 0.2–0.4 K [10]. This is the first s-metal superconductor which exhibits a T_c at ambient pressure. It is worth noting that C12A7 composed of CaO and Al_2O_3 can be transformed from an insulator to a superconductor through a semiconductor by appropriate electron doping. These findings encourage the exploration of TCO materials using only abundant constituents by creating novel nanostructures [48].

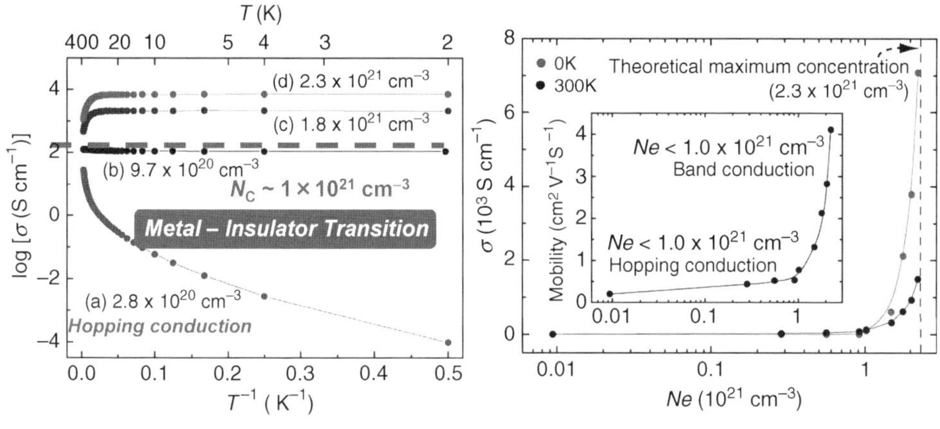

Figure 2.14 *Metal–insulator transition in C12A7:e⁻. Metallic conduction occurs at $>1 \times 10^{21}$ cm^{-3} electrons in the cages. Mobility is rapidly increased beyond this critical concentration*

2.5.5 Embedded Quantum Dots in C12A7 [49, 50]

The electronic structure of C12A7 has been studied by ultraviolet photoelectron spectroscopy (UPS) [50] and using an *ab initio* code WIEN2k. Figure 2.15 shows a calculated band structure of C12A7:e⁻ and the DOS measured by photoemission spectroscopy. The Fermi level was located at 1–2 eV below the edge of the fundamental band gap of the cage framework, which was in the energy bands formed by nearly free-electron states confined in

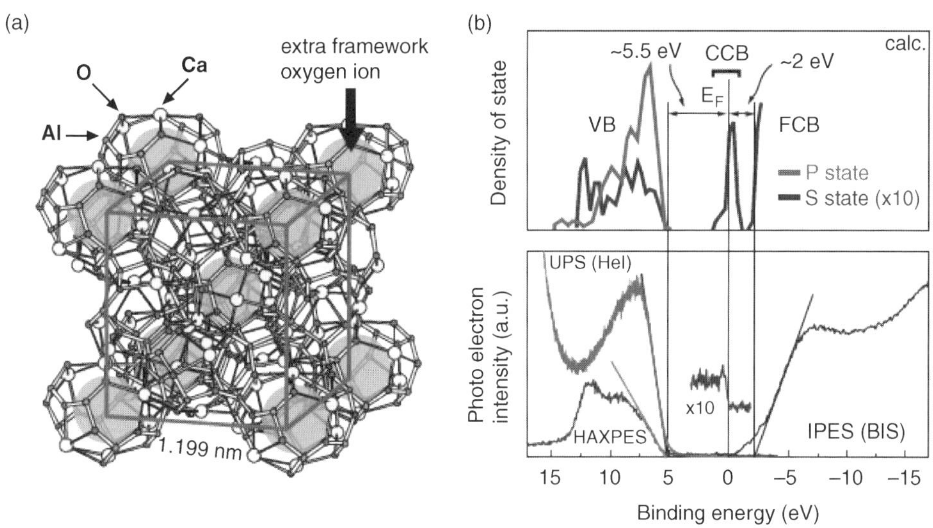

Figure 2.15 *Cage conduction band in C12A7:e⁻. (a) Crystal structure of $[Ca_{24}Al_{28}O_{64}]^{4+}(4e^-)$. (b) Calculated DOS and observed DOS by photoemission spectroscopy. The CCB is observed below the CBM*

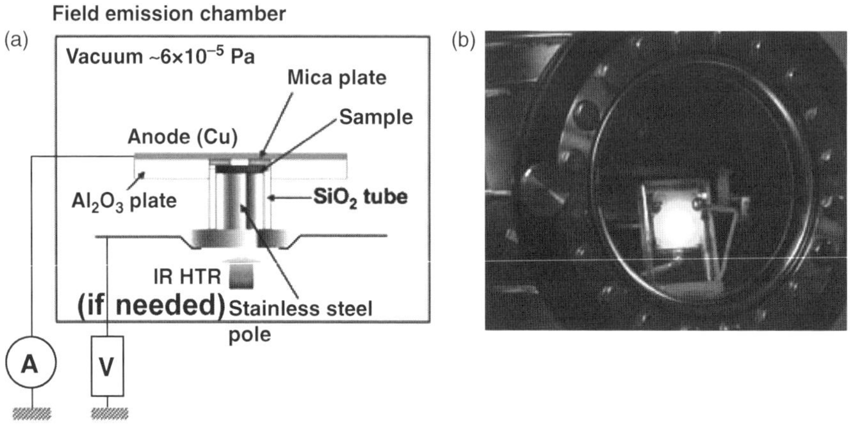

Figure 2.16 *Electron field emission using C12A7:e⁻ as the cold electron source. Device structure (a) and photograph (b) demonstrating the luminescence*

the cages. It is visually seen in the density map of the clathrated electrons. The clathrated electron shows a plateau around the cage center in C12A7:e⁻, showing that the electrons are loosely confined in the cages, while the electrons are attracted to the clathrated H^0 atoms to form H^- anions in C12A7:H^-, yielding more localized states. The above results suggest that the electronic states of the clathrated electrons are likely to be free electrons, which suggests the cages in C12A7:e⁻ can be regarded as quantum dots. Indeed, simple simulations based on the free-electron approximation and a coupled quantum dot model provided good quantitative agreement with optical transition energies and oscillator strengths estimated from the optical absorption spectra, validating the coupled quantum dot view for the cages in C12A7:e⁻ [50].

2.5.6 Device Application: Field Emission of Clathrated Electrons [49]

From the above consideration of electronic structure, we expected that C12A7:e⁻ might have a small work function (2.4 eV [51]) and exhibit good electron emission properties since the clathrated electrons are loosely confined in the cages. We used a mirror-polished surface of a C12A7:e⁻ single crystal as electron emitter to examine fundamental material properties. Field emission characteristics were measured in a vacuum chamber with an emitter surface–extraction electrode distance of 0.05 mm. Figure 2.16 shows the field emission luminescence at RT from a fabricated device using C12A7:e⁻. A clear bright emission was observed even in daylight.

2.6 TAOSs and their TFT Applications

2.6.1 TAOSs in Amorphous Semiconductors

The most important feature of semiconductors is in the controllability of carrier concentration over several orders of magnitude. A unique advantage of amorphous materials over crystalline materials is their capability of large-area deposition of uniform thin films at low

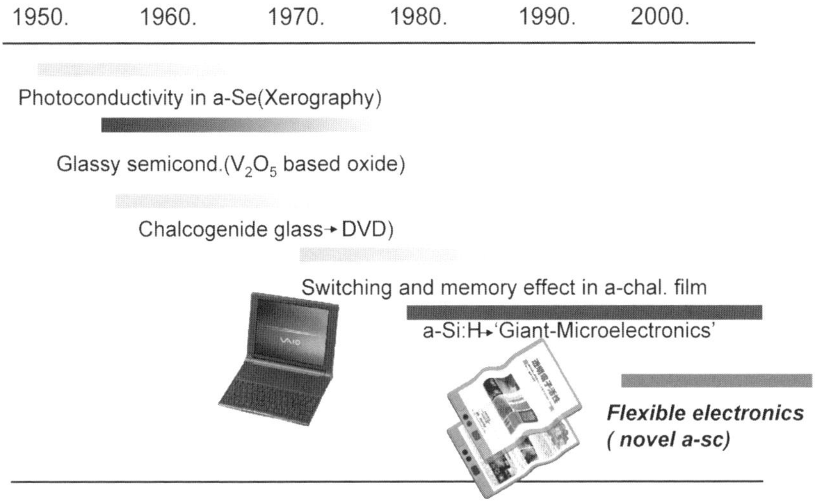

Figure 2.17 *History of amorphous semiconductors and applications*

temperatures. These process advantages make amorphous semiconductors extremely favorable for large-sized electronic devices fabricated on plastic substrates. Research on amorphous semiconductors started in the 1950s to seek materials which have both of these advantages. Figure 2.17 summarizes the history of amorphous semiconductors. The largest impact on electronics was the discovery of hydrogenated amorphous silicon (a-Si:H) in 1975 by Spear and LeComber [11]. This was the first material which could control both carrier type and concentration by impurity doping as in crystalline Si, and it opened a new frontier called 'Giant Microelectronics', which means electronics based on circuits fabricated on a large-area substrate. A TFT substrate a-Si:H on glass was first reported by LeComber *et al.* [12]. Although the field effect mobility was very a small value, such as $\sim$0.5 cm^2 (V s)$^{-1}$, this TFT met the demand for a backplane of liquid crystal displays.

A new area of electronics is rapidly emerging for applications which cannot be fabricated by Si complementary metal oxide semiconductor (Si-C-MOS) technology. This area, named 'flexible electronics', is characterized by electronic circuits fabricated on organic plastic (soft) substrates instead of inorganic (hard) glasses. This area was created to meet a strong demand for large-area light flexible displays because glass substrates, which are heavy and fragile, are obviously inconvenient. Amorphous semiconductors are preferable to crystalline semiconductors for flexible electronics. So far, organic semiconductors have been almost exclusively examined for such applications [5] but their performance and chemical/electrical instability are still insufficient for practical applications: e.g. field-effect mobilities of organic TFTs are too low to drive high-resolution, high-speed organic light emitting diode (OLED) displays and performance stability under electrical stress and an ambient atmosphere is rather poor.

Figure 2.18 shows a location map of the various types of amorphous materials on a plane constituted of a chemical bonding nature axis and a band gap axis. From Figure 2.18 it can be seen that conventional amorphous materials composed of chemical bonds having high covalency and large band gap (transparent) are electrically insulating. It is seen that a

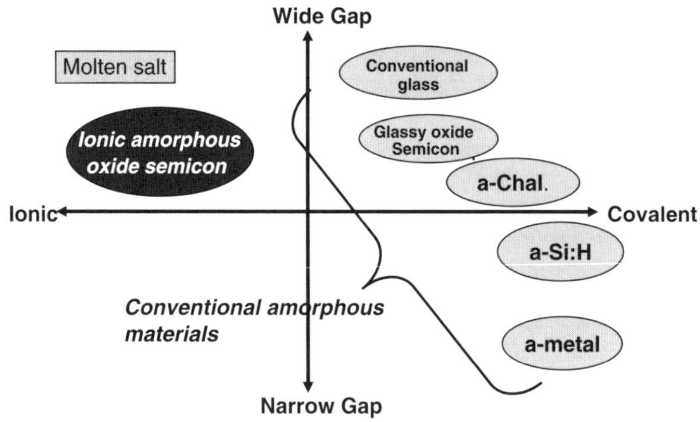

Figure 2.18 *Ionic amorphous oxide semiconductors on the location map of amorphous materials*

transparent, ionic, amorphous semiconductor is an unexploited class of amorphous semiconductor. As an extremely high quenching rate is needed to obtain an amorphous ionic material compared with a conventional oxide glass, physical deposition techniques from vapor phase on a substrate at RT are appropriate for this purpose. TAOSs belong to the family of ionic amorphous oxides in which the Fermi level is controllable.

2.6.2 Material Design for Transparent TAOSs with Large Electron Mobility [5, 52]

In ionic materials, the nature of the CBM which works as an electron pathway totally differs from that of the VBM which acts as a hole pathway. The CBM in ionic oxides is primarily composed of vacant s orbitals of a cation, and the contribution of oxygen 2p orbitals, which are dominant at the VBM, is rather small. The spatial spread of this vacant s orbital is so large that direct overlap between the s orbitals of the neighboring cations is possible in heavy metal oxides, and therefore an effective electron mass is small in these oxides, as described before. What happens if these TCO materials when they become amorphous? In an amorphous state, structural randomness concentrates on an energetically weak structural unit. In most amorphous materials structural randomness appears prominently as the bond angle distribution. When the bond angle has a large distribution, how is the effective mass (in other words, the transfer rate between neighboring cation s orbitals) modified for carrier electrons? We considered the two cases of covalent semiconductors and ionic semiconductors. In the former case, the magnitude of the overlap between the vacant orbitals of the neighboring atoms is very sensitive to the variation in the bond angle. As a consequence, rather deep localized states would be created at somewhat high concentrations and thereby the drift mobility would be largely degraded due to carrier scattering with these defects. However, the magnitude of the overlap in the latter case is critically varied by the choice of metal cations: when the spatial spread of the s orbital is larger than the inter-cation distance, the magnitude should be insensitive to the bond angle distribution because the s orbitals are isotropic in shape. As a consequence, we may anticipate that these ionic amorphous materials have large mobility comparable with that in the corresponding crystalline form. This is the amorphous material we have been seeking.

In the opposite case, such a favorable situation cannot be expected. The spatial spread of the s orbital is primarily determined by the principal quantum number (n) of a cation and is modified by the charge state of the cation. Here, we take the value of n as a measure of the spatial spread of the metal cation s orbitals, candidates for transparent amorphous semiconductors having large electron mobilities comparable with those of the corresponding crystals, which are transparent oxides consisting of post transition metal cations with an electronic configuration $(n-1)d^{10}ns^0$, where $n \geq 5$. Note that transition metal cations with an open shell structure are ruled out as candidates because they are not transparent due to absorptions arising from d-d transitions. In the case of crystalline oxide semiconducting oxides, this requirement is relaxed to be $n \geq 4$ as exemplified by ZnO; Zn^{2+} has the (Ar) $(3d)^{10}(4s)^0$ configuration, because crystalline materials have much more regular and compact structures than amorphous oxides.

Figure 2.19 compares the orbital drawings of a CBM and a post transition metal oxide (PTMO) for the crystalline and amorphous states. The drastic reduction in electron mobility in the amorphous state from the crystalline state may be understood intuitively from Figure 2.19, whereas medium mobility in c-PTMO is reserved even in the amorphous state. In a sense, the situation of CBM in PTMO is similar to that in amorphous metal alloys. The conductivity of amorphous metal alloys remains slightly lower compared with the corresponding crystalline phases, as illustrated in Figure 2.19. The structure of amorphous metal is modeled by dense random packing of metal spheres and occupation of metalloid in the interstitial positions. We may consider that in a-PTMO the vacant ns orbitals of post transition metal cations (PTMCs) work as metal elements in amorphous metal.

Electron dopability in semiconductors is determined by the stability of doped electrons and ease of counter defect formation in the hosts. If a higher valence state of candidate

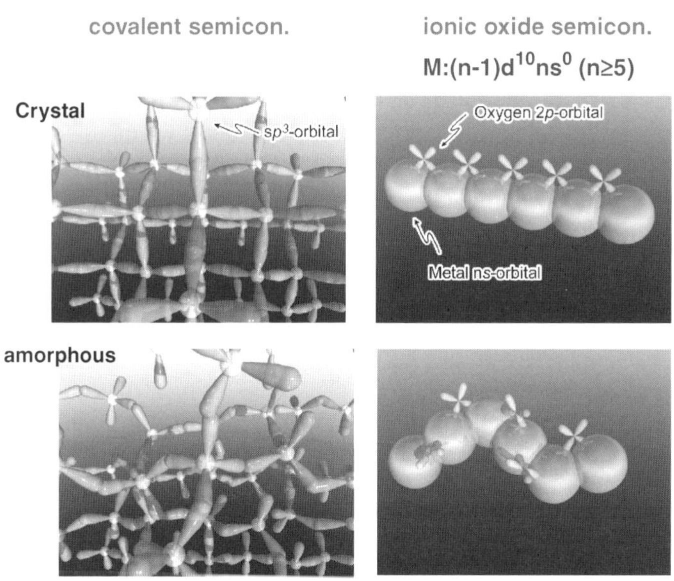

Figure 2.19 Comparison of orbital drawings in the vicinity of the CBM between covalent semiconductor and post transition metal oxides

PTMCs is not so stable, the doped electron is captured by the low valence state. Tin ion, which meets the criteria as a PTMC, has this tendency, i.e. Sn^{4+} is a cation suitable as an a-TOS but the low valence state Sn^{2+} is easily formed compared with In^{3+} with the same electron configuration. This valence stability of the PTMC is reflected in the easiness of fabrication process optimization.

As for the stability of doped electrons, the electron-neutrality level of the candidate metal oxides is close or above the CBM due to a large energy dispersion of CBM reflecting a large overlap in *ns* orbitals between the neighboring metal cations [53]. As a consequence, the above hypothesis predicts that transparent amorphous oxides are capable of electron doping and have a large electron mobility comparable with the corresponding crystalline phases.

2.6.3 Electron-Transport Properties

There are so many composition varieties for TAOSs. Among them amorphous In_2O_3-Ga_2O_3-ZnO (IGZO) has been extensively studied as the semiconducting channel layer of transparent TFTs since the first report [33] in 2004 [6]. Figure 2.20 summarizes the electrical properties] Hall mobility (μ_{Hall}) and carrier concentration (N_e)] for films in the IGZO system.

Both the Hall mobility and the carrier concentration rapidly decrease with increasing Ga^{3+} ion content. Hall motilities in the a-IGZO films decreased from $\sim$25 cm^2 (V s)$^{-1}$ at $N_{ew} \sim 10^{20}$ cm^{-3} to $\sim$1 cm^2 (V s)$^{-1}$ at $\sim 10^{18}$ cm^{-3} as the Ga^{3+} ion content increased from 30 to 50%. However, we should note that the Hall mobility values here are not the maximum potential of these materials because Hall mobility largely depends on carrier concentration in TAOSs due to structural randomness. Carrier mobility strongly depends on carrier concentration, and large mobilities are obtained at carrier concentrations larger than a threshold value (e.g. $\sim 10^{18}$ cm^{-3} for a-IGZO). However, introduction of high-density carriers (e.g. $>10^{20}$ cm^{-3}) became much difficult in the larger Ga content films. This result indicates that large mobility is not easily obtained in the a-IGZO films with large Ga contents if one tries to dope carriers by impurity doping or introducing oxygen vacancies. However, it would not be a

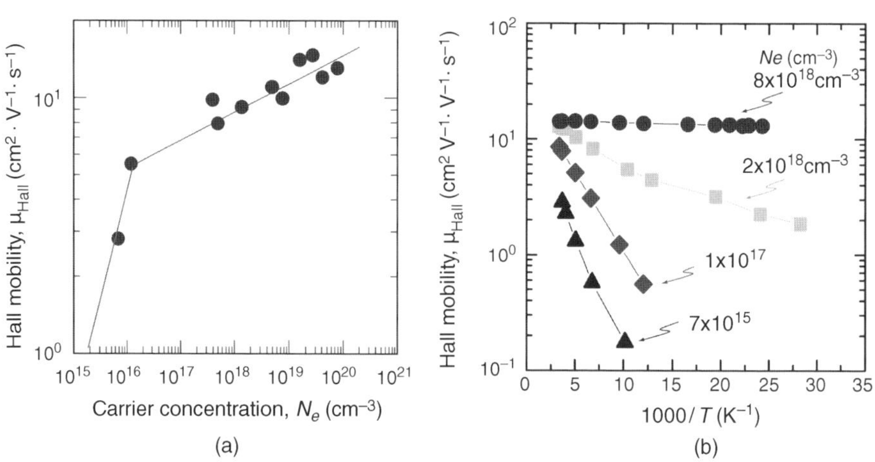

Figure 2.20 Hall mobility in a-In-Ga-ZnOx thin films as a function of (a) carrier concentration and (b) temperature

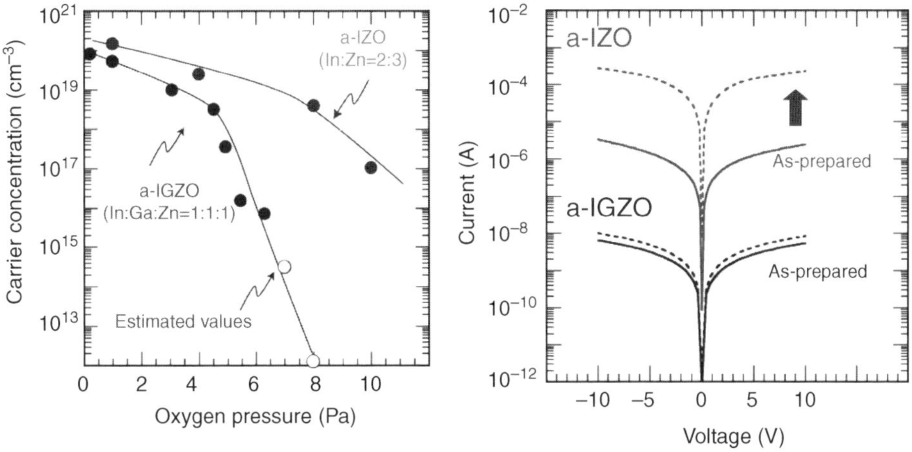

Figure 2.21 *Effect of oxygen pressure during the deposition and annealing at RT for 1 day on the conductivity of the resulting a-In-Ga-ZnO (a-IGZO) and a-In-ZnO (a-IZO)*

disadvantage for semiconductor device applications because the difficulty in carrier doping by oxygen vacancy suggests better controllability and stability of carrier concentration, especially at low concentrations. Even if high-density doping is difficult by choosing the deposition conditions, it is still possible to induce high-density carriers by external electric field if TFT structures are employed, which may make it possible to utilize the potential large mobilities that may be available at large carrier concentrations.

Hall mobilities larger than 10 cm^2 (V s)$^{-1}$ are obtained also in the In_2O_3-ZnO (a-IZO) and the In_2O_3-Ga_2O_3-ZnO (a-IGZO) systems. However, the controllability and stability of carrier concentrations are not satisfactory in a-IZO films. Figure 2.21 shows controllability of the carrier concentration of the a-IGZO (nominal chemical composition in atomic ratio was In:Ga:Zn = 1:1:1) and a-IZO (In:Zn = 2:3) films, in which carrier concentration is plotted against oxygen partial pressure during the film deposition. The carrier concentration was well controlled from $<10^{15}$ to 10^{20} cm^{-3} by varying oxygen pressures from 0.1 to 7 Pa for a-IGZO. In contrast, it is hard to reduce carrier concentration down to 10^{17} cm^{-3} for a-IZO. It was not impossible to further reduce carrier concentration if oxygen pressure was further increased. However, the properties of such films were not stable: excess carriers were easily generated in the a-IZO films even if they were kept in ambient air or even if photolithography processes were used to fabricate TFTs. Note that similar effects to suppress unintentional carrier generation were confirmed using Al^{3+} ions instead of Ga^{3+} ions by keeping reasonable mobilities of ~10 cm^2 (V s)$^{-1}$ [6]. However, fabrication of large-sized and dense sputter target for practical application is difficult for such a composition because such a ternary compound does not exist.

2.6.4 TAOS-TFTs

Recent research and development involving display applications of TAOS-TFTs is leading towards practical applications. The main drivers in this area are major display related companies and SID meetings are now the main forum of presentation including exhibitions.

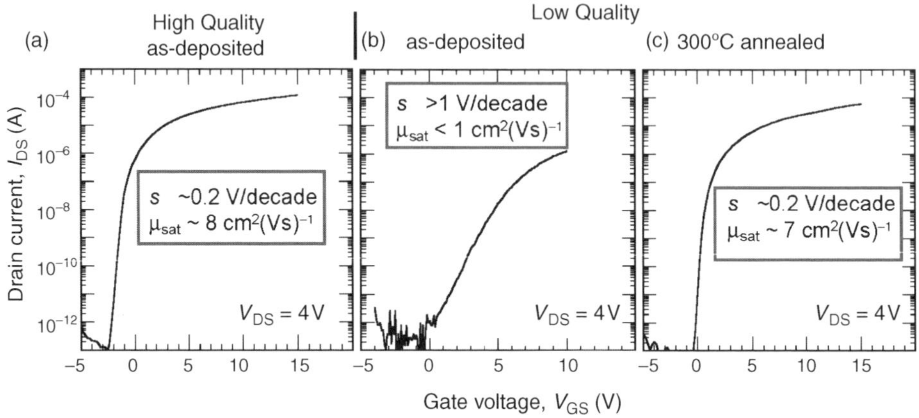

Figure 2.22 *Improvement of a-IGZO-TFT performance by post annealing. (a) High and (b, c) low quality samples mean the thin films were deposited under an optimized and intentionally unoptimized conditions*

Many papers on TAOS-TFTs have been published and the University of Oregon group led by Wager recently published a book [54]. Here recent advances in display applications using TAOS-TFT arrays are reviewed to avoid overlap with Wager *et al.*'s book.

2.6.4.1 Unique Features of TAOS-TFTs

TAOS-TFTs have three unique characteristics compared with other TFTs. The first is high field mobility [$>10\,cm^2\,(V\,s)^{-1}$]. The second is easy fabrication at low temperature using conventional DC sputtering. The third is a large process allowance. The TFTs fabricated at unoptimized conditions exhibits poor performance, but the TFT performance can be much improved to that prepared under optimized conditions just by annealing at an appropriate temperature far below the crystallization temperature of TAOSs. Figure 2.22 is an example of a-IZGO TFTs showing the effectiveness of post annealing to improve the TFT performance. The annealing temperature is 250–300 °C which is much lower than the crystallization temperature (>500 °C). No distinct structural change around each metal cation was noted before and after annealing. Pronounced annealing effects are observed commonly for TAOS-TFTs [53].

Figure 2.23 shows the performance histograms of a-IGZO TFTs which were fabricated on a glass substrate by conventional sputtering with subsequent annealing [55]. About 100 TFTs were fabricated from a 1 cm × 1 cm area of a-IGZO thin film. The TFT exhibits excellent uniformity and high average performance. The saturation mobility (μ_{sat}) value resides within a range of $0.5\,cm^2\,(V\,s)^{-1}$ and σ is $0.11\,cm^2\,(V\,s)^{-1}$ (0.76% of the average value), demonstrating the excellent uniformity of a-IGZO TFTs [54]. It strongly suggests that the a-IGZO TFTs essentially have a good short-range uniformity and are advantageous in integrated circuits and large areas.

2.6.4.2 Novel Display Structure

An innovative electronic paper display structure called 'front drive' type was recently proposed by Ito *et al.* of Toppan Printing [18]. Alignment of TFT array to color filter array is

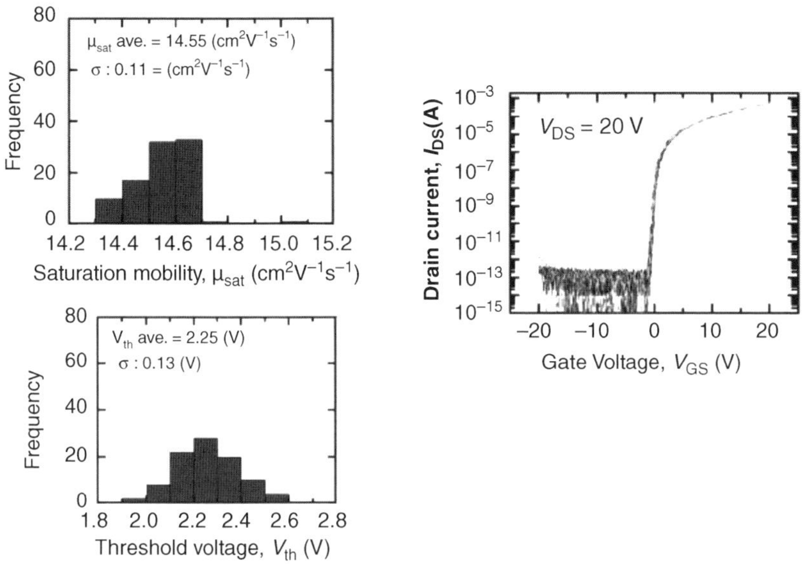

Figure 2.23 *a-IGZO-TFT performance distribution. TFTs are fabricated on SiO_2/Si substrates*

a troublesome process in the display assembly because a-Si:H is nontransparent and there is variation of substrate dimension with aging. Their idea to avoid this difficulty was to directly deposit the TAOS-TFT arrays on the color filter arrays utilizing the low temperature process and optical transparency simultaneously. This is the first demonstration of a device structure benefitting from the optical transparency of TAOSs. Figure 2.24 shows the front-drive structure applied to an electronic paper based on electrophoretic ink imaging film. Taking

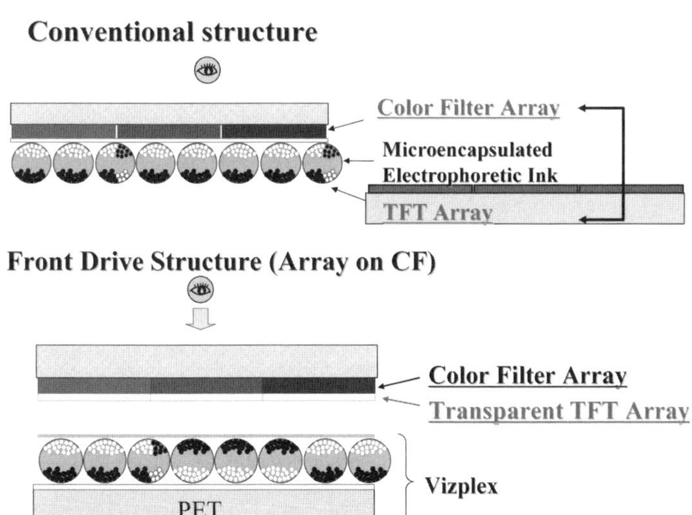

Figure 2.24 *Front drive structure. TFT arrays are fabricated on the color filter array directly*

advantage of the transparency of TAOS, the TFT and the color filter arrays can be positioned at the viewing side of displays. This display structure is applicable to other displays, facilitating the alignment.

2.6.4.3 Driving Backplane of OLED and Liquid Crystal Display (LCD) Panels

Conventional active-matrix (AM) flat panel displays are based on amorphous or polycrystalline silicon TFT technology. Limitations of the amorphous silicon (a-Si:H) include visible light sensitivity and a low field-effect mobility, which reduce the pixel aperture ratio and driving ability for some applications.

A typical example is an OLED which drives by current. Since the luminous intensity of OLEDs is proportional to the flow current, higher mobility TFTs are required. Although polycrystalline silicon TFTs have a larger field-effect mobility, its uniformity over large area is not acceptable for high yield manufacturing. In recent years, there has been a great interest in TFTs made of TOSs. This is mainly due to metal oxide semiconductor TFTs having unique advantages, such as visible light transparency, a large-area uniform deposition at low temperature, and a high carrier mobility. However, conventional metal oxide semiconductors, such as zinc oxide (ZnO), are polycrystalline in nature even at RT. The grain boundaries of such polycrystalline metal oxide could affect device properties, uniformity and stability over large areas.

Recently, OLED and LED panels driven by a-IGZO-TFT backplane have been presented in SID and related meetings. At SID '07 LG Electronics presented a 4 in. size AM full color OLED using this backplane [56]. This is the first demonstration of an OLED based on oxide semiconductor TFTs. Subsequently, they presented a flexible OLED at IMID '07 by depositing a-IGZO on a thin stainless plate [57]. Figure 2.25 shows the photo of a flexible a-IGZO TFT array and OLED. SID '07 was memorable for TAOS-TFTs because a technical session on oxide TFTs was launched for the first time at this conference, which is the largest

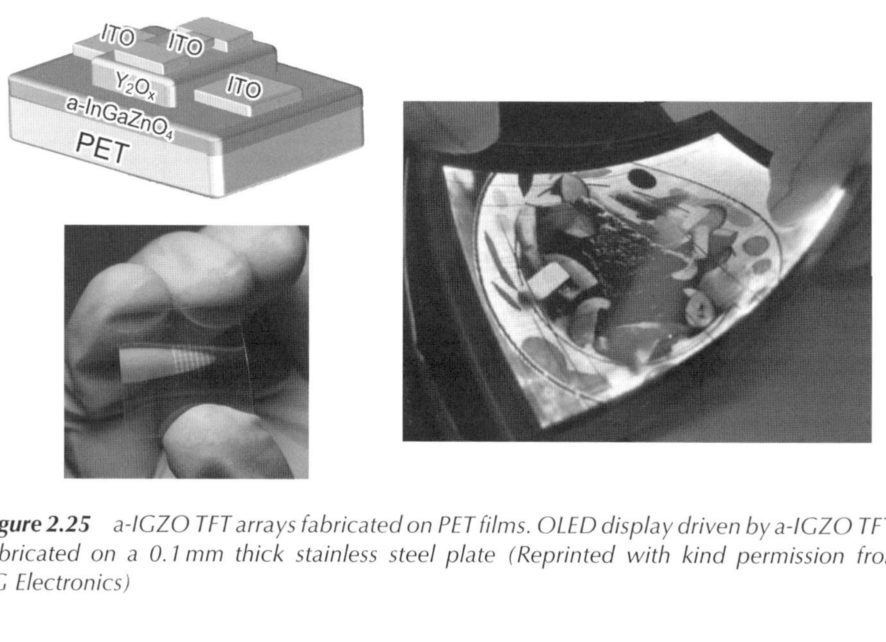

Figure 2.25 *a-IGZO TFT arrays fabricated on PET films. OLED display driven by a-IGZO TFTs fabricated on a 0.1 mm thick stainless steel plate (Reprinted with kind permission from LG Electronics)*

Figure 2.26 *12 in. OLED display driven by a-IGZO-TFT arrays (Reprinted with permission from Samsung SDI)*

and most important in the display area. In SID '08 oxide TFTs were highlighted as the backplane of LCDs as well as OLEDs. Samsung SDI demonstrated a 12 in. WXGA AMOLED prototype display fabricated by an a-IGZO TFT backplane (Figure 2.26). They successfully fabricated an a-IGZO THF array by the same lithographic process as a-Si:H TFTs. TAOS-TFTs are now attracting interest as a backplane for next generation LCDs. Larger size and high scanning frequency operation are requirements for next generation LCDs. When the display size is increased to >60 in., the number of TFTs required is higher to keep the same pixel density. In such a case 'on'-resistance of TFT arrays cannot be negligibly small. Furthermore, frame frequency quarupling (60 to 240 Hz) is now spreading. Thus, the performance of a-Si:H TFTs is insufficient to meet these requirements (Figure 2.27). This is one reason why a-IGZO TFTs are seriously considered for LCDs. Samsung Electronics and SAIT presented a 15 in. AM-LCD panel using a-IGZO backplane [56]. An excellent review article [7]on OLED panels using a TAOS-TFT backplane was published by a group from Samsung.

2.7 Perspective

Transparent oxide semiconductors (TOSs) were created from transparent conductive oxides (TCOs). The requirements for TOSs are rather differ from those of TCOs. Control of carrier concentration and carrier type is essentially important for the former. The current status of TOSs is far from the ideal situation, in particular for carrier polarity control. Although many papers have reported p-type TOSs, including p-ZnO, no p-channel TFTs with a field-effect mobility of $>0.1\ cm^2\ (V\ s)^{-1}$ had been realized by

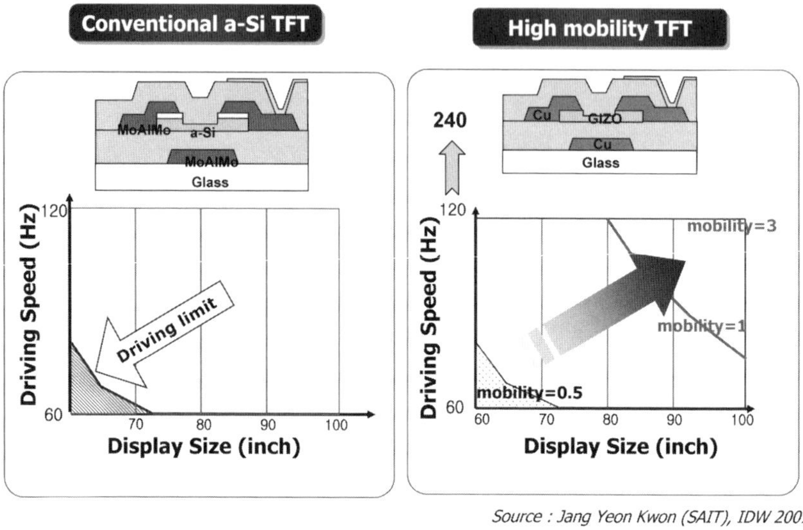

Figure 2.27 Future LCD and TFT mobility required (Reprinted with kind permission from Jang Yeon Kwon (SAIT) IDW Copyright (2007) Jang Yeon Kwon)

2007. It is considered that instability and/or high gap state density is the primary reason. For example, Cu_2O is a well know p-type semiconductor and has acted as the active layer since the first TFT proposed by Heil in 1935. We fabricated epitaxial thin films and obtained Hall mobility of $\sim 100\,cm^2\,(V\,s)^{-1}$ at a hole concentration of $\sim 10^{13}\,cm^{-3}$ [57], which is comparable to that in single crystalline Cu_2O. However, Cu_2O-based TFTs did not operate sufficiently and the estimated field-effect mobility remained $\sim$0.1. This striking difference comes from large tail state densities. Such a situation appears to be similar for other p-type oxide semiconductors.

In 2008, Ogo *et al.* [22] reported a p-channel TFT with a mobility of $1.4\,cm^2\,(V\,s)^{-1}$ employing SnO (not SnO_2) as the active layer. This is the first demonstration of a p-channel oxide TFT with a mobility $>1\,cm^2\,(V\,s)^{-1}$, which was a long-standing target. Figure 2.28

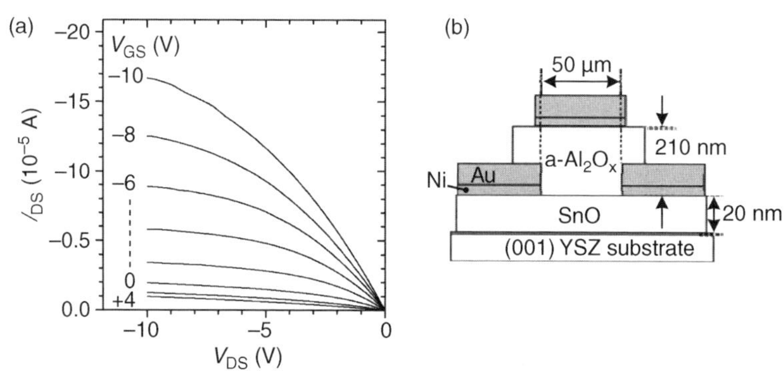

Figure 2.28 p-channel TFT using SnO as the active layer. (a) Output characteristics and (b) device structure

shows the device structure and TFT performances. In 1996 Mizoguchi *et al.* [58] proposed a p-block metal cation with ns^2 electronic configuration, whose energy level is expected to be higher or comparable with that of the O 2p level, as a promising candidate for p-TOSs. They tried to fabricate p-type conduction in PbO where Pb^{2+} has a $6s^2$ electronic configuration but were unsuccessful. The next goal is the fabrication of C-MOSs by combining p-channel and n-channel oxide TFTs. Although a monopolar channel is enough for TFTs for the backplane of displays, C-MOSs are applicable for logic circuits. Exploiting bipolar semiconductive oxides with low tail state densities which can be fabricated at low temperatures is essential to achieve this goal.

Oxide semiconductors are easy to fabricate by conventional sputtering and are robust to oxygen and radiation, in general. If oxide-based C-MOS structures can be fabricated on various types of substrates, including plastics, flexible electronic circuits would be promising. Of course, the formation of heterojunctions between a TOS and an organic semiconductor is a practical and promising way to achieve applications such as photo-sensors, C-MOSs and solar cells.

References

[1] G. Rupperecht, *Z. Phys.* **139**, 504 (1954).
[2] H. Kawazoe, M. Yasukawa, H. Hyoudou, M. Kurita, H. Yanagi, and H. Hosono, *Nature* **389**, 939 (1997).
[3] H. Ohta, K. Kawamura, N. Sarukura, M. Orita, M. Hirano, and H. Hosono, *Appl. Phys. Lett.* **77**, 475 (2000).
[4] H. Hosono, T. Kamiya, and M. Hirano, *Bull. Chem. Soc. Jpn.* **79**, 1 (2006).
[5] H. Hosono, M. Yasukawa, and H. Kawazoe, *J. Non-Cryst. Sol.* **203**, 338 (1996); H. Hosono, *J. Non-Cryst. Sol.* **352**, 851 (2006).
[6] K. Nomura, H. Ohta, A. Takagi, T. Kamiya, M. Hirano, and H. Hosono, *Nature* **432**, 488 (2004); *Jpn. J. Appl. Phys.* **45**, 4303 (2006).
[7] J. K. Jeong, H. J. Chung, Y. G. Mo, and H. D. Kim, *Inf. Display* **24**(9), 20 (2008).
[8] Y. Furubayashi, T. Hitoshigi, Y. Yamamoto, K. Inada, G. Kinoda, Y. Hirose, T. Shimada, and T. Hasegawa, *Appl. Phys. Lett.* **86**, 252 101 (2005).
[9] K. Hayashi, S. Matsuishi, T. Kamiya, M. Hirano, and H. Hosono, *Nature* **419**, 462 (2002).
[10] M. Miyakawa, S. W. Kim, M. Hirano, Y. Kohama, H. Kawaji, T. Atake, H. Ikegami, K. Kono, and H. Hosono, *J. Am. Chem. Soc.* **129**, 7270 (2007).
[11] W. E. Spear and P. G. LeComber, *Solid State Commun.* **17**, 1193 (1975).
[12] P. G. LeComber, W. E. Spear, and A. Ghaith, *Electron. Lett.* **15**, 179 (1979).
[13] O. Heil, UK Patent No. 439 457 (1935).
[14] O. M. Chapin, C. S. Fuller, and G. L. Perason, *J. Appl. Phys.* **25**, 676 (1954).
[15] C. W. Tang, and S. A. Van Slyke, *Appl. Phys. Lett.* **51**, 913 (1987).
[16] H. Yanagi, K. Ueda, H. Ohta, M. Orita, M. Hirano, and H. Hosono, *Solid State Commun.* **121**, 15 (2002).
[17] A. Tsukazaki, A. Ohtomo, T. Omuma, M. Ohtani, T. Makino, M. Sumiya, S. Chichibu, S. Fuke, Y. Segawa, H. Ohno, H. Koinuma, and M. Kawasaki, *Nature Mater.* **4**, 42 (2005).
[18] M. Itoh, M. Kon, C. Miyazaki, C. Ikeda, M. Ishigaki, Y. Urajin, and N. Sekine, *IEICE Trans. Electron* E90-c, 2105 (2007).
[19] L. Wang, M.-H. Yoon, G. L. Yang, A. Facchetti, and T. J. Mark, *Nature Mater.* **5**, 83 (2006).
[20] H. N. Lee, J. W. Kyung, S. K. Kang, D. Y. Kim, M. C. Sung, S. J. Kim, C. N. Kim, H. G. Kim, and S. T. Kim, *Proc. IDW '06* 663 (2006).

[21] T.-C. Fung, C.-S. Chuang, K. Nomura, H.-P.D. Shieh, H. Hosono, and J. Kanicki, *Information Display* **9**, 21 (2008).
[22] Y. Ogo, H. Hiramatsu, K. Nomura, H. Yanagi, T. Kamiya, M. Hirano, and H. Hosono, *Appl. Phys. Lett.* **93**, 032 113 (2008).
[23] H. Kawazoe, H. Yanagi, K. Ueda, and H. Hosono, *MRS Bull.* **25**, 28 (2000).
[24] J. Robertson, K. Xiong, and S. J. Clark, *Thin Solid Films* **496**, 1 (2006).
[25] A. Zunger, *Appl. Phys. Lett.* **83**, 57 (2003).
[26] H. Hosono, *Thin Solid Films* **615**, 6000 (2007).
[27] H. Yanagi, T. Hase, S. Ibuki, K. Ueda, and H. Hosono, *Appl. Phys. Lett.* **78**, 1583 (2001).
[28] G. Thomas, *Nature* **389**, 907 (1997).
[29] T. Kamiya and M. Kawasaki, *MRS Bull.* **33**, 1061 (2008).
[30] Y. R. Ryu, T. S. Lee, J. H. Leem, and H. W. White, *Appl. Phys. Lett.* **83**, 4032 (2003).
[31] Ü. Özgür, Ya.I. Alivov, C. Liu, A. Teke, M. A. Reshchikov, S. Doan, V. Avrutin, S.-J. Cho, and H. Morkoç, *J. Appl. Phys.* **98**, 041 301 (2005).
[32] H. Hiramatsu, H. Kamioka, K. Ueda, H. Ohta, T. Kamiya, M. Hirano, and H. Hosono, *Phys. Status Solidi A* **203**, 2800 (2007).
[33] K. Ueda, H. Hiramatsu, H. Ohta, M. Hirano, T. Kamiya, and H. Hosono, *Phys. Rev. B* **69**, 155 305 (2004).
[34] H. Ohta, K. Nomura, M. Orita, M. Hirano, K. Ueda, T. Suzuki, Y. Ikuhara, and H. Hosono, *Adv. Funct. Mater.* **13**, 139 (2003).
[35] H. Kamioka, H. Hiramatsu, H. Ohta, K. Ueda, M. Hirano, T. Kamiya, and H. Hosono, *Appl. Phys. Lett.* **84**, 879 (2004).
[36] H. Kamioka, H. Hiramatsu, H. Ohta, K. Ueda, M. Hirano, T. Kamiya, and H. Hosono, *Appl. Phys. Lett.* **84**, 879 (2004).
[37] Y. Kamihara, T. Watanabe, M. Hirano, and H. Hosono, *J. Am. Chem. Soc.* **130**, 3296 (2008).
[38] H. Hosono, *J. Phys. Soc. Jpn.* 77SC, 1 (2008), *Idem, Physica C*, **469**, 314 (2009).
[39] H. Hosono, K. Hayashi, and M. Hirano, *J. Mater. Sci.* **42**, 1872 (2007).
[40] S.-W. Kim, S. Matsuishi, M. Miyakawa, K. Hayashi, M. Hirano, and H. Hosono, *J. Mater. Sci.: Mater. Electron.* **18**, S5-14 (2007).
[41] K. Hayashi, P. V. Sushko, D. M. Ramo, A. L. Shluger, S. Watauchi, I. Tanaka, S. Matsuishi, M. Hirano, and H. Hosono, *J. Phys. Chem. B* **111**, 1946 (2007).
[42] P. V. Sushko, A. L. Shluger, K. Hayashi, M. Hirano, and H. Hosono, *Appl. Phys. Lett.* **86**, 092 101 (2005).
[43] P. Sushiko, A. Shluger, K. Hayashi, M. Hirano, and H. Hosono, *Phys. Rev. Lett.* **91**, 126 401 (2003).
[44] K. Hayashi, P. V. Sushko, A. L. Shluger, M. Hirano, and H. Hosono, *J. Phys. Chem. B* **109**, 23 836 (2005).
[45] J. L. Dye, *Inorg. Chem.* **36**, 3816 (1997).
[46] S. Matsuishi, Y. Toda, M. Miyakawa, K. Hayashi, T. Kamiya, M. Hirano, I. Tanaka, and H. Hosono, *Science* **301**, 626 (2004).
[47] S.-W. Kim, S. Matsuishi, T. Nomura, Y. Kubota, M. Takata, K. Hayashi, T. Kamiya, M. Hirano, and H. Hosono, *Nano Lett.* **7**, 1138 (2007).
[48] J. E. Medvedeva, A. J. Freeman, M. I. Bertoni, and T. O. Mason, *Phys. Rev. Lett.* **93**, 016 408 (2004).
[49] Y. Toda, S. Matsuishi, K. Hayashi, K. Ueda, T. Kamiya, M. Hirano, and H. Hosono, *Adv. Mater.* **16**, 685 (2004).
[50] T. Kamiya, and H. Hosono, *Jpn. J. Appl. Phys.* **44**, 774 (2005).
[51] Y. Toda, H. Yanagi, E. Ikenaga, J. J. Kim, M.-I. Kobata, S.-I. Ueda, T. Kamiya, M. Hirano, K. Kobayashi, and H. Hosono, *Adv. Mater.* **19**, 3564 (2007).
[52] S. Narushima, M. Orita, M. Hirano, and H. Hosono, *Phys. Rev. B* **66**, 035 203 (2002).
[53] J. Robertson, *J. Vac. Sci. Technol, B* **18**, 1785 (2000).
[54] J. F. Wager, D. A. Keszler, and R. E. Presley, *Transparent Electronics*, Springer, New York, (2006).

[55] R. Hayashi, M. Ofuji, N. Kaji, K. Takahashi, K. Abe, H. Yabuta, M. Sano, H. Kumomi, K. Nomura, T. Kamiya, M. Hirano, and H. Hosono, *J. Soc. Information Display* **15**, 915 (2007).
[56] J. K. Jeong, H.-J. Chung, Y.-G. Mo, and H. D. Kim, *Information Display* **9**, 20 (2008).
[57] K. Matsuzaki, K. Nomura, H. Yanagi, T. Kamiya, M. Hirano, and H. Hosono, *Appl. Phys. Lett.* **93**, 202 107 (2008).
[58] H. Mizoguchi, H. Kawazoe, and H. Hosono, *Chem. Mater.* **8**, 2769 (1996).

3

p-Type Wide-Band-Gap Semiconductors for Transparent Electronics

Janet Tate[1] *and Douglas A. Keszler*[2]
[1]*Department of Physics, Oregon State University, USA*
[2]*Department of Chemistry, Oregon State University, USA*

3.1 Introduction

Semiconductor electronics and photonics are an integral part of our lives, allowing us to communicate and direct information almost anywhere and anytime with ever-increasing speeds. Silicon and III-V compound semiconductors dominate in this narrow-band-gap semiconductor landscape. If some of the functions of these semiconductors were extended to visibly transparent materials, a host of new applications could be realized as discussed in detail later in this volume.

To realize the full functionality and capabilities of semiconductor electronics, both electron (*n*) and hole (*p*) type conductivity, i.e. bipolarity, is required. Although many useful unipolar devices can be made. Simultaneous transparency and conductivity is possible in all semiconductors, but it is rarer in *p*-type as compared with *n*-type semiconductors. Only a few wide-band-gap inorganic materials have been demonstrated to exhibit the necessary electronic and structural features for realization of effective *p*-type doping. In many of the known *p*-type materials, realizing high *p*-type conductivity can be problematic, largely because the mobility of valence-band-derived carriers is generally lower than that of conduction-band-derived carriers, as in the case of *n*-type conductivity. Indeed, the transparent circuits reported to date [1] are exclusively based on *n*-type materials (oxides of Zn, Sn, In, Ga, and others as well), and high-conductivity transparent conductors in

Transparent Electronics: From Synthesis to Applications Edited by Antonio Facchetti and Tobin J. Marks

commercial use (ITO, SnO_2:F, ZnO:Al) are also all *n*-type materials. The search for high-mobility *p*-type wide-gap semiconductors to complement these *n*-type materials has now been a major research thrust for more than 10 years. As many of the proposed applications for these materials entail large-area coverage or low-cost production, the investigated semiconductor materials sets differ considerably from those associated with conventional wafer-based technologies. As a result, the search for new materials is also spawning new techniques for their synthesis, film deposition and device fabrication.

Inorganic *p*-type transparent semiconductors are found among oxides, chalcogenides, nitrides, silicides, and others. Much recent attention has focused on Cu-based semiconductors, of which the delafossite family $CuMO_2$ is the primary oxide example, while LaCuOCh and BaCuChF (Ch = chalcogenide) are important representatives of the chalcogenides. The simple binary oxides ZnO and NiO have been extensive studied as *p*-type semiconductors. An extensive literature exists on organic *p*-type semiconductors, some of which are transparent. Organics offer the advantage of well-developed low temperature processing, but they suffer from thermal instability and inherently low carrier mobilities. There is also a growing canon on carbon-based electronics, particularly nanotubes, which may prove to be important in transparent electronics.

From a processing standpoint, it is highly desirable to selectively induce *n*- or *p*-type conductivity in a single host, or at least in compatible materials systems, so the pressing quest is for high-mobility *p*-type oxides to match the successful *n*-type oxides like ZnO, In-Ga-O, In-Ga-Zn-O, etc., where carrier mobilities are of order 10–100 $cm^2 V^{-1} s^{-1}$. To date, the search has yielded low mobility ($\approx$1 $cm^2 V^{-1} s^{-1}$) in *p*-type Cu-based oxides, and moderate-mobility (3–10 $cm^2 V^{-1} s^{-1}$) in *p*-type chalcogenide-based systems. The more promising hole mobility in the chalcogenides is offset by the lack of corresponding *n*-type semiconductors in similar materials. While considerable progress has been made, barriers to more rapid development are largely associated with the high-temperature processing needed to produce films with excellent crystalline quality and hence optimal mobility, the resulting inability to deposit them via low-temperature solution techniques on flexible substrates, and the lack of readily available photolithographic methods to pattern them. Some of these barriers have recently begun to crack, at least for *n*-type transparent oxides.

The discussion in the remainder of this chapter is focused mainly on inorganic *p*-type transparent oxide and chalcogenide semiconductors. To set the stage, applications that already feature wide-gap *p*-type semiconductors are presented. A discussion of general scientific issues associated with *p*-conductivity ensues, followed by consideration of specific materials systems and an overview of materials synthesis. The chapter ends with a summary and outlook.

3.2 Applications

A *p*-type transparent semiconductor would find application as the *p*-channel in a transparent thin film transistor (TTFT), the *p*-*n* diode operating as a rectifier, light emitter, or a window electron reflector in a solar cell. For band gaps smaller than 3.1 eV, such semiconductors are also likely to be of interest as absorbers in a variety of solar-cell configurations.

3.2.1 *p*-Channel TTFT

Realization of a *p*-channel TTFT would be an enormous step towards achieving the full functionality of transparent electronics, comparable with the development of CMOS relative to NMOS. Transparent field effect transistors (FETs) based on several *n*-channel transparent oxide semiconductors have been demonstrated with channel mobilities greater than $10\,cm^2\,V^{-1}\,s^{-1}$ and approaching $100\,cm^2\,V^{-1}\,s^{-1}$ [2]. Typical channel materials should also feature low native carrier concentration ($<10^{16}\,cm^{-3}$), and carrier injection must be possible. These all represent significant challenges for *p*-type transparent semiconductors. In addition, charge balance with the *n*-type materials will remain a significant issue because of the mobility difference. Although *I-V* characteristics of candidate *p*-channel FETs show gate-modulated behavior, a true *p*-channel TTFT has not yet been realized in transparent oxide or chalcogenide semiconductors. Organic semiconductors and carbon nanotubes are examples of materials in which such devices have been realized, as discussed later in this chapter, although with low drain currents and mostly with low mobility.

3.2.2 *p-n* Junctions

Fully transparent *p-n* and *p-i-n* diodes and bipolar junction transistors are of course not possible without both *n*- and *p*-type transparent semiconductors.

There have been many reports of transparent heterojunctions with rectifying characteristics. However, this is also a characteristic of Schottky diodes, so rectifying behavior alone is not evidence for a true bipolar junction. Compelling evidence for a true *p-n* junction comes from the *p*-$SrCu_2O_2$:K/*n*-ZnO structure of Ohta and coworkers [3]. This device exhibits rectifying *I-V* characteristics and ultraviolet (UV) emission under forward bias commensurate with the band gaps of the constituents, and the UV emission intensity can be modulated by the injection current above 2 mA. The carrier concentration is estimated at $5 \times 10^{17}\,cm^{-3}$. Despite very good heteroepitaxy, the efficiency is low (10^{-4}), illustrating the need to further improve growth techniques and material properties.

To date, the only transparent homojunction reported in a Cu-based oxide is for the $CuInO_2$:Mg/$CuInO_2$:Sn system [4]. The junction exhibits rectifying *I-V* characteristics with a turn-on voltage of 1.8 V, which is lower than expected for a wide-gap material. However, $CuInO_2$ actually has an indirect gap that is lower than the optical gap of 3 eV, so this junction may indeed be performing as expected. There are also reports of ZnO homojunctions, and blue electroluminescence attributed to donor-acceptor pair recombination in the *p*-layer of a *p-i-n* ZnO structure. A highly asymmetric carrier concentration between the *n* and *p* layers was inferred [5].

3.2.3 *p++* Contacts

Injection of holes into materials used in transparent electronics is challenging. Band-edge alignments must be carefully considered to ensure efficient injection from a contact (metal or degenerate semiconductor) into a semiconductor that forms one of the active device layers. For ohmic contacts to a *p*-type (*n*-type) semiconductor, the contact work function should be greater (less) than that of the semiconductor, or at least the carrier injection

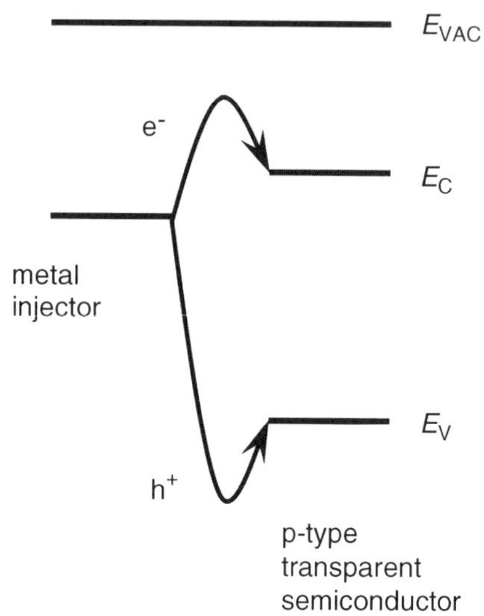

Figure 3.1 *Metal/semiconductor interface*

barriers should be less than a few tenths of an eV. This is readily achieved for an *n*-type semiconductor, where the conduction band minimum (CBM) is typically about 4–5 eV or less below the vacuum level, an energy easily comparable with the work function of most metals (Mg = 3.66 eV, In = 4.02 eV, Al = 4.28 eV [6]). As shown in Figure 3.1, if the same semiconductor were *p*-doped, its valence band maximum (VBM) would be lower by the band-gap energy, which is greater than 3 eV in a transparent material, putting the VBM well out of the range of even the highest work-function metals (Au = 5.1 eV, Pt = 5.65 eV [6]). Even for the narrower-gap semiconductors used in solar cells, the VBM can be 5 or 6 eV below the vacuum level, making injection problematic for the *p*-type materials. The usual solution is to make heavily doped (p++) contacts that allow injection via tunneling through the resulting Schottky barrier. For solar cells, this p++ contact should be transparent in the visible, so heavily doped *p*-type transparent semiconductors may well find application here. Similar issues apply in contacts to organic semiconductors, where the valence band language is replaced by highest occupied molecular orbital (HOMO). Recently, the Cu-based *p*-type transparent conductor LaCuOSe:Mg was demonstrated to have a low resistance to hole injection into the organic semiconductor *N,N′*-diphenyl-*N,N′*-bis (1,1′-biphenyl)-4,4′-diamine (NPB) [7].

3.2.4 Solar Cells

Solar cell materials require absorber layers to harvest solar energy and produce electrons and holes and also *n*- and *p*-type windows that will extract one type of carrier while blocking the other (and remaining transparent to the radiation in question). Particularly, when tandem solar cells are considered, rather specific band gaps and band offsets will be necessary for efficient functioning, and *p*-type wide band-gap semiconductors may be important elements.

3.2.5 Passive Applications

Highly conductive, highly transparent semiconductors are commonly used for passive applications where transparency is critical. These might include heat-reflecting window coatings, contacts for touch screens, and heating elements for windshields or windows on refrigerated displays. In these applications, the conductivity of p-type materials does not rival the present n-type materials for the same transparency, and therefore they are unlikely to compete. One exception may be p-type materials with larger effective masses and relatively large carrier concentrations having plasma edges that are tunable in the infrared (IR); these materials may find applications in IR-transparent electronics [8].

3.3 Challenges Associated with *p*-Type Wide-Gap Semiconductors

3.3.1 Band Structure and Dopability

p-Type conduction is associated with carrier movement through a valence band. An important feature of wide-gap p-type semiconductors is the relatively low mobility of these carriers in comparison with those in a conduction band. Indeed, most semiconductors have lower hole than electron mobility [9]. A large carrier mobility allows faster response, a critical issue for transistors, and it also allows larger current for a given carrier concentration, which is important for diodes and other high-current, high-power applications. Carrier mobility (μ) is inversely proportional to the band effective mass, which is in turn inversely proportional to the band curvature, so that $\mu \sim \nabla_k^2 E(k)$. Low mobility, then, arises from narrow (flat) energy bands. Valence bands, in contrast to conduction bands, are often derived from tightly bound states. For instance, it is common to find valence bands derived from highly directional p or d orbitals, while conduction bands are often derived from spherical and diffuse s orbitals. As a result, the spatial overlap of atomic orbitals that form the valence band is smaller than in the case of conduction bands, and valence bands are consequently narrower and correspondingly less disperse. Another factor that determines mobility is carrier scattering from neutral and ionized impurities. In principle, such scattering is similar for p- and n-type materials, but if the mobility μ is inherently small, a higher carrier concentration p is required to achieve a particular conductivity $\sigma = pe\mu$ (e is the electron charge, and the equation assumes a simple 1-band model). This results in a higher concentration of ionized and, probably, neutral impurities that increase scattering and further lower the mobility. In addition, increased optical absorption from free carriers or tail states compromises transparency.

The carrier mobility in Cu-based p-type transparent oxides is typically $<1\,cm^2\,V^{-1}\,s^{-1}$, but in Cu-based chalcogenides, mobilities as high as $8\,cm^2\,V^{-1}\,s^{-1}$ have been reported [10]. These values are generally lower than the carrier mobilities in n-type transparent oxides, which are in the range of $10\text{–}120\,cm^2\,V^{-1}\,s^{-1}$ [11, 12], and this mobility asymmetry is a problem for device applications that require balanced current injection from n- and p-type elements.

Doping of wide-gap semiconductors without compromising transparency is a nontrivial task and p-type wide-gap semiconductors present special challenges. The dopants should form shallow defect states so that band-to-defect absorption occurs at energies either above or below the visible range. The ability of acceptor dopants to generate hole carriers is compromised by the formation of compensating donor states. The cost of compensation is the formation energy of the donor state, but this is offset by the energy gain that results when a donor electron relaxes to the acceptor state. Formation energies are similar for most materials (some fraction of an eV), but the energy reduction from hole-filling usually scales with the gap energy. Wide-gap materials are therefore more susceptible to compensation. In more detail, one must consider the formation enthalpies of various defects as a function of the Fermi-level position and chemical potential to assess whether a particular doping strategy is likely to be successful [13, 14].

The introduction of defects shifts the Fermi level of a semiconductor. As discussed by Zunger [13], as the Fermi energy moves towards the valence band due to the introduction of acceptors or the injection of holes, the formation enthalpy of donors decreases to the point where spontaneous generation of donors renders further attempts to introduce acceptors ineffective for the purposes of carrier generation. This depends further on the chemical potential of the system. In the case of a compound semiconductor, the enthalpy of formation of acceptor states is favored when the system is anion rich, for example. The spontaneous formation of these 'killer defects' leads to the concept of a Fermi level 'pinning'. If the pinning level corresponding to spontaneous generation of hole killers lies well above the valence band, then it is impossible to p-dope a system, while if it is near or even below the valence band, the material is easily p-doped.

If semiconductor band edges are aligned according to their band offsets, then the observed Fermi pinning levels in different materials within classes of semiconductors are remarkably similar relative to the vacuum level [15, 16] (see Figure 3.2). This suggests that the positions of the VBM and CBM relative to the vacuum level also (separately) influence the dopability of a semiconductor. A large electron affinity lowers the CBM towards the upper pinning level or even below it, which in turn promotes n-dopability. Conversely, a small ionization potential pushes the VBM upwards towards or above the lower pinning level, which implies p-dopability. This general scheme explains the doping trends in many wide-gap semiconductors. For example, the position of the ZnO VBM shows that it is far more difficult to dope it p-type, than say, ZnSe, while n-doping in ZnO is considerably more favorable.

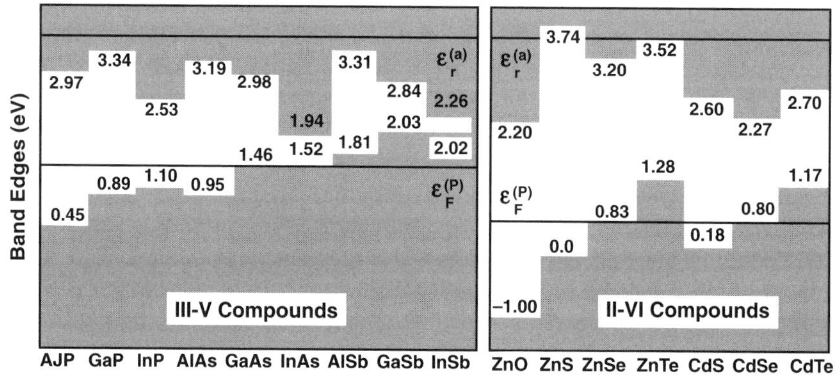

Figure 3.2 Pinning energies for p- and n-type semiconductors relative to the band edges of selected semiconductors. (Reprinted with permission from [13] © 2003, American Institute of Physics)

Identification of the defects in *p*-type wide-gap semiconductors is therefore critical to advancing the science and technology of the materials and their applications. While limited experimental data are available in this area, theoretical work can provide important insights. For example, Cu vacancies are expected on theoretical grounds to be the native defect in many Cu-based semiconductors [17], but it has yet to be experimentally verified that this defect is responsible for the conductivity in most transparent Cu-based *p*-type semiconductors.

3.3.2 Transport

Conductivity in wide-gap semiconductors results from carriers induced by defect levels that are either native to the material or deliberately introduced. The band gap is far too large for intrinsic conductivity to be important. Extrinsic doping by substitutional cations and anions affords control, but native defects are usually equally important and different native defects may dominate under different processing conditions (cation/anion rich, etc.). If the defect levels are within tens of meV of the band edge, they produce carriers at levels high enough to allow significant conductivity; degenerate doping produces free carriers at densities about two orders of magnitude lower than in metals. Degenerate conduction has been achieved in BaCuTeF and LaCuOSe:Mg at carrier densities approaching 10^{20} to $10^{21}\,cm^{-3}$. The highest conductivities observed are about $100\,S\,cm^{-1}$, about an order of magnitude smaller than commercial grade ITO, with mobilities $\leq 10\,cm^2\,V\,s^{-1}$, as discussed earlier.

Degenerate conduction is not the norm, and many of the *p*-type wide-gap materials show strongly activated conductivity, i.e. lower conductivity with decreasing temperature following a dependence modeled as $\sigma = \sigma_\infty e^{-(T_0/T)^n}$. When $n = 1$, as in the classical case for semiconductors exhibiting band conductivity, the activation energy of carriers from defects into the band (or across the band in the case of intrinsic conductivity) can be easily extracted from a semi-logarithmic plot of the conductivity against inverse temperature. In classic band conduction, it is the number of carriers that is exponentially sensitive to temperature, while the carrier mobility, determined by the band curvature and scattering mechanisms, is more weakly temperature dependent. When $n = 1/4$, the conductivity is known as variable range hopping, and carrier motion is considered to be a combination of thermally activated hopping and carrier tunneling from site to site within an impurity band or low-mobility states in band tails. Such behavior is particularly striking in $CuScO_{2+y}$, as shown in Figure 3.3. When strong interactions of the carriers with the lattice cause local distortions, the effective inertia of the carriers is increased, and the mechanism is known as polaron conductivity [18].

Commensurate with generally lower electrical conductivity, wide-gap *p*-type semiconductors exhibit large Seebeck coefficients (S) and may therefore be considered candidates for thermoelectric applications despite their large band gaps. The usual figure of merit $ZT = \sigma S^2 T/\kappa$ is of order 1 for Bi_2Te_3-Sb_2Te_3 alloys, the best thermoelectrics [19]. If transparent thermal elements could be realized, they could provide on-chip heating and cooling capability for transparent electronics. For example, BaCuTeF has $S \approx 100\,\mu V\,K^{-1}$ and $\sigma \approx 170\,S\,cm^{-1}$, which gives $S^2\sigma = 1.7 \times 10^{-4}\,W\,m^{-1}\,K^{-2}$. The thermal conductivity κ is unknown but a value of $0.1\,W\,m^{-1}\,K^{-1}$ would give $ZT \approx 0.5$ at room temperature.

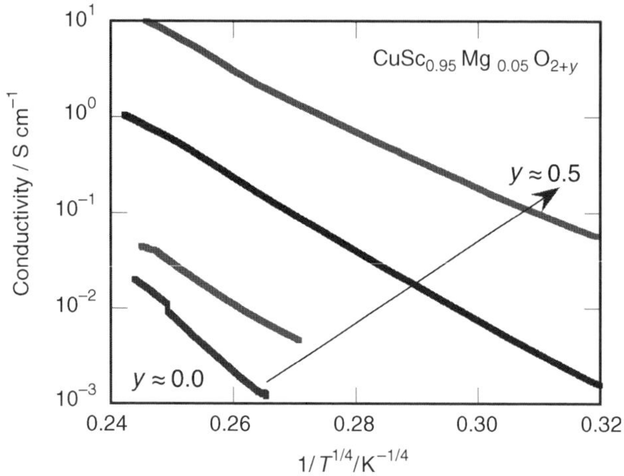

Figure 3.3 *Temperature-dependent conductivity of the delafossite $CuSc_{0.95}Mg_{0.05}O_{2+y}$, showing a good fit to variable-range hopping. The arrow indicates increasing doping by oxygen interstitials*

3.3.3 Optical Properties

Many of the *p*-type wide-gap semiconductors investigated for transistor-type applications have direct band gaps and exhibit light emission via band edge luminescence, including narrow exciton emission and sub-gap photoluminescence. Such properties are typically harnessed to produce light emitting diodes (LEDs), light sensors, or even lasers, and these elements might be incorporated into transparent circuits as well as being useful photonic elements independent of transparent circuitry.

The basic optical property is the band gap, whose signature is a sharp increase in absorption α as the photon energy E approaches the band gap energy E_g: $\alpha \propto (E - E_g)^{1/n}$, where n determines the nature of the gap, e.g., $n = 2$ is allowed and direct, and the energy intercept at $\alpha = 0$ gives the gap energy [20]. Most often, the gap is determined from a plot of $(\alpha E)^n$ vs. E with n chosen to make the data linear near E_g. This scheme works well for sharp transitions, but for the broad transitions typically found in newly developed materials, the method is fraught with inaccuracy. The linearity is subjective, the data commonly yield comparable linearity with different values of n, and, for a given n, the range of data used can change the intercept considerably. The defect-induced broadening of the absorption by thermal and static disorder is an additional complicating factor, although in clean materials, the Urbach edge can be distinguished by its exponential rather than linear dependence of α on E.

Many of the chalcogenide *p*-type transparent semiconductors have stable excitons, which give rise to sub-gap absorption that is usually narrow at low temperatures in highly crystalline material and broadens to merge into the normal band-to-band absorption under conditions of thermal and static disorder. In many cases the excitons are stable to room temperature, and they emit tunable near-band-edge light as discussed in a later section of this chapter. Mid-gap defect states also play a role in the optical properties, in some cases giving rise to visible photoluminescence under UV excitation. Although the visible luminescence might be useful in some cases, deep defects are in general detrimental to both conductivity and transparency, serving as compensators for shallow carrier acceptors and contributing to sub-gap absorption.

In degenerate semiconductors, there is a blue shift in the fundamental absorption caused by carriers occupying states within the band. This Moss-Burstein effect, readily observed when electrons occupy highly dispersed conduction bands, is less pronounced in wide-gap *p*-type semiconductors. This is because the valence band is less disperse and has a large density of states at the VBM so that even at high hole density, the highest hole-occupied states in the valence band remain close to the VBM. At the red end of the visible spectrum, transparency can be reduced because of free carrier absorption associated with intra-band transitions (and inter-band transitions in the case of nearly degenerate bands) and by strong reflection at wavelengths longer than the plasma edge. This occurs at about 1.5 μm in degenerately doped ITO ($n \approx 10^{20}\,\text{cm}^{-3}$). Larger effective masses push the plasma edge further into the IR, effectively extending the range of 'transparent' electronics for some *p*-type semiconductors. It is also worth pointing out that *p*-type transparent semiconductors have relatively large indices of refraction at visible frequencies, so that reflection from thin films can exceed 20% in some cases. This reduces transmission unless anti-reflection measures are implemented.

3.4 Materials

This section highlights some of the more significant wide-gap *p*-type semiconductor systems. An exhaustive review of every semiconductor is not the objective here, but rather we seek to highlight the features of the different classes of materials with particular emphasis on oxides and chalcogenides.

3.4.1 Oxides

3.4.1.1 Cu(I) Oxides

In 1997, Kawazoe *et al.* of the Tokyo Institute of Technology reported in *Nature* [21] a surprising result of *p*-type conductivity with high mobility in a transparent thin film of $CuAlO_2$. This result sparked considerable interest and research into *p*-type transparent conducting oxides; for example, more than 220 papers addressing the preparation and properties of $CuAlO_2$ have appeared since the initial report. While this was not the first report of *p*-type conductivity in a transparent oxide, cf. $SrTiO_3$ and NiO, it did offer a rationale for the design of a whole new family of potential *p*-type wide-gap semiconductors on the basis of the d^{10} electron configuration of Cu(I) [22].

Following the initial report on $CuAlO_2$, *p*-type conductivity was quickly reported in several derivative compositions $CuMO_2$ (M = Ga [23], In [24], Sc [25], Y [26]). Each of these materials crystallizes in the structure type of the mineral delafossite $CuFeO_2$. To gain insight into the electrical transport and optical properties of the wide-gap derivatives, it is very important to appreciate the unique aspects of the crystal structure. We will consider this structure type and then review the properties that have been reported for the various compositions. We will then return to a consideration of the

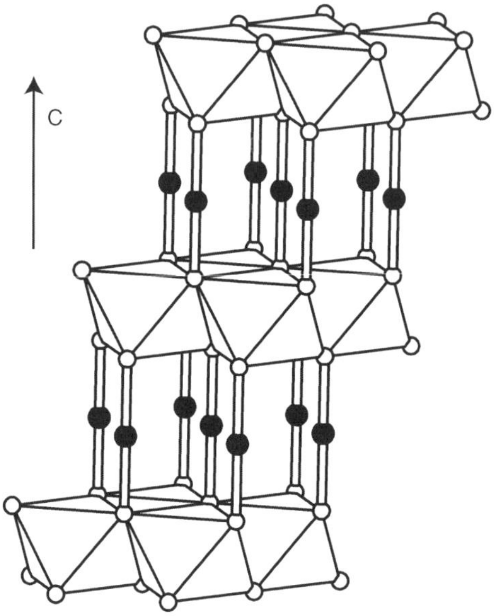

Figure 3.4 Crystal structure of the delafossite family $CuMO_2$. Cu atoms (black spheres) are linearly coordinated by two oxygen atoms (white spheres). Oxygen atoms form the vertices of tilted octahedra that contain the M atoms (hidden from view)

salient characteristics of the crystal structure as a means to rationalize the observed properties.

As shown in Figure 3.4, the delafossite materials $CuMO_2$ adopt an anisotropic trigonal/hexagonal structure [27]. The Cu and M atoms occupy discrete layers extending in the *ab* plane that are alternately stacked along the *c* axis. The Cu atoms are coordinated by two O atoms to form linear CuO_2 sticks, while the M atoms are coordinated by six O atoms in slightly distorted MO_6 octahedra. In this arrangement, the CuO_2 units are essentially isolated one from the other, i.e. there are no Cu-O-Cu interactions; the O atoms at the ends of the CuO_2 units are shared with six M atoms, occupying a distorted tetrahedral environment. Each Cu and M atom has three nearest neighbors of the same type with Cu···Cu and M···M distances scaling with the radii of the M atoms; representative distances are 2.86, 2.98 and 3.21 Å for M = Al, Ga and Sc, respectively, which compare with the Cu···Cu distance of 2.7 Å in Cu metal.

Many of the delafossites have been prepared as nearly white powders or visibly transparent thin films, features that are consistent with large band gaps, the closed shell d^{10} electron configuration of Cu^+, a large energy separation between the Cu 3d and Cu 4s levels, and the isolated nature of the CuO_2 structural units. Independent of structure, hole conductivity in Cu(I) oxides is likely to be realized via direct Cu···Cu interactions and mixing of Cu 3d and O 2p orbitals near the top of the valence band, which can provide broader bands than realized with O 2p orbitals alone. The presence or introduction of defects or appropriate acceptor dopants can then lead to *p*-type conductivity. Band structure calculations do indeed reveal Cu-O orbital mixing near the top of the valence band [28–30]; the position of the charge neutrality level slightly above the VBM in $CuAlO_2$ is consistent with hole doping [31]. From experimental studies [28], $CuAlO_2$ is found to exhibit an indirect gap near 1.8 eV and a direct gap near 3.5 eV. The band structure calculations are largely consistent with the indirect nature, and corrected band gaps [32] are quantitatively comparable with those observed experimentally. The calculations are also consistent with decreasing gaps in the series M = Al → Ga → In, where greater M orbital mixing at the Γ point serves to depress the energy of the conduction band. The direct gaps ($E_g > 3$ eV) of the delafossites are significantly larger than the gap ($E_g = 2.2$ eV) of Cu_2O, which also crystallizes in a structure characterized by the presence of linear CuO_2 sticks. Cu_2O, however, adopts a three-dimensional structure with a larger number of Cu···Cu interactions, which have been correlated to the smaller gap [33, 34] and a higher hole mobility. The nature of the Cu···Cu interactions and their contribution to the conductivity, however, has been questioned [35].

The highest hole mobility reported for the delafossites has been 10.4 $cm^2 V^{-1} s^{-1}$, reflecting the original work on $CuAlO_2$. Unfortunately, a mobility of this magnitude has not been reproduced in thin films of $CuAlO_2$ or any related delafossite. Mobilities in these oxides are always <1 $cm^2 V^{-1} s^{-1}$, and *p*-conductivity is usually confirmed on the basis of the Seebeck effect rather than the Hall effect. Seebeck coefficients are always positive, exhibiting values in the range 160–660 $\mu V K^{-1}$, typical of semiconductors. The highest reported conductivities are in the range of 1–30 $S cm^{-1}$, which in combination with the low mobilities, translate into relatively high carrier concentrations of 10^{19}–10^{20} cm^{-3}. For use in TTFTs, mobilities >1 $cm^2 V^{-1} s^{-1}$ and carrier concentrations $<10^{16}$–10^{17} cm^{-3} are preferred, so the delafossites have yet to be successfully employed as channel layers in *p*-type or hybrid TTFTs.

Control of conductivity by extrinsic doping has been found to be effective only when the radius of the M atom corresponds to approximately that of Sc or larger; this size induces sufficient expansion of the structure so that dopants such as Mg and especially interstitial O can be incorporated. $CuScO_2$:Mg and $CuScO_{2+y}$ both exhibit *p*-conductivity with a maximum value of 30 $S cm^{-1}$ for the composition $CuScO_{2.5}$ [25]. At this high O loading, however, the transparency is reduced from >85 to <25% at 550 nm. Among the derivatives with M = Al, Ga, In, only the In material allows extrinsic doping. In addition, $CuInO_2$ has the distinction of being the only transparent conducting oxide to exhibit bipolar character. Substitution of Sn and Ca doping for In in $CuInO_2$ produces *n*- and *p*-type carriers, respectively [24]. Ca serves as an acceptor to produce holes in the Cu-O valence band, while Sn donates electrons into the In 5s conduction band. The doping levels are not reported, but the conductivity is activated near room temperature with activation energies of 77 meV (*n*-type) and 190 meV (*p*-type). Bipolar doping in $CuInO_2$ was exploited to produce the homojunction described in Section 3.2 [4]. This is rather a special case; $CuInO_2$ actually has a much narrower gap than is suggested by its optical absorption (it is transparent across the visible spectrum) because of symmetry-suppressed transitions in the region of the Γ point, i.e., the location of the smallest direct gap [36].

The ubiquity of *p*-type behavior for Cu^+-containing materials strongly suggests that the defects producing conductivity are related to Cu^+ and its chemical oxidation. In the delafossites, it may be anticipated that hole concentrations could then be set through the formation of Cu vacancies, substitution of acceptors on the Cu or M sites, incorporation of O interstitials into the Cu planes, or some combination of these features. On the basis of an analysis of powder samples prepared by hydrothermal and conventional solid-state methods, the defect complex $(Al_{Cu}^{**}2O_i'')''$ has been postulated for $CuAlO_2$ [37]. Here, Al atoms substitute on the Cu sites, and excess O atoms associated with Al occupy interstitial positions in the Cu planes, producing materials that are compositionally rich in both Al and O. The complex oxidizes Cu^+ to produce holes, but the holes are effectively trapped at lower temperatures, which is consistent with a polaron model of conductivity. Higher conductivities are observed in $CuScO_2$ and $CuYO_2$, where the structure has expanded on the basis of the sizes of the Y and Sc atoms to allow ready insertion of O atoms into the Cu planes and acceptor dopants such as Ca_Y^* and Mg_{Sc}^*. At low O concentrations, the conductivity is characterized by an activated process and variable range hopping [38]. At higher O concentrations, nearly degenerate behavior and a conductivity of $\sim$30 S cm^{-1} are observed for $CuScO_{2+\delta}$ [25]; the conductivity of $CuScO_2$, in contrast to that of $CuAlO_2$, can also be enhanced by doping with the acceptor Mg. Degenerate behavior is similar to that of a metal, i.e., conductivity increases with decreasing temperature, indicative of the fact that that the number of carriers is essentially constant, and that the temperature dependence is dominated by scattering of the carriers by defects such as phonons and structural disorder. It is such degenerate conductivity ($\sim$1000 S cm^{-1}) in the *n*-type wide-gap oxides like ITO, ZnO:Al and SnO_2:F that has earned them the moniker 'transparent conductors'. In $CuScO_{2+\delta}$, however, the mobility is too low to provide a comparable conductivity, despite the presence of a high carrier concentration.

To summarize the work on the delafossites, it is clear that materials exhibiting both *p*-type behavior and transparency can be produced, but they do not exhibit sufficient mobility to be useful as conventional transparent conductive oxides. Moreover, as hole concentrations are increased, transparency is severely compromised. The low mobility, especially at low dopant concentrations, is largely related to the crystal structure. There are no Cu-O-Cu linkages in the structure, so while Cu-O orbital mixing may occur at the top of the valence band, there is simply no significant conduction pathway in the absence of sufficient Cu···Cu interactions. In essence, the CuO_2 sticks are behaving as isolated molecular units. For those materials containing larger M atoms, O acceptors can be readily inserted into the Cu planes. In the early stages of O incorporation, Cu(I) is locally oxidized to Cu(II), and samples and films become intensely colored. At higher O concentrations, many holes are introduced and a sufficient number of Cu-O-Cu in-plane interactions form to provide a conduction pathway, but even here carrier activation is required. In addition, the samples are no longer transparent. Overall, the delafossites have limited use for development of highly conducting and visibly transparent materials. Their continued development will therefore likely be focused on electro-optical properties, fundamental defect and anisotropic transport issues, and applications where polaron-type conductivity is useful.

Shortly after the report of *p*-type conductivity in $CuAlO_2$, similar behavior was reported for the oxide $SrCu_2O_2$ [39]. Like the delafossites, the structure of this Cu(I) material is characterized by the presence of CuO_2 linear sticks. In this case, the sticks share O vertices to form zig-zag one-dimensional chains. The band gap ($E_g = 3.3$ eV) and hole mobility ($\mu = 0.46$ $cm^2 V^{-1} s^{-1}$) are similar to those of the delafossites. Carrier transport is characterized by an activated process [39, 40], and the highest conductivity, 4.8×10^{-2} S cm^{-1}, is achieved with K doping. The Sr compound, however, is unique relative to the delafossites in exhibiting a direct band gap, which has stimulated efforts to develop electro-optical devices [41, 42].

3.4.1.2 Binary Oxides – p-ZnO and p-NiO

Among non-Cu-based *p*-type oxides, NiO and ZnO are the most important. NiO has been known as a *p*-type semiconductor for several decades, whereas the prospect of achieving *p*-type conductivity in ZnO, long a well-known *n*-type wide-gap semiconductor, has only recently been seriously considered.

NiO has a band gap of about 3.5–3.8 eV, and it is insulating when stoichiometric. *p*-Conductivity can be induced by the introduction of Li, which results in the formation of Li^+-Ni^{3+} acceptor complexes, or by making nonstoichiometric $Ni_{1-x}O$ in highly oxidizing conditions [43]. Conductivity seems to arise from hopping in a narrow d-band, but there are also arguments for band conductivity. Regardless of the mechanism, the carrier mobility is $<1\ cm^2\ V^{-1}\ s^{-1}$, which means NiO is unlikely to be useful as a channel layer in transistors, but there have been several reports of (semi)transparent diodes using NiO as the *p*-layer. Conductive thin films of NiO were reported by Sato *et al.* [44]. Films with $\sigma = 7\ S\ cm^{-1}$, $p = 1.3 \times 10^{19}\ cm^{-3}$ and hence $\mu \approx 0.3\ cm^2\ V^{-1}\ s^{-1}$ were, at 110 nm, 40% transparent over the visible range; futhermore heat treatment at 300–400 °C reduced the conductivity and improved the transparency to close to 80%, a typical trade-off for *p*-type conducting semiconductors. Such films were used in *p-i-n* junctions with configuration *p*-NiO/*i*-NiO/*i*-ZnO/*n*-ZnO to obtain rectifying characteristics. Ohta *et al.* constructed a UV detector based on a *p*-NiO/*n*-ZnO/ITO trilayer, and reported comparable UV sensitivity to GaN-based sensors [45].

NiO has also been used as a hole-transporting/electron-blocking anode interfacial layer in organic solar cells, improving the power conversion efficiency of the cells and recovering lost open-circuit voltage relative to control cells made without the NiO layer [46]. Typical films are about 55–60% transparent over the visible range.

p-Conductivity in ZnO is a highly desirable goal given its environmentally friendly composition, excellent optical transparency, the ease of inducing complementary *n*-conductivity, and its large exciton binding energy. Based on the principles discussed by Zunger, *p*-ZnO is not easy to achieve because the VBM lies too far below the vacuum energy. Despite many reports of *p*-conductivity in ZnO, the choice of dopant and method of growth remain controversial. It appears that low-carrier-density *p*-type conductivity is achievable by doping with N and P on the O site and that this *p*-conductivity dominates if the abundant native donors can be eliminated by careful growth methods. An example of such a high quality system came from Tohoku University where researchers reported an all-ZnO *p-i-n* homojunction with hole carrier concentration about $10^{16}\ cm^{-3}$ and hole mobility of 5–$8\ cm^2\ V^{-1}\ s^{-1}$ [47]. The broad electroluminescence observed in that device was later narrowed in a higher quality device, and blue electroluminescence was observed near 440 nm, attributed to electron-hole (*e-h*) recombination in the *p*-ZnO layer [48]. Limiting factors included the low concentration of holes and series resistance in the junction.

3.4.1.3 $ZnRh_2O_4$ and p-Type Spinels

An interesting class of non-Cu-based *p*-type oxides is ZnM_2O_4 (M = Ir, Rh, Co). The band gap originates from ligand-field-split M^{3+} *nd* states within the MO_6 octahedra. In this scenario, the valence band derives from the lower energy t_{2g} orbitals, while the conduction band derives from the e_g orbitals. The 6 M^{3+} d-electrons occupying the t_{2g} orbitals in a pseudo-closed-shell configuration then translates to a filled valence band. There is some question as to whether the crystal-field splitting would be large enough to achieve transparency in the visible, but a band gap of 3 eV has been reported for $ZnIr_2O_4$ [49], and 2.1–2.74 eV for $ZnRh_2O_4$ [49–51]. The *p*-type nature of the carriers has been confirmed by a positive Seebeck coefficient measured in polycrystalline thin films, but the Hall coefficient is too small to measure, which, coupled with a relatively large (thermally activated) conductivity of $0.7\,S\,cm^{-1}$ (M = Rh) and $2\,S\,cm^{-1}$ (M = Ir), means the mobility is likely to be small. $ZnRh_2O_4$ may be the only *p*-type oxide to be reported in the amorphous form [52]. In amorphous $ZnRh_2O_4$, the long-range order is destroyed, but local order probably remains in the form of the edge-sharing RhO_6 octahedra of the crystalline parent, which results in a good hole conductivity path. The disorder may be induced by corner-sharing octahedra similar to those found in the perovskite $LaRhO_3$. Amorphous $ZnRh_2O_4$ films maintain the same band gap as the crystalline parent, and the conductivity is $2\,S\,cm^{-1}$ at 295 K. $ZnRh_2O_4$ has been incorporated into an all-amorphous diode structure with an *n*-type amorphous oxide, a-$InGaZnO_4$ ($E_g \approx 3.0\,eV$). The diode shows rectifying characteristics with a threshold voltage of 2.1 eV. If some means could be found to strengthen the ligand field, and hence widen the band gap, then $ZnRh_2O_4$ might be a very promising material for transparent electronics. The caveat, of course, is that Rh is a precious metal whose price presently approaches US$ 10 000 per ounce [53].

The mixed oxide $Ni_{1-x}Co_xO_y$ is conductive over the entire range of x, exhibiting *p*-type conductivity close to the spinel composition, $x = 2/3$, in the range of $10–330\,S\,cm^{-1}$, depending on deposition conditions [54]. Such films exhibit moderate to good IR transparency with the higher transparency characteristic of more resistive films. Indications are that $NiRh_2O_4$ is more conductive than $NiCo_2O_4$ prepared under the same conditions [55] and more transparent in the visible range [56].

3.4.2 Chalcogenides, Chalcogenide Fluorides and Chalcogenide Oxides

3.4.2.1 BaCuChF, LaCuOCh and $BaCu_2S_2$

The design rationale for *p*-type semiconductors based on the d^{10} electron configuration of Cu(I) demonstrated by the delafossites can also be extended to the chalcogenides Ch = S, Se and Te. Relative to the oxides, the heavier group 16 chalcogenides have higher-lying p orbitals that can be expected to mix with the Cu 3d orbitals to form more disperse valence bands. In addition, the increasingly covalent nature of the Cu···Ch bond with heavier chalcogens favors increased mobility over the oxides. The prototype is Cu_2S, which has long been known as a *p*-type semiconductor. Although reported properties vary widely because $Cu_{2-x}S$ exists in various stoichiometries and structures, it is generally considered a high-mobility degenerate semiconductor ($\mu = 25\,cm^2\,V^{-1}\,s^{-1}$ [57]), but its band gap of about 1.2 eV is too small for transparency in the visible region. The basic structure, tetrahedral coordination of Cu(I) by S, also exists in other materials where wider band gaps are possible. In α-$BaCu_2S_2$, for example, the arrangement of the edge- and vertex-sharing CuS_4 tetrahedra decreases the Cu···Cu distances and widens the band gap to 2.3 eV, and thin films of this material showed $\mu = 3.5\,cm^2\,V^{-1}\,s^{-1}$ and $\sigma = 17\,S\,cm^{-1}$ [58]. In the P4/nmm structure of BaCuSF and LaCuOS (Figure 3.5), the edge-sharing CuS_4 tetrahedra are arranged in what can be considered as hole-conducting $[Cu_2S_2]^{2-}$ layers separated by, and only weakly coupled to, insulating $[Ba_2F_2]^{2+}$ or $[La_2O_2]^{2+}$ layers, leading to a highly anisotropic structure with a band gap above 3 eV [59, 60] and correspondingly smaller band gaps for the Se and Te analogues [61, 62] (Table 3.1). The structure also permits several

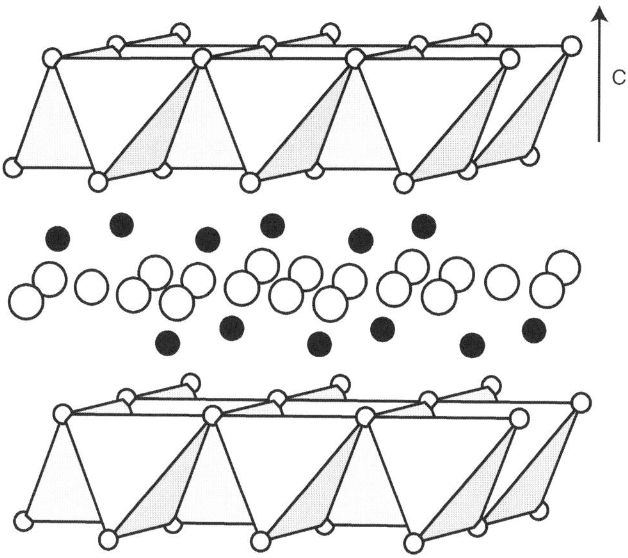

Figure 3.5 *P4/nmm crystal structure adopted by BaCuChF and LaCuOCh. The structure consists of layers of* $[Cu_2Ch_2]^{2-}$ *edge-sharing tetrahedra separated by layers of* $[Ba_2F_2]$ *or* $[La_2O_2]^{2+}$*. The large white spheres represent the anion and the smaller black spheres the cation*

Table 3.1 *Properties of BaCuChF and LaCuOCh at room temperature*

Material	Form	E_g (eV)	μ ($cm^2\,V^{-1}\,s^{-1}$)	σ ($S\,cm^{-1}$)	n (cm^{-3})	S ($\mu V\,K^{-1}$)	Ref.
LaCuOS (:Mg)	epi films	3.1 (–)	0.5 (0.25)	0.66 (7)	$10^{19} \times 1$ (2)		[61]
LaCuOSe (:Mg)	epi films	2.8 (–)	8 (4)	24 (140)	$10^{20} \times 2$ (2)		[61]
LaCuOSe:Mg	epi film		3.5	910	$10^{21} \times 1.7$		[71]
LaCuOTe	bulk	2.31	80.6	1.65	1.3×10^{17}		[65]
BaCuSF	poly film	3.2		0.8			
BaCuSF (:K)	bulk	3.2 (3.2)		0.09 (82)		56	[60]
BaCuSeF	film						
BaCuSeF (:K)	bulk	3.0 (3.0)		0.06 (43)		32	[60]
BaCuTeF	epi film	2.3 (weak) 3.0 (strong)	6	167	$10^{20} \times 1$	90	[10]
BaCuTeF	poly films	3.0	0.2	6.6	$10^{20} \times 1.7$	130	[10]
BaCuTeF	bulk	2.3		6–8		25–55	[73]

derivative compositions. For example, Bi, Y, Nd and Pr can all replace La in LaCuOCh, and Sr can replace Ba in BaCuSF. They are all *p*-type, but so far none has shown superior electrical or optical properties for transparent electronics. Although tellurides have the lowest band gaps, BaCuTeF exhibits fairly low absorption over the entire visible region [10]. It is possible that the transitions corresponding to the smallest gap are symmetry forbidden, resulting in greater optical transparency than expected.

FLAPW band structure calculations and XPS measurements [63–66] show significant mixing of the antibonding Cu 3d and S 3p orbitals at the top of the valence band. Taken together, these calculations and measurements paint a picture of a mixed Cu 3d/Ch *n*p valence band with the chalcogen contribution increasing with heavier chalcogenide, and pushing the top of the valence band higher, thus enhancing the *p*-dopability in Zunger's scheme. The conduction band states are also largely Cu- and Ch-derived with both Cu 4s and Ch *n*p character near the CBM, but also some La 5d or Ba 6s. The 'insulator' anions, O or F, make contributions well away from the band edges.

The band structure calculations indicating increased Cu-Ch mixing, larger dispersion and increasing VBM in the series O, S, Se, Te support the observations of the enhanced *p*-dopability and larger mobilities of the chalcogenides (Table 3.1). The largest mobility so far reported in a transparent thin film is $8\,cm^2\,V^{-1}\,s^{-1}$ in undoped, epitaxial LaCuOSe. Similar values can be obtained in amorphous *n*-type oxides [67]. Polycrystalline films reported thus far have mobilities of order $0.1\,cm^2\,V\,s^{-1}$ [10], but this regime is largely unexplored, and there is certainly potential for improvement. A very large mobility of $80\,cm^2\,V^{-1}\,s^{-1}$ was reported for bulk LaCuOTe [65], but this yet to be reproduced in films.

In undoped or lightly doped chalcogenides, the conductivity is activated, just as in the delafossites. However, degenerate conductivity can be achieved by doping, and it has less drastic consequences for transparency than in the oxides. For example, in the nominally undoped polycrystalline BaCuChF system, the conductivity in the sulfide is activated and is small, while it is degenerate and 10^2–10^3 times higher at room temperature in the telluride. *p*-Doping, in this case with K on the Ba site, increases the conductivity and changes it from activated to degenerate in both the sulfide and the selenide [68]. This general behavior is also observed in thin films, best illustrated for the case of epitaxial $La_{1-y}Mg_yCuOS_{1-x}Se_x$ [69] and shown in Figure 3.6. Here, degenerate *p*-type conductivity of $140\,S\,cm^{-1}$ was observed in $La_{0.9}Mg_{0.1}CuOSe$ at room temperature with $p = 2 \times 10^{20}\,cm^{-3}$ and $\mu = 4\,cm^2\,V^{-1}\,s^{-1}$. The mobility is about a factor of 10 larger than observed in the corresponding doped sulfide, again indicating that the heavier chalcogenides provide an improved path for hole conduction. Typical carrier concentrations are of order 10^{18}–$10^{19}\,cm^{-3}$ in undoped (activated) and $10^{20}\,cm^3$ in doped (degenerate) materials.

The degenerate conductivity characteristic of high carrier concentration is important because it allows current injection in diodes, exemplified by *p*-LaCuOSe:Mg(epi)/*n*-$InGaZnO_3$(amorph). This device exhibits electroluminescence that is controlled by current bias [70]. Even higher room-temperature conductivity ($910\,S\,cm^{-1}$ at $3.5\,cm^2\,V^{-1}\,s^{-1}$) and carrier concentration ($1.7 \times 10^{21}\,cm^{-3}$) have been demonstrated in a very thin (40 nm) epitaxially grown $La_{0.8}Mg_{0.2}CuOSe$ film [71]. Such high hole densities are difficult to obtain in other *p*-type materials such as GaN:Mg, ZnSe:N and ZnO:N.

The band gaps of the layered chalcogenide fluorides and oxides can be continuously tuned from the largest value in the sulfide to the smallest in the telluride by preparing S/Se or Se/Te solid solutions [72–74]. This feature is useful because the band gaps of the materials are

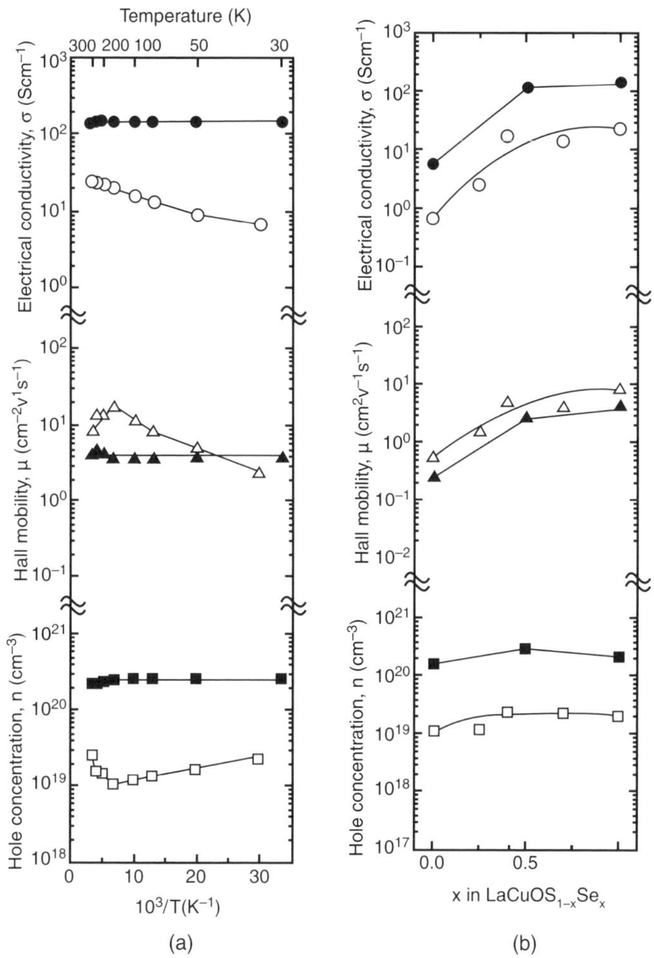

Figure 3.6 *Hole transport properties of undoped (open symbols) and Mg-doped (closed symbols) $LaCuOS_{1-x}Se_x$ epitaxial films (b), and temperature dependence of the properties for LaCuOSe films (a). (Reprinted with permission from [69] Copyright (2003) Elsevier Ltd)*

direct, with the possible exception of the oxide-telluride [65], and since most of them show band-edge excitonic absorption and also relatively narrow-band emission under UV excitation, the emission wavelength can therefore be controlled. The excitonic features remain stable to room temperature, unlike in conventional semiconductors like GaAs where they are observed only at low temperatures. The binding energies are estimated at above 50 meV, comparable with the well-known exciton in ZnO at 60 meV. Since such excitons are not seen in bulk Cu_2Ch, it may be that the anisotropic nature of the layered chalogenide-fluorides and oxides produces an effect akin to that of a two-dimensional quantum well, namely to confine the excitons and increase the binding energy. However, similar step-like absorption with excitonic features that persist to room temperature is seen in the cubic materials BaO, BaS and BaSe [75].

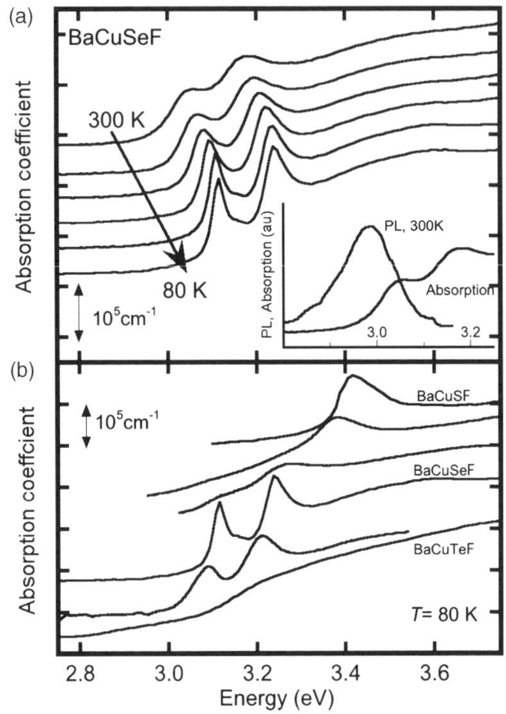

Figure 3.7 *Exciton absorption and emission in layered chalcogenide fluorides. Spectra are offset for clarity. (a) Spin-orbit-split exciton absorption in BaCuSeF narrows as the temperature is decreased. Inset shows the photoluminescence emission and absorption at 300 K. (b) Shift in exciton absorption as the band edge moves in solid solutions* $BaCuS_{1-x}Se_xF$ *(x = 0, 0.2, 0.8, 1) and* $BaCuSe_{1-y}Te_yF$ *(y = 0, 0.5, 1)*

As an example, Figure 3.7 shows absorption spectra of $BaCu(S_{1-x}Se_x)F$ and BaCu $(Se_{1-x}Te_x)F$. They all show excitonic peaks, which are narrow in the sulfides through the selenide, begin to broaden in selenide/telluride solutions, and are almost completely washed out in the telluride. The absorption spectra also show evidence of spin-orbit splitting, which is resolved in the selenide (89 meV), but unresolved in the sulfide (<22 meV). The peak splitting is in quantitative agreement with relativistic density functional theory (DFT) calculations that reveal the degeneracy at the top of the valence band is lifted when spin orbit coupling is taken into consideration, and that the splitting increases, as expected, with the heavier chalcogenides [76, 77]. In BaCuChF, room temperature emission at (slightly below) the excitonic absorption energy is seen only in the pure selenide. Such excitonic features were first reported in the LaCuO(S,Se) system, where control of the emission is achieved not only by partial substitution of Se for S in $LaCuOS_{1-x}Se_x$, but also by substitution of Pr and Nd for La in LaCuOS. At room temperature, the emission range is 386 nm (3.21 eV) to 407 nm (3.05 eV) [78, 79].

To summarize, the chalcogenide oxides and fluorides are the most promising of the Cu-based wide-gap semiconductors for transparent electronics applications. High conductivity and high carrier concentration have been achieved in transparent materials, useful mobilities are possible, and their optical properties are also potentially useful. So far only epitaxial films have been thoroughly explored, and the potential of polycrystalline or amorphous films has not been determined. The low carrier concentrations that would be necessary for use in a *p*-channel TTFT have yet to be reported. Potential *p*-TTFT structures typically show gate-modulated resistance, but it is difficult to turn the devices off, and hard saturation of the *I-V* characteristics is also not realized. An additional concern is the lack of complementary *n*-type conductivity in a process-compatible transparent material, although *n*-type conductivity (and indeed superconductivity at 26 K) has been observed in the isostructural, but semimetallic, $LaO_{1-x}F_xFeAs$ [80].

3.4.2.2 Cu_3TaCh_4 (Ch = S, Se, Te)

Cu_3TaCh_4 is another example of a Cu(I)-based materials system that exhibits *p*-type conductivity in a Cu 3d – Ch *n*p valence band. The band gap is large enough for Cu_3TaS_4 to be transparent (2.75 eV), while Cu_3TaSe_4 (2.35 eV) absorbs in the blue [81, 82]. The room-temperature Seebeck coefficient of both Cu_3TaS_4 and Cu_3TaSe_4 powders is about $+25\,\mu V\,K^{-1}$, confirming the *p*-type nature. This class of compounds is unusual among transparent Cu-based *p*-type conductors in that it is cubic, adopting the *P*-43*m* sulvanite structure. This isotropy is important for device applications, since precise orientation is less critical for good transport properties. Cu_3TaS_4 is readily doped on the Ta site with Zr (an acceptor) and W (a donor). Ta-doping increases and W-doping decreases the *p*-conductivity relative to the undoped powders.

Visible photoluminescence at 540 nm in Cu_3TaS_4 decreases in intensity when excess Cu is added, suggesting that a V_{Cu} acceptor is the final state in the photoluminescence emission. If so, it is located 450 meV above the valence band, and is unlikely to be linked to the conductivity.

3.4.3 Organic Semiconductors

Organic semiconductors are discussed in detail elsewhere in this volume, and here it suffices to comment briefly on the *p*-channel TTFT application. Organic semiconductors also exhibit *p*-conductivity, but they generally exhibit carrier mobilities that are several orders of magnitude lower than those of typical inorganic semiconductors. An exception is the acene family, where field effect mobilities of $3\,cm^2\,V^{-1}\,s^{-1}$ in pentacene-based *p*-channel TFTs [83] and an intrinsic mobility of $35\,cm^2\,V^{-1}\,s^{-1}$ in single-crystal pentacene [84] have been measured. Pentacene ($E_g = 2.5$ eV [85]), is unfortunately not transparent over the entire visible range. *p*-TFTs have been made with tetracene ($E_g = 3.1$ eV), which is slightly orange, but the characteristics are not as good as *p*-channel pentacene TFTs. Complementary *n*-channel TFTs are also unlikely to be realized within the acenes, and in general, *n*-channel organic TFTs are much more problematic, being susceptible to 'air doping' which tends to promote electron traps [86].

3.4.4 Nanomaterials

Carbon nanotubes (CNTs) represent a unique system in which a range of electronic functionality can be found. Single-walled CNTs of a particular chirality are semiconductors, but their band structure differs radically from that in the preceding discussion, giving rise to extremely large mobilities. Given the problem of low mobility in conventional p-type semiconductors, it is ironic that the highest mobility demonstrated in a TFT comes from a p-type material in the form of a single CNT [87], where extremely large values ($>10^5\,cm^2\,V^{-1}\,s^{-1}$) were found or inferred for the field effect, saturation, and intrinsic mobilities. The difficulty of assembling such a device, along with its low current-carrying capacity (μA in this case), of course spurs the search for a TFT based on a network of CNTs, i.e. thin films, and an all-CNT TFT-like structure has been reported [88]. In the network scenario, device manufacture becomes simple, current capacity increases, and although the full mobility of a single CNT is unlikely to be realized because of the percolative current path, even a very large reduction is acceptable. However, very thin films must be made because CNTs are not inherently transparent (band gaps depend on the radius of the CNTs, but are generally $\approx$1 eV), and contamination of the p-type semiconductors with metallic CNTs and amorphous carbon is a problem yet to be solved. Reports of p-TFTs using CNT networks on plastic substrates indeed indicate much lower p-mobility ($1\,cm^2\,V^{-1}\,s^{-1}$ or less), and the transistor-like behavior suffers from the same problems as other inorganic p-TFTs, namely lack of saturation, low on/off ratios, or inability to turn off the devices [88–90]. The predominance of p-CNT-TFTs reflects the lack of suitable low-work function metals to inject n-carriers. However, Sc has recently been identified as a good electron injector into CNTs and a Sc-contacted n-CNT-FET has been reported [91].

3.4.5 Materials Synthesis

Materials synthesis is obviously critically important to produce higher quality materials so as to understand and address the limits of their performance. Unlike ITO, which has seen 50 years of development, the p-type wide-gap semiconductors are relative newcomers. For electronics applications, it is the thin-film deposition that is most important. Many of the best p-type wide-gap semiconductors have multiple components, so deposition methods have to be carefully evaluated for fidelity of stoichiometry. Most of the methods have been met with reasonable success, but process optimization takes time. In the case of p-type wide-band-gap semiconductors, many physical vapor deposition methods have been reported with pulsed laser deposition (PLD) and sputtering being the most common, and chemical vapor deposition (CVD) methods having been used mostly for oxides. An exhaustive review can be found in Banerjee and Chattopadhyay [92].

PLD has been a common choice because it has been a reliable method to achieve reasonable stoichiometry transfer from target to film in oxides, and also because *in-situ* epitaxial films can be achieved in many cases. A common wavelength used for PLD is 248 nm (5 eV), and most of the wide-gap materials in question have significant absorption in this regime. Even with reliable stoichiometric transfer, process optimization is tedious, and often the parameter space for successful film production is narrow. Much of the early work on delafossites came from the Hosono group at the Tokyo Institute of Technolgy, and almost all of the materials were produced by PLD with *in-situ* processing. High quality, epitaxial films of single materials and of multilayers with excellent interfaces were reported. *In-situ* formation of layered chalcogenide fluorides by PLD has been achieved in the case of BaCuChF. BaCuTeF has been deposited at 650 °C on MgO with both in-plane and out-of plane order, resulting in mobilities of $8\,cm^2\,V^{-1}\,s^{-1}$ and conductivities of $165\,S\,cm^{-1}$ [10].

In some cases, the formation temperatures of the materials are too high to achieve *in-situ* formation. In such cases, the technique of reactive solid phase epitaxy (R-SPE) has proven useful [93]. For example, epitaxial growth of *p*-type LaCuOS was achieved by deposition of amorphous LaCuOS on a 5-nm seed layer of Cu on single crystal YSZ at room temperature. In the subsequent 1000 °C anneal of the film in the presence of LaCuOS powder, full in-plane and out-of-plane alignment are realized and extremely narrow diffraction peaks are obtained. The critical component for success is the nucleation of epitaxial LaCuOS growth at the interface of the substrate, seed layer and amorphous precursor. The technique has also been extended to produce high-quality films of other conductive oxides [94]. The high temperatures involved for some systems are prohibitive for all but the highest-performance devices.

p-ZnO presents a particularly challenging system. The low *p*-carrier densities and the ease with which *n*-type defects are formed mean that *p*-ZnO is very susceptible to degradation in a dirty deposition environment. An example of some of the best material comes from a laser-MBE technique adopted by Tsukazaki *et al.* [47] in a process they dubbed 'repeated temperature modulation epitaxy'. Epitaxial ZnO is grown at high temperature (950 °C) in an ultra high vacuum environment, and then a lower temperature growth (400 °C) is employed to allow incorporation of N donors from N_2 gas. This process is repeated layer-by-layer, resulting in films that appear to reproducibly exhibit *p*-type conductivity. The high-temperature growth phase may be essential for annealing out H, which likely deactivates acceptors.

High temperature physical vapor deposition methods are critical for producing the best quality material to understand the basic physical principles underlying the materials systems. These methods are expensive, and large-scale production becomes feasible only with low-temperature, high-throughput methods involving inexpensive substrates. This means that polycrystalline or amorphous materials result. If *p*-type conductivity could be achieved in amorphous optically transparent semiconductors, this would allow low-temperature film deposition by a variety of methods and integration with the already-demonstrated *n*-conductivity in amorphous multicomponent oxides [95]. Although X-ray amorphous $ZnRh_2O_4$ with a band gap of 2.1 eV has been reported [52], there are no amorphous *p*-type semiconductors with good visible transparency, nor reports of low-cost spin-coating methods for the *p*-type semiconductors discussed here.

The exploitation of the relatively high mobility of amorphous oxide semiconductors opens up low-temperature processing routes, including new solution methods. High-quality *n*-channel TTFTs with solution processed dielectrics [96] and amorphous oxide semiconductors have been produced [97]. Solution-based precursors containing hydrazine have also been used to fabricate transistors containing small-gap *p*-type semiconductors [98] and thin films of the conductor $BaCu_2S_2$ [99].

3.5 Outlook and Prospects

Over the past decade, considerable progress has been made in the development of new *p*-type transparent semiconductors. Hole mobilities near $10\ cm^2\ V^{-1}\ s^{-1}$ have been achieved with high visible transparency, and conductivities as high as $900\ S\ cm^{-1}$ have been realized in epitaxial films. Numerous electronic and optical devices have been fabricated on the basis of these unique properties.

Looking forward, there remains a strong need to identify new oxides exhibiting high hole mobilities to complement the known and widely used *n*-type TCOs. While the delafossites and $SrCu_2O_2$ have only limited mobilities, new Cu(I)-rich compositions exhibiting structural characteristics more closely related to those of Cu_2O are worthy synthesis targets for realizing the desired performance. Materials containing atoms with ns^2 valence electron configurations are also potential targets for generation of new wide-gap *p*-type materials, but their propensity to adopt distorted environments that drop the s orbitals below the top of the valence band remains an inherent limitation.

Several studies on epitaxial systems produced via high-end vapor deposition processes (most at rather high temperatures) have provided means for fabrication of a variety of devices and elucidation of the science governing physical behavior. In these systems, continued attention will be given to understanding defect and doping chemistries, particularly for achieving strict and reproducible control of carrier concentrations.

In the future, more attention is likely to be directed to the study of polycrystalline and amorphous materials prepared via low-cost methods that allow large-area coverage. Here, the demonstrated capabilities for tuning the luminescence colors of *p*-type chalcogenides across the visible spectrum, the rapidly narrowing gap in processing complexity between inorganic and organic materials, and the long projected timeline for realizing useful and efficient organic-based lighting and displays, make an expanded effort to develop new types of stable inorganic (or hybrid) light emitting devices on the basis of these materials and methods a potentially rewarding venture. At the same time, similar methods could be used to examine more complex mixed-metal systems that afford new opportunities for realizing simultaneously high dispersity in both the valence and conduction bands. In such systems, charge neutrality levels are likely to appear near the middle of the band gap, promoting bipolar behavior, i.e. selective *p*- or *n*-type conduction in a single material.

In this brief overview, we have attempted to illustrate the progress, challenges, and opportunities for research on transparent *p*-type materials. Hopefully, it can be a source of stimulation for continued forward movement of the field.

References

[1] R. E. Presley, D. Hong, H. Q. Chiang, C. M. Hung, R. L. Hoffman, J. F. Wager, Transparent ring oscillator based on indium gallium oxide thin-film transistors, *Solid State Electron.*, **50**, 500–503 (2006).

[2] K. Nomura, H. Ohta, K. Ueda, T. Kamiya, M. Hirano, H. Hosono, Thin-film transistor fabricated in single-crystalline transparent oxide semiconductor, *Science*, **300**, 1269–1272 (2003).

[3] H. Ohta, M. Orita, M. Hirano, H. Hosono, Fabrication and characterization of ultraviolet-emitting diodes composed of transparent *p-n* heterojunction, *p*-$SrCu_2O_2$ and *n*-ZnO, *J. Appl. Phys.*, **89**, 5720–5725 (2001).

[4] H. Yanagi, K. Ueda, H. Ohta, M. Orita, M. Hirano, H. Hosono, Fabrication of all oxide transparent *p–n* homojunction using bipolar $CuInO_2$ semiconducting oxide with delafossite structure, *Solid State Commun.*, **121**, 15–17 (2001).

[5] A. Tsukazaki, M. Kubota, A. Ohmoto, T. Onuma, M. Ohtani, H. Ohno, S. F. Chichibu, M. Kawasaki, Blue light-emitting diode based on ZnO, *Jpn. J. Appl. Phys.*, **44**, L643–L645 (2005).

[6] H. B. Michaelson, The work function of the elements and its periodicity, *J. Appl. Phys.*, **48**, 4729–4733 (1977).
[7] H. Yanagi, M. Kikuchi, K.-B. Kim, H. Hiramatsu, T. Kamiya, M. Hirano, H. Hosono, Low and small resistance hole-injection barrier for NPB realized by wide-gap *p*-type degenerate semiconductor, LaCuOSe:Mg, *Org. Electr.* (2008) doi: 10.1016/j.orgel.2008.03.004.
[8] L. F. Johnson, M. B. Moran, T.L. Cole, Conductive, infrared-transparent intermetallic boride, carbide and oxide coatings, *41st Annual Technical Conference Proceedings of the Society of Vacuum Coaters*, 232–237 (1998).
[9] S. J. Fonash, *Solar Cell Device Physics*, pp. 76–80, Academic Press, New York, 1981.
[10] R. Kykyneshi, D.H. McIntyre, J. Tate, C.-H. Park, D.A. Keszler, Electrical and optical properties of epitaxial transparent conductive BaCuTeF thin films deposited by pulsed laser deposition, *Solid State Sci.* (2007) doi: 10.1016/jsolidstatesciences.2007.10.005.
[11] H. Ohta, M. Orita, M. Hirano, H. Hosono, Surface morphology and crystal quality of low resistive indium tin oxide grown on yittria-stabilized zirconia, *J. Appl. Phys.*, **91**, 3547–3550 (2002).
[12] C.-H. Park, D. A. Keszler, J. Tate, P. S. Nyholm, High electron mobility W-doped In_2O_3 thin films by pulsed laser deposition, *Appl. Phys. Lett.*, **87**, 112 108 (2005).
[13] A. Zunger, Practical doping principles, *Appl. Phys. Lett.*, **83**, 57–59 (2003).
[14] S. B. Zhang, J. E. Northrup, Chemical potential dependence of defect formation energies in GaAs: application to Ga self diffusion, *Phys. Rev. Lett.*, **67**, 2339–2342 (1991).
[15] W. Walukiewicz, Intrinsic limitations to the doping of wide-gap semiconductors, *Physica B*, **302–302**, 123–134 (2001).
[16] S. B. Zhang, S.-H. Wei, A. Zunger, A phenomenological model for systematization of and prediction of doping limits in II-VI and I-III-VI_2 compounds, *J. Appl. Phys.*, **83**, 3192–3196 (1997).
[17] H. Raebiger, S. Lany, A. Zunger, Origins of the *p*-type nature and cation deficiency in Cu_2O and related materials, *Phys. Rev., B*, **76**, 045 209 (2007).
[18] B. J. Ingram, M. I. Bertoni, K. R. Poeppelmeier, T. O. Mason,Point defects and transport mechanisms in transparent conducting oxides of intermediate conductivity *Thin Solid Films*, **486**, 86–93 (2005).
[19] T. K. Reynolds, R. G. Bales, R. F. Kelley, F. J. DiSalvo, The synthetic search for better thermoelectrics, in *Chemistry, Physics and Materials Science of Thermoelectric Materials: Beyond Bismuth Telluride*, M. G. Kanatzidis, S. D. Mahanti, T. P. Hogan (Eds), Kluwer Academic/Plenum, New York, 2003.
[20] P. Y. Yu, M. Cardona, *Fundamentals of Semiconductors*, Springer, Berlin, 2001.
[21] H. Kawazoe, M. Yasukawa, H. Hyodo, M. Kurita, H. Yanagi, H. Hosono, *p*-Type electrical conduction in transparent thin films of $CuAlO_2$, *Nature*, **389**, 939–942 (1997).
[22] H. Kawazoe, Chemical design of transparent *p*-type conducting oxides, *Kotai Butsuri*, **33** (11), 937–943 (1998).
[23] H. Kawazoe, *p*-Type oxide transparent conductive films, *Tomei Dodenmaku no Shintenkai*, 47–57 (1999).
[24] H. Yanagi, T. Hase, S. Ibuki, K. Ueda, H. Hosono, Bipolarity in electrical conduction of transparent oxide semiconductor $CuInO_2$ with delafossite structure, *Appl. Phys. Lett.*, **78**, 1583–1585 (2001).
[25] N. Duan, A. W. Sleight, M. K. Jayaraj, J. Tate, Transparent *p*-type conducting $CuScO_{2+\delta}$ films, *Appl. Phys. Lett.*, **77**, 1325–1326 (2000); R. Nagarajan, N. Duan, M. K. Jayaraj, J. Li, K. A. Vanaja, A. Yokochi, A. Draeseke, J. Tate, A. W. Sleight, *p*-Type conductivity in the delafossite structure, *Int. J. Inorg. Mater.*, **3**, 265–270 (2001).
[26] R. L. Hoffman, J. F. Wager, M. K. Jayaraj, J. Tate, Electrical characterization of transparent *p-i-n* heterojunction diodes, *J. Appl. Phys.*, **90**, 5763–5767 (2001).
[27] W. Staehlin, H. R. Oswald, Polytypes of the double oxides, $AgCoO_2$, $AgFeO_2$, and $AgCrO_2$, *Z. Anorgan. Allg. Chem.*, **373**, 69–72 (1970).
[28] H. Yanagi, S. Inoue, K. Ueda, H. Kawazoe, H. Hiroshi, H. Hosono, N. Hamada, Electronic structure and optoelectronic properties of transparent *p*-type conducting $CuAlO_2$, *J. Appl. Phys.*, **88**, 4159–4163 (2000).

[29] B. J. Ingram, T. O. Mason, R. Asahi, K. T. Park, A. J. Freeman, Electronic structure and small polaron hole transport of copper aluminate, *Phys. Rev., B*, **64**, 155 114/1–155 114/7 (2001).
[30] J. Robertson, P. W. Peacock, M. D. Towler, R. Needs, Electronic structure of *p*-type conducting transparent oxides, *Thin Solid Films*, **411**, 96–100 (2002).
[31] B. Falabretti, J. Robertson, Electronic structures and doping of SnO_2, $CuAlO_2$, and $CuInO_2$, *J. Appl. Phys.*, **102** (12), 123 703/1–123 703/5 (2007).
[32] J. Robertson, K. Xiong, S. J. Clark, Band gaps and defect levels in functional oxides, *Thin Solid Films*, **496**, 1–7 (2005).
[33] A. Buljan, M. Llunell, E. Ruiz, P. Alemany, Color and conductivity in Cu_2O and $CuAlO_2$: a theoretical analysis of $d^{10}\cdots d^{10}$ interactions in solid-state compounds, *Chem. Mater.*, **13**, 338–344 (2001).
[34] S. Boudin, C. Felser, F. Studer, Cu-Cu interactions in the transparent *p*-type conductors: $CuAlO_2$ and $SrCu_2O_2$, *Solid State Sci.*, **5**, 741–744 (2003).
[35] J. Li, K. A. Vanaja, A. Yokochi, A. Draeske, J. Tate, A. W. Sleight, *p*-Type conductivity in the delafossite structure, *Int. J. Inorg. Chem.*, **3**, 265–270 (2001).
[36] X. Nie, S.-H. Wei, S. B. Zhang, Bipolar doping and band-gap anomalies in delafossite transparent conductive oxides, *Phys. Rev. Lett.*, **88**, 066 405 (2002).
[37] B. J. Ingram, G. B. González, T. O. Mason, D. Y. Shahriari, A. Barnabè, D. Ko, K. R. Poeppelmeier, Transport and defect mechanisms in cuprous delafossites. 1. Comparison of hydrothermal and standard solid-state synthesis in $CuAlO_2$, *Chem. Mater.*, **16**, 5616–5622 (2004).
[38] W. W. Warren, A. Rajabzadeh, T. Olheiser, J. Liu, J. Tate, M. K. Jayaraj, K. A. Vanaja, Nuclear quadrupole resonance studies of transparent conducting oxides, *Solid State Nuclear Magn. Reson.*, **26**, 209–214 (2004).
[39] A. Kudo, H. Yanagi, H. Hosono, H. Kawazoe, $SrCu_2O_2$: a *p*-type conductive oxide with wide band gap, *Appl. Phys. Lett.*, **73**, 220–222 (1998).
[40] C. C. B. Lynch, R. G. Egdell, D. S. L. Law, High resolution X-ray photoemission of $SrCu_2O_2$, *Chem. Phys. Lett.*, **401**, 223–226 (2005).
[41] H. Ohta, H. Hosono, Transparent oxide electronics, *Materials Today*, **7**, 42–51 (2004).
[42] H. Hosono, H. Ohta, K. Hayashi, M. Hirano, Near-UV emitting diodes based on a transparent *p-n* junction compound of heteroepitaxially grown *p*-$SrCu_2O_2$ and n-ZnO, *J. Cryst. Growth*, **237–239**, 496–502 (2002).
[43] Z. M. Jarzebski, *Oxide Semiconductors*, p. 150, Pergamon, Oxford, 1973.
[44] H. Sato, T. Minami, S. Takata, T. Yamada, Transparent conducting *p*-type NiO thin films prepared by magnetron sputtering, *Thin Solid Films*, **236**, 27–31 (1993).
[45] H. Ohta, M. Kamiya, T. Kamiya, M. Hirano, H. Hosono, UV-detector based on pn-heterojunction diode composed of transparent oxide semiconductors, *Thin Solid Films*, **445**, 317–321 (2003).
[46] M. D. Irwin, D. B. Buchholz, A. W. Hains, R. P. H. Chang, T. J. Marks, *p*-Type semiconducting nickel oxide as an efficiency-enhancing anode interfacial layer in polymer bulk-heterojunction solar cells, *PNAS*, **105**, 2783–2787 (2008).
[47] A. Tsukazaki, A. Ohtomo, T. Onuma, M. Ohtani, T. Makino, M. Sumiya, K. Ohtani, S. F. Chichibu, S. Fuke, Y. Segawa, H. Ohno, H. Koinuma, M. Kawasaki, Repeated temperature modulation epitaxy for *p*-type doping and light emitting diode based on ZnO, *Nature Mater.*, **4**, 42–46 (2005).
[48] A. Tsukazaki, M. Kubota, A. Ohtomo, T. Onuma, K. Ohtani, H. Ohno, S. F. Chichibu, M. Kawasaki, Blue light emitting diode based on ZnO, *Jpn. J. Appl. Phys.*, **44**, L643–L645 (2005).
[49] M. Dekkers, G. Rijnders, D. H. A. Blank, $ZnIr_2O_4$, a *p*-type transparent oxide semiconductor in the class of spinel zinc-d^6-transition metal oxide, *Appl. Phys. Lett.*, **90**, 021 903 (2007).
[50] H. Mizoguchi, M. Hirano, S. Fujitsu, T. Takeuchi, K. Ueda, H. Hosono, $ZnRh_2O_4$: a *p*-type semiconducting oxide with a valence band composed of a low spin state of Rh^{3+} in a $4d^6$ configuration, *Appl. Phys. Lett.*, **80**, 1207–1209 (2002).
[51] D. J. Singh, R. C. Rai, J. L. Musfeldt, S. Auluck, N. Singh, P. Khalifah, S. McClure, D. G. Mandrus, Optical properties and electronic structure of spinel $ZnRh_2O_4$, *Chem. Mater.*, **18**, 2696–2700 (2006).

[52] S. Narushima, H. Mizoguchi, K. Shimizu, K. Ueda, H. Ohta, M. Hirano, T. Kamiya, H. Hosono, A *p*-type amorphous oxide semiconductor and room temperature fabrication of amorphous oxide *p*-n heterojunction diodes, *Adv. Mater.*, **15**, 1409–1413 (2003).
[53] http://www.kitco.com/charts/rhodium.html, accessed 28 Mar 2008. The price has increased by about a factor of 30 since 2003.
[54] C. F. Windisch Jr., K. F. Ferris, G. J. Exarhos, S. K. Sharma, Infrared transparent spinel films with *p*-type conductivity, *Thin Solid Films*, **398–399**, 45–52 (2001).
[55] C. F. Windisch Jr., G. J. Exarhos, K. F. Ferris, M. H. Engelhard, D. C. Stewart, Conducting spinel oxide films with infrared transparency, *Thin Solid Films*, **420–421**, 89–99 (2002).
[56] G. J. Exarhos, C. F. Jr. Windisch, K. F. Ferris, R. R. Owings, Cation defects and conductivity in transparent oxides, *Appl. Phys. A*, **89**, 9–18 (2007).
[57] S. J. Fonash, *Solar Cell Device Physics*, p. 78, Academic Press, New York, 1981.
[58] S. Park, D. A. Keszler, M. M. Valencia, R. L. Hoffman, J. P. Bender, J. F. Wager, Transparent *p*-type conducting $BaCu_2S_2$ films, *Appl. Phys. Lett.*, **80**, 4393–4395 (2002).
[59] K. Ueda, S. Inoue, S. Hirose, H. Kawazoe, H. Hosono, Transparent *p*-type semiconductor: LaCuOS layered oxysulfide, *Appl. Phys. Lett.*, **77**, 2701–2703 (2000).
[60] H. Yanagi, J. Tate, S. Park, C.-H. Park, D. A. Keszler, *p*-Type conductivity in wide-band-gap BaCuQF (Q = S,Se), *Appl. Phys. Lett.*, **82**, 2814–2816 (2003).
[61] H. Hiramatsu, K. Ueda, H. Ohta, M. Hirano, T. Kamiya, H. Hosono, Degenerate *p*-type conductivity in wide-gap $LaCuOS_{1-x}Se_x$ (x = 0–1) epitaxial films, *Appl. Phys. Lett.*, **82**, 1048–1050 (2003).
[62] K. Ueda, H. Hosono, Band gap engineering, band edge emission, and *p*-type conductivity in wide-gap $LaCuOS_{1-x}Se_x$ oxychalcogenides, *J. Appl. Phys.*, **91**, 4768–4770 (2002).
[63] K. Ueda, H. Hosono, N. Hamada, Valence-band structures of layered oxychalcogenides, LaCuOCh (Ch = S, Se, and Te), studied by ultraviolet calculated by the full-potential linearized augmented plane wave method, *J. Phys.: Condens. Matter*, **16**, 5179–5186 (2004).
[64] K. Ueda, H. Hosono, N. Hamada, Energy band structure of LaCuOCh (Ch = S, Se, and Te), studied by ultraviolet photoemission spectroscopy and energy-band calculations, *J. Appl. Phys.*, **98**, 043 506 (2005).
[65] M. L. Liu, L. B. Wu, F. Q. Huang, L. D. Chen, J. A. Ibers, Syntheses, crystal and electronic structure, and some optical and transport properties of LnCuOTe (Ln = La, Ce, Nd), *J. Solid State Chem.*, **180**, 62–69 (2007).
[66] H. Yanagi, J. Tate, S. Park, C.-H. Park, D. A. Keszler, M. Hirano, H. Hosono, Valence band structure of BaCuSF and BaCuSeF, *J. Appl. Phys.*, **100**, 083 705 (2006).
[67] H. Q. Chiang, J. F. Wager, R. L. Hoffman, J. Jeong, D. A. Keszler, High mobility transparent thin-film transistors with amorphous zinc tin oxide channel layer, *Appl. Phys. Lett.*, **86**, 013 503 (2005).
[68] H. Yanagi, J. Tate, S. Park, C.-H. Park, D. A. Keszler, *p*-Type conductivity in wide-band-gap BaCuQF (Q = S, Se), *Appl. Phys. Lett.*, **82**, 2814–2816 (2003).
[69] H. Hiramatsu, K. Ueda, H. Ohta, M. Hirano, T. Kamiya, H. Hosono, Wide gap *p*-type degenerate semiconductor, Mg-doped LaCuOSe, *Thin Solid Films*, **445**, 304–308 (2003).
[70] H. Hiramatsu, K. Ueda, H. Ohta, T. Kamiya, M. Hirano, H. Hosono, Excitonic blue luminescence from *p*-LaCuOSe/n-$InGaZn_5O_8$ light-emitting diode at room temperature, *Appl. Phys. Lett.*, **87**, 21107 (2005).
[71] H. Hiramatsu, K. Ueda, H. Ohta, M. Hirano, M. Kikuchi, H. Yanagi, T. Kamiya, H. Hosono, Heavy hole doping of epitaxial thin films of a wide gap *p*-type degenerate semiconductor, LaCuOSe, and analysis of the effective mass, *Appl. Phys. Lett.*, **91**, 012 104 (2007).
[72] C.-H. Park, D. A. Keszler, H. Yanagi, J. Tate, Gap modulation in $MCu[Q_{1-x}Q'_x]F$ (M = Ba, Sr; Q, Q′ = S, Se, Te) and related materials, *Thin Solid Films*, **445**, 288–293 (2003).
[73] C.-H. Park, R. Kykyneshi, A. Yokochi, J. Tate, D. A. Keszler, Structure and physical properties of BaCuTeF, *J. Solid State Chem.*, **180**, 1672–1677 (2007).
[74] H. Hiramatsu, K. Ueda, K. Takafuji, H. Ohta, M. Hirano, T. Kamiya, H. Hosono, Fabrication of heteroepitaxial thin films of layered oxychalcogenides LnCuOCh (Ln = La-Nd; Ch = S-Te) by reactive solid-phase epitaxy, *J. Mater. Res.*, **19**, 2137–2143 (2004).

[75] R. J. Zollweg, Optical absorption and photoemission of barium and strontium oxides, sulfides, selenides, and tellurides, *Phys. Rev.*, **111**, 113–119 (1958).
[76] A. Zakutayev, R. Kykyneshi, J. Kinney, D. H. McIntyre, G. Schneider, J. Tate, Excitonic absorption and emission in thin-film BaCu(Q,Q′)F [Q, Q′ = S, Se, Te], unpublished.
[77] T. Kamiya, K. Ueda, H. Hiramatsu, H. Kamioka, H. Ohta, M. Hirano, H. Hosono, Two-dimensional electronic structure and multiple excitonic states in layered oxychalcogenide semiconductors, LaCuOCh (Ch=S, Se, Te): optical properties and relativistic ab initio study, *Thin Solid Films*, **486**, 98–103 (2005).
[78] H. Hiramatsu, H. Kamioka, K. Ueda, M. Hirano, H. Hosono, Electrical and photonic functions originating from low-dimensional structures in wide-gap semiconductors LnCuOCh (Ln = lanthanide, Ch = chalcogen): a review, *J. Ceram. Soc. Jpn.*, **113**, 10–16 (2005).
[79] H. Hiramatsu, K. Ueda, K. Takafuji, H. Ohta, M. Hirano, T. Kamiya, H. Hosono, Degenerate electrical conductive and excitonic photoluminescence properties of epitaxial films of wide gap *p*-type layered oxychalcogenides, LnCuOCh (Ln=La, Pr and Nd; Ch=S or Se), *Appl. Phys. A*, **79**, 1521–1523 (2004).
[80] Y. Kamihara, T. Watanabe, M. Hirano, H. Hosono, Iron-based layered superconductor $La[O_{1-x}F_x]FeAs$ (x = 0.05–0.12) with Tc = 26 K, *J. Am. Chem. Soc. Commun.*, **130**, 3296–3297 (2008).
[81] P. F. Newhouse, P. A. Hersh, A. Zakutayev, A. Richard, H. A. S. Platt, D. A. Keszler, J. Tate, Thin film preparation and characterization of wide band gap Cu_3TaQ_4 (Q = S or Se) *p*-type semiconductors, *Thin Solid Films*, **517**, 2473–2476 (2008).
[82] J. Tate, P. F. Newhouse, R. Kykyneshi, P. A. Hersh, J. Kinney, D. H. McIntyre, D. A. Keszler, Chalcogen-based transparent conductors, *Thin Solid Films*, **516**, 5795–5799 (2008).
[83] H. Klauk, M. Halik, U. Zschieschang, G. Schmid, W. Radlik, W. Weber, High-mobility polymer gate dielectric pentacene thin film transistors, *J. Appl. Phys.*, **92**, 5259–5263 (2002).
[84] O. D. Jurchescu, J. Baas, T. T. M. Palstra, Effect of impurities on the mobility of single crystal pentacene, *Appl. Phys. Lett.*, **84**, 3061–3063 (2004).
[85] E. A. Silinsh, V. Capek, *Organic Molecular Crystals*, AIP, New York, 1994.
[86] N. Koch, Organic electronic devices and their functional interfaces, *ChemPhysChem*, **8**, 1438–1455 (2007).
[87] T. Dürkop, S. A. Getty, E. Cobas, M. S. Fuhrer, Extraordinary mobility in semiconducting carbon nanotubes, *Nano Lett.*, **4**, 35–39 (2004).
[88] Q. Cao, S.-H. Hur, Z.-T. Zhu, Y. Sun, C. Wang, M. A. Meitl, M. Shim, J. A. Rogers, Highly bendable, transparent thin-film transistors that use carbon-nanotube-based conductors and semiconductors with elastomeric dielectrics, *Adv. Mater.*, **18**, 304–309 (2006).
[89] E. Artukovic, M. Kaempgen, D. S. Hecht, S. Roth, G. Grüner, Transparent and flexible carbon nanotube transistors, *Nano Lett.*, **5**, 757–760 (2005).
[90] X. Han, D. C. Janzen, J. Vaillencourt, X. Lu, *Micro Nano Lett.*, **2**, 96–98 (2007).
[91] Z. Zhang, X. Liang, S. Wang, K. Yao, Y. Hu, Y. Zhu, Q. Chen, W. Zhou, Y. Li, Y. Yao, J. Zhang, L.-M. Peng, Doping-free fabrication of carbon nanotube based ballistic CMOS devices and circuits, *Nano Lett.*, **7**, 3603–3607 (2007).
[92] A. N. Banerjee, K. K. Chattopadhyay, Recent developments in the emerging field of crystalline *p*-type transparent conducting oxide thin films, *Progr. Cryst. Growth Characteriz. Mater.*, **50**, 52–105 (2005).
[93] H. Hiramatsu, H. Ohta, T. Suzuki, C. Honjo, Y. Ikuhara, K. Ueda, T. Kamiya, M. Hirano, H. Hosono, Mechanism for heteroepitaxial growth of transparent *p*-type semiconductor: LaCuOS by reactive solid-phase epitaxy, *J. Crystal Growth Design*, **4**, 301–307 (2004).
[94] Y. Ogo, K. Nomura, H. Yanagi, H. Ohta, T. Kamiya, M. Hirano, H. Hosono, Growth and structure of heteroepitaxial thin films of homologous compounds $RAO_3\ (MO)_m$ by reactive solid-phase epitaxy: applicability to a variety of materials and epitaxial template layers, *Thin Solid Films*, **496**, 64–69 (2006).
[95] H. Hosono, M. Yasukawa, H. Kawazoe, Novel oxide amorphous semiconductors: transparent conducting amorphous oxides, *J. Non-Cryst. Solids*, **203**, 334–344 (1996).

[96] J. T. Anderson, C. L. Munsee, C. M. Hung, T. M. Phung, G. S. Herman, D. C. Johnson, J. F. Wager, D. A. Keszler, Solution-processed HafSOx and ZircSOx inorganic thin-film dielectrics and nanolaminates, *Adv. Funct. Mater.*, **17**, 2117–2124 (2007).
[97] J. T. Anderson, D. A. Keszler, S. T. Meyers, H. Q. Chiang, D. Hong, R. E. Presley, J. F. Wager, Solution-processed oxide films, devices, and integrated circuits, *MRS Proc.*, **998E**, QQ01-06 (2006).
[98] D. J. Milliron, D. B. Mitzi, M. Copel, C. E. Murray, Solution-processed metal chalcogenide films for *p*-type transistors *Chem. Mater.*, **18**, 587–590 (2006); D. B. Mitzi, M. Copel, C. E. Murray, High-mobility *p*-type transistor based on a spin-coated metal telluride semiconductor, *Adv. Mater.*, **18**, 2448–2452 (2006).
[99] Y. Wang, M. Liu, F. Huang, L. Chen, H. Li, X. Lin, W. Wang, Y. Xia, Solution-processed *p*-type transparent conducting $BaCu_2S_2$ thin film, *Chem. Mater.*, **19**, 3102–3104 (2007).

4

Lead Oxides: Synthesis and Applications

Dale L. Perry
Lawrence Berkeley National Laboratory, Berkeley, USA

4.1 Introduction

While the binary oxides of lead might seem to be a rather mundane topic, there perhaps exist only a few other metal oxide systems that are as (or more) complex and interesting than this group of compounds. Lead oxides, several members of which are in the class of conducting metal oxides, are used in a wide array of applications in the field of electronics. Other representative examples of their applications include the use of lead oxide as a direct conversion material for X-ray imaging detectors [1], transparent conducting films [2] that exhibit optical transmission in the visible region and reflectance in the infrared region of the spectrum, and lead oxide-graphite composite electrodes [3]. As a result of their electronic properties and ability to be synthesized and fabricated in many forms by a variety of methods, there is great potential for lead oxides to be used for many more applications in the future. Additionally, combining them with other materials may easily result in many more applications as composites.

This chapter discusses the synthetic chemistry, structural forms, physical properties, and some of the representative applications of lead oxides. The various synthetic schemes that are discussed represent a very wide variety and range of synthetic approaches. Several experimental techniques are described for their analyses, from a chemical content, structural, and bonding standpoint. Methods are discussed for their synthesis and growth

Transparent Electronics: From Synthesis to Applications Edited by Antonio Facchetti and Tobin J. Marks

into powders, crystals, and films, while several techniques are described for fabrication of the lead oxides into forms that are useful in various devices.

While this work focuses only on the dominant, principal oxides of lead, minor oxides sometimes mentioned in the research literature are mentioned, too, with comments and references. The complexity of the Pb-O system is such that, while a principal lead oxide being sought is often obtained, it may be contaminated with both other major phases in various amounts as well as minor phases, too. Even though much research effort has been expended in the chemistry of lead and its oxides, there still remain many aspects of the field that have not been fully examined.

4.2 Overview of Synthetic Methods and Approaches

The synthesis of lead oxides can be effected by a wide range of synthetic methods and approaches. The bulk syntheses of the lead oxides can be as straightforward as simple oxygenation of the metal, but varying the reaction conditions, for example, can alter what specific oxide is obtained even by this simple oxygenation. The oxides can be synthesized and fabricated in many different forms other than bulk powders, such as thin films, large crystals, nanoparticles and wires. Other synthetic routes include tribology, electrochemistry, thermal decomposition, and solution chemistry.

The various chemical forms of lead oxides represent an extremely complex reaction chemistry by which they are formed, including interconversion among different chemical and structural forms as a function of both chemical and physical conditions. The parameters that effect transitions between different chemical and structural forms of the oxides include heat, time, initial reactants, tribological effects, the presence of dopants during their synthesis, and pH conditions in the case of aqueous syntheses.

Two other aspects of the syntheses of lead oxides that are present somewhat complicate the syntheses of other elemental oxides. First, because of the two oxidation states, the different state oxides can form a mixed oxidation species containing both the Pb(II) and Pb(IV) states, i.e. Pb_3O_4. In several cases, such transitions from one single oxide to another oxidation state oxide or mixed oxidation state oxide may be effected by merely heating. Secondly, either an impurity-free product is difficult to obtain for a specifically desired product, or impurities present may determine the phase of the oxide as discussed below. Both concepts are discussed below with respect to the syntheses for PbO and PbO_2. Also, when attempting the preparation of all members of the lead oxide family, *it is prudent to compare the obtained product with accepted X-ray diffraction patterns found in the literature.*

Finally, while the syntheses of bulk lead oxides can be achieved by a wide variety of routes, more exotic morphological forms of the oxides can be obtained; for example, forms such as films, rods, large crystals, and much smaller nanocrystals can be produced. While by no means an exhaustive list, several examples will be given to illustrate the large range of possibilities.

Tables 4.1 and 4.2 list the principal lead oxides, representative commercial synonyms, physical and chemical properties, and *Chemical Abstract* (American Chemical Society) registry numbers for the parent lead oxides themselves.

Table 4.1 *Principal lead oxides, synonyms, chemical literature registry numbers, and molecular weights*

Chemical name and formula	Synonyms and trade names [42]	Chemical Abstract Service (CAS) registry number [43]	Molecular weight
Lead oxide (PbO)	Lead monoxide (PbO); lead oxide yellow; lead protoxide; lead(2+) oxide; lead(II) oxide; Litharge S; Litharge Yellow L-28; plumbous oxide; yellow lead ochre; massicot; C. I. Pigment Yellow 46	[1317-36-8]	223.2
Lead dioxide (PbO_2)	Lead oxide; lead brown; lead(II) lead oxide brown; lead peroxide; lead superoxide; lead(IV) oxide; plumbic oxide, lead(4+), scrutinyite; plattnerite, Thiolead A	[1309-60-0]	239.2
Lead trioxide (Pb_2O_3)	Lead sesquioxide; plumbous plumbate lead(II,IV); dilead trioxide; lead(II, IV); lead(2+, 4+)	[1314-27-8]	462.4
Lead tetraoxide (Pb_3O_4)	IC oxide (Pb_3O_4); Azarcon; C. I. 77578; C. I. Pigment Red 105; Entan Gold Satinobre; Heuconin 5; lead orthoplumbate; triplumbic tetroxide; lead oxide (3:4); lead oxide red; lead tetroxide; Mennige; Mineral Orange; Mineral red; Minium; Minium Non-Setting RL 95; Minium red; Orange Lead; Paris Red; red lead; red lead oxide; Sandix; Saturn Red; trilead tetraoxide; trilead tetroxide; plumboplumbic oxide	[1314-41-6]	685.6

Table 4.2 *Physical and chemical properties of lead oxides [43]*

Chemical name and formula	Physical form/melting point	Density ($g\,cm^{-3}$)
Lead oxide (PbO) (Litharge)	Tetrahedral red crystals/alters to massicot at 489 °C	9.35
Lead oxide (PbO) (Massicot)	Orthorhombic yellow crystals/897 °C	9.64
Lead dioxide (α-PbO_2) (Scrutinyite)	Orthorhombic red-brown crystals	9.1–9.4
Lead dioxide (β-PbO_2) (Plattnerite)	Tetragonal brown-black crystals	9.1–9.4
Lead trioxide (Pb_2O_3)	Monoclinic black crystals or red amorphous powder/decomposes at 530 °C	10.05
Lead tetraoxide (Pb_3O_4) (Minium)	Tetrahedral red crystals. Decomposes at 550 °C, melts at $\sim$830 °C under partial oxygen pressure	8.92

4.3 Synthesis of Lead Oxides

4.3.1 Synthesis of PbO

PbO exists in two structural polymorphs, the α and β forms. α-PbO, referred to, mineralogically, as *litharge*, is the red tetragonal form [4], while β-PbO, *massicot*, is the yellow orthorhombic form [5] (Figure 4.1). Depending on the conditions of storage and thermal and mechanical treatments, they will interconvert between one another. Also, there are a number of synthetic routes which will effect the synthesis of different forms of PbO, such as particles, nanorods, etc.

The two phases of PbO can be made in bulk quantities by a wide variety of methods, including elevated temperature combination with oxygen, solution chemical methods, and thermal decomposition. Litharge, for example, may be made by heating elemental lead in air, although unless great care is taken to control the temperature, the final product can be contaminated with other lead oxide phases. Alternately, a very pure compound can be made by an aqueous reaction of lead(II) acetate trihydrate reacting with NaOH at elevated temperature in a Teflon beaker with vigorous stirring [6].

$$\mathrm{Pb(NO_3)\cdot 3H_2O + NaOH(aqueous) \rightarrow \alpha\text{-}PbO}$$

During this synthetic procedure, the yellow form of PbO is formed first, then transformed into the red litharge form; this process is easily followed visually due to the color change between the two forms. This same reaction can be used in a regular glass laboratory beaker to make the massicot form due to the very slight dissolution of silicon from the glass walls of the beaker that stops the phase transformation.

Grinding the product with a mortar and pestle results in both nanoparticles and clusters of nanoparticles. After allowing the crop of crystals to stand for several months at room temperature, the deep red color of the litharge begins to turn lighter in color, indicative of its

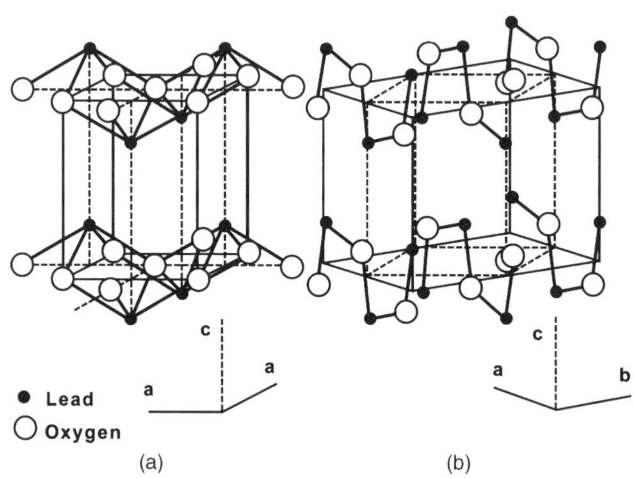

Figure 4.1 Comparative schematic structures of the tetragonal α-PbO (a) and orthorhombic β-PbO (b) phases of lead(II) oxide. The unit cell frameworks of the molecules are represented by the solid lines. From Perry and Wilkinson [6]. Copyright (2007) Springer Science + Business Media

very slow conversion to the yellow massicot form. Further grinding of the litharge, however, accomplishes two purposes. First, the small clusters of the litharge are broken into many more single nanoparticles than are formed in the initial synthetic processes. Secondly, when samples of the several-month-old material that are beginning to convert into massicot are ground, the material is converted back into litharge, with its tell-tale change to the deep red color. Other parameters also affect the phase formed when PbO is formed by evaporative techniques [7].

One must take precautions in the use of commercial sources of PbO. Many of these commercial sources of PbO use a material, that, because of the bulk industrial processes used to make it, is usually a mixture of both the α- and β-PbO forms. Experience has shown that with PbO, several batches of 'pure' massicot or litharge can be seen by X-ray diffraction to be a mixture. The mixture also can be due to the effects mentioned above, including ageing of the commercial sample.

Several synthetic approaches for obtaining PbO revolve around grinding. One [8] combines the grinding and decomposition of $PbCO_3$ to form various ratios of the massicot/litharge in PbO mixtures. Another technique involves the thermal decomposition of lead(II) precursors such as lead(II) diaqua 3,6-dicarboxylatopyridazine to form both α-PbO and β-PbO phases as a function of controlled heating [9]. Another synthesis of PbO (both structural phases) is centered on spray pyrolysis (aerosol decomposition) to derive either single phase or mixed products which can be produced [10] with respect to careful heating schemes. In addition to its phase studies, the product's morphology can easily be studied by scanning electron microscopy (SEM).

In addition to bulk lead monoxide, several other products can be made that have unique morphology or form. α-PbO films, for example, can be made by the spray pyrolysis [11] of an aqueous solution of $Pb(NO_3)_2$. Other films can be made by evaporatively produced layers that result [12] in an initial seeding layer of mostly yellow orthorhombic massicot, followed by the deposition of the subsequent layer growth of the red tetragonal litharge phase. Highly photoactive films of α-PbO can be formed by electrochemical methods [13].

There is still yet a third phase (several intermediate ones, too, actually) of PbO which can be generated under high pressure conditions. This high-pressure phase (γ-PbO) is obtained [14] at room temperature at a pressure of $\sim$0.7 and $\sim$2.5 GPa. Other rare low temperature phases (α'-PbO) of the compound have been described in this same work. Another form of PbO, 'black lead oxide', is, in reality, a mixture of either α-PbO or β-PbO and finely powdered metallic lead [15].

4.3.2 Synthesis of PbO_2

Several methods of synthesizing PbO_2 have been reported in the research literature, although PbO_2, like PbO, presents the same problems with respect to obtaining a truly pure compound. The compound – like PbO – exhibits two structural polymorphic phases. The α-PbO_2 phase (mineralogically, *scrutinyite*) has the orthorhombic structure, while β-PbO_2 (mineralogically, *plattnerite*) has the tetragonal rutile structure. The traditional syntheses for the two phases focus on both regular chemical and electrochemical approaches, with one general approach being the oxidation of Pb(II) species to the tetravalent lead oxide by high-temperature oxidation with oxygen, by melts, or by solution reactions. The majority of the syntheses involve the solution reaction approach.

Methods of synthesis for PbO_2 that might seem obvious do not necessarily yield the desired product or a pure product. Oxidation of either PbO or Pb_3O_4 by oxygen at elevated temperatures, for example, does *not* yield a pure PbO_2 product but rather a material that is of an intermediate Pb-O stoichiometric compound between PbO and PbO_2 [16]. Many of these syntheses directed at the formation of PbO_2 lead to products that are either deficient in oxygen or are nonstoichiometric phases. Several other researchers have also attempted the direct combination of lead with oxygen at high pressure and temperature without success [17].

There are several methods for making bulk α- and β-PbO_2, with the experimental details and pitfalls having been extensively discussed [18]. For α-PbO_2, they include the repeated formation of a flux of yellow PbO (massicot) with $NaClO_4$ and $NaNO_3$ at 340 °C for 10 min, followed by a washing and a drying of each cycle to remove any remaining soluble salts. The resulting solid is then suspended in 3 M HNO_3 overnight to remove any lead(II) ions from the lattice. The solution is heated to 60 °C, filtered, and washed with water. If the temperature is allowed to exceed 340 °C, the product is reduced to minium, Pb_3O_4.

A second method for making bulk α-PbO_2 is the oxidation of sodium plumbite by chlorine dioxide. β-PbO is reacted with an aqueous NaOH, with a flow of a chlorine dioxide/air mixture bubbled through it for 4 h. The mixture of PbO_2-$NaClO_2$ is filtered, washed with water, and boiled with 3 M HNO_3 for 45 min. The final product is then washed with water and dried.

A third way for making α-PbO_2 is the ammonium persulfate oxidation of lead(II) acetate in a strong ammonia solution of ammonium acetate according to the reaction:

$$4NH_4OH + Pb(C_2H_3O_2)_2 + (NH_4)_2S_2O_8 \rightarrow \alpha\text{-}PbO_2 + 2(NH_4)C_2H_3O_2 + 2(NH_4)_2SO_4 + 2H_2O$$

This synthetic approach proceeds slowly, with additional amounts of ammonium persulfate being added and the solution being stirred for 24 h. It is then heated to 70 °C. The precipitate is filtered, washed with ammonium acetate solution, followed by a wash with water, and then dried.

β-PbO_2 can be synthesized by several methods, including the electro-oxidation of lead perchlorate [19] by the treatment of PbO with 2 M perchloric acid in a Pt-Pb anode-cathode set. The β-PbO_2 material is removed, ground, and washed with water. A similar electrodeposition approach uses the same electrode pair immersed in a solution of lead acetate in a 0.5 M acetic acid solution. The deposit of β-PbO_2 is then collected, washed, and dried.

High-pressure phases of PbO_2 also are known. A fluorite-type polymorph PbO_2 of PbO_2 has been reported by several groups. The initial report [20] of this phase described its preparation above 70 kb making use of a tetrahedral anvil-type of high-temperature high-pressure press. A later report expanded the X-ray crystallographic aspects of the high-pressure phase [20] in much greater detail. One of the sample products from a reaction to produce this material is shown in Figure 4.2. It consists of a very well crystallized dark brown powder with grains that are 10–30 μm in size. Prismatic crystals are 40–50 μm in width and 250–750 μm in length along the *c*-axis. For the single crystal growth of α-PbO_2, a hydrothermal synthesis can be used, including a high-pressure chamber. Tetragonal PbO_2 and water can be used as starting materials in sealed platinum ampoules.

PbO_2 also can be synthesized in different forms, many of which have practical applications. Highly oriented layers of α-PbO_2 can be formed on silicon and quartz by successive ionic layer deposition (SILD) techniques [22]. The thickness of the layers is defined by the number

Figure 4.2 *Single crystal of α-PbO_2. From Filatov et al. [21]. Copyright (2005) Elsevier Ltd*

of deposition cycles, with the mechanism for the formation of the layers apparently involving the oxidation of Pb(II) in solution with the permanganate, MnO_4^-. This process results in very high quality α-PbO_2 layers with respect to their crystalline and chemical characteristics.

PbO_2 can be produced [23] as a transparent conducting film, one of its most important forms. The films exhibit the tetragonal β-PbO_2 phase. These films have conducting properties that are comparable with other transparent metal oxide semiconductor films. The pristine, undoped PbO films are found to exhibit optical transmission in the visible region and reflectance in the infrared. Vacuum annealing of these films increases the infrared reflectance, visible transmission, and optical band gap.

Submicrometer-sized PbO_2 hollow spheres, along with Pb_3O_4 microtubes [24] can been prepared by a selected-control synthesis using poly(vinyl pyrrolidone) (PVP), $Pb(NO_3)_2$, and $(NH_4)_2S_2O_8$ in NaOH solutions in an autoclave at 90 °C for 10 h. The reaction sequence results in a brown β-PbO_2 product (Figure 4.3), which is then washed with distilled water, followed by a wash with ethanol, and finally dried under vacuum at 50 °C for 4 h.

4.3.3 Synthesis of Pb_2O_3

Black crystals of Pb_2O_3 [16] can be made by the decomposition of PbO_2 in a hydrothermal bomb at temperatures of 300–500 °C and 1 kbar of pressure. The microscopic appearance of

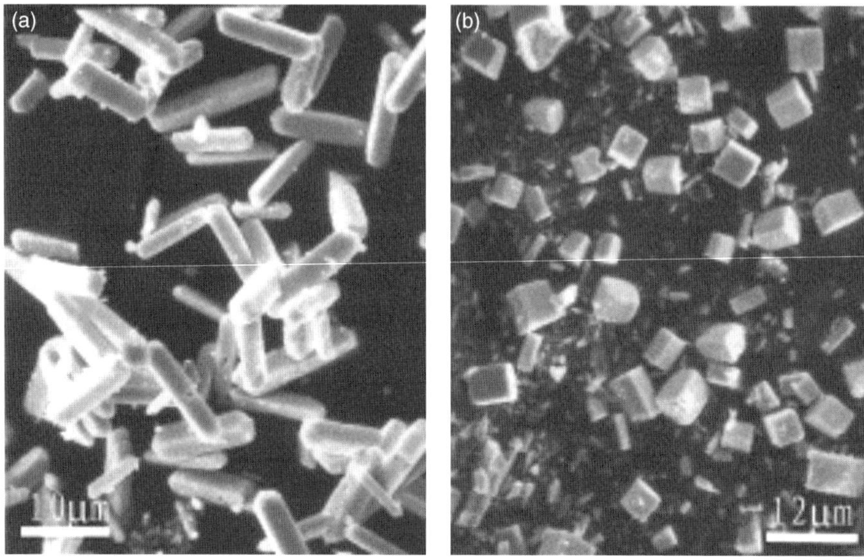

Figure 4.3 *(a) SEM image of PbO_2 microrods; (b) SEM image of Pb_3O_4 microtubes. Reprinted with permission from Xi et al. [24] Copyright (2004) Elsevier Ltd*

the crystals made in this fashion are identical to those that had been made earlier [25]. Pb_2O_3 can be made without solvent by sealing PbO_2 in a gold tube at 400 °C at a pressure of 1 kbar or higher. With this same pressure regime, it can be prepared by the oxidation of PbO or the reduction of PbO_2 in the 580–620 °C range [16].

4.3.4 Synthesis of Pb_3O_4

Pb_3O_4, or, mineralogically, *minium*, which can also be represented by the formula $2PbO{\cdot}PbO_2$ that indicates its lead(II,IV) nature, manifests itself as red tetrahedral crystals consisting of $Pb^{IV}O_6$ octahedra joined at their edges [26]. The interatomic Pb^{IV}-O and Pb^{II}-O are interesting in that the oxygen octahedra around Pb^{IV} atoms reveal bonds that are atypically long, and the Pb^{II} atoms are atypically short. It can be made using several techniques, including the rerystallized product of $K_2Pb(OH)_4$ and $K_2Pb(OH)_6$ [27] or by the reaction between PbO_2 and aqueous NaOH in a steel bomb at 375 °C [25]. Other procedures for preparing it [28] include the heating of PbO in air.

$$6PbO + O_2 \rightarrow 2Pb_3O_4$$

For an ultra-pure product, the solid should be washed well with an aqueous solution of KOH to remove any contaminant PbO. Other preparations [29] include reactions of the types:

$$6PbCO_3 + O_2 \rightarrow 2Pb_3O_4 + 6CO_2$$
$$3Pb_2CO_3(OH)_2 + O_2 \rightarrow 2Pb_3O_4 + 3CO_2 + 3H_2O$$

Pb_3O_4 microtubes can be made using the β-PbO described in the synthesis above [24] as a starting material. The only difference in the synthetic procedure is the running of the reaction scheme at 180 °C. The material produced in this synthesis can be seen in Figure 4.3.

4.3.5 Other Minor Lead Oxides

There are several reports of other lesser known and much more poorly chracterized minor lead oxides, with varying degrees of documention of their properties and characteristics. Perhaps the most rigorous summaries of all of them can be found in the study of phase relations in the lead-oxygen system [16, 30] which also includes exhaustive X-ray diffraction analyses, studies of pressure–temperature univariant equilibrium curves, and comparisons with other Pb-O phases that have been reported in the research literature. Another study is that of Anderson and Sterns [31] that, too, has reported very rigorous comparisons of solid phases of Pb-O with other reported lead oxides.

4.4 Applications of Lead Oxides

Partially because lead oxides exist as both lead(II) and lead(IV) oxides in addition to mixed oxidation state oxides, and because they represent very complex electronic systems, there are many other applications of the oxides. They include semiconductors, transparent conductors, batteries, electrocatalysts, gas sensors, and other electronic or electronic-related materials.

The applications of the lead oxides are many, especially when it comes to their applications in electronics and related technologies. Thin films of the tetragonal PbO, for example, that are formed by the reactive sputtering of lead in an oxygen atmosphere exhibit dielectric anomalies that may make them possible candidates for potential use in electronics [32]. Also, these films are photosensitive [11], making them possibly useful as optical sensors. PbO also is used as an X-ray imaging material in detectors [1] (Figure 4.4), as a solar

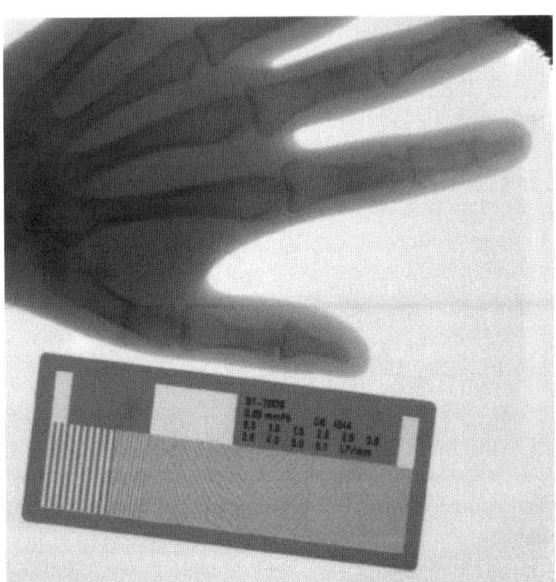

Figure 4.4 PbO detector. Reprinted with permission from Sellin [1] Copyright (2006) Elsevier Ltd

energy conversion material [33], and as a material useful as an X-ray sensitive photoconductor [34]. In conjunction with other elements, PbO has important electro-optic applications [35].

In addition to PbO_2 being useful as a transparent conducting oxide [23], it also shows high activity in anodic oxygen transfer reactions [36], especially when it is doped with metal ions such as iron(III). The resulting doped PbO_2 exhibits catalytic properties. PbO_2 can be incorporated into matrix composites [37] by the electrolysis of Pb(II) solutions that contain other metal oxides suspended in solution.

The electrodeposition of PbO_2 from solutions of soluble lead(II) salts also can be used in the study of the effects of ultrasound on the surfaces of electrodes [38]. The initially proposed mechanism [39] indicates the features shown below but subsequent researchers [40] have introduced other modifications.

$$OH^- \leftrightarrow OH_{ads} + e^-$$
$$Pb^{2+} + OH_{ads} + OH^- \rightarrow Pb(OH)_{2,ads}{}^{+} \text{ (slow)}$$
$$Pb^{2+} + OH_{ads} + OH^- \rightarrow Pb(OH)_{2,ads}{}^{2+} + e^-$$
$$Pb(OH)_{2,ads}{}^{2+} \leftrightarrow PbO_2(s) + 2H^+$$

These modifications include the generation of soluble intermediates which are capable of being removed from the electrode surface by the use of convective-diffusional processes.

Pb_3O_4 is used in paints and as a component in the litharge-minium pair for use in storage batteries. It is also used as a component in glasses, vitreous enamels and glazes [41].

There are potentially many more applications of lead oxides, both in their normal molecular form and possible bimetallic oxides. Due to the large size of the lead cations, many other large metal ions can substitute into a parent lead oxide lattice. Other contaminants that are present in the synthesis stream and the final product, such as the silicon mentioned above in the synthesis of the structural polymorphs of PbO [6], can be replaced with other contaminants to form yet still new products. This phenomenon of 'fuzzy chemistry' [6], in which very slight concentrations of reaction dopants can change a material's chemistry, its reactions relative to the formation of new products, or different structural phases of the material, could very well lead to new areas of chemistry for lead oxides and many new applications.

4.5 Summary

The oxides of lead present a complex set of chemical and physical properties, chemistries, and a wide variety of synthetic approaches. In the attempt to produce a particular oxide of either lead(II) or lead(IV), one must use very precise synthetic approaches to produce the desired oxide. This is because that while there are only a handful of lead oxides, there are literally dozens of ways to obtain them. Also, there are many synthetic schemes which may result in mixed products.

Perhaps the most important message from the study of lead oxides is their complexity. Literally hundreds of papers directed at the actual stoichiometric composition of lead oxides can be found in the research literature, along with even more studies regarding their electronic structures. In addition to their bulk chemistry, they may be fabricated into many forms such as films, powders, single cubes, nanorods, and other different shapes and morphologies.

On account of the unique electronic aspects of the various lead oxides, they can be employed in a wide range of applications, ranging from the somewhat classical use in batteries to advanced uses in domains such as transparent metal oxide semiconductors.

Acknowledgement

The author wishes to acknowledge support of the US Department of Energy under Contract Number DE-AC02-05CH11231.

References

[1] P. J. Sellin, Thick film compound semiconductors for X-ray imaging applications, *Nucl. Instrum. Methods Phys. Res. A*, **563**, 1–8 (2006).
[2] D. Raviendra, Transparent conducting PbO_2 films prepared by activated reactive evaporation, *Phys. Rev. B, Condens. Matter*, **33**, 2660–2664 (1986).
[3] B. Sljukic, C. E. Banks, A. Crossley, and R. G. Compton, Lead(IV) oxide-graphite composite electrodes: application to sensing of ammonia, nitrite, and phenols, *Anal. Chim. Acta*, **587**, 240–246 (2007).
[4] J. Leciejewicz, On the crystal structure of tetragonal (red) PbO, *Acta Cryst.*, **14**, 1304 (1961).
[5] R. J. Hill, Refinement of the structure of orthorhombic PbO (massicot) by Rietveld analysis of neutron powder diffraction data, *Acta Cryst. C*, **41**, 1281–1284 (1985).
[6] D. L. Perry and T. Wilkinson, Synthesis of high-purity α- and β-PbO and possible applications to synthesis and processing of other lead oxide materials, *Appl. Phys. A: Mater. Sci. Process.*, **89**, 77–80 (2007).
[7] D. U. Wiechert, S. P. Grabowski, and M. Simon, Raman spectroscopic investigation of evaporated PbO layers, *Thin Solid Films*, **484**, 73–82 (2005).
[8] J. M. Criado, F. Gonzalez, M. Gonzalez, and C. Real, Influence of the grinding of $PbCO_3$ on the texture and structure of the final products of its thermal decomposition, *J. Mater. Sci.*, **17**, 2056–2060 (1982).
[9] S. Sobanska, J.-P. Wignacourt, P. Conflant, M. Crache, M. Lagrenee, and E. M. Holt, Synthesis, thermal analysis and crystal structure of lead(II) diaqua 3,6-dicarboxylatopyridazine. Evaluation of performance as a synthetic precursor, *New J. Chem.*, **23**, 393–396 (1999).
[10] S. W. Lyons, Y. Xiong, T. L. Ward, and T. T. Kodas, Role of particle evaporation during synthesis of lead oxide by aerosol decomposition, *J. Mater. Res.*, **7**, 3333–3341 (1992).
[11] B. Thangaraju and P. Kaliannan, Optical and structural studies on spray deposited α-PbO thin films, *Semiconduct. Sci. Technol.*, **15**, 542–545 (2000).
[12] D. U. Wiechert, S. P. Grabowski, and M. Simon, Raman spectroscopic investigation of evaporated PbO layers, *Thin Solid Films*, **484**, 73–82 (2005).
[13] P. Veluchamy and H. Minoura, Simple electrochemical method for the preparation of a highly oriented and highly photoactive α-PbO film, *Appl. Phys. Lett.*, **65**, 2431–2433 (1994).
[14] D. M. Adams, A. G. Christy, J. Haines, and S. M. Clark, Second-order phase transition in PbO and SnO at high pressure: implications for the litharge-massicot phase transformation, *Phys. Rev. B*, **46**, 11358–11367 (1992).

[15] E. G. Rochow and E. W. Abel, *The Chemistry of Germanium, Tin and Lead*, Pergamon Texts in Inorganic Chemistry, Vol. **14**, Pergamon Press, Oxford, 1973, p. 119.
[16] W. B. White and R. Roy, Phase relations in the system lead-oxygen, *J. Am. Ceram. Soc.*, **47**, 242–249 (1964).
[17] J. A. Duisman and W. F. Giaque, Thermodynamics of the lead storage cell. The heat capacity and entropy of lead dioxide from 15 to 318 K, *J. Phys. Chem.*, **72**, 562–573 (1968).
[18] J. P. Carr and N.A. Hampson, The lead dioxide electrode, *Chem. Rev.*, **72**, 679 (1972).
[19] N. E. Bagshaw, R. L. Clarke, and B. Hallwell, The preparation of lead dioxide for X-ray diffraction studies, *J. Appl. Chem.*, **16**, 180–184 (1966).
[20] Y. Syono and S. Akimoto, High pressure synthesis of fluorite-type PbO_2, *Mater. Res. Bull.*, **3**, 153–158 (1968).
[21] S. Filatov, N. Bendeliani, B. Albert, J. Kopf, T. Dyuzeva, and L. Lityagina, High pressure synthesis of α-PbO_2 and its crystal structure at 293, 203, and 113 K from single crystal diffraction data, *Solid State Sci.*, **7**, 1363–1368 (2005)
[22] V. P. Tolstoi and E. V. Tolstobrov, Synthesis of highly oriented α-PbO_2 layers on the surfaces of single-crystal silicon and quartz by successive ionic layer deposition, *Russ. J. Appl. Chem.*, **75**, 1529–1531 (2002).
[23] D. Raviendra, Transparent conducting PbO_2 films prepared by activated reactive evaporation, *Phys. Rev. B*, **33**, 2660–2664 (1986).
[24] G. Xi, Y. Peng, L. Xu, M. Zhang, W. Yu, and Y. Qian, Selected-control synthesis of PbO_2 submicrometer-sized hollow spheres and Pb_3O_4 microtubes, *Inorg. Chem. Commun.*, **7**, 607–610 (2004).
[25] G. L. Clark, N. C. Schieltz, and T. T. Quirke, A new study of the preparation and properties of the higher oxides of lead, *J. Am. Chem. Soc.*, **59**, 2305–2308 (1937).
[26] J.-R. Gavarri and D. Weigel, Oxydes de plomb. I. Structure crystalline du minium Pb_3O_4, a temperature ambiante (293 K), *J. Solid State Chem.*, **13**, 252–257 (1975).
[27] M. Baudler, in *Handbook of Preparative Inorganic Chemistry*, 2nd Edn, G. Brauer (Ed.), Vol. 1, Academic Press, New York, 1963, p. 755.
[28] E. G. Rochow and E. W. Abel, *The Chemistry of Germanium, Tin and Lead*, Pergamon Texts in Inorganic Chemistry, Vol. **14**, Pergamon Press, Oxford, 1973, p. 121.
[29] Lead tetroxide, *Wikipedia*, http://en.wikipedia.org/wiki/Minium
[30] W. B. White, F. Dachille, and R. Roy, High-pressure-high-temperature of polymorphism of the oxides of lead, *J. Am. Ceram. Soc.*, **44**, 170–174 (1961).
[31] J. S. Anderson and M. Sterns, Intermediate oxides of lead, *J. Inorg. Nucl. Chem.*, **11**, 272–285 (1959).
[32] K. Wasa and S. Hayakawa, Dielectric properties of PbO thin films, *Jpn. J. Appl. Phys.*, **8**, 276–276 (1969).
[33] P. Veluchamy and H. Minoura, Simple electrochemical method for the preparation of a highly oriented and highly photoactive α-PbO film, *Appl. Phys. Lett.*, **65**, 2431–2433.
[34] J. R. Clarke, A. K. Weiss, J. L. Donovan, J. E. Greene, and R. E. Klinger, Ion-plated lead oxide, an x-ray sensitive photoconductor, *J. Vac. Sci. Technol.*, **14**, 219–222 (1977).
[35] L. R. P. Kassab, R. D. Monsano, L. da S. Zambom, and V. D. D. Cacho, Semiconductor characteristics of Nd doped PbO-Bi_2O_3-Ga_2O_3 films, *Braz. J. Phys.*, **36**, 451–454 (2006).
[36] A. B. Velichenko, D. V. Girenko, S. V. Kovalyov, A. N. Gnatenko, R. Amadelli, and F. I. Danilov, Lead dioxide electrodeposition and its application: influence of fluoride and iron ions, *J. Electroanal. Chem.*, **454**, 203–208 (1998).
[37] U. Casellato, S. Cattarin, P. Guerriero, and M. M. Musiani, Anodic synthesis of oxide-matrix composites. Composition, morphology, and structure of PbO_2-matrix composites, *Chem. Mater.*, **9**, 960–966 (1997).
[38] J. Gonzalez-Garcia, J. Iniesta, A. Aldaz, and V. Montiel, Effects of ultrasound on the electrodeposition of lead dioxide on glassy carbon electrodes, *New J. Chem.*, 343–347 (1998).
[39] M. Fleischmann and M. Liler, The anodic oxidation of solutions of plumbous salts, *Trans. Faraday Soc.*, **54**, 1370–1381 (1958).

[40] A. B. Velichenko, D. V. Girenko, and F. I. Danilov, Electrodeposition of lead dioxide at an Au electrode, *Electrochim. Acta*, **40**, 2803–2807 (1995).
[41] Minium, *Britannica Online Encyclopedia*, http://www.britannica.com/EBchecked/topic/384272/minium
[42] International Agency of the Evaluation of Carcinogenic Risks to Humans, Vol. 87, Inorganic and Organic Lead Compounds (2006), p. 42. http://monographs.iarc.fr/ENG/Monographs/vol87/mono87-6.pdf
[43] D. R. Lide (Ed.), *CRC Handbook of Chemistry and Physics*, CRC Press, Boca Raton, 2004.

5

Deposition and Performance Challenges of Transparent Conductive Oxides on Plastic Substrates

Clark I. Bright

Condor Group Technical Leader, 3M Corporate Research Process Laboratory, USA

5.1 Introduction

Transparent conductive oxides (TCO) have undergone significant commercialization and these films are used by most of us virtually everyday in applications ranging from digital watches, to cell phones, to computer screens or other types of displays. The most important TCO today, typically called indium-tin-oxide or ITO, is tin doped indium oxide, In_2O_3:Sn. ITO on glass is used as a transparent electrode in nearly all flat panel displays (FPDs), and this application represents the largest annual dollar value for TCO thin film coatings. The other major application of TCO is as low-emissivity (low-e) coatings for energy efficient windows. The spectrally selective properties of SnO_2:F coatings, pyrolitically deposited on float glass, provides windows with high visible and near infrared transmission along with thermal insulation by limiting radiative heat loss from the interior. Low-e coatings account for the largest deposited area of TCO coatings used annually. The growth of the energy generation market and the demand for solar photovoltaic (PV) devices is also driving larger consumption of TCO for transparent electrodes. All of the above applications are served by TCO typically deposited or post-treated at high temperatures ($\sim$200–600 °C).

The application of TCO transparent oxide electronic devices also has been reported in the literature [1]. Typically, these transparent active electronic devices also are made at high

Transparent Electronics: From Synthesis to Applications Edited by Antonio Facchetti and Tobin J. Marks

process temperatures. However, there is considerable interest in depositing TCO and transparent electronic devices at low temperatures on flexible plastic film substrates. One reason for this interest is that using a roll-to-roll process can enable low cost, high volume production. Examples of commercial products using roll-to-roll vacuum deposition processes are multilayer coatings for antireflection [2], solar control and energy efficient windows [3], color-shifting security coatings [4] and ITO coatings for various transparent electrode applications [5].

The largest applications for ITO coated plastic film are as transparent electrodes for touch panels and electroluminescence lamps. While ITO on plastic film is currently a relatively small market (approximately US$100 million per year), compared with ITO on glass (>US$100 billion since 2006), this market is expected to grow significantly in mobile devices, flexible electronics and displays, photovoltaic devices and transparent electronics applications. For transparent electronics applications, a single transparent electronic device can require up to a total of four layers of TCO and transparent semiconductive oxides (TSO).

The term TCO historically embraces highly transparent and degenerate *n*-type semiconductors. For this chapter, we define TCO in this way. Nondegenerate transparent conductive oxides used in transparent electronics for their semiconductive properties are discussed later in this chapter. When referring to nondegenerate transparent semiconductive oxides the abbreviation TSO will be used.

For example, a simple *p*-*n* junction device requires both an *n*-type highly conductive TCO contact electrode and an *n*-type semiconductive layer for one side of the junction and two *p*-type TCO and TSO layers to form the other half of the junction. For field effect transistors (FET), also called thin film transistors (TFT) when made from thin films, only one type of TCO and TSO material is required. Because only *n*-type oxide materials can be used, which are more highly developed than *p*-type oxide materials, progress has been more rapid than with oxide junction devices. Fully transparent TFT require three conductive layers, two TCO and one TSO.[1]

In this Chapter, the basics of electrical conductivity and transparency in TCO are briefly reviewed to illustrate the differences between TCO on glass (high temperature process) and on plastics (low temperature process). The electrical and optical (E/O) performance achieved with high and low temperature processes are compared.

The E/O performance of TCO is shown to be very process dependant and process sensitive. For transparent oxide electronics to be successful, a robust industrial deposition process, which is low cost and high volume, will be required. For TCO deposition on plastic film, today that means vacuum deposition primarily by D.C magnetron sputtering but occasionally by thermal (boats and e-beam) evaporation. These two vacuum deposition methods are briefly reviewed.

Understanding the origin of conductivity, aids in designing and controlling the TCO deposition process to achieve the desired E/O properties. A very simplified, qualitative model of the TCO doping process is presented [6] to aid practitioners in designing and defining their TCO deposition process and in understanding their process results. A step-by-step procedure for developing the TCO deposition process is also given in this Chapter [6].

[1] For a fully transparent TFT, typically one TCO is patterned to form the source and drain contact areas. The TSO forms the channel layer separated from and unpatterned 'continuous' TCO contact by the transparent dielectric gate layer.

5.2 Challenges with Plastic Substrates

5.2.1 Temperature Limitation

There is significant interest in depositing transparent oxide electronics on flexible plastic substrates. However, the limited temperature capability of common plastic substrates, e.g. polyethylene terephthalate (PET), requires process temperatures for the TCO well below optimum (for ITO ≥200 °C). With temperature-sensitive plastic substrates, process temperatures must be much lower, typically ~20 °C to <100 °C, than on glass. As discussed later, these low process temperatures yield TCO (ITO) films with significantly reduced conductivity.

The characteristics of several commercially available plastic substrates as a function of temperature have been reviewed [7]. Polymer films other than the traditional PET offer higher glass transistion temperature (T_g) and service (continuous use) temperatures. The T_g and melting point temperature (T_m) for several commercially available plastic substrates are shown in Figure 5.1.

5.2.2 Mechanical Limitation

Several of these candidate plastic substrates would appear to be suitable for deposition of ITO (TCO) at or near the desired temperature of ~200 °C. However, along with the temperature capabilities of the plastic substrate, its shrinkage or expansion with temperature and its coefficients of (linear) thermal expansion (CTE) must be considered for suitability of use with TCO [8] and transparent oxide electronic devices. If substrate shrinkage occurs or there is a mismatch in thermal expansion, the TCO or device will become strained during the temperature cycling of the deposition process. Typically, a plastic substrate with a CTE < 20 ppm °C^{-1} is preferred to limit the mismatch in thermal expansion with the deposited layers, which can become strained and cracked under thermal cycling [9].

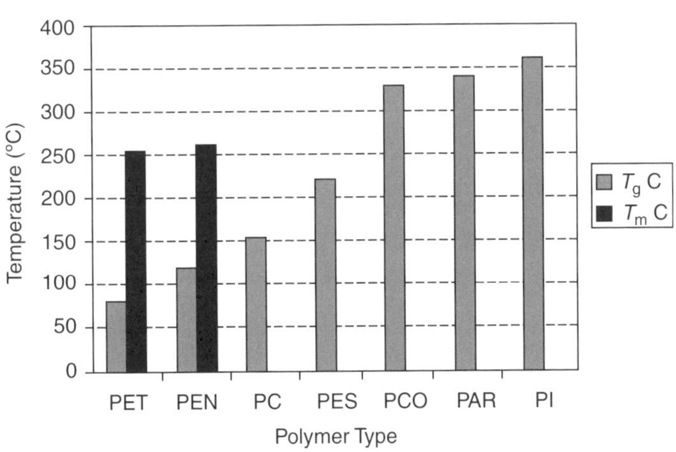

Figure 5.1 *Glass transition temperature (T_g) and melting point temperature (T_m) for several commercial plastics. (Reprinted with permission from [7] Copyright (2004) Royal Society of Chemistry)*

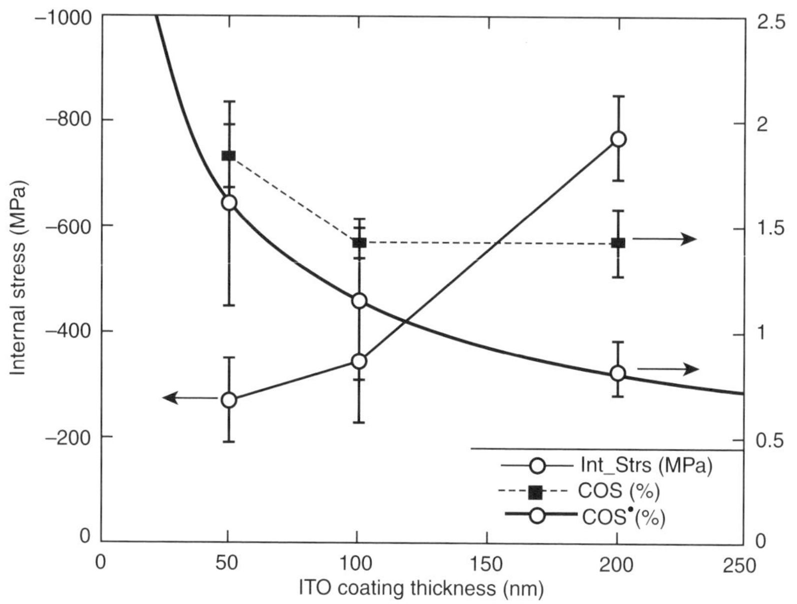

Figure 5.2 *Internal stress and crack onset strain versus ITO coating thickness. (Reprinted with permission from [10] Copyright (2003))*

Recent work on thin film mechanical properties has measured the critical strain-to-failure of ITO [$\sim$100 nm thin films on 100 μm thick aromatic polyester (Arylite™ – ARY)], in both bending and tensile testing, to be about 1.5% [10]. This measured crack onset strain (COS) value includes the influence of internal stress and internal cohesion of the ITO coatings. The intrinsic crack onset strain value, COS*, i.e. a COS strain-to-failure value corrected for these factors, shows the expected classical square root of thickness dependence (Figure 5.2 [10]). The COS* value for these $\sim$100 nm thick ITO films was $\sim$1.1%. As the fundamental material properties of metal oxides, e.g., TCO, primarily determine the COS and COS* values, other TCO (metal oxides) thin films of similar thickness are expected to have strain-to-failure values near 1.0–1.5%. These mechanical limitations, of course, affect TCO on both rigid and flexible substrates. For flexible transparent oxide electronics the effects of both bending/tension and expansion mismatch at processing/operating temperatures must be accommodated.

One commercial (DuPont Teijin) plastic substrate candidate for flexible transparent oxide electronics is heat-stabilized polyethylene naphthalate (PEN). Even though PEN has a T_g of $\sim$120 °C, after it is heat-stabilized well above the T_g, it becomes dimensionally stable up to 200 °C [7]. However, even if significant shrinkage does not occur, the CTE for PEN must be considered for suitability of use with TCO. For example, with ITO deposited on PEN at 200 °C, the mismatch in thermal expansion, during the temperature cycling for the deposition process, must not exceed the crack onset strain value of $\sim$1.5%.

Some properties of commercial polymer substrates, including PEN, are listed in Tables 5.1 and 5.2 [7]. Note in Table 5.1 that PEN has the lowest CTE value, 13 ppm/ °C, of the listed plastic substrate materials, and it is well below the cited acceptable value of <20 ppm/°C. However in Table 5.2, the CTE value for PEN in the temperature range 100 °C–150 °C

Table 5.1 Basic properties of high temperature plastic film substrates. (Reprinted with permission from [7] Copyright (2004) Royal Society of Chemistry)

Base polymer	PET	PEN	PC	PES	PAR	PCO	PI
CTE (−55 to 85 °C) (ppm °C^{-1})	15[a]	13[a]	60–70	54	53	74	30–60
%Transmission (400–700 nm)	>85	0.85	>90	90	90	91.6	Yellow
Water absorption (%)	0.14	0.14	0.4	1.4	0.4	0.03	1.8
Young's modulus (GPa)	5.3	6.1	1.7	2.2	2.9	1.9	2.5
Tensile strength (MPa)	225	275	NA[b]	83	100	50	231

[a]CLTE.
[b]Not available.
N.B. The information in this table in taken from different datasheets and should only be taken as illustrative.

Table 5.2 Coefficient of linear thermal expansion (CLTE) for heat stabilized PEN. (Reprinted with permission from [7] Copyright (2004) Royal Society of Chemistry)

	CLTE PPM −50–0 °C	CLTE PPM 0–50 °C	CLTE PPM 50–100 °C	CLTE PPM 100–150 °C
Machine direction	13	16	18	28
Transverse direction	8	11	18	29

is 25–29 ppm and values at 200 °C are not given. The highest temperature range with CTE values of <20 ppm/°C is 50 °C–100 °C where heat-stabilized PEN exhibits a CTE of 18 ppm. Therefore, depositing ITO on PEN at 200 °C may be problematic, and the substrate temperature may have to be reduced to ≤100 °C [11].

5.3 TCO Performance Comparison – Glass Versus Plastic Substrates

5.3.1 Typically Achieved E/O Properties

The ranges of E/O performance, typically achieved in industry for ITO deposited at high (≥200 °C) and low ('room temperature', RT, ~23 °C) process temperatures, on glass and plastic substrates, respectively, are summarized in Tables 5.3 and 5.4[2] [12]. Tables 5.3 and 5.4 show that ITO deposited at RT on typical PET plastic substrates has a resistivity, ρ ($=1/\sigma$, see Section 5.4.1), range, that is ~4 times higher than on glass (>200 °C). The reasons for the significantly reduced ITO (TCO) conductivity with deposition at low process temperatures are discussed in detail in Section 5.4.

5.3.2 Baseline ITO (90 wt% In_2O_3/10 wt% SnO_2) E/O Properties

The average visible transmittance values (%T) given in Tables 5.3 and 5.4 include the substrate losses (mainly reflection).

[2] For example, Colorado Concept Coating, $\rho = 1.61 \times 10^{-4}$ ohm cm, $\mu = 29$, $N = 1.3 \times 10^{21}$; see Appl. Phys. Lett., 87, 032 111 (2005).

Table 5.3 *Performance for high temperature process (glass). (Reprinted with permission from [12] Copyright (2008) Society of Vacuum Coaters)*

Resistivity range (ohm cm)	$1-2\times10^{-4}$
Visible transmittance (%T)	%T $\geq$ 85% (average 400–700 nm)
R_s (ohm per square)	10–20
Film thickness d (nm)	120–135
ρ (ohm cm)	1.4×10^{-4}–1.8×10^{-4}
N (cm^{-3})	1.3×10^{21}–2×10^{21}
μ (cm^2 V^{-1} s^{-1})	20–35

Table 5.4 *Performance for low temperature process (PET plastic). (Reprinted with permission from [12] Copyright (2008) Society of Vacuum Coaters)*

Resistivity range (ohm cm)	$5–8\times10^{-4}$
Visible transmittance (%T)	%T $\geq$ 80% (average 400–700 nm)
R_s (ohm per square)	40–60
Film thickness d (nm)	125–140
ρ (ohm cm)	5×10^{-4}–8×10^{-4}
N (cm^{-3})	2×10^{20}–3×10^{20}
μ (cm^2 V^{-1} s^{-1})	40–45

This increase in ITO (TCO) resistivity of $\geq$4 times is a serious problem not only for flexible transparent oxide electronics but also for many other applications like organic light-emitting devices (OLEDs), PVs, and large-area flexible displays and electronics, where a low sheet resistance, transparent electrode is needed.

However, the issues of mechanical properties mismatch and resistivity/sheet resistance are not independent. For example, increasing the deposition temperature can reduce resistivity but will increases strain in the TCO thin film. The possibility of lowering sheet resistance by increased the TCO thickness, altering the sputter deposition process or using a more temperature-resistant plastic substrate to allow increasing the deposition temperature, to meet specific E/O requirements was analyzed [13]. The conclusions were that none of these approaches alone could meet the electrical (ohm per square) requirement and the effectiveness of combining several techniques is unproven. One promising potential solution presented was to use multiple thin layers of ITO (TCO) between flexible optical polymer layers [14]. However, specialized vacuum deposition equipment which is not yet commercially available was used to produce these films.

Some of the conclusions of these studies were the following:

- Mechanical properties limit maximum ITO thickness to ~300 nm with a sheet resistance of ~25–30 ohm per square.
- ITO/substrate mechanical properties limit maximum process temperature even with a high temperature plastic substrate.
- Modifications to the deposition process yields only small (~10%) improvements in sheet resistance.
- A multilayer (thin ITO/polymer)^n stack may meet the electrical/optical requirements.
- No single approach meets all requirements for low sheet resistance/high transmittance.

These two main issues of mismatched mechanical properties for TCO on plastic substrates and higher resistivity due to temperature limitations of plastic substrates remain substantial issues to commercializing a roll-to-roll process for flexible transparent oxide electronics [11, 13].

5.4 Conductivity Mechanisms in TCO

A phenomenological discussion of the causes of TCO conductivity and the parameters on which it depends in both high temperature and low temperature processes are discussed in this section. This discussion explains the reasons for the significantly reduced conductivity of TCO (ITO) when deposited at low process temperatures. A very simplified, qualitative model of the TCO doping process also is presented [6]. This qualitative model and knowledge of the conductivity mechanisms can be used to design and control the deposition process to achieve the desired E/O properties in the TCO thin film.

5.4.1 Metallic Conductivity

Recall that a degenerate *n*-type semiconductor behaves electronically like a metal so TCO have the same expression for conductivity, σ, as do metals: $\sigma = Ne\mu$, where N is the carrier density, the number of conduction electrons per unit volume (usually stated in cm^{-3}), μ, defined below, is the electron mobility, and e is the value of its electronic charge. Therefore, conductivity, σ, and its reciprocal, resistivity, ρ, are determined by only two variables, N and μ.

The electron mobility will be discussed first. In a conductor the motion of conduction electrons is limited by very frequent collisions, primarily with phonons and occasionally with lattice defects. Even in an applied electric field, the motion of the electrons is still dominated by collisions related to thermal energy effects and defects. The electrons travel, on average, a distance between collisions, the mean free path, in a time interval, τ, the scattering time. The electron mobility, μ, is related to the scattering time by $\mu = e\tau\, m^{*-1}$, where m^* is the electron (effective) mass. Thus, a longer time between collisions, i.e. a larger τ, means a higher mobility, μ, and therefore higher conductivity.

Next consider N, the carrier density. The more a TCO is doped (effective) the larger is the number of conduction electrons per unit volume, and the higher the conductivity. However, impurity metal doping at substitutional sites in the TCO also creates charge defects in the TCO lattice resulting in ionized impurity scattering of the electrons, lowering the mobility. Eventually, excess impurity metal doping, (and interstitials) results in greater loss in mobility (scattering of the electrons) than gain from the increased number of electrons, resulting in a net loss in conductivity [15].

5.4.2 Optical Properties

Both, N, the electron concentration and μ, their mobility, affect the optical properties of TCO (and TSO) but each affects them differently. The value of N affects both the short wavelength transmittance cut-off limit (band gap) and the long wavelength transmittance cut-off (plasma wavelength) of a TCO (and TSO). Typically, both of these transmittance limits move to shorter wavelengths with increasing N value. Generally, the change in the band gap transmittance limit is proportional to $N^{2/3}$ (Burstein–Moss effect), while the plasma wavelength limit is proportional to $N^{1/2}$. The value of μ does not affect the location of the short wavelength transmittance cut-off limit or the plasma wavelength, although the slope of the change in transmittance (and reflectance) near the plasma wavelength is increased by a higher μ value. Achieving high conductivity (low resistivity) with a high μ value and a low N value is preferred [12].

With the above brief review of the fundamentals as a basis, the following discussion begins with examinations of conductivity in the archetypical TCO, ITO, but while the details are specific to ITO, the discussion conclusions are exemplary of any impurity (cationic) doped TCO.

5.4.3 Impurity Doping

Early work (1972) by Fraser and Cook [16] and others [17] studied the effect of tin doping in indium oxide. The results of Fraser and Cook for ITO thin films with different Sn concentrations deposited by DC diode sputtering on substrates at ~420 °C are shown in Figure 5.3. As explained above, at first the resistivity was lowered with increased Sn doping then eventually increased with further Sn doping. The lowest ITO resistivity was achieved at about 10 mol% SnO_2 doping. As shown in Figure 5.3, the typical minimum resistivity was $\sim 3 \times 10^{-4}$ ohm cm; however, the best value achieved was 1.77×10^{-4} ohm cm.

A qualitative explanation for these results is presented next based on a simplified impurity doping model for ITO and the schematic atomic structure representation shown in Figure 5.4 [18]. Initially as Sn^{+4} is added, some Sn^{4+} atoms replace In^{3+} atoms substitutionally in the In_2O_3 lattice (Figure 5.4) and each donates one electron to increase the carrier concentration, lowering the resistivity. Of course, the tin doped substitutional sites also act as charge defects in the TCO lattice and cause ionized impurity scattering of the electrons.

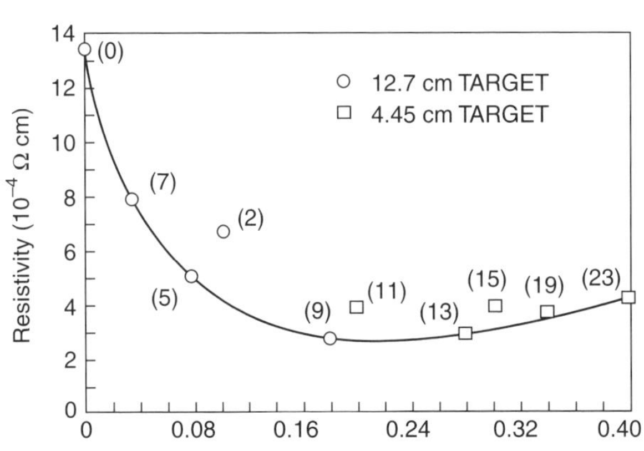

Figure 5.3 *Variations of ITO resistivity with Sn/In ratio (mol%). (Reprinted with permission from [16] Copyright (1972) Electrochemical Society)*

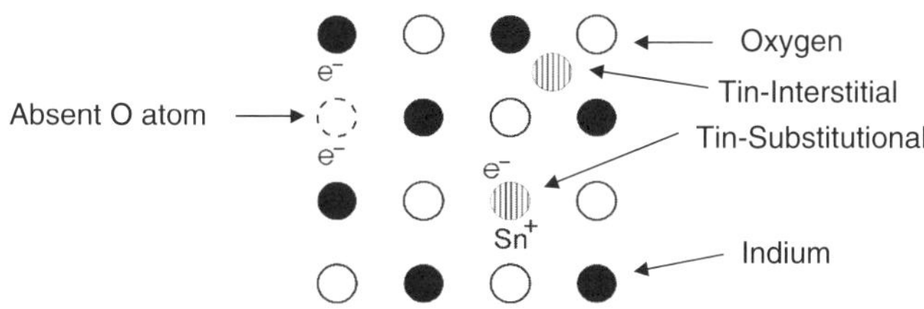

Figure 5.4 *Simplified crystal structure and doping model for ITO. (Adapted from [18])*

Also, not all of the Sn atoms are incorporated substitutionally; some are in interstitial positions (Figure 5.4), which also scatter electrons but do not directly contribute any conduction electrons.[3] As the concentration of Sn^{4+} doping is increased, eventually the excess Sn^{4+}, interstitial and substitutional, result in greater losses from scattering of the electrons (lowering mobility) than increase in conductivity from the contributed conduction electrons. This condition causes the rise in the curve in Figure 5.3, shown at high Sn/In ratios, i.e. high at% or wt% Sn [6].

5.4.4 Defect Doping

In addition to creating conduction electrons by impurity doping of a TCO, e.g. In_2O_3:Sn, carriers (conduction electrons) can be created by altering the film composition from stoichiometry. Lack of stoichiometry (defects) from oxygen vacancies can occur during the deposition process or during post-processing in vacuum or a reducing atmosphere. Oxygen vacancies also are represented schematically in Figure 5.4. According to the cluster model, at high oxygen partial pressure, ITO conductivity can be limited by the formation of a neutral defect cluster composed of two substitutional Sn^{4+} cations and one interstitial O^{2-} anion. However, some of these clusters can be broken at low oxygen partial pressure, by loss of the interstitial O^{2-}, which results in two additional conduction electrons from the now active Sn^{4+} dopants on In^{3+} sites.[4]

The cluster model also showed that with highly reducing conditions (lower oxygen partial pressure, pO_2) oxygen vacancies become the important contribution to the conductivity. This carrier generation mechanism was proposed first by Frank and Kostlin [19] with a detailed model for the cluster process. Experimental support [20] for the cluster model explanation for defect doping in IO was recently reported.

An absent oxygen atom (or removing oxygen from a cluster), would leave two unbound electrons available for conduction. Thus, oxygen vacancies are fundamentally twice as effective in creating conduction electrons as a cationic impurity substitution with a valence difference of one from the host lattice, e.g. Sn^{4+} for In^{3+}. In fact, in many TCO, including ITO, deposited on substrates at high temperature, the conduction electrons are usually created primarily by oxygen deficiencies (vacancies and tin interstitials) not tin doping

[3] Sn^{+4} can form clusters which can affect conductivity. This mechanism is discussed in Section 5.4.4.
[4] The model assumes the neutral defect cluster is composed of one interstitial O^{2-} anion, and two substitutional Sn^{4+} cations, which would typically require a high temperature process.

(substitutional). This observation was pointed out even in the early work of Fraser and Cook discussed above. They found, for their experimental conditions, only about a factor of 4 improvement in resistivity from tin doping.

5.4.5 TCO Microstructure

Another difference in TCO thin films deposited on substrates at low and high temperatures is the film microstructure. Consider the reported difference in conductivity of amorphous (and textured) versus polycrystalline ITO (TCO) films [15]. The higher conductivity of ITO (TCO) in polycrystalline films often has been attributed to significantly lower electron scattering, i.e. higher mobility, than similar TCO films with an amorphous microstructure. One intuitively assumes that as with other semiconductors, like silicon, the mobility in the amorphous material is substantially, often several orders of magnitude, less than in the crystalline (and polycrystalline) state.

However, Bellingham *et al.* [15] showed that the conductivity of amorphous films is at most reduced by a factor of 2 compared with polycrystalline ITO ($\rho \sim 2 \times 10^{-4}$ ohm cm) and IO thin films. Bellingham argued that even if the entire factor of 2 were ascribed to structural disorder due to the amorphous microstructure, this would result in a lower limit of ~10 nm for the electron mean free path (MFP). Further, the MFP of electrons at the Fermi surface in IO is much smaller, ~3 nm, and therefore electrons would not be significantly scattered by the microstructure. Rather than the amorphous structure, the correct explanation for the conductivity difference Bellingham states is unavoidable ionized impurity scattering, which alone can account for nearly all of the conductivity loss in IO (and ITO) films. The conclusion that scattering by the amorphous structure, i.e. low mobility, is not the cause of the lower conductivity is supported by unpublished Hall effect measurements made in the author's laboratory[5] and more recently by similar measurements reported by others [21]. Both groups found similar mobilities of ~40–45 $cm^2 V^{-1} s^{-1}$ in amorphous (and textured) ITO thin films deposited at RT on PET film substrates. These amorphous ITO mobilities are as high as or higher than found in very good quality polycrystalline, low resistivity ITO deposited at high temperatures.[6]

Therefore, as shown by Bellingham *et al.* [15], the higher resistivity of amorphous films deposited at low substrate temperatures is not caused by low mobility. Instead, it is caused by the inability to activate the impurity dopant, i.e. Sn^{4+} is not activated into an In^{3+} site, which therefore does not contribute conduction electrons in ITO. The cause of low conductivity is thought to be similar in many other TCO, e.g. IZO [6, 22]. The conductivity of TCO (ITO) films deposited at low substrate temperatures is only a result of oxygen vacancies (and perhaps interstitial Sn^{4+}).

With the background discussed above, the reasons for the conductivity difference (Tables 5.3 and 5.4) between TCO deposited at high and low substrate temperatures are now clear.

[5] The author's ITO (10 wt% SnO_2) films were ~140 nm thick and had a $\rho \sim 5 \times 10^{-4}$. These mobility values (40–45 $cm^2 V^{-1} s^{-1}$) are nearly equivalent to the best found in (degenerate) ITO films on glass. (Of course, the higher carrier concentration value can affect the mobility achieved.)

[6] The approximately equal mobilities were not achieved in thin films with equal carrier concentrations, and generally mobility is reduced as carrier concentration increases.

5.5 Qualitative TCO Doping Model [6]

For any TCO doping process, there is an optimum doping level, which results in minimum resistivity. At lower doping levels, the film resistivity is increased primarily by lack of conduction electrons. At higher doping levels, increased electron scattering from many sources, e.g. impurity atoms, cations and interstitials, from defects, oxygen vacancies, phonons, film microstructure and grain boundaries, can lower mobility causing the TCO resistivity to increase. However, one conclusion is that in highly degenerate TCO (10^{20}–10^{21} carriers cm^{-3}), deposited or post annealed at high temperatures ($\geq 200\,°C$), ionized impurity scattering is the dominant mechanism. Even for TCO thin films deposited onto low temperature substrates like plastics, with amorphous microstructure, ionized impurity scattering is the dominant scattering mechanism [15].

The effective mobility, μ, in a given TCO film is the results of all the various scattering mechanisms present, and is approximated by

$$\frac{1}{\mu} = \frac{1}{\mu_i} + \frac{1}{\mu_p} + \frac{1}{\mu_{gb}} + \frac{1}{\mu_n} + \ldots$$

where μ_i is the scattering effect of the ionized impurity dopant atoms, μ_p is the reduction in mobility due to phonons scattering, μ_{gb} is the result of scattering at grain boundaries and μ_n is the reduction caused by scattering of electrons from neutral impurities, etc.

The minimum TCO resistivity occurs when the doping level has increased the carrier concentration to a level where scattering caused by doping, i.e. ionized impurity scattering, is comparable with scattering from all other mechanisms.

This conductivity concept is shown schematically in Figure 5.5 [6].

This qualitative doping model will be used extensively to discuss developing a TCO deposition process and understanding and controlling the results.

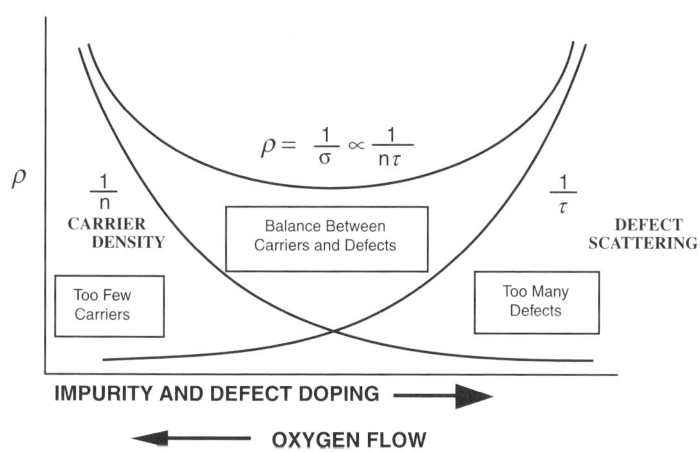

Figure 5.5 Qualitative TCO doping model. (Reprinted with permission from [6] Copyright (2009) Society of Vacuum Coaters)

5.6 Industrial TCO Deposition Methods on Plastic Substrates

TCO films have been deposited by many different methods. Chemical methods such as pyrolysis, chemical vapor deposition (CVD) and metal organic chemical vapor deposition (MOCVD), require substrate temperatures well above those compatible with common plastic substrates. However, many physical vapor deposition (PVD) methods such as various vacuum evaporation and sputtering techniques also have been used on temperature tolerant substrates like glass, and most can be compatible with low substrate temperatures and plastic substrates. The attributes and limitations of evaporation and sputtering technologies are discussed briefly below. In addition, the use of these PVD methods for coating flexible plastic substrates is explained as background to permit the detailed discussion of the effect of process parameters on the properties of the TCO thin film deposited.

5.6.1 Evaporation

Thermal evaporation is one of the simplest vacuum deposition methods and is widely used in industry, particularly to produce metalized paper and plastics for food packaging and labeling applications. Metal deposition by thermal evaporation is high rate and wide web, roll-to-roll coating line speeds exceeding 1000 m min^{-1} are typical.

However, depositing TCO films by thermal evaporation of the constituent metal has challenges not faced when depositing metal films. For example, if an indium-tin metal alloy is evaporated, oxygen must be provided and react with the indium and tin atoms to form the indium and tin oxides. The amount of added oxygen in the vacuum chamber must be kept low, typically creating a pressure of $\sim 10^{-4}$ Torr (0.013 Pa), which also may limit the surface reaction rate forming the TCO. Typically, this requires a substrate temperature of $\sim 200\,°C$ or more, which is incompatible with common transparent plastics. Further, the speed of this surface reaction is proportional to the substrate temperature and much higher temperatures are needed to achieve substantial reaction rates.

Alternatively, instead of starting with the metals, the metal oxides, for example, indium and tin oxides, can be vacuum evaporated, preferably from separate evaporation sources for better control to achieve the desired In-Sn composition in the oxide film.

However, generally some oxygen is lost from the oxides during the evaporation process, and so again some oxygen background gas must be provided to achieve the desired oxide composition. As discussed above, a high substrate temperature is needed to achieve a high reaction rate, although because the starting materials are the oxides rather than the base metals, far less reaction is needed, which allows higher deposition rates.

Another form of vacuum evaporation, which solves the problem of heating of the substrate by radiation from the hot evaporation source(s), is electron beam (e-beam) evaporation. Both metals and metal oxides can be readily evaporated by heating with an e-beam. Using an e-beam to impart the energy needed to vaporize the TCO material to be deposited, however, does not generally solve the issue of needing an oxygen background during the evaporation process. If an oxygen background is not used during the evaporation process the TCO (ITO) typically will be oxygen deficient, with high carrier concentration, low mobility and very low visible transmittance [6, 22].

On account of these limitations, a substrate-temperature limited reaction and the requirement for an oxygen gas background, the vacuum evaporation method is not commonly used for depositing TCO coatings on plastic film. Ion plating, plasma activation or other energetic techniques may be used with evaporation, however, to increase the reaction rates and reduce the substrate temperature requirements.

Optical coatings of metal oxide thin films are produced commercially at high rates by high power, e-beam evaporation on a plastic film substrate (PET) in a roll-to-roll (R2R) process [4]. Although the author is not aware of TCO being deposited commercially by e-beam evaporation on plastic film substrate (PET) in a R2R process, it certainly is practical. The advantages (throughput, scaleable, etc.) of an e-beam deposition system, particularly with ion assist for improved film microstructure, are often overlooked.

Pulsed laser deposition (PLD) is basically a highly efficient, flash evaporation technique. Its use for TCO research is rapidly growing in universities and laboratories. The highest reported conductivity, lowest resistivity value ($\sim 10^{-5}$ ohm cm) for several TCO, e.g. AZO $8.54 \times 10^{-5}\ \Omega$ cm [23] and ITO $7.2 \times 10^{-5}\ \Omega$ cm [24], was achieved in films deposited by this method (although on nonplastic, heated substrates). One benefit of PLD is the very high energy imparted to the evaporated material in a very short pulse, which can result in the composition of the deposited film matching the composition of the target. A disadvantage of the PLD method is the nonuniformity in the spatial distribution of deposited material (and composition) on the substrate.

PLD is not generally considered a practical industrial technique suitable for applications requiring coating of a large area, e.g. wide rolls of plastic film. The cited limitations of PLD for this conclusion are that the laser only illuminates a very small area of the source material, and this 'point source' only deposits on a small substrate area. Further, if optical or scanning techniques are used to increase the illuminated TCO source material area, and therefore deposit on a larger substrate area, the incident laser energy density is reduced, reducing the TCO deposition rate. For larger area substrates the achieved deposition rate is too low to be cost-effective, and the reduced incident laser energy density can result is depositing TCO with poorer properties. However, some commercial PLD systems are being offered [25], multipoint techniques have been demonstrated, and multisource methods as used with e-beam sources are possible.

5.6.2 Sputtering

Sputtering is the ejection of material from a target when bombarded by gas ions from a plasma. These ions are attracted to the target by an applied electric field. The most common type of sputtering uses a DC electric field. Magnets placed behind the sputtering target (magnetrons) trap electrons increasing the ion bombardment of the target, hence increasing the deposition rate and also reducing substrate heating.

Sputtering is fundamentally a more energetic deposition process than evaporation. However, it is not the arrival energy of the sputtered material on the substrate but the accelerated ions and neutrals from the plasma bombard the growing film, which promote adatom surface mobility and film densification resulting in a superior microstructure compared with evaporated films [26]. A typical DC magnetron process has an argon pressure of ~ 1–$10\ \mu$ (mTorr) or 0.13–1.3 Pa.

In addition to achieving a better microstructure, another attribute of sputtering is that the atomic composition (metallic) of deposited film is equal (approximately) to that of the target. Thus depositing metal oxide compounds with a dopant from a ceramic target, for example made of 90% In_2O_3/10% SnO_2 (wt%), or from a metal alloy target of 90% In/10% Sn (wt%) will result in ITO of the correct mixture ratio (atomic metal ratio).

There are many implementations of the magnetron sputtering process, such as direct current (DC), alternating current (AC), mid frequency (MF), radio frequency (RF), and in planar, cylindrical, inverse-cylindrical, rotatable and dual magnetron formats. Many of these magnetrons may be used with a metal target in a reactive process or may use a ceramic (conductive or insulating) target to form the TCO. However, DC magnetron sputtering is the dominant commercial ITO (TCO) deposition method.

The details of magnetron sputtering sources, implementations and targets have been discussed previously in numerous articles and books [26, 27]. The interested reader is referred to the voluminous literature for details. The purpose here is only to provide enough background on sputtering (and evaporation) to permit the detailed discussion of the effect of process parameters on the properties of the deposited TCO thin film.

Four major benefits of magnetron sputtering sources are that (1) it scales easily in length permitting production coating on very large substrates, (2) its compatibility with temperature-sensitivities substrates, e.g. plastic, (3) its ability to operates at lower pressure (than diode sputtering) and (4) the atomic composition (metallic) of deposited film is equal (approximately) to that of target.

In-line coaters with magnetron sputtering of metal and metal oxide thin films on glass substrates >4 m wide for energy efficient architectural windows is routine in industry. Similar coatings and TCO are deposited by magnetron sputtering on plastic.

5.7 Developing a TCO Deposition Process

Process conditions have the largest effect on the TCO thin film properties, and generally the most important process variable is oxygen partial pressure (flow rate). As most TCO conductivity is primarily a result of oxygen vacancies, precise control of oxygen partial pressure (flow rate) is crucial to achieving the desired TCO electrical and optical properties in both high and low temperature processes.

As pointed out in Section 5.4, substrate temperature also is a very important deposition process parameter for influencing impurity (cationic) doping and the microstructure of a TCO thin film. Typical process temperatures for many TCO deposited on a temperature-resistant substrate, like glass, are ~200–500 °C. Usually the TCO deposition process (substrate) temperature is chosen empirically to achieve activation of the cationic dopant, e.g. Sn in ITO, and suitable film microstructure.

Of course, deposition geometry, e.g. source-to-substrate distance, deposition rate, vacuum level (pressure) and gas or plasma conditions all influence TCO film properties as in any thin film vacuum deposition process.

The purpose of the following discussion is to provide a procedural outline for developing a vacuum coating process for a TCO. While for discussion convenience a R2R coating system and magnetron sputtering are assumed, this procedural outline is readily adapted to an in-line, batch or other coating systems, and to other types of deposition sources [6]. The details of actual vacuum coaters are readily available in commercial literature, e.g. see supplier listings on SVC.org website, and not discussed here. The procedural outline also is applicable to high temperature deposition processes with the simple addition of the step of selecting the substrate temperature.

5.7.1 TCO Deposition Process Procedural Outline [6]

A procedural outline for developing a vacuum coating process for a TCO by magnetron sputtering is given below. The discussion here will focus on the deposition of a TCO on a plastic substrate and the process control of TCO properties. However, some of the steps, e.g. 1–3, are the same as used for depositing thin films other than TCO.

1. Evacuate the coating chamber to its base pressure $\leq 10^{-5}$ Torr (0.001 Pa).
2. Admit process gas, e.g. argon, to bring the chamber pressure up to the desired sputtering pressure (~1–10 mTorr/0.13–1.3 Pa).
3. Apply power to the magnetron source and pre-sputter the target (metal or ceramic) (typically ~5–10 min while creeping the web) to remove unwanted oxides or other contamination from its surface.
4. Plastic film line-speed, magnetron power and reactive gas flow interactively affect the deposited TCO film thickness and composition. During process development, the line-speed and sputtering power are typically (initially) fixed at convenient values (~1 ft min^{-1} and 2–3 W cm^{-2}). Often these values are selected based on prior experiments.
5. Reactive gas, oxygen, is slowly introduced into the chamber to react with the sputtered material and form a thin film of the TCO on the plastic film substrate.
6. While holding the other parameters fixed, the oxygen flow rate is adjusted through a range of values, preferably starting at zero, to achieve variations in the deposited TCO film properties; hopefully including the desired film properties.[7]
7. The incremented values of oxygen flow should be sufficiently small and adequate in number to well define the relationship between ρ and O_2 flow rate, i.e. the resistivity-well curve shape (discussed below), which includes the minimum resistivity (ρ) value. The location of each of these TCO samples deposited on the plastic film at each of these incremented values of oxygen flow is marked or recorded (footage).
8. The coating system preferably includes on-line electrical and optical measurement instrumentation to guide the choice of oxygen flow values, and to record the corresponding sheet resistance (R_s), also called surface resistivity (ohm per square) and optical transmittance (%T) values.
9. After completing the coating run and removing the coated roll, sheet samples are cut from the roll at the location corresponding to the various incremental O_2 values and (additional) measurements of important performance parameters, e.g. optical transmittance (%T), reflectance (%R), absorptance (%A) and sheet resistance (R_s), are taken.
10. The resistivity (ρ) values calculated from the measured sheet resistance and TCO film thickness[8] of the samples are plotted (y-axis) versus the oxygen flow rate (x-axis). One or more selected performance parameter, e.g. %T, also may be plotted versus the oxygen flow rate using a second y-axis.

[7] If the film properties achieved do not include the desired electrical and optical (E/O) performance, the procedure must be repeated (from step 4) with new settings for the starting conditions for sputtering power and/or line-speed and sometimes sputtering pressure.

[8] $\rho = d \times R_s$, i.e. the sample thickness × surface resistivity (ohm per square); however, if the sample thicknesses are unknown, and hence ρ can not be determined, an approximate well can be plotted using the R_s values assuming that all samples are approximately equal in thickness.

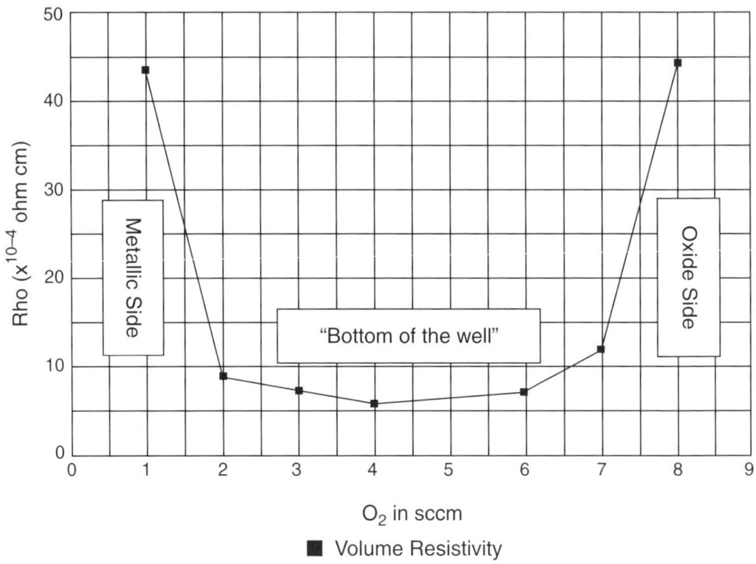

Figure 5.6 *Resistivity-well for ITO thin films (from data in Table 5.5). (Reprinted with permission from [6] Copyright (2009) Society of Vacuum Coaters)*

The above procedure was followed using DC magnetron with an indium tin oxide ceramic target (90 wt% In_2O_3/10 wt% SnO_2) was used to sputter deposit ITO films on PET film with an in-line (conveyor) coater. Figure 5.6 and Table 5.5 show measured data for these ITO thin films [6]. For ease of discussion and illustrative purposes, the concaved curve shown in Figure 5.6 has been named, colloquially, the 'resistivity-well' [6], abbreviated for convenience from here on as 'ρ-well'. This well-shaped curve has the lowest resistivity at the 'bottom of the well'. Figure 5.6, generated by following the above procedural outline, the accompanying process, and the sample performance data in Table 5.5, are the most important data for selecting process parameters and controlling the deposition process to achieve the desired TCO film properties.

Table 5.5 *Measured data for ITO films deposited following the process procedural outline (Reprinted with permission from [6] Copyright (2009) Society of Vacuum Coaters)*

Target: 90 wt% In_2O_3/10 wt% SnO_2
Substrate: Glass slide/PET
Conditions: No heat
Power 0.50 kW
Speed 5.75 in. min^{-1}
Total flow 50 sccm (20% H_2)
Pressure 1.8–1.9 mTorr (μ), ~0.24–0.25 Pa

O_2 (sccm)	1	2	3	4	6	7	8
R (ohm per square)	425	91.3	64.9	61.1	75	119	443
%T vis (luminous)	66.8	80.6	84.9	87.1	87.8	86.3	83.3
Thickness (nm, slide)	102.6	96.3	107.5	94.3	93.3	99.9	100.5
ρ ($\times 10^{-4}$ ohm cm)	43.6	8.8	7.0	5.8	7.0	11.9	44.5

5.7.2 Interpreting Results

Examine the effect of varying the oxygen flow rate (oxygen partial pressure) in the above sputtering deposition process on the achieved TCO film properties. Initially[9] with only the process gas, argon, flowing, the deposited ITO film (not shown in Figure 5.6 and Table 5.5) will be 'metallic'. It is typically dark in appearance with poor transmission and poor conductivity (high resistivity). On the metallic-side of the (ρ-well, adding a small increment (relative to the process gas) of reactive gas, oxygen (here, 1 sccm) causes the surface resistivity/sheet resistance to decrease significantly, to ~425 ohm per square and the transmittance to increase to 67%T (Table 5.5). Increasing the reactive gas flow by adding another increment of oxygen again decreases the surface resistivity/sheet resistance, ~91 ohm per square at 2 sccm, and increases the transmittance, 81%T. This trend of lowering the resistivity with increasing oxygen flow continues until the surface resistivity/sheet resistance (and ρ) minimum (the 'bottom of the well') of ~61 ohm per square ($\rho = 5.8 \times 10^{-4}$ ohm cm) is reached at 4 sccm, and then the trend reverses with the surface resistivity/sheet resistance increasing with additional oxygen, e.g. ~75 ohm per square at 6 sccm on the oxide-side of the well. Adding still more oxygen causes a steeper rate of increase in the resistivity with an equal incremental increase in oxygen flow, e.g. 1 sccm change from 7 to 8 sccm causes an incremental ohm per square increase of 324 ohm per square. On the oxide side of the ITO ρ-well, the film refractive index (real part) also is increasing, slightly raising reflections and decreasing %T. If additions of oxygen are continued, the resistivity will continue to increase and eventually a stoichiometric, nonconductive metal oxide of In_2O_3 and SnO_2 will result.

Compare the ρ-well of Figure 5.6 with the simplified crystal structure of Figure 5.4 discussed in Section 5.5. The two figures are representations of the same doping phenomenon. Figure 5.4 is the qualitative physical explanations for the experimental results shown in Figure 5.6. Thus, ρ rises on the metallic side of the ρ-well because the large number of defect carriers (electrons) N, causes too much ionized impurity scattering, which results in a small scattering time, τ, and hence, a low mobility, μ. Conversely, on the oxide side, ρ rises because N is low and there are too few electrons.

5.8 Controlling TCO E/O Properties

The determined ρ-well shown in Figure 5.6 is only unique for that specific set of process parameters. Actually, there is a family of ρ-wells, like Figure 5.6, one for each different set of process parameters and the minimum resistivity achieved for a given set is generally a local minimum not a global minimum. If the minimum resistivity achieved with a given set of conditions is too high, change the operating conditions (increase power, lower or raise argon pressure, decrease oxygen flow rate and increment), and repeat the procedure to generate a new resistivity well. Of course, many additional factors affect the minimum ρ achieved or achievable, e.g. target surface oxidation state, kW-hours of use, erosion groove condition, nodules, etc.

[9] Starting with a substoichiometric ceramic target (homogeneous throughout its bulk).

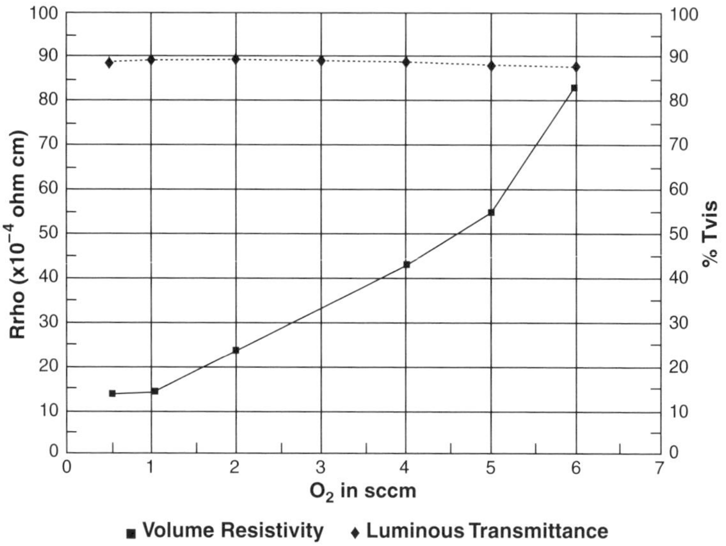

Figure 5.7 Resistivity-well for AZO (ZnO_x) thin films. (Reprinted with permission from [6] Copyright (2009) Society of Vacuum Coaters)

To develop a vacuum coating process for a different TCO than ITO, the same procedure described above is again followed. As an example, a ZnO target doped with 2 wt% Al_2O_3 (AZO), was used to DC magnetron sputter thin films and generate the ρ-well shown in Figure 5.7 [6]. Notice however, that only the oxide side of the ρ-well is obtained even when using no reactive gas, i.e. only argon. The reason that the metallic side of the ρ-well cannot be obtained is because of the target's oxidation state and/or background oxygen in the deposition chamber. When coating plastic film, the substrate itself is often a significant source of oxygen (water). Because the Zn-O bond is formed even more readily than the In-O bond, control of a ZnO_x or AZO deposition process is more difficult than with ITO. This issue will be discussed further in a later section when ZnO_x based semiconducting films are addressed.

As shown in Figure 5.7, the lowest resistivity ('bottom of the ρ-well') is observed with zero oxygen flow. Having a process where the operating-point on the ρ-well is determined by an uncontrolled source(s) of oxygen is not desirable from a process control standpoint. Also, depending on the TCO film surface resistivity/sheet resistance (ohm per square) target and tolerance values, such a process may cause a yield issue in a production.

For a viable industrial TCO deposition process, coating system variables such as the dynamics of pumping system, outgassing from the chamber walls and substrate, the condition of the sputtering target, etc., all must be accounted for in a practical way. Using the step-by-step procedure for developing a TCO ρ-well described above, understanding the ρ-well data and using on-line E/O and vacuum system monitoring, along with regular prescribed maintenance are all important parts of achieving a robust TCO process.

TCO are not a single material like a stoichiometric compound or an elemental metal that have fixed atomic and electronic properties.[10] Each time one deposits, e.g. gold (following good vacuum practices), the film will have a carrier concentration (conduction electrons) determined by the material. However, TCO are engineered materials [6] with a carrier concentration determined by the amount of (active) doping. Actually, each data point on a ρ-well curve, represents process conditions, where a different material was formed.[11] In addition, the optical properties of TCO are strongly dependent on the electrical properties, and the electrical properties are primarily determined by the processing conditions. Widely different E/O film properties are easily achieved from the same starting material. For a given TCO starting material, the properties achieved in the deposited film are the result of the (final) operating point on the ρ-well.

5.9 TSO for Transparent Oxide Electronics

As stated above, the term TCO in this chapter only means highly transparent and degenerate *n*-type (*p*-type are called *p-TCO*) semiconductors, which electrically are metals. Transparent and nondegenerate semiconductive oxides, used for their *n*-type semiconducting properties in transparent electronics, are discussed in this section. When referring to these *n*-type nondegenerate transparent semiconductive oxides, the abbreviation TSO will be used in this chapter (*p*-type are called *p-TSO*).

The focus of this section is the discussion of various TSO candidate materials and their deposition on plastics. Much of the discussion can be independent of the specific electronic device application. However, to cover situations where this is not the case, and for ease and clarity of discussion, transparent oxide electronic devices will be divided into two basic categories:

1. field effect transistors (FET) made with thin films, often called thin film transistors (TFT);
2. junction-type devices, i.e. *p–n* junctions and their many variations

The *p–n* junction and its many device variations have been widely discussed in the literature and are reviewed in other chapters of this book. Similarly, transparent TFT, also called TTFT, are reviewed in other chapters and discussed in the literature. Therefore, no separate discussion of these electronic devices is included in this chapter, except in commentary related to the performance achieved with and material suitability of the TSO and its compatibility with deposition on plastic substrates.

In this chapter any reference to transparent electronic devices whether junction-type devices, TFT or TTFT will mean devices made with oxide materials unless stated otherwise. The term TFT in this chapter will mean a FET with only some layers transparent, typically the channel and the dielectric layers. In these TFT devices the drain and source, and sometimes the gate electrode, are typically metals and not transparent. The term TTFT will be reserved for devices made with *all* transparent layers because this is required for many

[10] Ignoring thin film effects and just considering material effects.
[11] However, when the operating point is far up the oxide side of the (ρ-well a fully oxidized, insulating, stoichiometric 'TCO' compound is formed. Further oxygen additions to the process do not change the composition which remains stoichiometric.

envisioned applications. Another reason for this distinction is that fabricating TTFT devices requires deposition of TCO and TSO compatible coatings.

Transparent inorganic electronic devices will in general, require both TCO and TSO types of thin films. The TCO (*p-TCO*) are used as transparent electrodes to make low resistance ohmic contacts to the TSO (*p-TSO*) layer(s) of the electronic device. The deposition of TCO on plastics and the control of their conductive and optical properties were reviewed above. Discussed below are the deposition of TSO on plastic substrates (where available) and the control of its semiconductive and optical properties.

5.9.1 TSO for TTFT/TFT Devices

TSO deposition and resultant film properties are discussed first to parallel the similar TCO discussion given above. The E/O characteristics and suitability of these TSO coatings for TTFT (TFT) applications are next examined.

Generally, for TCO thin films the objective is achieving the highest combined conductivity and transparency. However, when fabricating electronic devices with a TSO, the primary objective is to deposit the TSO film with a controlled, stable and relativity high resistivity. Therefore, TSO process conditions and sometimes even the starting materials may be quite different from those used for TCO. Interestingly, when fabricating TSO thin films using TCO materials deposited at low substrates temperatures, e.g. on plastics, the penalty of 4–6 times higher resistivity which results is actually a benefit.

Deposition and resulting device performance issues which occur when traditional TCO materials are used as active TSO layers in amorphous oxide electronic devices are described below. However, the focus in this chapter will be on the issues to be overcome in making a TSO suitable for device applications rather than the performance of the device per se. Depositing a TSO with a known TCO material first requires achieving control of carrier concentration to produce a nondegenerate film. In addition, in making a TTFT/TFT the TSO channel should, preferably, have a low background carrier concentration to achieve an enhancement (device off at $V_{gate} = 0$ V) rather than a depletion type of device. Often, poor TTFT/TFT performance can result from layer incompatibility, a poor semiconductor/ dielectric interface or because of a poor quality dielectric layer resulting from the low substrate temperature used during deposition. These are examples of device performance issues not addressed here because they are not caused by the deposition of the TSO or its E/O characteristics but rather by other effects such as the low substrate temperature needed to deposit on a plastic film.

The approximate electrical requirements for TSO used as an active *n*-channel material in a TTFT are control of carrier concentration[12] from $<10^{14}$ cm^{-3} to $>10^{18}$ cm^{-3} and achieving 'high' Hall mobility ($\gtrsim 10$ cm^2 V^{-1} s^{-1}) [28]. Exemplary binary TCO materials[13] that also have been deposited as TSO include IO (In_2O_{3-x}), ITO (In_2O_3:Sn), ZnO_z, AZO (ZnO:Al), IZO (In_2O_3-ZnO) and TO (SnO_{2-y}).[14] However, most reports of TFT/TTFT devices

[12] The necessary carrier concentration range is material dependant but a low carrier value is critical to achieving low off-current and a high on-off current ratio.

[13] In this chapter, the addition of a dopant, e.g. ITO (In_2O_3:Sn) with In_2O_3 90 wt%/SnO_2 10 wt% or a dopant like amount of another (stabilizing) oxide like IZO with In_2O_3 90 wt%/ZnO 10 wt% is still considered a binary oxide material, not a ternary, because the E/O properties are dominated by the binary host material. (See also Sections 5.9.5, 5.9.8 and footnote 19.)

[14] Unfortunately, the nomenclature and abbreviations for TCO/TSO are imprecise and not standardized. Often the stoichiometric compound formula is written when the reduced oxide is intended, e.g. ITO is represented as In_2O_3:Sn to indicate Sn doping of In_2O_3 but nearly always the TCO ITO with Sn as a dopant (deposited at high temperature) is actually In_2O_{3-x}:Sn.

incorporating a TSO channel made using these binary TCO materials were deposited on temperature-tolerant substrates at a high temperature.

Fundamentally, all of these TCO materials could be or have been deposited on plastic substrates (as TCO) with low temperature deposition processes. However, to the author's knowledge, there have been only a few reports of binary TSO, e.g. one with an indium oxide host, deposited on plastic substrates [29]. While there are more reports [30–35] of these TSO deposited by low temperature processing, which potentially could be used with plastic substrates, recall that depositing a TCO/TSO at 'RT', i.e. which usually means not intentionally heated, or at a low temperature (often not reported) on glass, silicon or other temperature-tolerant material is not identical to depositing it on a plastic substrate, even at that temperature [8, 11–13].

Typically in the literature, the TSO was deposited not as a single layer but in a multilayer structure to form a device, and the characteristics of the TSO were derived from/or dependent on the device performance. In addition, reported TFT electrical performance with TSO based on these TCO materials, in general, was only fair and often a high temperature process for at least one step in the fabrication was required [31, 32, 34, 35].

One exception is the recent report [29] of TTFT with TSO IO channel layer, TCO IO source and drain layers and TCO ITO gate electrode layer. Two unusual attributes of this work are (1) the deposition methods for the TSO/TCO layers and the gate insulator layer and (2) the gate insulator material. The gate insulator was an organic dielectric layer produced by solution spin coating (in air) and crosslinking of a polymer blend. The TSO/TCO layers were fabricated by ion-assisted deposition (IAD), an energetic process, which fundamentally is a vacuum ion sputter deposition process with a separated ion bombardment source for enhancement of film microstructure (see, for example, [27]).

Devices were successfully deposited on heat-stabilized or organic hard-coated PET substrates at or near RT, i.e. not intentionally heated; however, some heating would occur from the energetic IAD process (no temperature measurements or substrate cooling was reported). The IO TSO/TCO layers were reported to have significant crystallinity (texture) due to the IAD process. The IO TCO source and drain were reported to have high conductivity of 1400 S cm^{-1} ($\rho = 7 \times 10^{-4}$)[15] ascribed to the crystallinity (texture).

TFT devices constructed as described above, except with Au source/drain electrodes, fabricated on PET exhibited field effect mobilities of $\sim$20 cm^2 V^{-1} s^{-1} and I_{on}/I_{off} ratios of 10^3. Similar TTFT devices but with transparent IO source/drain, IO channel layer, crosslinked polymer gate insulator and a thin (20 nm) SiO_2 layer on both sides of the PET, exhibited field effect mobilities of $\sim$10 cm^2 V^{-1} s^{-1} and I_{on}/I_{off} ratios $> 10^2$.

Most significant were similar TFT devices fabricated on n^+-Si substrates (Au source/drain electrodes), with crosslinked polymer gate insulator and IO channel layer, which exhibited field effect mobilities of $\sim$160 cm^2 V^{-1} s^{-1} and I_{on}/I_{off} ratios of 10^4 with a threshold voltage of $\sim$0.2 V. This exceptional performance was ascribed, primarily, to use of the IAD process for the IO channel deposition, which allowed the determination of the semiconducting IO composition by controlling the ion-assist power providing oxygen ions to the film, and to the dielectric quality of the crosslinking polymer gate insulator.

[15] This resistivity (conductivity) value is consistent with the resistivity range achieved by industrial sputtering processes given in Section 5.3.2, Table 5.4.

Due to the lack of reports for most of these exemplary TCO materials deposited as TSO by traditional sputtering processes on plastics, selected examples of results on temperature-tolerant substrates will be reviewed later in this section to demonstrate issues common to both high and low temperature processes.

5.9.2 Binary TCO Materials for TSO

What are the issues to be overcome when using TCO materials for TSO? How does one select the TCO material and develop a process to make a TSO? What are the consequences of a low substrate temperature and of a plastic substrate? These questions and issues will be discussed in this section.

Developing a robust deposition process to make TSO is quite challenging. One way to develop a TSO deposition process is to follow the procedural outline given previously (Section 5.8) for developing a vacuum coating process for depositing a TCO. As an example, consider using one of the classic binary TCO materials (CdO, SnO_2, In_2O_3 and ZnO) to generated a TSO ρ-well.

5.9.3 Tin-doped Indium Oxide

Choosing ITO as an example, the TSO ρ-well generated following the given procedure will be conceptually, similar to the ITO TCO ρ-well of Figure 5.6. However, because for a TSO a low conduction electron density ($N \leq 10^{18}\ \text{cm}^{-3}$) is desired, a higher oxygen flow rate (pO_2), compared with a TCO process, will be used to remove many oxygen vacancies. With sputter deposition of a TSO, the process gas pressure, e.g. argon, also typically will be higher ($\geq$10 mTorr) to promote gas collisions to reduce possible damage of the growing film by bombardment of negatively charged oxygen ions and neutrals. The TSO process conditions are varied similarly to those shown for TCO in Figure 5.6 and Table 5.5 but scaled for the higher operating settings, to generate the characterizing ρ-well data but with additional oxide side, high resistivity, points.

Next the generated ρ versus O_2 flow rate (pO_2) results are plotted. However, compared with Figure 5.6 for an ITO TCO process, which is a linear plot and shows about one order of magnitude change in ρ, the TSO plot typically will be a log plot and show many orders of magnitude change in ρ versus pO_2. The process operating point, high on the oxide side of the ρ-well, is chosen to make the TSO film with the desired low N (high ρ). However, the steep slope of the ρ-well in the region of operation will make control of the process very difficult, even when sputtering with a ceramic target. Further, depending on other process and vacuum pumping system parameters, adequate manual control of the oxygen partial pressure to reliably achieve the target N value may not be possible. Conceptually, this will not be a robust process and a fast feedback control system for the oxygen partial pressure will likely be necessary but may not be sufficient to control the process.

From the qualitative discussion above the difficulty of making a TSO with controllable electrical properties from an In_2O_3 host (ITO) binary material was explained. Consider next a different classic binary TCO material as a candidate for making a TSO thin film on plastics. On account of toxicity concerns with CdO, it will not be considered or discussed. The remaining two common binary oxides, host ZnO and SnO_2, as candidates for TSO are discussed below.

5.9.4 Zinc Oxide

ZnO-based films can be water-white and highly transparent at visible wavelength including the short wavelength blue region. Optically, ZnO-based films are good TSO candidates. What about the electrical properties and process considerations for ZnO-based TSO films?

For a qualitative understanding of process issues with making ZnO-based TSO films, reexamine the ρ-well shown in Figure 5.7, for ZnO_x TCO films sputter deposited from an AZO target. In Figure 5.7, as discussed in the TCO examples, the metallic side of the ρ-well is missing primarily because of the oxidation state of the target but background oxygen, e.g. from the plastic substrate, also adds oxygen (water) to the process. However, for TSO films this is not a concern because only the oxide side of the ρ-well, per se, will be of interest.

To obtain a TSO thin film with low N, the partial pressure of oxygen is increased and the process operating point on the ρ-well curve must be far up right side (oxide side) of the ρ-well curve (off the top of the TCO curve in Figure 5.7). To sputter deposit a low N TSO film from this AZO target, the argon pressure generally is increased also raising the system pressure. Common values for the total sputtering pressure are 5–10 mTorr or more, compared with 2–3 mTorr for the TCO process when the resistivity 'minimum' also is included.

As with the ITO (In_2O_{3-x}) TSO qualitative example above, the ZnO_x ρ-well curve in the region of desired operation will exhibit a very steep slope. As the Zn-O bond is formed even more readily than the In-O bond, control of oxygen vacancies in a ZnO-based material deposition process is difficult. Further, control of the deposition process to achieve a repeatable, specific, low N value for a ZnO-based TSO is very difficult. Therefore, one would expect that fabricating TFT/TTFT devices with a ZnO-based TSO active layer would not be a reliable and reproducible process.

However, many groups [30–32, 34, 36, 37] have reported fabrication of TFT/TTFT with ZnO_x-based layer as the active channel material. Often a high substrate deposition or post-process annealing temperature (300–700 °C) was used [36] to achieve improved electrical properties, e.g. improved channel mobility, gate dielectric, off-on ratio, etc. Generally, these TFT with ZnO-based channels were depletion-mode devices exhibiting a high background carrier concentration of $> 10^{16}\,cm^{-3}$, and required special measures to suppress leakage current, along with application of a negative gate voltage to turn off the devices [37].

An example from the literature [30], which will demonstrates the issues discussed above in depositing a ZnO_x TSO film is that of TFT devices made with a ZnO_x TSO channel deposited at near RT[16] (but not on plastic film) on heavily doped patterned Si substrates with a thermally grown (high temperature process) oxide dielectric layer (~100 nm thick). The source and drain electrodes and a backside common gate electrode on the Si were of Ti-Au. Thin Si and glass (Corning 7059) substrates were also included for TSO layer deposition to permit thin film stress and %T measurements. This work provides a specific quantitative example of the process issues encountered when making ZnO-based TSO films for which a qualitative understanding was provided above and will be discussed in detail below.

These ZnO_x channel TFT were made by RF magnetron sputtering at two different total $Ar + O_2$ process pressures, 10 mTorr and 20 mTorr. The partial pressure of oxygen (pO_2) was varied (10^{-8}–10^{-5} Torr) to determine the corresponding electrical resistivity. The plotted

[16] The actual temperature was not reported but the substrates were placed on a water cooled table parallel to the sputtering target.

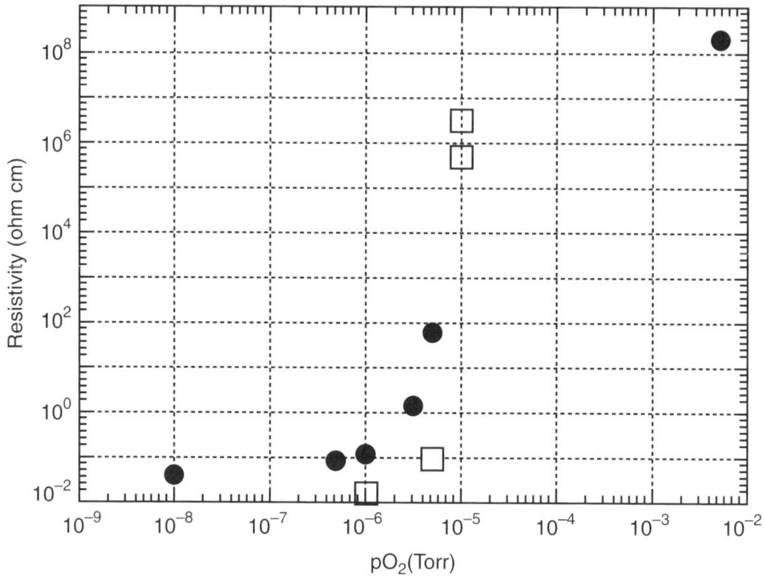

Figure 5.8 *Resistivity-well for ZnO_x TSO thin films. Total sputtering pressures ($Ar + O_2$): 10 mTorr (solid circles) and 20 mTorr (open squares). (Reprinted with permission from [6] Copyright (2003) American Institute of Physics)*

experimental data represent two partial (oxide side) ρ-wells, one for each $Ar + O_2$ process pressure (Figure 5.8). At both of these sputtering pressures (10 mTorr or 20 mTorr), the electrical properties varied over orders of magnitude with very limited changes in oxygen partial pressure (see Figure 5.8); as would be expected from knowledge of a characteristic ρ-well.

RF magnetron sputtering at 10 mTorr total $Ar + O_2$ process pressures and for very low oxygen 'flow' ($pO_2 \sim 10^{-8}$ Torr), semiconducting films of ZnO_x with a resistivity of $\rho \sim 0.03$ ohm cm were deposited near the bottom of the ρ-well, for these conditions. At higher oxygen 'flows' ($pO_2 \sim 10^{-5}$ Torr) on the oxide side of the ρ-well, the resistivity changed abruptly to semi-insulating with $\rho \sim 10^6$–10^8 ohm cm. Similar behavior of the resistivity was observed when sputtering at 20 mTorr total pressure. However, the ρ-well characteristics at 20 mTorr, as would be expected, exhibited an even steeper slope than at 10 mTorr.

TFT devices were made at 20 mTorr of total pressure with three different pO_2 settings of 7.5, 10 and 20 μTorr near the abrupt transition in resistivity (on the oxide side of the well). Resistivities measured 15, 6.5×10^4 and 2×10^6 ohm cm, respectively (not shown in Figure 5.8). TFT made at these three settings operated in the charge accumulation mode. One of the better devices made at the best pO_2 compromise (10 μTorr) had a calculated field effect mobility of 1.2 $cm^2\,V^{-1}\,s^{-1}$, an on-off ratio of 1.6×10^6 with drain current $> 10^{-5}$ A and a threshold voltage of $\sim$0 V. The reported field effect mobility and on-off ratio varied inversely and widely for the three pO_2 settings. The carrier concentrations corresponding to the three resistivities (at the three pO_2 settings) were not stated.

The TFT properties obtained were rather poor even though patterned Si substrates with a high temperature thermally grown oxide dielectric layer and metalic deposited source, drain and gate electrodes were used. Clearly, the poor TFT properties were related to the inferior ZnO_x TSO channel layer properties.

This work demonstrates some of the process difficulties in depositing a ZnO_x TSO thin film with controlled electrical properties. A $pO_2 \sim 10^{-5}$ Torr was used for the process operating point but with a small variation of $0.75–2.0 \times 10^{-5}$ Torr the measured resistivities changed from 15 to 2×10^6 ohm cm, more than five orders of magnitude! These achieved resistivity versus pO_2 values do not demonstrate a process with reasonable control suitable for an industrial process to deposit ZnO_x TSO thin films.

5.9.5 Indium Zinc Oxide

Another popular TCO material with good performance when deposited at low temperature and on plastic substrates is amorphous indium zinc oxide, IZO (In_2O_3 90 wt%/ZnO 10 wt%) A relatively low resistivity of $\sim 3 \times 10^{-4}$ Ohm cm, typically better than ITO, can be achieved with IZO on plastic substrates at RT [38]. As zinc is of lower valence than indium it is not truly an impurity dopant in the In_2O_3 host. However, the addition of dopant like amounts of zinc results in a stabilized indium oxide-zinc oxide material that can remain amorphous up to high temperatures, ~500 °C [39].

Amorphous IZO has also been used as a TSO for the channel semiconductor in TTFT devices [35, 40–43]. However, TTFT with TSO channels based on IZO generally exhibited the same issues (device and process) as discussed above for ZnO-based devices, i.e. they are depletion-mode devices exhibiting a high background carrier concentration $> 10^{16}$ cm^{-3}, required special measures to suppress leakage current, and application of a negative gate voltage to turn off the devices. A deposition at a high substrate temperature or post-process annealing temperature (~300–600 °C) is needed to achieve improved electrical properties, e.g. channel mobility, on-off ratio, gate dielectric properties, etc. [43]. High carrier concentration causes large negative threshold voltage that requires very thin channel layers to allow pinch-off of the channel.

5.9.6 Tin Oxide

Of the four classic binaries, tin oxide (TO) is perhaps the preferred TSO candidate, electrically and from a process control standpoint. For example as a TCO, TO films DC sputtered on plastics at RT from a 98 wt% SnO_2/2 wt.% Sb_2O_3 target achieved a minimum resistivity of $\sim 5 \times 10^{-3}$ ohm cm (Figure 5.9) [6], which is approximately an order of magnitude higher than for IO or ITO deposited under similar conditions (at the respective ρ-well minimums). Further, a resistivity of $\sim 8 \times 10^{-2}$ ohm cm was obtained (Table 5.6 and Figure 5.9) with an easily manually controllable operating point on the oxide side of the TCO ρ-well (10 sccm O_2 flow rate, ~2 mTorr total pressure). Thus, developing a TSO process should be less sensitive than those already discussed (IO, ITO, ZnO_x, AZO and IZO).

Optically, however, a TCO of reduced SnO_{2-x} may not be preferred because it is generally more absorbing than ITO, particularly at short visible wavelengths, which tends to cause a light yellow color in relatively thick TO films (~300 nm). However, a TSO of SnO_2 will be much less reduced than when deposited as a TCO, i.e. the film is much more oxidized, lowering absorption. The TSO film thickness likely also will be smaller than the TCO film, further reducing the absorption. Therefore, the SnO_2 film optical transmittance and color may not be an issue in a TTFT application.

Table 5.6 *Measured data for tin oxide thin films deposited on PET. (Reprinted with permission from [6] Copyright (2009) Society of Vacuum Coaters)*

Target: 98 wt% SnO_2/2 wt% Sb_2O_3
Substrate: DuPont 725 PET
Conditions: No heat
Power 0.50 kW
Speed 5.75 in. min^{-1}
Total flow 50 sccm
Pressure 1.8–1.9 μ

O_2 (sccm)	4	5	6	7.5	9	10
R (ohm per square)	4710	420	183	214	714	3000
%T vis (luminous)	67.2	69.4	76.6	83	83.2	80.1
Thickness (nm, slide)	311	298.0	298.7	294.3	283.1	252.6
ρ ($\times 10^{-3}$ ohm cm)	146.5	12.5	5.5	6.3	20.2	75.8

TFT/TTFT made with TO active channels have been reported [44, 45] but all were made by high temperature processes on temperature resistant substrates. The ferroelectric TFT (unidentified opaque source and drain) in Prins *et al.* [44] was fabricated on a crystalline substrate by pulsed laser deposition (PLD) with a 110 nm thick SnO_2:Sb (ATO) film semiconductor channel layer (resistivity ~1 Ohm cm), with $PbZr_{0.2}Ti_{0.8}O_3$ as a ferroelectric insulator, and $SrRuO_3$ (opaque) as a gate electrode. Hall measurements on the ATO film

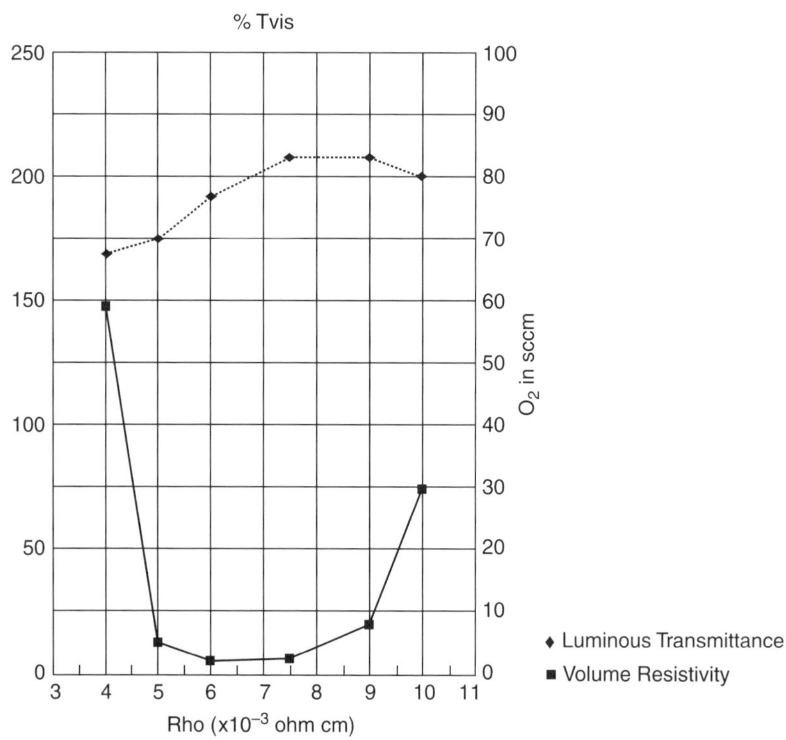

Figure 5.9 *Resistivity-well for TO (SnO_x) thin films (from data in Table 5.6). (Reprinted with permission from [6] Copyright (2009) Society of Vacuum Coaters)*

indicated values for N of $\sim 10^{18}$ cm^{-3} and μ of ~ 5 cm^2 V^{-1} s^{-1}. Substrate temperature during deposition was not given but the carrier density was reported to be 'entirely from Sb dopant atoms', indicating a high temperature was used.

More recently, a bottom-gate TTFT with a SnO_2 film channel was reported [45]. Glass substrates commercially coated with ITO and superlattice AlO_x and TiO_x were used at the gate electrode and dielectric, respectively. ITO was used for source and drain electrodes and the average visible (400–700 nm) transmittance through the entire device was $\sim 75\%$.

The SnO_2 channel was deposited by three different methods, RF magnetron sputtered (5 mTorr with Ar/O_2 ratio of 97%/3%), thermal evaporation ($\sim 10^{-6}$ Torr) and activated reactivated evaporation ($\sim 5 \times 10^{-4}$ Torr with microwave activated O_2 or N_2) with no intentional substrate heating. The as-deposited SnO_2 channel layer was highly insulating and was subsequently annealed (furnace and rapid thermal annealing, RTA) at 600 °C in oxygen. The post annealing created polycrystalline SnO_2 (RF sputtered) films but typically, with too much conductivity (note there was no intentional impurity doping) for TTFT channel layer applications. The report stated that it was very difficult to achieve enhancement-mode devices. Only by making a very thin layer SnO_2 channel layer could the resistance be raised sufficiently for TTFT application. A 10–20 nm channel layer was used to achieve a TTFT device that operated in the enhancement mode with a maximum field effect mobility of 0.8 cm^2 V^{-1} s^{-1} and drain current on-off ratio of 10^5. Depletion mode devices of similar construction exhibited maximum field effect mobility of 2.0 cm^2 V^{-1} s^{-1}.

In this example, the as-deposited SnO_2 channel layer was highly insulating but after post-annealing at 600 °C in oxygen, polycrystalline 'SnO_2' films with too much conductivity were created. The oxygen pressure was not stated but it was apparently insufficient to prevent forming SnO_x films through loss of oxygen and the generation of carriers through oxygen vacancies. Indium acceptor doping also was tried to reduce the SnO_x conductivity but a controllable process was not achieved. As emphasized in Section 5.8 and discussed in the previous TSO examples, achieving the desired TSO film properties is highly dependant on control of the process details.

5.9.7 Ternary and Multicomponent (TCO) Materials for TSO

Combinations of binary TCO materials forming ternary and multicomponent oxide compounds and mixtures have been investigated as TCO for decades. In general, the E/O properties of these materials as TCO made at high temperatures (>300 °C) are rarely equivalent and not superior to what is achieved with ITO[17] [46]. Not surprisingly, these TCO alternatives have not found general application.

5.9.8 Zinc Indium Oxide and Zinc Tin Oxide

Ternary oxide mixtures and compounds have been successfully used as TSO in TFT and TTFT devices. For example, ZIO[18] (also called 'IZO')[19] thin films of various $ZnO\text{-}In_2O_3$

[17] Ternary Cd_2SnO_4 is an exception but has the substantial disadvantage of containing toxic cadmium.
[18] ZIO here designates examples of the $ZnO\text{-}In_2O_3$ family where the ZnO concentration is >10 wt%.
[19] The abbreviations for TCO/TSO, like their chemical nomenclature, are not standardized. Multiple material compositions have been named by the same letters. IZO has been used to designate ZnO:In, indium doped ZnO and 'In_2O_3:Zn, Zn doped In_2O_3' but zinc is not truly an impurity dopant in the In_2O_3 host (see Section 9.5).

(molar) mixture ratios (1 : 1 [47] and 2 : 1 [48]) have been reported. TTFT and TFT devices with RT ZIO TSO channel and ITO source and drain electrodes were deposited with no intentional heating. Some ZIO (1 : 1) TFT depletion-mode devices (transparent except for Cr/Au bottom gate electrode) were deposited on PEN plastic substrates with a maximum process temperature of 175 °C [47]. Other ZIO (1 : 1) TFT devices on Si wafer substrates were processed or annealed at temperatures from 150 to 400 °C. The best device properties were produced with ZIO on Si and with deposition/annealing at high temperatures but the results on PEN are encouraging.

The earlier work [48] produced TTFT with RF sputtered ZIO (2 : 1) TSO channel layer and ITO source and drain electrodes, deposited with no intentional heating. Glass substrates commercially coated with ITO and a superlattice of AlO_x and TiO_x were used at the gate electrode and dielectric, respectively. Both depletion mode and enhancement mode devices were fabricated with high temperature (600 °C) and medium temperature (300 °C) air annealing, respectively. The ZIO layer was amorphous with annealing up to 500 °C.

Both enhancement-mode and depletion-mode TTFT devices also were fabricated with RT deposited ZIO channels without post annealing. Some process parameters for the vacuum RF sputtering TSO depositions were listed and differed for the three annealing temperatures (600 °C, 300 °C and RT) but details of the deposition and process control were not discussed. The listed process parameters for the RF sputtering of the ZIO TSO depositions at RT without annealing were: Ar flow 15 sccm, O_2 flow 5 sccm, pressure 1–2 mTorr, RF power density 4.9 W cm^{-2} and with a target–substrate distance of 10 cm.

The best devices exhibited peak incremental mobilities of 8 and 17 $cm^2\,V^{-1}\,s^{-1}$ and drain current on/off ratios of 10^4 and 3×10^3, respectively [47]. For TTFT devices with a RT deposited TSO channel these device characteristics are quite respectable but recall that the gate dielectric, which also significantly affects performance, was atomic layer deposition (ALD) deposited superlattice oxides made by a commercial, high temperature process on glass (as was the ITO gate electrode).

Another example of ternary TCO materials used as TSO is zinc tin oxide (ZTO), i.e. ZnO: SnO_2. Flexible transistors [49] were made on polyimide (PI) sheets (backed with stainless steel) by first sputtering an Al gate electrode, followed by a silicon oxynitride dielectric layer deposited by PECVD (300 °C) then a TSO channel (50 nm thick) of ZTO was sputter deposited from a 1 : 1 molar (ZnO : SnO_2) ratio target without heating (<70 °C) but later was annealed at 250 °C for 10 min. The source and drain electrodes for good devices were indium tin oxide (ITO)/Au. The details of fabrication are given in Jackson *et al.* [50]. ZTO transistors fabricated as described above on PI plastic substrates exhibited mobilities of 14 $cm^2\,V^{-1}\,s^{-1}$ and on/off ratios of 10^6.

5.9.9 Indium Gallium Zinc Oxide and Cadmium Indium Antimony Oxide

It has been pointed out that amorphous transparent semiconducting oxides (ATSO) with high mobilities are better suited than polycrystalline TSO films for use as active semiconducting layers in transparent electronic devices because they provide better device uniformity and allow lower processing temperature [51]. Amorphous materials with the $(n-1)\,d^{10}\,ns^0$ (where the principal quantum number $n \geq 4$) structure and with post-transition-metal (heavy-metal) cationic dopants were shown to be particularly preferred [51, 52]. Examples of this class of materials are the amorphous multicomponent oxides of the In-Ga-Zn-O system, e.g. $InGaO_3[ZnO]_{1-4}$ and $InGaO_3[ZnO]_5$, abbreviated a-IGZO. Note that ternary oxides of the In-Zn-O and Zn-Sn-O systems, discussed above, also belong to this class of materials [40]. Further, these ternary and multicomponent oxide systems also appear attractive as TSO candidates from a process stand point.

Typically, amorphous ternary and multicomponent oxides formed from combinations of binary TCO materials appear to be advantageous as TSO candidates. One reason multicomponent oxides like a-IGZO are preferred for TFT on temperature-tolerant substrates, is because the amorphous phase is stabilized up to $\sim$600 °C allowing high temperature processing to obtain superior channel (and gate dielectric) characteristics.

For low temperature processing, for example on plastic substrate, amorphous thin films are the most common result. The a-IGZO electronic structure is insensitive to chemical bond distortion and carrier transport is maintained in the amorphous structure (like ZIO and IZO). In a-IGZO the electron MFP is much larger than the chemical bond distances much like amorphous IO, ITO and IZO [15, 51]. The theoretical explanation for these materials with heavy-metal cations and $(n-1)d^{10}ns^0$ ($n \geq 4$) structures is that they have extended, overlapping spherical s orbitals (of the heavy-metal cations) which provide conduction paths even in an amorphous structure [51, 52].

Amorphous multicomponent oxides, e.g. the In-Ga-Zn-O system, actually appear attractive as TSO candidates from a process standpoint as well as because of their 'poorer performance' as a TCO. Recall that a-IGZO TCO, ($InGaO_3[ZnO]_{1-4}$) and ($InGaO_3[ZnO]_5$) with $\sim 10^{20}$ cm^{-3} carriers exhibited a conductivity of only $\sim$200–400 S cm^{-1} (with mobilities of $\sim$10–20 cm^2 V^{-1} s^{-1}) [51–53]. Thus, like TO on plastic, the difficulty of process control (pO_2) of conductivity (*N*, resistivity) is reduced by about an order of magnitude compared with TCO like ITO.

TTFT made with single crystal IGZO exhibited very good performance compared with devices made from binary TCO materials. Typical (over 100 devices) single-crystal, top gate devices exhibited field effect mobility of $\sim$80 cm^2 V^{-1} s^{-1}, with a threshold voltage of $\sim$+3 V, a on/off current ratio of $\sim 10^6$ and operated in the enhanced mode [1]. This performance, albeit in single crystal, is far superior to the performance of depletion mode devices made with typical binary TCO materials discussed above. Of course, one would expect far better performance from a single-crystal device versus amorphous FET devices.

Perhaps, the most successful TTFT device, deposited by PLD all at room temperature *and* on a plastic (200 μm thick PET) film substrate, used the same multicomponent oxide system, $InGaZnO_4$ (a-IGZO), as the channel material [28]. The top gate devices exhibited saturation mobility of $\sim$6–9 cm^2 V^{-1} s^{-1} (a Hall mobility of > 10 cm^2 V^{-1} s^{-1}) and field effect mobility of $\sim$5.6 cm^2 V^{-1} s^{-1} with a threshold voltage of $\sim$+1.6 V, an on/off current ratio of $\sim 10^3$ and operated in the enhanced mode. After bending the TTFT to a radius of curvature of 30 mm (tensile strain of $\sim$0.3%) the performance was almost unaffected (a slight decrease in saturation current) and was stable with further bending. This is the work that first showed the potential of transparent oxide electronics on flexible, plastic substrates and stimulated widespread interest in the field.

Since this work there are many other reports of TTFT/TFT devices made using a-IGZO as the active TSO layer. However, nearly all of the devices reported were made using more than one type of deposition process, have one or more layers made by a high temperature process (or anneal) and were not on a plastic substrate. These same process limitations can generally

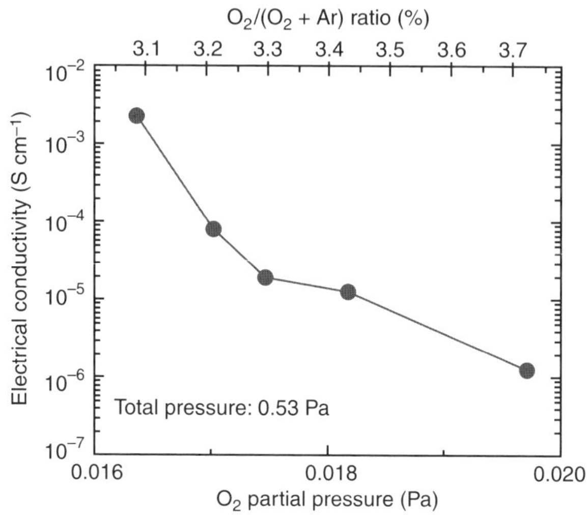

Figure 5.10 *'Conductivity-well' for $InGaZnO_4$ TSO (TFT channel) RF sputter deposited at RT (40 °C). (Reprinted with permission from [54] Copyright (2006) American Institute of Physics)*

be ascribed to TTFT/TFT made with all other TSO candidate materials whether binary or ternary oxides.

However, recently a top-gate TFT was deposited by RF sputtering on glass at 'RT', i.e. without intentional heating [54]. The maximum temperature measured during the a-IGZO channel deposition was 40 °C and during the Y_2O_3 gate insulator layer deposition the glass substrate reached 140 °C. TFT devices exhibited field effect mobilities of ~2 $cm^2\,V^{-1}\,s^{-1}$, with an on/off current ratio of 10^6–10^8, threshold voltage of ~.5 V and subthreshold swing of ~0.2 V $(decade)^{-1}$. Multiple TFT devices were deposited and a reproducible RF sputtering process with control of the pO_2 was demonstrated (Figure 5.10). While the TFT were not transparent because of the Au/Ti electrodes for source, drain and gate, and these were deposited by e-beam evaporation, the achievement of 'commercially useful' performance with a RT deposited channel and gate insulator by a reproducible RF sputtering process is quite a significant step toward realization of amorphous oxide TFT. The substrate was glass but the process temperatures were compatible with the temperature capability of commercial plastic film substrates, e.g. PET. However, recall that as pointed out in Section 5.2, there are additional hurdles to overcome when using a plastic substrate.

Another multicomponent amorphous oxide system recently reported is Cd-In-Sb-O (a-CISO) [55] derived from Cd_2InSbO_6, the crystalline phase of this material. This material system is promising because it has the preferred electronic configuration $4d^{10}\,5s^0$ similar to that discussed above for the In-Ga-Zn-O system.

TSO films of a-CISO (Cd:In:Sb:O ratio in the films of 2 : 1 : 1 : 6) were deposited at room temperature using RF magnetron sputtering on hardcoated PEN plastic film. The a-CISO thin films were vacuum deposited to a thickness of ~150 nm using either pure Ar or Ar + O_2 gas at a pressure of 7 mTorr. The maximum Hall mobility of ~26 $cm^2\,V^{-1}\,s^{-1}$ was measured for a-CISO film deposited in Ar only without oxygen, and its resistivity and carrier density were $\sim 1.5 \times 10^{-3}$ Ohm cm and $\sim 2.1 \times 10^{20}$ cm^{-3}, respectively. A Hall mobility as high as ~ 17 cm^2 V^{-1} s^{-1} was measured at a carrier density of $\sim 10^{17}$ cm^{-3}. These electrical characteristics are very comparable with those achieved with the a-IGZO films discussed above. The average visible transmittance of the a-CISO film in the visible region was $>75\%$ including the substrate losses.

TTFT/TFT were deposited at room temperature with a ~ 150 nm thick Cd-In-Sb-O TSO film channel layer having a carrier density of $\sim 10^{17}$ cm^{-3}, ITO (5 wt% SnO_2) source and drain electrodes (~ 200 nm) and an Al_2O_3 gate dielectric layer (~ 300 nm). Given the high Hall mobility, the TTFT/TFT enhancement mode performance was disappointing with a field effect mobility value of ~ 0.45 cm^2 V^{-1} s^{-1} and an on/off current ratio of $\sim 10^2$ at RT. However, all layers were deposited at RT on a plastic film using only RF magnetron sputtering and not optimized.

There are two major concerns when using a-Cd-In-Sb-O films. First, is the inclusion of toxic Cd, which will preclude the use of this multicomponent material in many circumstances. Second, is that the electrical properties of the deposited a-CISO TSO films were very sensitive to the pO_2 used during sputter deposition.

The maximum Hall mobility of ~ 26 cm^2 V^{-1} s^{-1} was obtained for TSO films deposited in Ar only. The carrier density dropped monotonically from $\sim 10^{20}$ to $\sim 10^{17}$ cm^{-3}, as oxygen was added (pO_2) from 0.0 to 2.0 mTorr, and at $pO_2 > 2.0$ mTorr, the carrier density was $<10^{16}$ cm^{-3}. Similarly, the Hall mobility also dropped monotonically with increasing pO_2. Thus, the a-Cd-In-Sb-O material system like many binary and ternary oxides, e.g. ZnO-based materials, suffers from pO_2 process control issues with the decrease of oxygen defects leading to channel conductivity variations.

5.10 *p*-Type TCO and TSO

The first wide band gap (transparent) *p*-type semiconducting oxide, $CuAlO_2$, with a conductivity of ~ 1 S cm^{-1}, from a hole concentration of 1.3×10^{17} cm^{-3} and a mobility of 10 cm^2 V^{-1} s^{-1}, was reported in 1997 [56]. Investigations of similar oxide materials for new *p*-type semiconductors were spurred by the success with $CuAlO_2$. Examples of some *p*-type oxides demonstrated include other delafossites, $CuFeO_2$, $AgCoO_2$, $CuYO_2$, Cu-GaO_2, $CuInO_2$ is *n*-type in natural state, spinels $NiCo_2O_4$, a-$ZnRh_2O_4$, $ZnIr_2O_4$, and $SrCu_2O_2$, and the materials with various dopants, e.g. $CuInO_2$-Ca [57]. Traditional n-type TCO material ZnO recently was made *p*-type by the codoping method with, for example, N-In, N-Al and N-Ga [58]. To deposit these different materials as *p*-TCO and *p*-TSO, a number of different processes, spray pyrolysis, PECVD, PLD and RF or DC sputtering, were used with substrate temperatures or post treatments (or both) ranging from ~ 400 to 700 °C.

5.10.1 Junction-type Devices

Of course, with useful *p*-type oxide materials demonstrated, making a *p*-*n* junction of transparent oxides became a possibility. A rectifying *p*-*n* junction diode fabricated on a glass substrate using only TCO materials with the structure, n^+-ZnO/*n*-ZnO/*p*-$SrCu_2O_2$/ITO was soon demonstrated [59, 60].

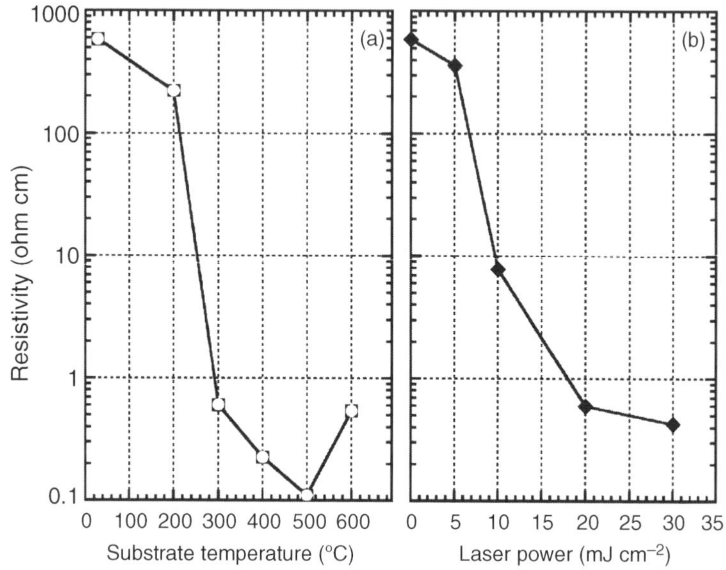

Figure 5.11 *Change in resistivity of p-$CuCrO_2$:Mg films (a) with substrate temperature and (b) with laser in situ annealing. (Reprinted with permission from [61] Copyright (2008) Elsevier Ltd)*

Several transparent oxide *p-n* junctions (and its variations) have been made to date. Most commonly PLD and sputtering processes were used but with substrate temperatures or post treatments (or both) that ranged from ~400 to 700 °C. Recently, *p*-$CuCrO_2$:Mg and *n*-type ZnO_x single layer films were investigated in preparation for making a transparent oxide *p-n* junction. Both the *p*-type and *n*-type films were made without intentional substrate heating by PLD in a vacuum chamber, followed by *in situ* laser annealing [61].

Figure 5.11 [61] shows the change in resistivity of *p*-$CuCrO_2$:Mg films with substrate temperature [Figure 5.11(a)] and with laser *in situ* annealing [Figure 5.11(b)]. The effect of the post deposition laser annealing was compared with other similar samples made on heated substrates at various temperatures. As shown both treatments have a similar initial effect of *decreasing* the resistivity as either the temperature or the laser power was increased. Films deposited at RT and without laser annealing, had an initial resistivity of ~600 ohm cm and were amorphous. Films deposited on a hot substrate or that were laser heated, activated the Mg dopant and became polycrystalline, thus lowering the resistivity. In this case, the PLD vacuum coating chamber was maintained at an oxygen pressure of only 0.27 Pa during deposition and annealing of the *p*-$CuCrO_2$:Mg films to prevent compositional changes.

Figure 5.12 shows the change in resistivity of *n*-type ZnO_x films with substrate temperature [Figure 5.12(a)] and with laser annealing [Figure 5.12(b)]. As shown, both treatments have a similar effect of *increasing* the resistivity as the temperature or the laser power is increased from ~0.005 to ~6–60 Ohm cm.

This general trend can be understood and would be expected from the standpoint of the qualitative doping process model and the ρ-well discussed earlier (Section 5.8). The PLD vacuum coating chamber during deposition and annealing of the ZnO_x files was maintained at an oxygen pressure of 1.3 Pa. In both the deposition onto a hot substrate and the laser

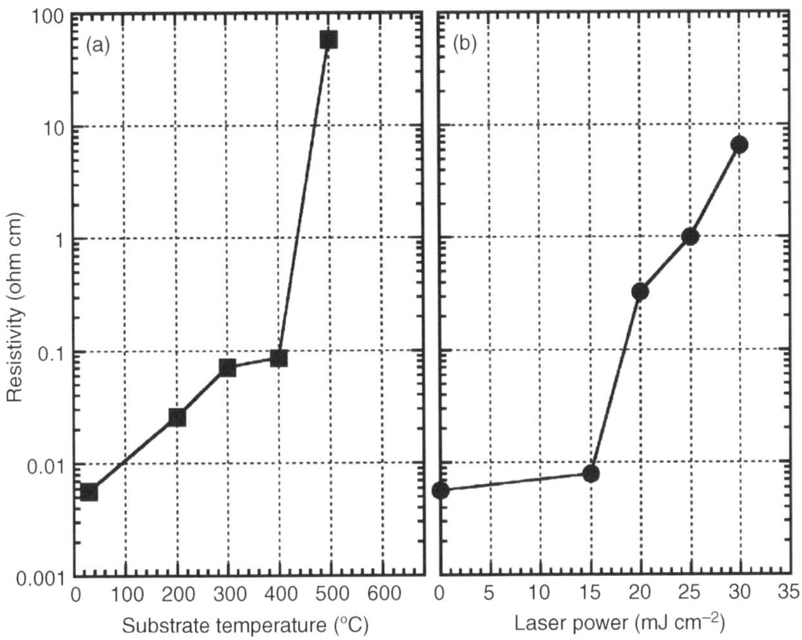

Figure 5.12 *Change in resistivity of n-type ZnO_x films (a) with substrate temperature and (b) with laser annealing. (Reprinted with permission from [61] Copyright (2008) Elsevier Ltd)*

heating processes, the composition of the ZnO_x films was becoming more oxidized. These processes are operating on the oxide side of the ρ-well increasing resistivity as oxygen vacancies (and interstitial Zn)[20] are lost.

The interest in this work is the hope that *p-n* junction devices can be made at low enough temperatures to be compatible with the temperature limitations of plastic film.

While no intentional substrate heating was used, however, thermal modeling suggested a peak surface temperature of ~440 °C in the ZnO thin film and of ~340 °C in the glass. The average temperatures are surely lower but probably still not suitable for common plastic substrates, e.g. PET or PEN. Another concern is that the processing times were too long for current practical application, e.g. the deposition time was 40 min and the annealing times were not given [61].

Previously reported work with excimer laser annealing to make polycrystalline Si from a-Si on polycarbonate and polyethersulfone plastic substrates and to crystallize low temperature sputtered ITO on PEN required heat buffering layers, e.g. SiO_2, to protect these higher temperature plastic substrates [62, 63].

To the author's knowledge no *p*-type transparent oxide coatings in a separate single layer in junction devices have been made on plastic film. The E/O performance of current *p*-type layers even when made at high temperatures is quite modest and with the current state of development useful *p*-type coatings probably cannot be deposited at RT/low temperature deposition.

[20] Zn is readily lost at >300 °C.

5.11 Key Points and Summary

Some of the key points discussed in this chapter on the challenges of depositing TCO and TSO on plastic substrates are reviewed in the bulleted listed below:

- The resistivity of ITO deposited at a low substrate temperature, e.g. on plastics, is ~4–6 times higher than with a high temperature process ($\geq$200 °C).
- The conductivity of TCO (ITO) deposited at a low substrate temperature is from oxygen vacancies not the impurity dopant.
- Deposition process (post-process) conditions, e.g. pO_2, are the most important parameters controlling TCO thin film properties.
- The major commercial ITO (TCO) deposition method is DC magnetron sputtering.
- Roll-to-Roll (R2R) vacuum deposition of ITO (TCO) on plastic film, e.g. PET, has been practiced commercially for more than three decades.
- The binary oxides, e.g. In_2O_3, SnO_2 and ZnO, are the best materials today for depositing TCO by both a high temperature process and at a low substrate temperature on plastics.
- However, none of these same materials, as a single binary oxide, makes the best TSO.
- A 'high resistivity' TCO material with low sensitivity to native defects (oxygen vacancies) is best for fabricating TSO.
- Ternary and multicomponent materials (combinations of binary oxides) are best for making TSO thin films and TTFT/TFT devices today, by both high and low temperature processes.
- Discovery of a *p*-type oxide material, $CuAlO_2$, made transparent junction oxide devices possible but they are still far from commercial quality.
- The big near term commercial opportunity for oxide TFT/TTFT devices (without a *p*-type material) appears to be in AMLCD applications.
- Few, (if any) *p*-type TCO have been deposited at low temperature on plastics; all require high temperature processing.
- Nearly all of the oxide TFT/TTFT devices reported to date, were made:
  - using more than one type of deposition process;
  - have one or more layers made by a high temperature process (or anneal);
  - and were not on a plastic substrate.

Significant progress has been made in the five years since the benchmark report of fabrication of an all amorphous oxide TTFT with TCO and TSO layers deposited on commercial plastic film by RT processing in 2004 [29]. Several reports have been made of TFT/TTFT demonstrations with materials of the a-IGZO system and with various ternary materials, particularly in the In_2O_3-ZnO (ZIO) and SnO_2-ZnO (ZTO) systems. However, most of the devices demonstrated used high temperature processing, incompatible with commercial plastic film substrate, and there is a paucity of recent reports of transparent oxide electronic devices on plastic films. While not on plastic, the demonstration of reproducible, 'commercially useful' performance, exceeding the benchmark performance [28], for a batch of a-IGZO channel TFT deposited by RF sputtering on glass at, plastic compatible, low maximum temperature is a significant step toward realization of amorphous oxide TFT [54].

No reports of R2R deposition of either TSO layers or TFT/TTFT devices were found. The benefits of R2R processing are well known and some progress toward this goal was reported [47] with the necessary but insufficient step, of lower temperature processing ($\leq$175 °C) of TFT/TTFT. Further, an announcement of plans for R2R processing for conventional backplanes [64] could bring the goal of R2R transparent oxide backplanes one step closer to commercialization.

References

[1] K. Nomura, H. Ohta, K. Ueda, T. Kamiya, M. Hirano and H. Hosono, Thin-Film Transistor Fabricated in Single-Crystalline Transparent Oxide Semiconductor, *Science*, **300**, 1269–1272 (2003).

[2] C. Bright, Conductive Antireflection Coatings on Plastic Films, *American Vacuum Society, Southern California Chapter, Vacuum Coating Workshop,* September 26–27, 1995; C. Bright, Development and Scale-up of a Multifunctional Antireflection Coating, Sputter Deposited on Plastic Film, *American Vacuum Society, 44th National Symposium, Vacuum Metallurgy Division,* October 20–24, 1997.

[3] S. Barth,CPFilms, Inc., Martinsville, VA; H. Memarian, Solar Control Window Films—A Historical and Technological Perspective, *43rd Annual Technical Conference Proceedings of the Society of Vacuum Coaters*, 329–334 (2000).

[4] R. W. Phillips and A. Argoitia, Using Vacuum Roll Coaters to Produce Anti-Counterfeiting Devices, *48th Annual Technical Conference Proceedings of the Society of Vacuum Coaters*, 145–152 (2005).

[5] R. L. Cormia, J. B. FennJr, H. Memarian, and G. Ringer, Roll-to-Roll Coating of Indium Tin Oxide—A Status Report, *41st Annual Technical Conference Proceedings of the Society of Vacuum Coaters*, 452–456 (1998).

[6] C. I. Bright, ITO and Other Transparent Conductive Coatings: *Fundamentals, Deposition, Properties, and Applications*, Society of Vacuum Coaters, Class C-304 Notes, Presented annually at the Technical Conference (1995–2009).

[7] W. A. MacDonald, Engineered Films for Display Technologies, *J. Mater. Chem.*, **14**, 4–10 (2004).

[8] C. I. Bright, Properties and Issues with ITO on Plastic Substrates, *Symposium on Functional Coatings and Surface Engineering,* Montreal, Quebec, Canada, June 5–8, 2005.

[9] N. D. Young, M. J. Trainor, S. -Y. Yoon, D. J. McCulloch, R. W. Wilks, A. Pearson, S. Godfrey, P. W. Green, S. Roosendaal and E. Hallworth, *Mater. Res. Soc. Symp. Proc.*, paper H2.1 769 (2003).

[10] Y. Leterrier, L. Medico, F. Demarco, J. -A. E. Manson, P. C. P. Bouten, J. de Goede, and G. Nisato, Mechanical Properties of Transparent Functional Thin Films for Flexible Displays, *46th Annual Technical Conference Proceedings of the Society of Vacuum Coaters*, 169–174 (2003).

[11] C. I. Bright, Issues with ITO Deposited on Temperature Sensitivity Plastic Substrates, *7th Symposium of European Vacuum Coaters*, Anzio, Italy, October 2–4, 2006.

[12] C. I. Bright, Alternative Transparent Conductive Oxides (TCO) to ITO, *51st Annual Technical Conference Proceedings of the Society of Vacuum Coaters*, 840–851 (2008).

[13] C. I. Bright, Fundamental Issues with TCO on Plastic Substrates, *1st International Symposium on Transparent Conductive Oxides, Crete, Greece, Conference Proceedings*, C6, 62 (2006).

[14] C. I. Bright, Multilayer ITO Coatings with High Transparency and Low Sheet Resistance, *Optical Interference Coatings Topical Meeting*, Optical Society of America, Banff, Alberta, Canada, July 15–20, 2001.

[15] J. R. Bellingham, W. A. Phillips, and C. J. Adkins, Electrical and optical properties of amorphous indium oxide, *J. Phys.: Condens. Matter*, **2**, 6207–6221 (1990).

[16] D. B. Fraser, and H. D. Cook, Highly Conductive Transparent Films of Sputtered $In_{2-x}Sn_xO_{3-y}$, *J. Electrochem. Soc.*, **119**, 1368–1374 (1972).

[17] J. L. Vossen, RF Sputtered Transparent Conductors. The System In_2O_3-SnO_2, *RCA Rev.*, **32**, 269 (1971).

[18] C. G. Granqvist, *Spectrally Selective Surfaces for Heating and Cooling Applications*, TT 1, SPIE 1989.

[19] G. Frank, and H. Kostlin, Electrical Properties and Defect Model of Tin-Doped Indium Oxide Layers, *Appl. Phys. A: Solids Surf.*, **27**, 197 (1982).

[20] G. B. Gonzalez, J. B. Cohen, J. -H. Hwang, T. O. Mason, J. P. Hodges, and J. D. Jorgensen, Neutron Diffraction Study on the Defect Structure of Indium–Tin–Oxide, *J. Appl. Phys.*, **89** (5), 2550–2555 (2001).

[21] B. Yaglioglu, Y. -J. Huang, H. -Y. Yeom, and D. C. Paine, A Study of Amorphous and Crystalline Phases in In_2O_3_10 wt.% ZnO Thin Films Deposited by DC Magnetron Sputtering, *Thin Solid Films*, **496**, 89–94 (2006).

[22] H. Han, and T. L. Alford, Influence of Metal Impurity Defects on the Electrical and Optical Properties of ITO Films on the PEN Substrates, *Mater. Res. Soc. Symp. Proc.*, **1012**, 1012–Y12-10 (2007).

[23] H. Aguraa, A. Suzukia, T. Matsushitaa, T. Aokia, and M. Okudab, Low Resistivity Transparent Conducting Al-Doped ZnO Films Prepared by Pulsed Laser Deposition, *Thin Solid Films*, **445**, 263–267 (2003).

[24] A. Suzuki, T. Matsushita, T. Aoki, and Y. Yoneyama, Pulsed Laser Deposition of Transparent Conducting Indium Tin Oxide Films in Magnetic Field Perpendicular to Plume, *Jpn. J. Appl. Phys.*, **40**, L401–L403 (2001).

[25] W. Waldhauser, and J. M. Lackner, Room-Temperature Industrially-Scaled Pulsed Laser Deposition of Coatings, *49th Annual Technical Conference Proceedings of the Society of Vacuum Coaters*, 50–54 (2006).

[26] W. D. Westwood, Sputter Deposition, in *AVS Education Committee Book Series, Vol. 2*, H. G. Thompkins (Ed.), AVS, New York, 2003.

[27] D. M. Mattox, *Handbook of Physical Vapor Deposition (PVD) Processing*, Noyes Publication, Park Ridge, NJ, 1998.

[28] K. Nomura, H. Ohta, A. Takagi, T. Kamiya, M. Hirano, and H. Hosono, Room-Temperature Fabrication of Transparent Flexible Thin-Film Transistors using Amorphous Oxide Semiconductors, *Nature*, **432**, 488–492 (2004).

[29] L. Wang, M. -H. Yoon, A. Facchetti, and T. J. Marks, Flexible Thin-Film Transistors Using All Transparent Component Materials, *Adv. Mater.*, **19**, 3252–3256 (2007).

[30] P. F. Carcia, R. S. McLean, M. H. Reilly, and G. NunesJr, Transparent ZnO Thin-Film Transistor Fabricated by rf Magnetron Sputtering, *Appl. Phys. Lett.*, **82**, 1117–1119 (2003).

[31] E. M. C. Fortunato, P. M. C. Barquinha, A. C. M. B. G. Pimentel, A. M. F. Goncalves, A. J. S. Marques, R. F. P. Martins, and L. M. N. Pereira, Wide-Bandgap High-Mobility ZnO Thin-Film Transistors Produced at Room Temperature, *Appl. Phys. Lett.*, **85**, 2541–2543 (2004).

[32] E. M. C. Fortunato, P. M. C. Barquinha, A. C. M. B. G. Pimentel, A. M. F. Goncalves, A. J. S. Marques, L. M. N. Pereira, and R. F. P. Martins, Fully Transparent ZnO Thin-Film Transistor Produced at Room Temperature, *Adv. Mater.*, **17**, 590–594 (2005).

[33] L. Wang, M. -H. Yoon, G. Lu, Y. Yang, A. Facchetti, and T. J. Marks, High-Performance Transparent Inorganic-Organic Hybrid Thin-Film n-type Transistors, *Nat. Mater.*, **5**, 893–900 (2006).

[34] B. -Y. Oh, M. -C. Jeong, M. -H. Ham, and J. -M. Myoung, Effects of the Channel Thickness on the Structural and Electrical Characteristics of Room-Temperature Fabricated ZnO Thin-Film Transistors, *Semicond. Sci. Technol.*, **22**, 608–612 (2007).

[35] D. C. Paine, B. Yaglioglu, Z. Beiley, and S. Lee, Amorphous IZO-Based Transparent Thin Film Transistors, *Thin Solid Films*, **516**, 5894–5898 (2008).

[36] R. L. Hoffman, B. J. Norris, and J. F. Wager, ZnO-based transparent thin-film transistors, *Appl. Phys. Lett.*, **82**, 733–735 (2003).

[37] S. Masuda, K. Kitamura, Y. Okumura, S. Miyatake, H. Tabata, and T. Kawai, Transparent Thin Film Transistors using ZnO as an Active Channel Layer and their Electrical Properties, *J. Appl. Phys.*, **93**, 1624–1630 (2003).

[38] H. Hara, T. Hanada, T. Shiro, and T. Yatabe, Properties of Indium Zinc Oxide Thin Films on Heat Withstanding Plastic Substrate, *J. Vac. Sci. Technol. A*, **22**, 4 (2004).
[39] B. Yaglioglu, H. Y. Yeom, and D. C. Paine, Crystallization of Amorphous In_2O_3_10 wt % ZnO Thin Films Annealed in Air, *Appl. Phys. Lett.*, **86**, 261 908 (2005).
[40] B. Yaglioglu, H. Y. Yeom, R. Beresford, and D. C. Paine, High-Mobility Amorphous In_2O_3_10 wt % ZnO Thin Film Transistors IZO, *Appl. Phys. Lett.*, **89**, 062 103 (2006).
[41] P. Barquinha, A. Pimentel, A. Marques, L. Pereira, R. Martins, and E. Fortunato, Influence of the Semiconductor Thickness on the Electrical Properties of Transparent TFTs Based on Indium Zinc Oxide, *J. Non-Cryst. Solids*, **352**, 1749–1752 (2006).
[42] P. Barquinha, A. Pimentel, A. Marques, L. Pereira, R. Martins, and E. Fortunato, Effect of UV and Visible Light Radiation on the Electrical Performances of Transparent TFTs Based on Amorphous Indium Zinc Oxide, *J. Non-Cryst. Solids*, **352**, 1756–1760 (2006).
[43] P. Barquinha, G. Gonçalves, L. Pereira, R. Martins, and E. Fortunato, Effect of Annealing Temperature on the Properties of IZO Films and IZO Based Transparent TFTs, *Thin Solid Films*, **515**, 8450–8454 (2007).
[44] M. W. J. Prins, K. -O. Grosse-Holz, G. Muller, J. F. M. Cillessen, J. B. Giesbers, R. P. Weening, and R. M. Wolf, A Ferroelectric Transparent Thin-Film Transistor, 3650-3652, *Appl. Phys. Lett.*, **68**, 25 (1996).
[45] R. E. Presley, C. L. Munsee, C. -H. Park, D. Hong, J. F. Wager, and D. A. Keszler, Tin Oxide Transparent Thin-Film Transistors, *J. Phys. D: Appl. Phys.*, **37**, 2810–2813 (2004).
[46] C. I. Bright, Alternative Transparent Conductive Oxides (TCO) to ITO, Society of Vacuum Coaters, Class C-321 Notes, *Presented annually at the Technical Conference* (2008–2009).
[47] T. Emery, R. Hoffman, B. Yeh, and T. Koch, Low-Temperature ZIO TFTs: Progress Toward Roll-to-Roll Manufacturing, *Mater. Res. Soc. Fall Symp.* (2008).
[48] N. L. Dehuff, E. S. Kettenring, D. Hong, H. Q. Chiang, J. F. Wager, R. L. Hoffman, C. -H. Park, and D. A. Keszler, Transparent Thin-Film Transistors with Zinc Indium Oxide Channel Layer, *J. Appl. Phys.*, **97**, 064 505 (2005).
[49] W. B. Jackson, R. L. Hoffman, G. S. Herman, C. Taussig, S. Braymen, F. Jeffery, and J. Hauschildt, Zinc Tin Oxide Transistors on Flexible Substrates, *J. Non-Cryst. Solids*, **352**, 1753–1755 (2006).
[50] W. B. Jackson, R. L. Hoffman, and G. S. Herman, High-Performance Flexible Zinc Tin Oxide Field-Effect Transistors, *Appl. Phys. Lett.*, **87**, 193 503 (2005).
[51] M. Oritay, H. Ohta, and M. Hirano, Amorphous Transparent Conductive Oxide $InGaO_3(ZnO)_m$ $m \leq 4$: a Zn 4s Conductor, *Phil. Mag. B*, **81**, 501–515 (2001).
[52] A. Takagi, K. Nomura, H. Ohta, H. Yanagia, T. Kamiya, M. Hirano, and H. Hosono, Carrier Transport and Electronic Structure in Amorphous Oxide Semiconductor, a-$InGaZnO_4$, *Thin Solid Films*, **486**, 38–41 (2005).
[53] K. Nomura, T. Kamiya, H. Ohta, K. Ueda, M. Hirano, and H. Hosono, Carrier Transport in Transparent Oxide Semiconductor with Intrinsic Structural Randomness Probed using Single-Crystalline $InGaO_3(ZnO)_5$ Films, *Appl. Phys. Lett.*, **85**, 11, 1993–1995 (2004).
[54] H. Yabuta, M. Sano, K. Abe, T. Aiba, T. Den, H. Kumomi, K. Nomura, T. Kamiya, and H. Hosono, High-Mobility Thin-Film Transistor with Amorphous $InGaZnO_4$ Channel Fabricated by Room Temperature rf-Magnetron Sputtering, *Appl. Phys. Lett.*, **89**, 112 123 (2006).
[55] H. Tetsuka, Y. -J. Shan, K. Tezuka, and H. Imoto, Transparent Amorphous Conductive Cd–In–Sb–O Thin Films for Flexible Devices, *Vacuum*, **80**, 1038–1041 (2006).
[56] H. Kawazoe, M. Yasukawa, H. Hyodo, M. Kurita, H. Yanagi, and H. Hosono, P-Type Electrical Conduction in Transparent Thin Films of $CuAlO_2$, *Nature (London)*, **389**, 939–942 (1997).
[57] H. Yanagi, T. Hase, S. Ibuki, K. Ueda, and H. Hosono, Bipolarity in Electrical Conduction of Transparent Oxide Semiconductor $CuInO_2$ with Delafossite Structure, *Appl. Phys. Lett.*, **78**, 1583–1585 (2001).
[58] M. Kumar, T. -H. Kim, S. -S. Kim, and B. -T. Lee, Growth of Epitaxial p-Type ZnO Thin Films by Codoping of Ga and N, *Appl. Phys. Lett.*, **89**, 112 103 (2006).

[59] A. Kudo, H. Yangi, H. Hosono, H. Kawazoe, and Y. Yano, Fabrication of Transparent p–n Heterojunction Thin Film Diodes Based Entirely on Oxide Semiconductors, *Appl. Phys. Lett.*, **75**, 18 2851–2853 (1999).

[60] H. Kawazoe, H. Yanagi, K. Ueda, and H. Hosono, Transparent p-Type Conducting Oxides: Design and Fabrication of p-n Heterojunctions, *MRS Bull.*, **25**, 28–36 (2000).

[61] T. -W. Chiu, K. Tonooka, and N. Kikuchi, Fabrication of ZnO and $CuCrO_2$:Mg Thin Films by Pulsed Laser Deposition with in situ Laser Annealing and its Application to Oxide Diodes, *Thin Solid Films*, **516**, 5941–5947 (2008).

[62] Y. -H. Kim, S. -K. Park, D. -G. Moon, W. -K. Kim, and J. -I. Han, Excimer Laser Crystallization of Sputter Deposited a-Si Films on Flexible Substrates, *Mater. Res. Soc. Symp. Proc.*, **814**, I7.5 (2004).

[63] W. Chung, P. Wickboldt, D. Toet, and P. G. Carey, Laser-Crystallized High Quality ITO on Plastic Substrates for Flexible Displays, *Mater. Res. Soc. Symp. Proc.*, **769**, H10.1 (2003).

[64] C. Taussig, B. Cobene, R. Elder, W. Jackson, M. Jam, A. Jeans, H. Luo, P. Mei, C. Perlov, F. Jeffrey, M. Almanza-Workman, K. Beacom, S. Braymen, B. Garcia, J. Hauschildt, H. -J. Kim, O. Kwon, and D. Larson, Progress on R2R Manufactured Backplanes using Self-Aligned Imprint Lithography (SAIL), *Proceedings of the 7th Annual Flexible Electronics and Displays Conference*, 14.1 (2008).

6

Oxide Semiconductors: From Materials to Devices

Elvira Fortunato, Pedro Barquinha, Gonçalo Gonçalves, Luís Pereira and Rodrigo Martins

CENIMAT/I3N, Materials Science Department, Faculdade de Ciências e Tecnologia, Universidade Nova de Lisboa, Portugal

6.1 Introduction

We are currently experiencing a fascinating scientific period in the area of thin film transistors (TFTs) using nonconventional materials, like oxide conductors and semiconductors. Conductive and semiconductive transparent oxides are a special class of materials, because transparency and conductivity are somewhat contradictory. There are two major conditions to be met for a material to be transparent to visible light: the band gap should be above 3 eV and the carrier concentration (free electrons or holes) should be below 20^{21} cm^{-3}. The band gap is related to how easily the material absorbs electromagnetic waves at certain frequencies. Carrier concentration, however, is related to the plasma frequency, which determines the boundary between reflected electromagnetic frequencies and those which are passed (transparency) or absorbed.

In general a wide band gap material means a low carrier concentration, which makes the material an insulator, like glass. ZnO is an exception because while it is highly transparent it is also a good conductor. Usually dopants are used to supply the carriers needed to make many oxides electrical conductors. ITO, for example, is doped with Sn to achieve carrier concentration of the order of 10^{21} cm^{-3}.

Transparent Electronics: From Synthesis to Applications Edited by Antonio Facchetti and Tobin J. Marks

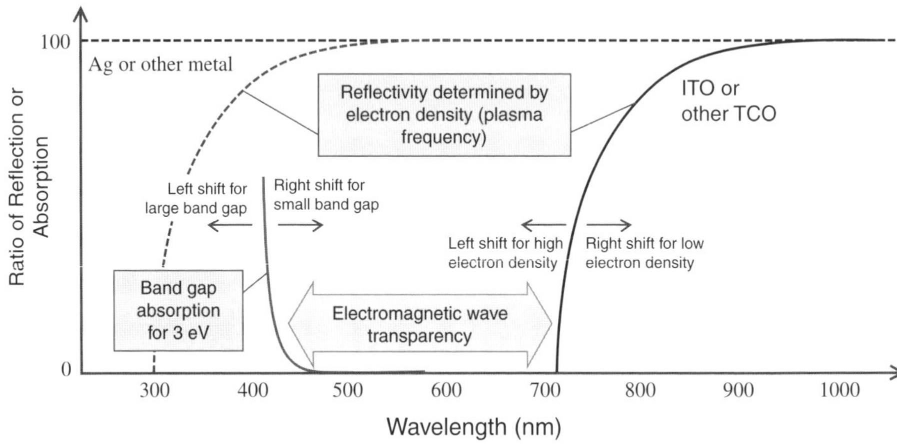

Figure 6.1 *Conditions for transparent materials: for a semiconductor, dielectric or other material to be transparent we must follow two conditions: (1) the band gap must be wider than 3 eV; and (2) the carrier concentration must be below 10^{21} cm^{-3}. If both conditions are met simultaneously, visible light will not be reflected or absorbed*

Amorphous $InGaZnO_4$ (a-GIZO) was created with the objective to increase the carrier mobility (the ease with electrons move) and decrease the carrier concentration, making possible the production of high performance TFTs.

There are two major conditions to be met for a material to be transparent to visible light: the band gap should be 3 eV and the carrier concentration (free electrons or holes) should be below 20^{21} cm^{-3} (see Figure 6.1).

In this chapter, we start with the history of TFTs along with the major developments achieved over nearly the last 80 years, focusing particularly on the different type of semiconductor materials used. After that we will present some results obtained in our laboratory on passive and active applications especially for amorphous IZO and GIZO films, respectively. We will finish this chapter with some emerging devices based on cellulose, namely the paper transistor and the nonvolatile memory paper transistor. We hope with this chapter to motivate new researchers to initiate work in this exciting field as well as to contribute one more step to the progress of Transparent Electronics.

6.2 Historical Background: From Field Effect Transistors (FETs) to TFTs

6.2.1 The Field Effect Invention

The history of TFTs is very interesting because their invention was around 30 years before the production of the first TFT. Even though the FET did not come into widespread use until the 1960s, its invention predated both the junction and point contact transistors by many years. As is normal with many innovations at that time, its practical realization was delayed until adequate materials and technologies were available for its fabrication. We can even say

(a)

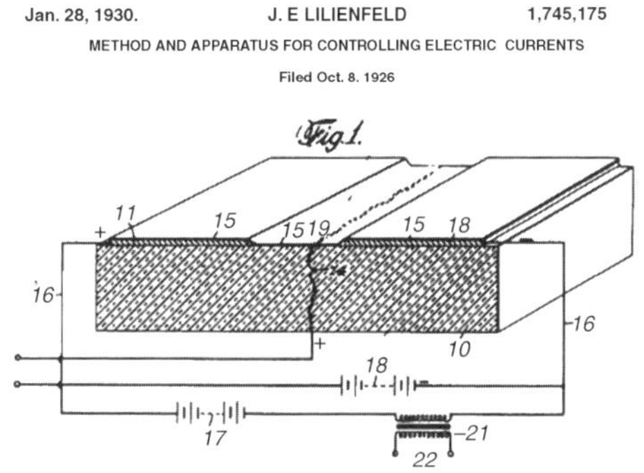

(b)

Figure 6.2 *(a) Photograph of Julius Edgar Lilienfeld and (b) Lilienfeld's proposal for a FET*

(and for most of us this could be a surprise) that the TFT was the first solid-state amplifier ever patented. The basic principle of the FET (what we now call a JFET) was proposed first by Julius Edgar Lilienfeld[1] in the USA [Figure 6.2(a)] as early as 1925 and patented in 1930 [1] [Figure 6.2(b)]. His dream was to control the conductivity of a semiconductor by an external electric field. The FET he described was probably the first successful solid-state amplifier invented.

The advantages of this device over the vacuum tubes, voltage controlled devices able to sustain high currents, the only alternative high frequency amplifier available at that time, are also described in this patent and in later ones. Lilienfeld's later patent from 1933 [2] described a new FET (now called a MISFET) were he specified the thickness of the dielectric layer, which insulates the metal control electrode from the copper sulfide channel, to about 10^{-5} cm (1000 Å). This is, in fact, the typical thickness used in the insulated-gate transistors which were developed many years later. However, since his experiments were conducted in near obscurity he was unable to draw serious attention to them; there are no published research articles about his devices and as a result they were ignored by industry.

Around 1935 in England, one of the first patents was issued to the German inventor Oskar Heil[2] for a field effect crystal amplifier (Figure 6.3) [3]. Basically its concept used a control

[1] Julius Edgar Lilienfeld (18 April 1881–28 August 1963) was an Austrian-Hungarian physicist. He was born in Lemberg in Austria-Hungary (now called Lviv in Ukraine). From 1900 to 1904 he studied at the Friedrich-Wilhelms-Universität in Berlin. In 1905 he started to work at the physics institute at the University of Leipzig. Lilienfeld attained habilitation in 1910. Among other things, he invented the FET in 1925 and the electrolytic capacitor in the 1920s. He filed several patents describing the construction and operation of transistors as well as many features of modern transistors. When Brattain, Bardeen and Shockley tried to get a patent on their device, most of their claims were rejected due to Lilienfeld's patents (from http://en.wikipedia.org/wiki/Julius_Edgar_Lilienfeld).

[2] Oskar Heil (1908, in Langwieden/Rhineland-Palatinate–1994) was a German electrical engineer and inventor. He studied physics, chemistry, mathematics, and music at the Georg-August University of Göttingen and was awarded his PhD in 1933, for his work on molecular spectroscopy. Oskar Heil worked on microwave vacuum tubes in Germany during the Second World War. He and his wife wrote a famous paper in 1935 about microwave vacuum tubes. This paper continues to be cited in the 21st century (from http://en.wikipedia.org/wiki/Oskar_Heil).

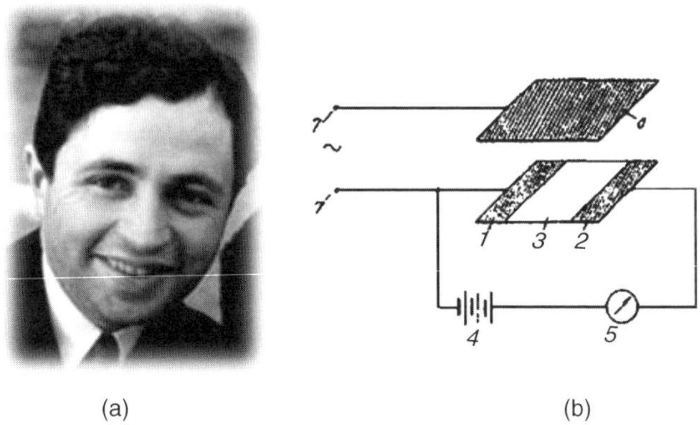

Figure 6.3 (a) Photograph of Oskar Heil and (b) part of the schematics from Heil's British patent of 1935

electrode to regulate the current flow through a thin semiconductor layer made with several types of materials like copper oxide, vanadium pentoxide, tellurium and iodine. In reality the Heil proposal for such a kind of crystal amplifier was the predecessor of the insulated-gate FET (metal oxide semiconductor FET, MOSFET), since the control electrode was isolated from the semiconductor substrate. We can say nowadays that Heil was the inventor of the MOS technology.

It is ironic that the concept of the FET, so marvellously simple and elegant (see Figure 6.4), only provides practical implementations after the invention of the far more complex bipolar transistor in 1947 [4].

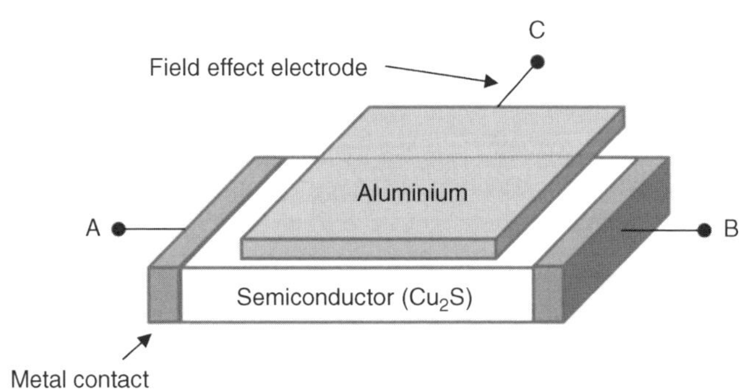

Figure 6.4 Idealization of the transistor structure proposed by Lilenfield and Heil, independently. The device worked on the principle that a voltage applied to the metallic plate modulated the conductance of the underlying semiconductor, which in turn modulated the current flowing between ohmic contacts A and B. This phenomenon, where the conductivity is modulated by an electric field applied normal to the surface of the semiconductor, has been named the field effect

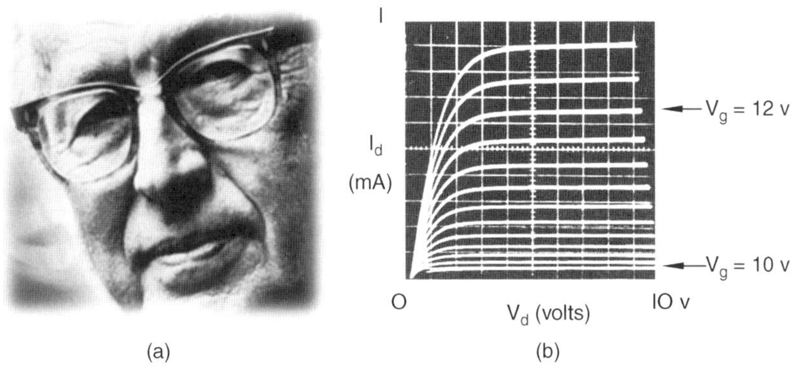

Figure 6.5 *(a) Photograph of Paul K. Weimer and (b) typical drain characteristics of a Weimer's TFT based on CdSe with an electron mobility of 200 cm^2 V^{-1} s^{-1} . Reproduced from http://en.wikipedia.org/wiki/Julius_Edgar_Lilienfeld*

6.2.2 The First Working TFT

The first functional working TFT was demonstrated by Weimer[3] in 1962 [Figure 6.5(a)] [5]. He used thin films of polycrystalline cadmium sulfide, similar to those developed for photodetectors, deposited onto glass substrates. In one pump down of his vacuum system, he would deposit a gold source and drain, then deposit polycrystalline semiconductor material over that and, finally, place a gold gate on top. This was a coplanar process that was similar to what he used in the tricolour vidicon. (It should be noted that Weimer's work depended on depositions, not photoresists, which were used in developing integrated circuits, just then getting under way.)

At first, these deposited transistors were not very good. However, he then placed an insulator between the gate and the semiconductor material and got what he called 'beautiful characteristics' [see Figure 6.5(b)]. His 1962 paper, 'The TFT – a new thin-film transistor', drew worldwide attention [5].

At first, Weimer used cadmium sulphide as the semiconductor material because it was a high-resistivity semiconductor with which he was somewhat familiar. He later used cadmium selenide. This made for an even better FET, not as good as silicon, Weimer recalled, but quite respectable.

Other TFT semiconductor materials like Te, InSb and Ge were investigated, but in the mid 1960s with the emergence of the MOSFET based on the crystalline silicon technology and the possibility to perform integrated circuits, led to a decline in TFT development activity by the end of the 1960s [6].

6.2.3 The (R)evolution of TFTs: Amorphous Silicon Thin Film

What dramatically changed the prospects for TFTs in the 1970s was the realization that with crystalline silicon, low cost was inseparable from miniaturization, whereas some

[3] Paul K. Weimer (5 November 1914–6 January 2005) was a noted contributor to the development of television and the TFT. Dr Weimer was born in Wabash, IN. He received a BA in mathematics and physics from Manchester College (IN) in 1936, an MA in physics from the University of Kansas in 1938, and a PhD in physics from Ohio State University in 1942. He then joined the RCA laboratory in Princeton, NJ, where he worked until retirement in 1981 (from http://en.wikipedia.org/wiki/Paul_K._Weimer).

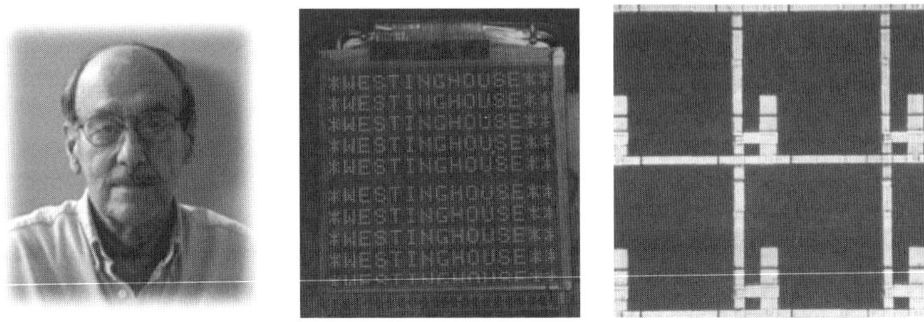

(a) (b) (c)

Figure 6.6 *(a) Photograph of Peter Brody, (b) operation of the 6″ × 6″ 20 lpi EL display and (c) detail of the pixel layout in the active matrix (Adapted from [8] Copyright (1975) IEEE)*

applications required large arrays of low cost electronics, like for example displays. By this time many researchers and engineers were engaged in improving the characteristics of liquid crystal displays (LCDs), which had been recently discovered.

The first so-called active matrix LCDs (AMLCDs) was successfully demonstrated by Peter Brody in 1973 [7], where a CdSe TFT was used as a switching element for each pixel of a 120 × 120 matrix (Figure 6.6).

Unfortunately for him and for the US display industry, Westinghouse ignored Brody's innovations. 'Westinghouse was very short-sighted,' Brody said. 'We were too far ahead.' However, Japanese manufacturers took note of Brody's papers and presentations, and in the early 1980s began heavily investing in development. When Westinghouse terminated his project in 1979, Brody formed a company called Panelvision and by 1983 was producing TFT LCD panels, but the company could not compete with the Japanese companies. Panelvision was sold to Litton Industries in 1985. US manufacturers were unwilling to invest in LCD development and TFT-LCD technology moved to the Far East. Today, the multibillionaire flat-panel LCD industry is dominated by Samsung, Sharp and LG Philips.

Thirty-plus years later, Brody is the president and CEO of a stealth-mode startup developing organic LEDs, which one day could replace TFT LCDs. Ironically, he is renting space in the same laboratory where he first developed the innovation that revolutionized display technology and paved the way for ubiquitous computing.

In spite of the many successful demonstrations of CdSe TFT LCDs, the industry did not enter this market. CdSe was not compatible with standard processing in the microelectronic industry, which uses mainly silicon as the semiconductor material. Advanced photolithographic and etching processes have been developed over the years for silicon devices and this technology was not readily applicable to CdSe TFTs.

In parallel to the early development of liquid crystal cell technology and CdSe TFTs, thin film amorphous silicon was investigated in the 1970s [9, 10]. The rationale behind this interest was initially not its potential for LCDs, but rather its promise for low-cost solar cells. A major development occurred at the University of Dundee in 1979 [11], when LeComber, Spear and Ghaith described a TFT using a-Si:H as the active semiconductor material, and

(a) (b)

Figure 6.7 *Photographs of (a) Peter LeComber and (b) Walter Spear*

suggested the active matrix LCD as one of its applications (Figure 6.7). Interestingly, a patent on the basic a-Si TFTs was never filed, since this work was performed at an academic institution. The University of Dundee had been a pioneer in the development and understanding of amorphous silicon materials and devices, but it was Japanese companies that reaped the profits.

After LeComber reported the first a-Si:H TFT, many laboratories started the development of AMLCDs formed on glass substrates. Although this result attracted much attention, the major disadvantage of a-Si:H TFT is its low electron mobility that limits the ultimate speed of devices. However, an adequate device speed for the switching applications in the LCD has been achieved.

Since the mid 1980s, silicon-based TFTs have become the most important devices for AMLCDs, and have successfully dominated the large area LCD product market [12]. In the mid 1990s Korean companies started mass production AMLCD modules followed a few years later by massive investments by several companies in Taiwan. For example, there is an expected increase of 219% between 2008 and 2012 in the worldwide demand for LCD displays, which will reach 14 million units in 2012 [13]. The success of AMLCD technology is the result of many years cooperation between scientists and engineers with different backgrounds, including organic chemists, physicists, electrical, electronic, mechanical, packaging, and manufacturing engineers, all supported by the increasing revenues from sales of LCDs. Figure 6.8 illustrates the exponential increase in the total market for AMLDCs. Based on an extrapolation of this plot, the AMLCD industry will achieve nearly $170 billion in annual revenue around the year 2015.

6.2.4 Looking for Higher Mobilities: Polycrystalline Silicon TFTs

As we indicated earlier, amorphous silicon TFTs are the dominant active matrix technology, and most of the production investments have been made in amorphous silicon. The main advantage is the low temperature process (that do not require the crystallization and doping processes) and the possibility to deposit over large areas (up to

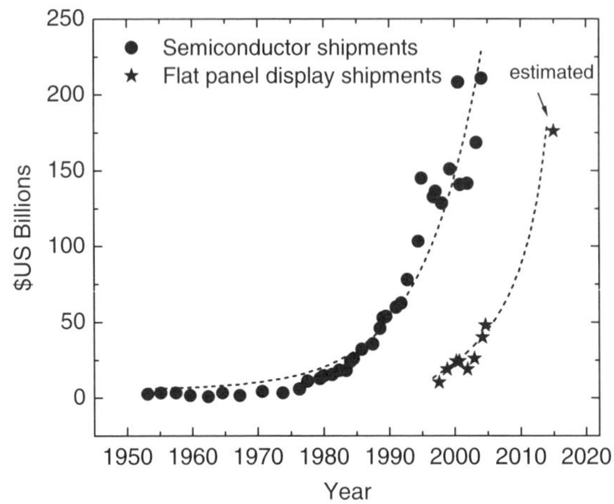

Figure 6.8 *Comparison between the semiconductor and flat panel display market (data adapted from [12])*

Gen 8). However, as the pixel density of displays increases, it is increasingly difficult to achieve reliability from all the connections necessary with the display drivers (for example, even a monochrome display with 480 × 640 pixels requires over a thousand connections). Much current development work is directed at integrating some, or all, of the driver circuitry onto the display panel. This requires switching speeds that are difficult to achieve with amorphous silicon TFTs.

In the late 1990s several companies succeeded in producing low-temperature poly-Si TFTs processed at temperatures below 600 °C, compatible with lower cost glass substrates, after the successful results obtained by Philips researchers [14] reporting off-currents of 0.1 pA, an on/off ratio of 10^{10} and electron mobility of $35\,cm^2\,V^{-1}\,s^{-1}$.

Various techniques have been developed to create high-mobility polycrystalline silicon. Most commonly, physical vapour deposition (PVD) or low pressure chemical vapour deposition (LPCVD) amorphous silicon films are used as the layer to be crystallized. The key processes in fabricating low temperature TFTs are the crystallization methods that convert amorphous silicon into polycrystalline silicon. These methods are based on non-laser crystallization and laser annealing. Besides non-laser crystallization, the simplest method is solid-phase crystallization (SPC), nevertheless SPC requires annealing at 600 °C for tens of hours, which makes it unsuitable for use on large area glass substrates [15]. Other non-laser methods employ metal seeds for crystallization, which may result in a large leakage current [16]. Among the laser methods available, excimer-laser annealing (ELA) has been the most widely used because of the resulting excellent crystallinity, fast crystallization speed, and high mobility [17]. In addition, ELA is already being employed in mass production. Finally, all low temperature poly-Si TFTs (LTPS TFTs), including the ELA technique, suffer from nonuniformity issues because of the existence of grain boundaries, which require the use of a complicated compensation unit pixel circuit such as a 5 transistor + 2 capacitor pixel circuit, leading to a loss in device yield.

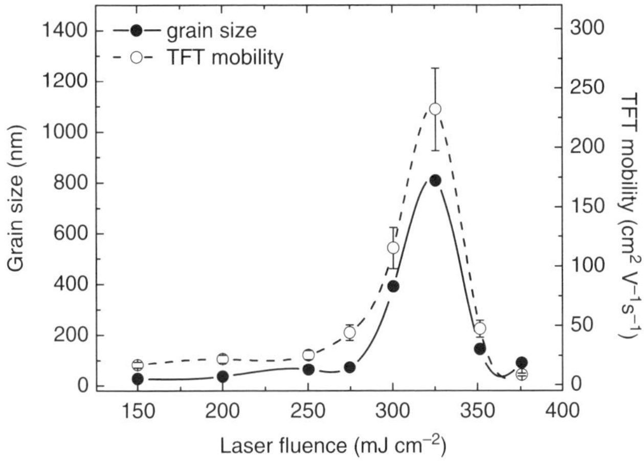

Figure 6.9 *Grain size dependence and corresponding poly-Si TFT mobility vs. laser fluence for the conventional Excimer Laser Crystallization (ELC) process (adapted from [18])*

By varying the parameters in the crystallization process it is possible to control the polycrystalline grain size, as can be observed in Figure 6.9. The large grain size corresponds to higher electron mobilities, resulting in better TFT performance.

Poly-Si can be used to make both *p*-channel and *n*-channel TFTs. On account of its relatively high mobility, both row and column drivers can be integrated on the glass, even D/A converters, DC/DC converters and (micro)processors can be integrated too, which significantly cuts the cost from external drivers and chips from other devices. However the off current poly-Si is much higher than a-Si, i.e. the off state is not stable because of the charge on the pixel capacitor cannot be maintained. In order to decrease the off current, a dual gate structure and a lightly doped drain (LDD) were proposed. Both methods can effectively lower the off current. Table 6.1 compares some of the pertinent properties of various switches, along with display applications in which they are common and

Table 6.1 *Different types of switches for AMLCDs and their main applications*

Switching device	Mobility ($cm^2 V^{-1} s^{-1}$)	Highest processing temperature (°C)	Major applications
a-Si TFT	0.3–1	~300 (glass)	Notebooks, flat panel monitors, LCD TVs
High-T poly-Si TFT	100–300	~100 (quartz)	Projection light valves, viewfinders
Low-T poly-Si TFT	10–200	~500 (glass)	PDAs, notebooks, projection light valves, viewfinders
Crystalline Si MOSFET	400	~1100 (c-Si)	Projection light valves, viewfinders

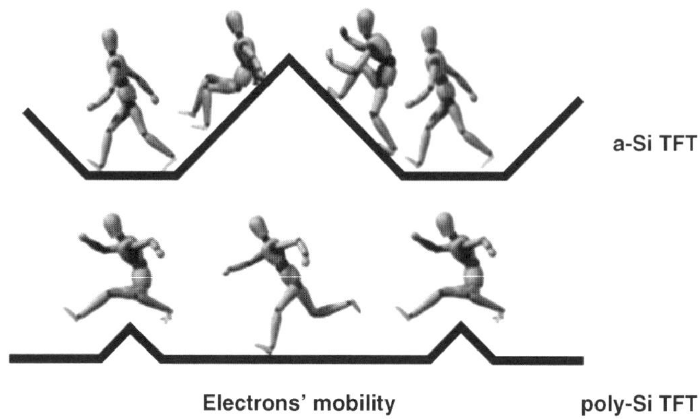

Figure 6.10 Comparison of the electron mobility between amorphous silicon and polycrystalline silicon TFTs

Figure 6.10 compares the electron mobility in amorphous silicon and polycrystalline silicon TFTs.

Modern displays have already reached the boundary of human perception. In 1988, IBM released the VGA standard, with resolutions at or under 60 PPI (pixels per inch). By the late 1990s, however, display technology had undergone a revolution: graphics hardware was a dedicated device instead of being a subcomponent of the main processor. With more memory and processing resources available, displays began to rapidly offer more resolution, with the industry passing 100 PPI in 2000 and exceeding 200 PPI by 2003 (Figure 6.11). Devices claiming over 300 PPI began to appear in 2005. The world of display technology will undergo a revolution in the early 21st century. The advent of true 3D vision is not far off, and its roots will be in our understanding of human vision today.

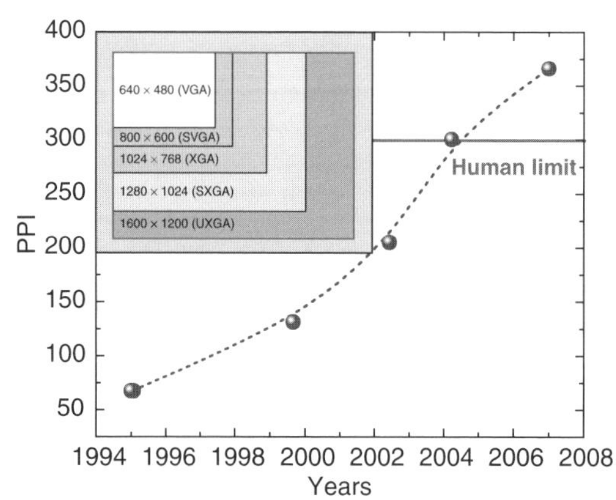

Figure 6.11 Evolution of the display PPI

6.2.5 The Organic Era

The 1990s marked the debut of a new class of TFTs, based upon an organic semiconductor active layer material [19], with electron mobilities similar to that of a-Si:H. This new class of TFTs are very promising candidates for integration onto flexible plastic substrates for a future generation of rugged, lightweight displays than can be rolled up like a map. Organic field effect transistors (OFETs) were first described in 1985 [20]. OFETs based on solution processing polymers, as well as small molecular semiconductors, have seen impressive improvements in their performance [21].

OFETs, which use thin films of organic semiconductors, are interesting because of their capability to make large-area flexible circuits on plastic substrates. In addition, their low processing temperatures and the wide variety of organic semiconductors enable them to be compatible with various flexible substrates.

Functional devices on flexible plastics, paper, and cloth have been demonstrated [22]. In addition, organic materials may be printed via ink-jet or roll-to-roll printing which can greatly reduce fabrication cost [23]. Due to these promising applications and the potential for low cost circuits, the number of research papers pertaining to OFETs is increasing rapidly, and new conferences that focus on OFETs have emerged [24]. Currently, research is simultaneously performed on all levels, from device physics to the design of functional circuits like decoders and AMLCD backplanes, but the focus is shifting from device fabrication towards circuit applications especially for low-cost and large-area electronic products. Figure 6.12 shows the evolution of the mobilities of organic FETs over the last 15 years, where improvements by more than five orders of magnitude have been obtained, for the particular case of pentacene [25].

In a similar way to what happens to poly-Si TFTs by increasing the grain size, an increase in the mobility is observed (Figure 6.13), since grain size in pentacene significantly influences the carrier transport.

Many pentacene TFTs with field effect mobilities of more than 1 $cm^2\ V^{-1}\ s^{-1}$ (comparable with those of amorphous silicon TFTs) have been achieved. This may indicate that

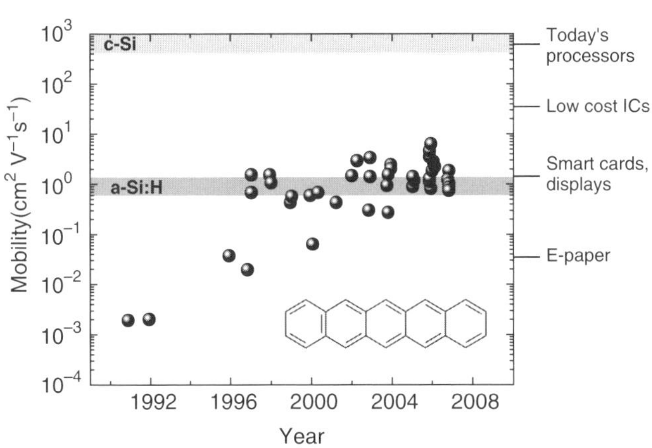

Figure 6.12 *Evolution of the field effect mobility of pentacene FETs with high mobilities (Adapted from [25] Copyright (2008) Institute of Physics Publishing)*

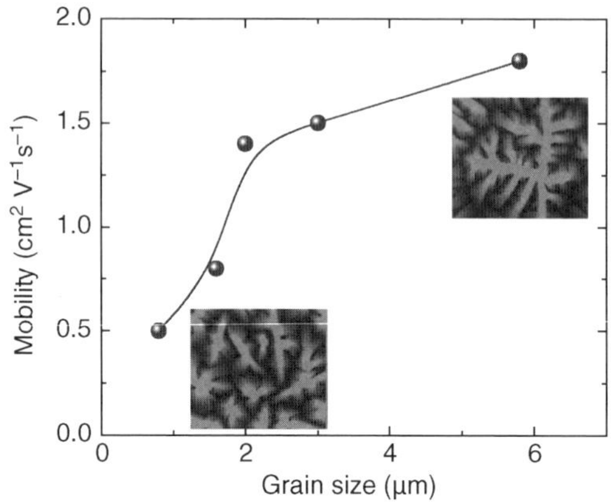

Figure 6.13 Relationship between mobility of pentacene FETs and grain size (Adapted from [25] Copyright (2008) Institute of Physics Publishing)

amorphous silicon TFTs can be technologically replaced by pentacene TFTs for some specific applications like large-area, mechanically flexible, lightweight and low-cost devices.

Pentacene TFTs have exhibited high mobilities of more than 1 $cm^2\ V^{-1}\ s^{-1}$, despite their polycrystalline films. This indicates that the intrinsic mobility is higher than those estimated from the characteristics of the TFTs and further improvement in the mobility of the TFT is possible. Obtaining pentacene films consisting of grains without defects leads to improvement in the mobility, which is very important for device applications, since the output current and cut-off frequency are proportional to the mobility.

6.2.6 The Future Generation of TFTs: Metal Oxide Semiconductors

A new generation of oxide semiconductors are being studied and applied as the active material for TFTs. The first TFT based on using an oxide material as a channel layer was reported in 1964 using evaporated SnO_2 as the semiconductor and a bottom gate staggered configuration, while the first ZnO based TFT was proposed in 1968. Although these initial attempts were done during the 1960s, it was 30 years later that Hosono's group [26], Wager [27] and Fortunato *et al.* [28] created a significant worldwide interest in active matrix organic light emitting diode (AMOLED) technology, both in industry and academia. The main advantages of this new technology are the high mobility, excellent uniformity in device parameters, and amorphous structure which lead to good scalability over large substrate sizes, associated with a low or even room temperature deposition process.

Several oxide based TFTs have been reported in the last few years. The first reports from 2003 had transistors based on polycrystalline ZnO [29–34]. In 2004 Fortunato *et al.* [28] and Nomura *et al.* [35] presented the first fully transparent TFTs produced at room temperature, using ZnO and amorphous Ga-In-Zn oxide, respectively. In the latter semiconductor the conduction mechanism is different from that of conventional semiconductors like, for example, silicon. In covalent semiconductors there are strongly directed sp^3 bonds in contrast to amorphous oxide semiconductors (AOSs) whose conduction band minima are

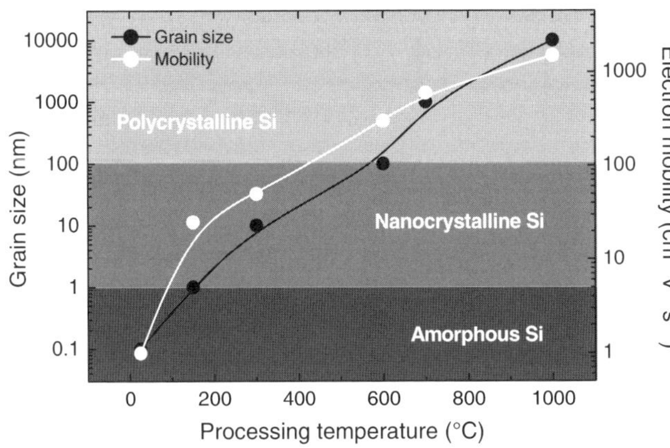

Figure 6.14 *Comparison of the grain size and the electron mobility as a function of the processing temperature for silicon*

derived from isotropic and spatially expanded s orbitals, presenting structural disorder but without any significant affect on the electrical properties of these semiconductors (Figure 6.14). This result was a significant milestone in the field of oxide based TFTs with the introduction of a new concept, that of AOSs.

AOSs are very attractive for TFT applications, from a manufacturing point of view, because they combine simultaneously the advantages of amorphous silicon and poly-Si based TFTs. They present stability and uniformity over large areas, as schematically shown in Figure 6.15.

Concerning the different oxide semiconductors there is a wide variety, ranging from binary oxides to quaternary oxides, as well as the different types of deposition techniques. These reports employ a variety of deposition techniques for fabrication, including radio frequency (RF) sputtering, ion beam sputtering, and pulsed laser deposition (PLD). Non-vacuum techniques, including spin coating and chemical bath deposition, have been also employed to

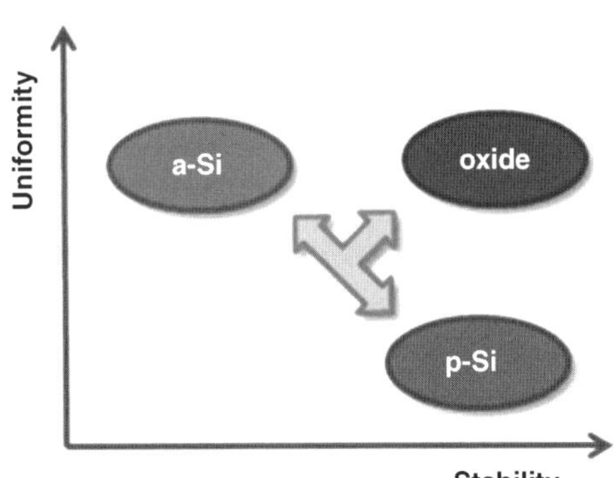

Figure 6.15 *Comparison of the stability and uniformity for different TFT technologies. a-Si, amorphous silicon; p-Si, polycrystalliine Si*

Table 6.2 *Survey of oxide semiconductors applied to TFTs as well as the technology used for the channel layer*

Oxide semiconductor	Technology	Reference
ZnO	Sputtering	[31]
	Pulsed laser deposition	[29]
	Solution based	[36]
SnO_2	Sputtering	[37]
	Evaporation	[38]
	VLS process	[39]
In_2O_3	Sputtering	[40]
TiO_2	MOCVD	[41]
	Solution based	[42]
Zn-Sn-O (ZTO)	Solution based	[43, 44]
	Pulsed laser deposition	[45]
	Sputtering	[46]
Zn-In-O (ZIO)	Sputtering	[47–50]
In-Zn-O (IZO)	Sputtering	[51, 52]
	Solution based	[53]
In-Ga-O (IGO)	Sputtering	[54]
Ga-In-Zn-O (GIZO)	Sputtering	[35, 55–58]
	Solution based	[59]
Ga-Sn-Zn-O (GSZO)	Sputtering	[60]

Note: VLS, vapour-liquid-solid; MOCVD, metal organic chemical vapour deposition.

fabricate TFTs. Table 6.2 shows some of the semiconductors (binary, ternary and quaternary materials) most used as channel layers in TFTs as well as the associated technology.

Some companies are working actively on the development of transparent electronics, mainly for display applications. Canon demonstrated in 2006 that a high performance transistor (mobility, $>10\,cm^2\,V^{-1}\,s^{-1}$; gate swing, $0.2\,V\,decade^{-1}$) can be achieved using RF sputtering and by using large-area deposition rather than PLD [61]. Major display producers such as LG and Samsung began performing research and development on oxide TFTs for AMOLEDs in 2006 [62, 63]. The first AMOLED display was released by LG Electronics in 2007 [64]. The fabricated InGaZnO transistor with a top-gate structure exhibited good device performance. This prototype of a full colour 3.5 in. QCIF+ AMOLED demonstrated the possibility of being used as a backplane for an OLED device. At SID's Display Week 2008, Samsung SDI showcased a full colour 12.1 in. AMOLED prototype (Table 6.3) that used InGaZnO TFTs; this is the world's largest AMOLED panel of any oxide TFT driven

Table 6.3 *Display prototypes with an a-GIZO active matrix*

LG	Toppan	Samsung SAIT	Samsung SID	Samsung Electronics
3.5″ QCIF AM-OLED 2T1C	4″ QVGA AM-e-paper 1T1C	4″ QVGA AM-OLED 2T1C	12″ WXGA AM-OLED 2T1C	15″ XGA AM-LCD 1T1C
SID 07 [64]	IDW 06 [68]	SID 08 [69]	SID 08 [70]	SID 08 [71]

OLED display. The WXGA high resolution (1280 × RGB × 765) is compatible with TFT-LCDs that are currently commercially available for notebook PCs. In addition, even though a 2 transistors + 1 capacitor pixel circuit was implemented a 'randomness-free' high-quality display was demonstrated, due to the excellent short-range uniformity of the threshold voltage of oxide TFTs. More recently, at IEDM 2008 Samsung presented the largest (40″) and best performance AMOLED display, which applied amorphous oxide TFTs with a 300% improvement in screen functionality. This new technology could slash the cost of OLED manufacturing and change the face of display production. Y. Kim, Managing Director of Samsung, said: 'In mid and long term, we will be able to expand its application to semiconductor peripheral circuits'. He added that it would become the key technology to semiconductor devices for the future [65]. The wide broad range of applications is not limited to TFTs but also applies to other semiconductor devices like the high speed ring oscillator (RO), memories resistance RAM and inverters [66]. The reported RO has a delay of 0.94 ns per stage, which is 75 times faster than previously reported. The Resistive Random Access Memory (RRAM) has the potential to make the footprint smaller than that of a 3D RRAM with a readout circuit arranged around the memory.

From a manufacturing point of view these semiconductors are very attractive, because they combine simultaneously the advantages of a-Si and poly-Si based TFTs (Figure 6.15). They can be produced at room low or even low temperatures (compatible with low cost polymeric substrates or even cellulose paper [67]) presenting very smooth surfaces without grain boundaries (an advantage for process integration). In addition, this new technology based on amorphous multicomponent oxides is 100 times more effective than the existing a-Si TFTs.

Table 6.3 shows some examples of recent display prototypes using an a-GIZO active matrix.

6.3 Transparent Oxide Semiconductors

Transparent Electronics is growing very fast and is one of the most advanced topics for a wide range of device applications [72] where the key components are wide band gap semiconductors.

As active material, they exploit the use of truly electronic semiconductors where the main emphasis is on transparent TFTs for display applications [73].

Two primary technologies which represent and underlie Transparent Electronics are TCOs transparent conductive oxides (TCOs) and TFTs. During the past two decades the techniques for oxide production have been dramatically enhanced. In addition, oxides have a variety of elements and structures providing great potential for integrating a diverse range of functions. Nevertheless, the application of TCOs has been restricted to 'transparent metals', notwithstanding the fact that TCOs are high conductive *n*-type degenerate semiconductors.

TCOs constitute an unusual class of materials possessing two physical properties – high optical transparency and high electrical conductivity – that are generally considered to be incompatible. This particular combination of physical properties is only achievable if a material has a sufficientlylarge energy band gap so that it is nonabsorbing or transparent to

Table 6.4 Electrical properties of common TCOs reported for the best polycrystalline films (Adapted from [72] Copyright (2008) Springer)

Materials	Bandgap (eV)	Conductivity ($S\,cm^{-1}$)	Electron concentration (cm^{-3})	Mobility ($cm^2\,V^{-1}\,s^{-1}$)
In_2O_3	3.75	10 000	$>10^{21}$	35
ZnO	3.35	8000	$>10^{21}$	20
SnO_2	3.60	5000	$>10^{20}$	15

visible light (>3 eV), and also possesses a high enough concentration of electrical carriers ($>10^{19}\,cm^{-3}$) with moderate mobility ($\sim 1\,cm^2\,V^{-1}\,s^{-1}$), that the material can be considered to be a 'good' conductor of electricity. The three most common TCOs used today are based on In_2O_3, SnO_2 and ZnO thin films. Table 6.4 shows the electrical properties of TCOs based on these materials (Figure 6.16).

The first large scale use of TCOs was during the Second World War, involving transparent heaters for de-icing applications in aircraft windshields. Since then TCOs have been used in a wide range of applications, including: automobile, airplane and marine window defrosters; LCDs, electrochromic, electroluminescence and plasma displays; solar cell electrodes; infrared reflectors for energy-efficient windows; transparent barrier layers for food, cigarette and other types of packaging; heated glass freezer doors; heating stages for optical microscopes; photoconductors for television camera vidicons; electromagnetic shielding; touch screens; abrasion- and corrosion-resistant coatings; and gas sensors [75].

In all these applications, TCOs are used electrically in a passive manner as conductors or resistors and typically these applications require high conductivities. The materials of choice are typically indium based, like indium tin oxide (ITO). Nevertheless, as volume production increases, industry is more and more keen to find an alternative material to the

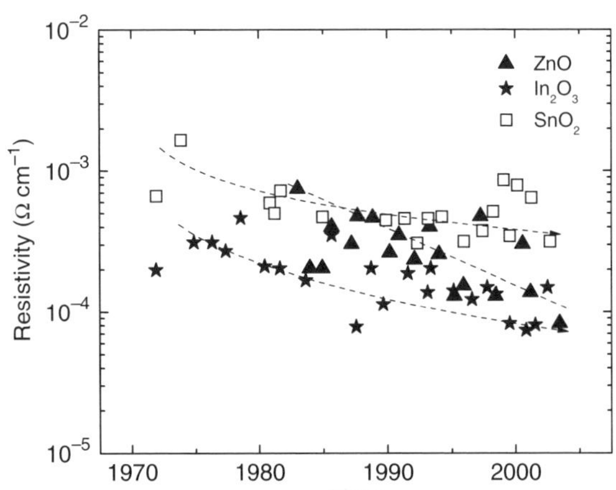

Figure 6.16 Electrical resistivity dependence over the last 30 years for TCOs based on doped In_2O_3, SnO_2 and ZnO (adapted from [74])

expensive indium, as well as to look for lower resistance materials and low temperature deposition methods (for low cost flexible substrates).

Nevertheless, TCO films based on zinc oxide are receiving a lot of attention because of the advantages over the more commonly used TCO, like low cost, resource availability (about a factor of 1000 more abundant than indium), nontoxicity and high thermal/chemical stability [76]. Non-doped zinc oxide usually presents a high resistivity due to a low carrier concentration. Al, In and Ga have been reported as an effective dopant for zinc oxide based films. Most of the work related to zinc oxide uses aluminium as a dopant. Nevertheless, aluminium presents a very high reactivity leading to oxidation during the film's growth, which may become a problem. Gallium is less reactive and more resistant to oxidation compared with aluminium [77, 78]. Recently it was demonstrated by the present authors that doping with gallium led to films with low resistivity associated with a high transmittance in the visible region [79–83], even when processed at room temperature. Several techniques have also been used like MOCVD, evaporation, magnetron sputtering, sol gel and plasma assisted molecular beam epitaxy [84]. However, most of these techniques need to use moderate temperatures to obtain low values of resistivity. In addition to the several advantages presented by RF magnetron sputtering, it is also possible to produce highly conductive and transparent GZO without heating the substrate, since additional energy is delivered from the plasma to the growing film – a characteristic of a plasma assisted process. Highly conducting and transparent GZO films have been produced by the present authors using RF sputtering at room temperature, $2.6 \times 10^{-4}\,\Omega\,\mathrm{cm}$ with growth rates above $280\,\text{Å}\,\mathrm{min}^{-1}$ [85].

In contrast to conventional *passive* TCO applications, the key issue of Transparent Electronics is to use TCOs (or other transparent conductors) in what we call *active* electronic applications. For this particular application when used for example as a channel layer in a TFT, the carrier should be as low as possible in order to permit the modulation of the channel current as a function of gate voltage: the well known *field effect*.

6.3.1 Passive Applications: Amorphous TCOs (a-IZO)

Amorphous oxide semiconductors are becoming one of the most promising semiconductor materials, for instance in *passive* and *active* electronic applications, due to the superior electrical performances and better uniformity over large areas, when compared with the conventional primary polycrystalline materials already mentioned (In_2O_3, SnO_2 and ZnO). In polycrystalline material, grain boundaries play a major role in determining a variety of performance characteristics in active electronic devices. Amorphous materials provide a means to eliminate the effects of these boundaries, especially for devices covering large areas. This is one of the reasons why amorphous silicon is massively used in active matrix flat panel displays. Amorphous oxides have been known for more than half a century for several transition metal oxides, with particular focus on V_2O_5 [86]. On account of the presence of the transition metals, these films are quite coloured and the conductivity is quite low, and the mobility is of the order of $10^{-4}\,\mathrm{cm^2\,V^{-1}\,s^{-1}}$, preventing their use in transparent electronics.

Recently Fortunato *et al.* [87] reported results on amorphous transparent conductors for *passive* electronic applications, based on IZO (In_2O_3:ZnO) with resistivities of $10^{-4}\,\Omega\,\mathrm{cm}^{-1}$ and Hall mobilities of $60\,\mathrm{cm^2\,V^{-1}\,s^{-1}}$ with the advantage of being produced at room

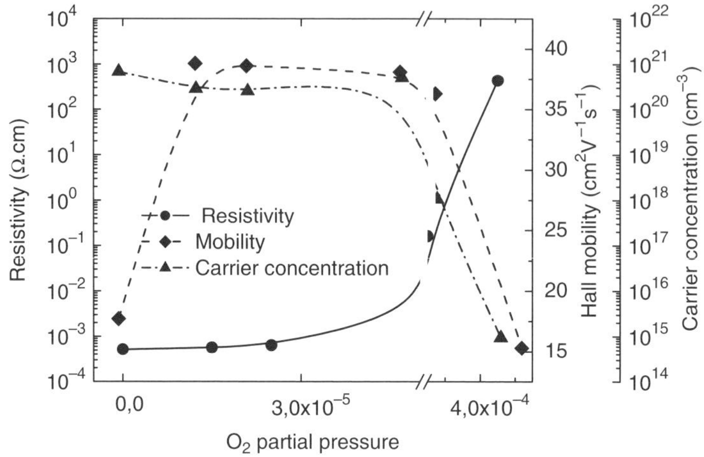

Figure 6.17 Dependence of the resistivity, Hall mobility and carrier concentration on the oxygen partial pressure during deposition (the target composition is In_2O_3:ZnO in the proportion of 75:25 wt%)

temperature by RF magnetron sputtering. Figure 6.17 shows how the electrical properties of the films (resistivity, mobility and carrier concentration) are affected by the oxygen partial pressure. The lower resistivities are achieved for films produced with an oxygen partial pressure lower than 2×10^{-5}. For higher oxygen partial pressures, we observe an increase in the resistivity mainly due to the annihilation of oxygen vacancies that are responsible for the conducting mechanism of the films.

From the optical plots (Figure 6.18) we can observe an increase in the transmittance on the near infrared region of the films as a result of the increase in carrier concentration, in accordance with the Burstein-Moss effect.

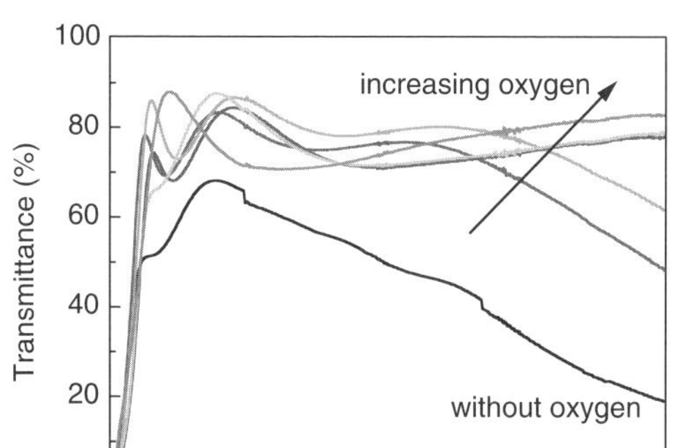

Figure 6.18 Optical transmittance as a function of the wavelength for IZO films deposited at different oxygen partial pressures

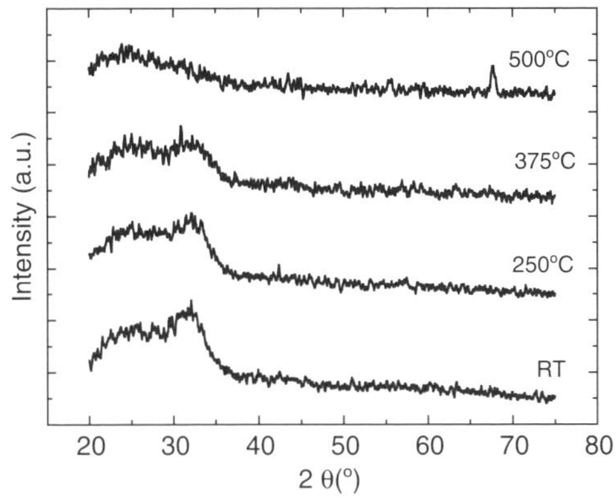

Figure 6.19 *X-ray diffraction analysis as a function of the annealing temperature in air for a-IZO films. Reprinted with permission from [90] Copyright (2008) Elsevier Ltd*

These materials present peculiar electrical properties when compared with traditional covalent semiconductors. It is possible to obtain high mobilities in amorphous structures due to their unique transport mechanism, primarily related to the large ns orbitals that compose the conduction band and to their overlap. This orbital overlap is nearly insensitive to the degree of disorder of the material [35, 88–90], which is not the case, for example, for silicon (Figure 6.14).

Since these are relatively new materials, is important to study the evolution of their properties when they are subjected to moderate-to-high temperatures and different atmospheres [91, 92], both from a material research perspective and also for the possible integration of these materials in electronic devices. This will be a further step in the direction of introducing TTFT technology, or more broadly, the amorphous oxide technology, into industrial processes and real world applications [93].

Figure 6.19 shows the diffraction patterns for a-IZO films annealed at different temperatures. For temperatures around 500 °C an important structural changes occurs, with a peak at $2\theta = 68^\circ$ and appears superimposed on the broad peak of the glass substrate. This suggests that crystallization occurs at this temperature, and consequently, the carrier concentration and the mobility decrease. This is corroborated by the high resolution transmission electron microscopy (TEM) analysis performed. For temperatures between room temperature (RT) and 375 °C, no changes are verified in the X-ray diffraction (XRD) data.

These results are also confirmed with electrical measurements. We can observe in Figure 6.20 that as the annealing temperature increases we start to observe deterioration in the macroscopic properties, namely the resistivity. Concerning the microscopic properties, it is clear that two regimes are observed: (I) up to ~400 °C; and (II) between 400 °C and 500 °C. From the Hall effect measurements we observe an increase in the mobility due to a rearrangment of the amorphous structure associated with stress release and/or chemical composition for regime I, while a drastic decrease is observed for regime II. This behaviour is in line with the carrier concentration, corroborated by the fact that the annealing was done

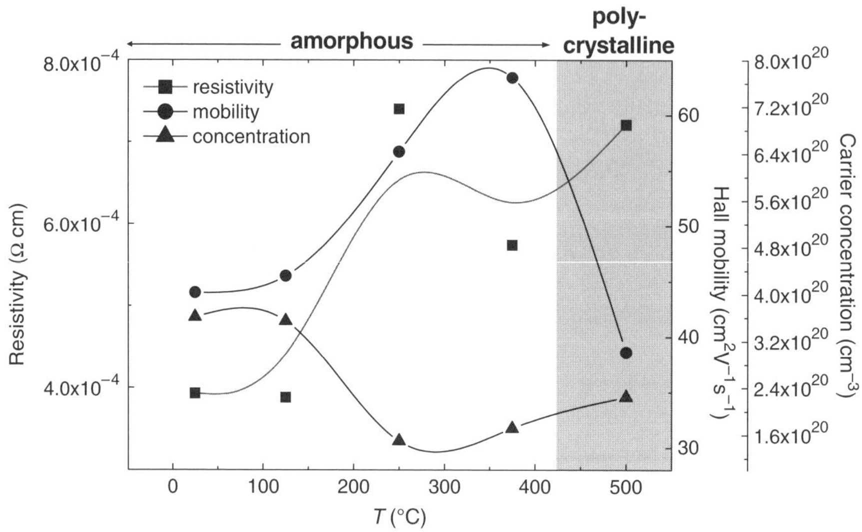

Figure 6.20 *Dependence of the resistivity, Hall mobility and carrier concentration on the annealing temperature*

in air, promoting the annihilation of oxygen vacancies associated with a decrease in the structural defects.

From the high resolution TEM observation it is observed that the crystallization of the IZO layer occurs near the SiO_2 interface (Figure 6.21). The crystalline part of the IZO layer comprises the first 50 nm and it is not continuous. However, the rest of the layer is amorphous. Several crystalline grains can be observed near the interface between the IZO layer and the SiO_2.

Figure 6.22 shows a photograph of an alphanumeric display using as TCO an a-IZO thin film. This work is part of the European Project Multiflexioxides (NMP3-CT-2006-032231) [94].

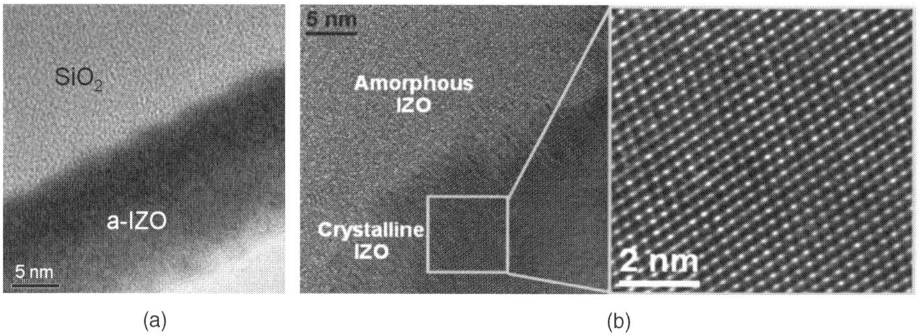

Figure 6.21 *High resolution cross-section TEM image for an a-IZO film (a) as produced (at room temperature) and (b) annealed at 500 °C in air*

Figure 6.22 Photograph of an alphanumeric display with commercial LEDs

6.3.2 Active Applications: Amorphous Oxide TFTs (a-IZO and a-GIZO)

A detailed study concerning the IZO [52, 93, 95–97] and GIZO [56, 98, 99] films to be used as active layers in TFTs was made with simple TFT structures using standard plasma enhanced chemical vapour deposition (PECVD) SiO_2 deposited on Si wafers. The results for these devices are presented in the following sections of this chapter. However, isolated films were also characterized, with some of the obtained results presented below. Figure 6.23 (a) and (b) show sthe resistivity (ρ) measured by the four point probe method on IZO 1:1 samples, with different p_{O_2} and p_{dep}, respectively, as a function of the annealing temperature (T_A). All the annealing steps were made in air for 1 h.

Besides the expected increase of ρ as p_{O_2} increases for nonannealed samples, related to the existence of more oxygen vacancies hence more free carriers for lower p_{O_2}, it can be seen that as T_A increases ρ tends to similar values for all the tested p_{O_2} [Figure 6.23(a)]. A similar

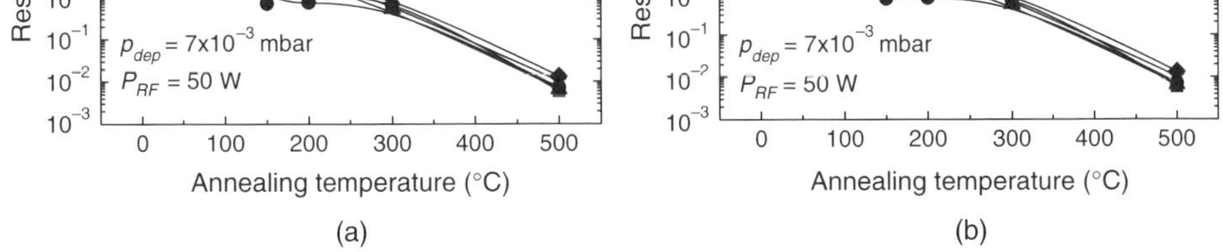

Figure 6.23 Variation of resistivity as a function of annealing temperature for IZO 1:1 films deposited with different (a) p_{o_2} and (b) p_{dep}

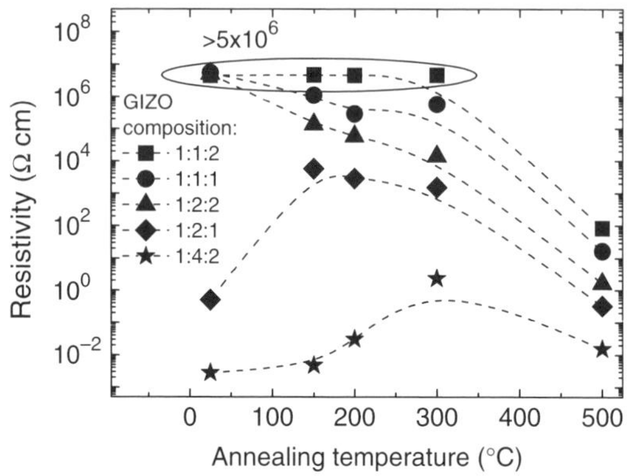

Figure 6.24 *Variation of resistivity as a function of annealing temperature for GIZO films deposited with different target compositions*

effect is verified for the evolution of p_{dep} [Figure 6.23(b)], with ρ increasing with T_A for the initially conductive films and decreasing for the initially more resistive films. The effect of T_A for GIZO films produced with ceramic targets with different compositions is shown in Figure 6.24.

Note that the sample geometry and measuring equipment used here only allowed values of ρ up to $5 \times 10^6\ \Omega\,\mathrm{cm}$ to be measured, with the more resistive films being expected to surpass this value. Even with this limitation it is clear that the same convergence of properties with T_A verified for IZO is also verified for GIZO, with the compositions 1:2:1 and 1:2:2 being the most adequate for TFT production, since they allow moderate ρ values (10^3–$10^5\ \Omega\,\mathrm{cm}$) to be obtained in the region of interest of T_A (150–300 °C). In addition, these compositions have small variations in ρ between 150 °C and 300 °C, contrary to what happens for instance for 1:4:2 compositions, where ρ changes around three orders of magnitude for the same T_A range. This will possibly lead to higher stability in the properties of TFTs with 1:2:1 and 1:2:2 compositions, which in fact is verified in actual devices.

In this section the influence of several processing parameters of the semiconductor layer on the electrical properties of TFTs produced with a simple structure (i.e. Si/SiO_2 substrates and patterning/deposition of active layer and Au source/drain electrodes only) is discussed. This study allowed the selection of the proper semiconductor deposition conditions for the production of the transistors.

6.3.2.1 Influence of Oxide Semiconductor Target Composition

Figure 6.24(a) shows the ceramic target compositions tested for the production of TFTs on the Ga_2O_3-In_2O_3-ZnO system. For each composition, the field effect mobility (μ_{FE}) and turn on voltage (V_{on}) values are shown in Figure 6.25(a) while the transfer characteristics used to extract those values are presented in Figure 6.25(b) and (c). All the TFTs were produced with $P_{RF} = 50\,\mathrm{W}$, $p_{dep} = 7 \times 10^{-3}$ mbar, $p_{O_2} = 1.5 \times 10^{-5}$ mbar, $d_s = 40$ nm and were annealed in air at 150 °C (the effect of the annealing treatment is discussed later). Note that

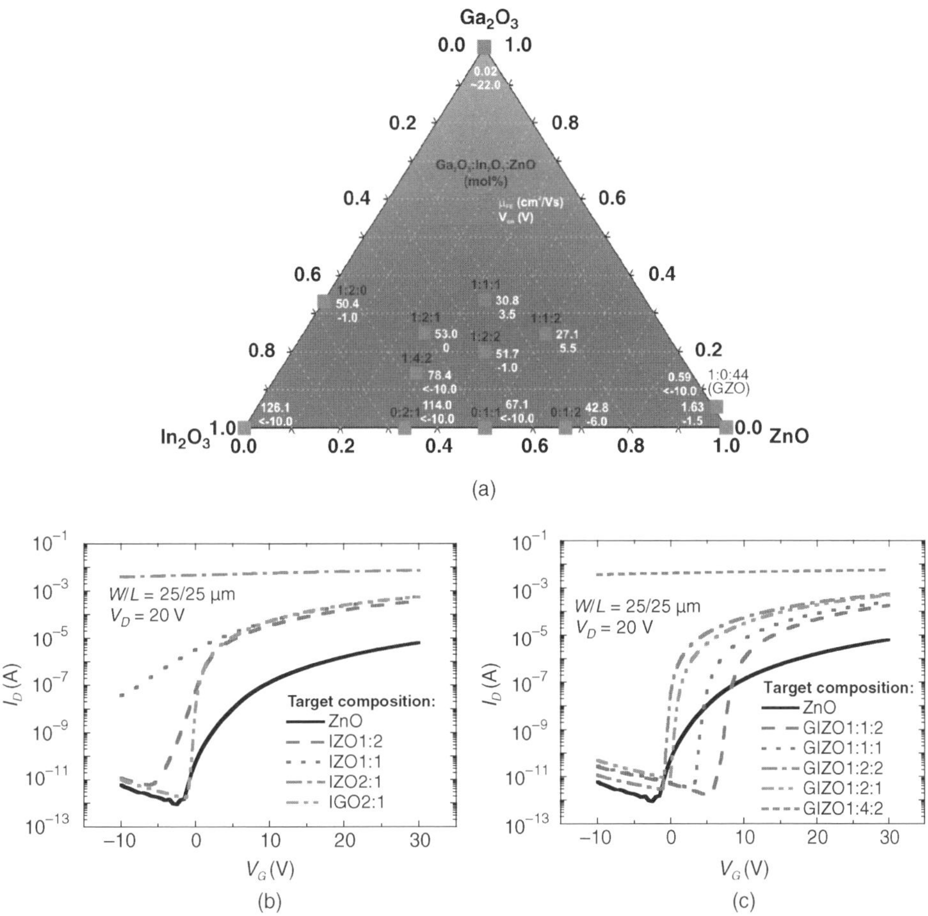

Figure 6.25 *Influence of oxide semiconductor target composition on the electrical properties of TFTs: (a) ternary diagram on the Ga_2O_3-In_2O_3-ZnO system, showing the studied compositions and the μ_{FE} and V_{on} obtained for each composition; (b) transfer characteristics for the ternary oxides; (c) transfer characteristics for the quaternary oxides*

these conditions were chosen since they reflect the best overall conditions for most of the devices presented herein, but for certain compositions enhanced properties can be obtained using other deposition conditions.

Several differences and trends can be verified analysing Figure 6.25:

1. ZnO, a binary oxide, leads to considerably worse properties than multicomponent oxides. This is justified by the polycrystalline structure of the binary oxide, where the grain boundaries make the movement of electrons difficult since they work as depletion regions with high potential barriers. For the multicomponent oxides, an amorphous structure is obtained, and charge transport is mostly limited by potential barriers associated with the structural randomness and shallow traps close to the bottom of the conduction band, which in materials with controlled free carrier density (N around

10^{15}–10^{17} cm^{-3}) can generally be surpassed by accumulating a higher charge density in the semiconductor/dielectric interface by increasing V_G.

2. μ_{FE} increases for In_2O_3 richer compositions, which is related to the higher N obtained for those cases. Note that in the multicomponent oxides μ generally increases with N, since as N is increased the Fermi level moves closer and actually penetrates the conduction band, where carriers are not prone to be affected by the shallow states located below the conduction band. However, as the initial (i.e. in the unbiased state of the device) N increases above a certain level the channel conductivity modulation is lost, leading to always-on devices. This is the case of the 0:2:1 and 1:4:2 compositions, for which $N \approx 10^{19}$ cm^{-3}.
3. The introduction of Ga_2O_3 into the IZO structure leads to a decrease in μ_{FE}, due to the strong bonds that gallium forms with oxygen, suppressing the generation of N. As an example, 0:1:1 and 1:2:2 compositions, which have similar $In_2O_3/(In_2O_3 + ZnO)$ ratios, can be compared: in spite of higher μ_{FE} being obtained for the former, only for the latter is it possible to suppress excessive N and thus be able to properly control channel conductivity modulation, achieving well defined on and off states. However, above a certain Ga_2O_3 content, dependent on the $In_2O_3/(In_2O_3 + ZnO)$ ratio, active layers with high ρ and low N are obtained, which are insufficient to compensate all the traps present on the material and its interfaces. This translates to degradation in TFT properties: compare, for instance, the trend obtained going from 1:2:2 to 1:1:1, where a significant decrease in μ_{FE} and increase in V_{on} are verified.

From all the analysed compositions, it seems that the ceramic target with composition close to 1:2:2 or 1:2:1 leads to the best overall properties, allowing a high μ_{FE}, close to 0 V_{on} and low S to be obtained.

6.3.2.2 *Influence of Oxide Semiconductor Oxygen Partial Pressure (p_{O_2})*

Oxygen content is expected to be the main parameter controlling the electrical properties obtained in the active layer of the TFTs. In fact, as already shown for the material characterization, oxygen vacancies are the main source of free electrons in oxide semiconductors. Figure 6.26 shows the influence of the p_{O_2} used to produce GIZO 1:2:2 layers on the electrical properties of TFTs. These devices were all produced with $P_{RF} = 50$ W, $p_{dep} = 7 \times 10^{-3}$ mbar and $d_s = 40$ nm and were annealed in air at $T_A = 150$ °C.

It is clearly visible that V_{on} (V_T) and the on/off ratio increase with p_{O_2} [Figure 6.26(a) and (b)], which is justified by the higher ρ and lower N obtained with increasing p_{O_2}. Essentially, as p_{O_2} increases (N decreases), higher V_G is necessary to accumulate enough free carriers to establish a conductive path between source and drain electrodes. The lower N as p_{O_2} increases also justifies the significant decrease in μ_{FE} for higher p_{O_2} [Figure 6.26(c)], as discussed above. Nevertheless, for high p_{O_2} another factor can also be important to justify the verified degradation in properties, which is the high substrate bombardment by highly energetic negative ions that can damage the dielectric's surface and also the growing semiconductor film, increasing its number of defects and decreasing its compactness. This is reflected in the higher S [Figure 6.26(c)] and in the trend verified for the μ_{FE} evolution as a function of V_G [Figure 6.27(a)], being noticeable not only as a decrease in the peak μ_{FE} value as p_{O_2} increases, but also as a trend for μ_{FE} saturation and even decrease for high V_G. Note that as predicted by the conventional field

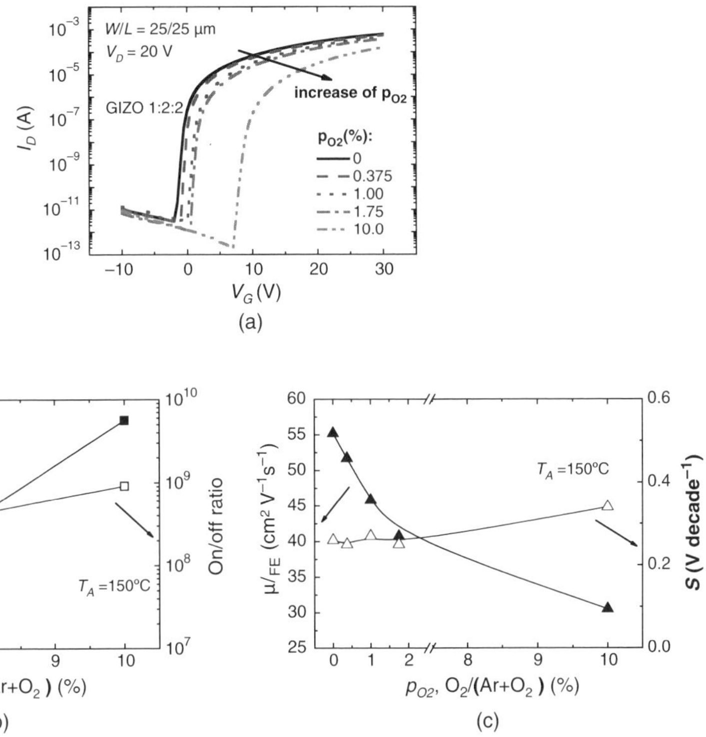

Figure 6.26 *Influence of p_{O_2} used for active layer production on the electrical properties of TFTs: (a) transfer characteristics; (b) variation of V_{on} and on/off ratio; (c) variation of μ_{FE} and S*

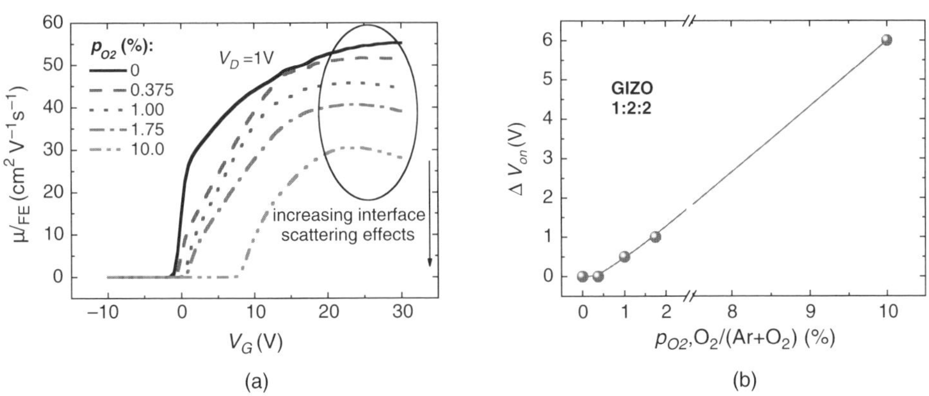

Figure 6.27 *(a) Variation of μ_{FE} with V_G for different p_{O_2}; (b) ΔV_{on} as a function of p_{O_2} after four consecutive transfer characteristics*

effect theory, electrons move closer to the dielectric/semiconductor interface as V_G increases, so if there is a large number of defects at that interface μ_{FE} is negatively affected.

Stability is also negatively affected as p_{O_2} increases [Figure 6.27(b)]: in fact, by performing a simple stability test consisting of measuring four consecutive transfer characteristics in one device, the V_{on} shift (ΔV_{on}) can be as high as 6 V for $p_{O_2} = 10.0\%$, while $\Delta V_{on} = 0$ V can be obtained for $p_{O_2} = 0$ or 0.375%. The high ΔV_{on} should be related to the high defect density and with the low *N*, which may be too low to compensate for all the defects.

The overall best conditions, in terms of performance and stability, seem to be $p_{O_2} = 0$ and 0.375%. However, generally the latter is preferable since devices produced without oxygen sometimes present highly negative V_{on} or even always-on properties, especially if no post thermal annealing process is performed.

6.3.2.3 *Influence of Annealing Temperature (T_A)*

Devices with the active layer processed under different conditions were annealed at temperatures ranging from 150 to 300 °C. Higher annealing temperatures, although important from a scientific viewpoint, were not tested here, since one of the main purposes was to produce the transistors entirely at low temperatures. Figure 6.28 shows some of the results obtained for devices produced with $P_{RF} = 50$ W, $p_{dep} = 7 \times 10^{-3}$ mbar and $p_{O_2} = 1.5 \times 10^{-5}$ mbar, using different target compositions and d_s. For the different situations presented it can be seen that at least until $T_A = 200$ °C the properties are improved, which is related to the following factors:

- ρ adjustment to an equilibrium value, being increased or decreased depending on whether the as-produced films are low (IZO) or highly (ZnO) resistive, respectively.

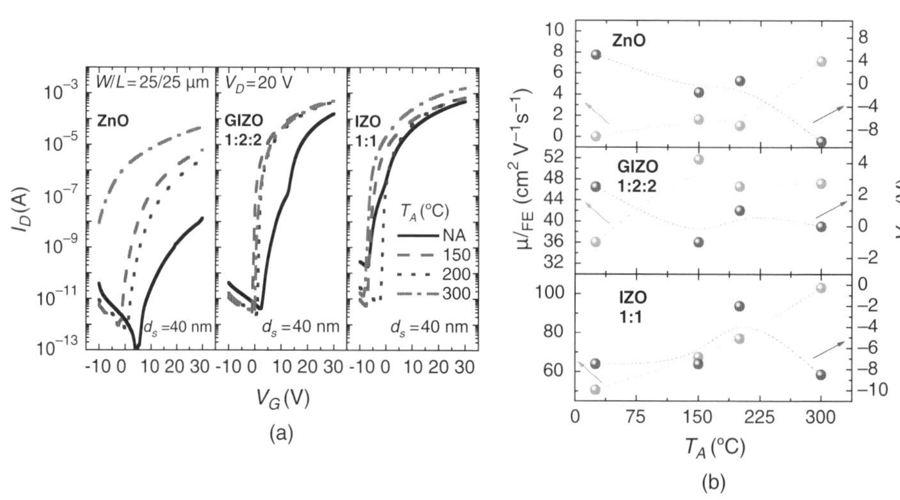

Figure 6.28 Influence of T_A on the electrical properties of TFTs: (a) transfer characteristics for ZnO, GIZO 1:2:2 and IZO 1:1 compositions; (b) μ_{FE} and V_{on} obtained for the devices presented in (a)

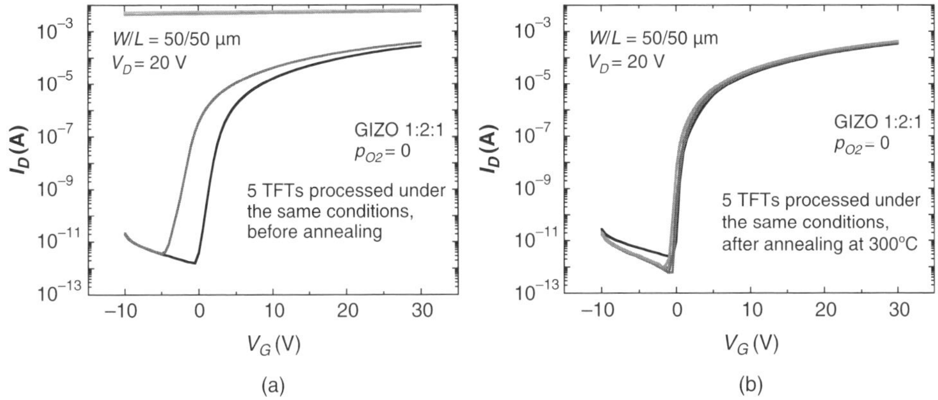

Figure 6.29 *Variation of the transfer characteristics of devices where the active layer is processed under the same conditions: (a) before annealing; (b) after annealing at 300 °C*

- Local atomic rearrangement, allowing reduction of the potential barriers and the defect density associated with the random distribution of the metallic cations close to the bottom of the conduction band. Note that in the case of IZO and GIZO, this rearrangement does not lead to the film's crystallization, with the amorphous structure preserved until $T_A > 500\,°C$.
- Improvement of dielectric/semiconductor interface properties.

Concerning the three different situations presented here, it can also be seen that the lower variation in properties as T_A increases is verified for the GIZO 1:2:2 composition: for this case, a significant improvement is only observed from nonannealed to $T_A = 150\,°C$. This fact reinforces that this composition should lead to the most stable properties in TFTs.

Besides the highly important role in improving the electrical properties of devices, annealing proves also to be extremely important to attenuate differences in the properties among devices processed under the same conditions. For critical conditions, for instance when the active layer is processed without the introduction of oxygen in the sputtering chamber, large variations in the properties of the TFTs can be obtained before annealing [Figure 6.29(a)], but these variations are considerably attenuated after annealing at 300 °C [Figure 6.29(b)]. Note that even lower temperatures (150–200 °C) can be used for this standardization process, depending on whether the semiconductor is processed or not under (close to) ideal conditions.

Another example of the effectiveness of annealing to attenuate differences in different devices is shown in Figure 6.30, this time for devices where the active layer was intentionally processed with different p_{O_2}. If before annealing the properties are considerably affected by p_{O_2}, after an annealing treatment at 300 °C all the TFTs present similar properties (despite the best properties still being obtained with low p_{O_2}), meaning that annealing starts to be the predominant processing parameter dictating the final properties obtained in the devices.

In summary, annealing is a key processing parameter for the production of oxide semiconductor based TFTs, allowing not only the improvement of their electrical properties

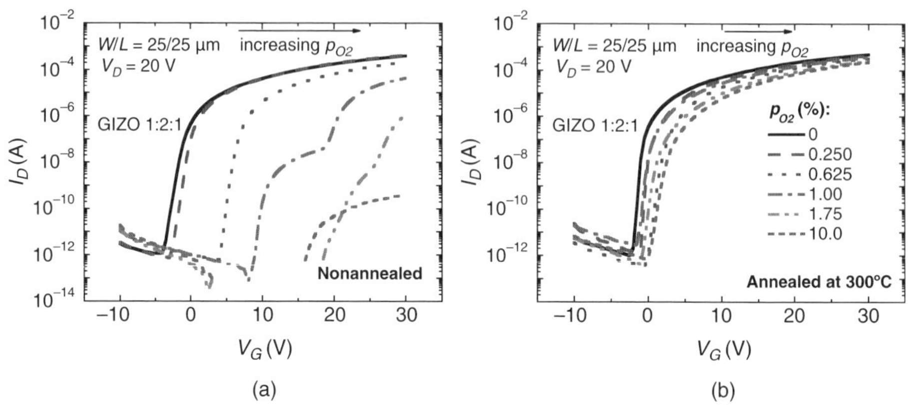

Figure 6.30 *Transfer characteristics of devices where the active layer is intentionally processed under different p_{O_2}: (a) before annealing; (b) after annealing at 300 °C*

but also to considerably reduce the differences verified in unintentionally and intentionally processed devices. This should be seen as a remarkable characteristic of these materials/devices if they are to be transferred to mass production and integration in electronic circuits.

6.3.2.4 TFT Stability Over Time

For proper integration in electronic circuits, the electrical properties of the transistors need to be stable over a long period of time, otherwise the lifetime of the end product would be too short to be viable. TFTs with the active layer processed under optimal and nonoptimal conditions (based on the study presented above) were tested over 6 months. During that time, the devices were stored in typical ambient conditions, i.e. normal indoor light, temperature of 20–25 °C and relative humidity of 40–45%. Figure 6.31

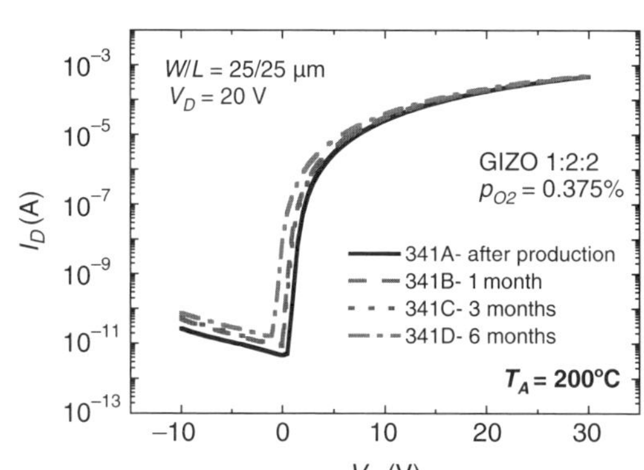

Figure 6.31 *Transfer characteristics evolving over 6 months for TFTs with the active layer processed under optimized conditions, annealed at 150 °C*

Table 6.5 Variation of electrical properties of TFTs over 6 months extracted from the transfer characteristics presented in Figure 6.31

Month	μ_{FE} (cm^2 V^{-1} s^{-1})	V_{on} (V)	ΔV_{on} (V)
0	52.0	0.5	0.5
1	53.0	0	0.5
3	53.2	0	0.5
6	53.7	–1.0	0

and Table 6.5 show the results obtained for optimized devices, annealed at a low $T_A = 150\,°C$.

Even with this low T_A, compatible with a large number of plastic substrates, it can be seen that the variations are not very significant. V_{on} is the most affected parameter, but even after 6 months its final value was maintained close to 0 (the fact that this final value is -1 V instead of the ideal 0 V can be justified mainly by work-function differences between the gate electrode and the semiconductor and by charges existent on the dielectric's bulk). Interestingly, after storage over 6 months the devices evolve to a more stable state, with $\Delta V_{on} = 0$ V instead of the 0.5 V previously obtained. Additionally, note that the semiconductor surface of these devices was not passivated, being always exposed to ambient air and light.

For devices processed under nonideal conditions, like high p_{O_2} (10.0%), the variation in properties is considerably higher, as expected [Figure 6.32(a)]. Nevertheless, the same trend of increasing stability with time is verified for this case, with V_{on} shifting towards values close to 0 V and nonidealities observed in transfer characteristics having the tendency to be eliminated. The effectiveness of the annealing treatment is once again verified here: if $T_A = 300\,°C$ is used for these nonideal semiconductor conditions, very good stability is obtained right from the beginning [Figure 6.32(b)].

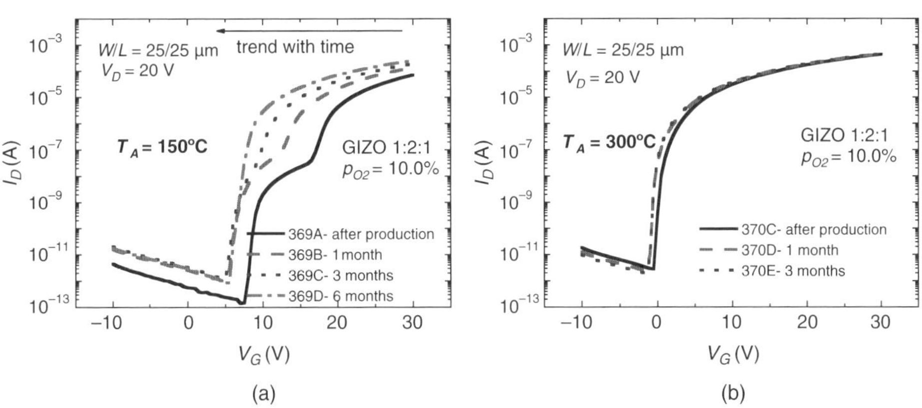

Figure 6.32 Transfer characteristics evolving over 6 months for TFTs with the active layer processed with $p_{O_2} = 10.0\%$, after annealing at (a) 150 °C and (b) 300 °C

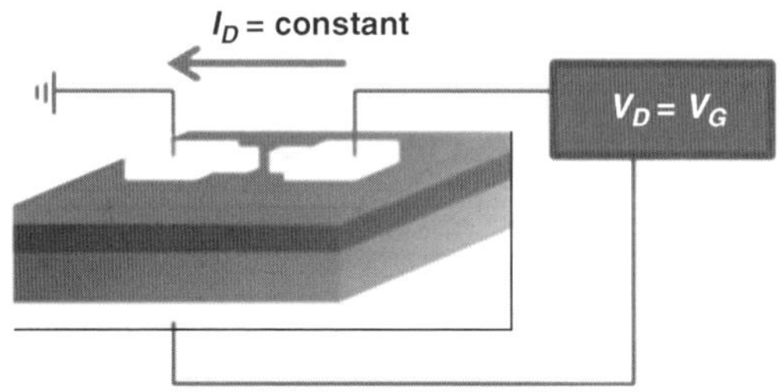

Figure 6.33 Schematic representation of the current stress set-up

6.3.2.5 TFT Stability Under Constant Current Stress

Regarding current stress measurements, they are based on forcing a constant I_D over time, while having the drain and gate short-circuited (Figure 6.33). In this way, the circuit will continuously adjust the $V_D = V_G$ (thus V_T) needed to maintain the forced I_D. This is a critical test for OLED application, since OLEDs are current driven devices.

GIZO 1:2:1 TFTs annealed at 300 °C were tested for 100 consecutive hours and the results are presented in Figure 6.34.

It can be seen that a $\Delta V_T = 0.69$ V was obtained after the 100 h stress. This variation was partly recovered after 4 h, where a $\Delta V_T = 0.36$ V in relation to the initial value was measured, meaning that much of the induced damage can be easily recovered. Another aspect that is worth mentioning is that most of the final ΔV_T is verified after the first 24 h of stress and the data do not show a trend to a continuous increase in this value with time, so it might be expected that even with longer stress times the obtained ΔV_T should not be very different from the 0.69 V obtained here. It is also interesting to compare these values with the ones typically found in the literature for a-Si:H TFTs: ΔV_T exceeding 6 V (one order of magnitude higher) is obtained under the same testing procedure in

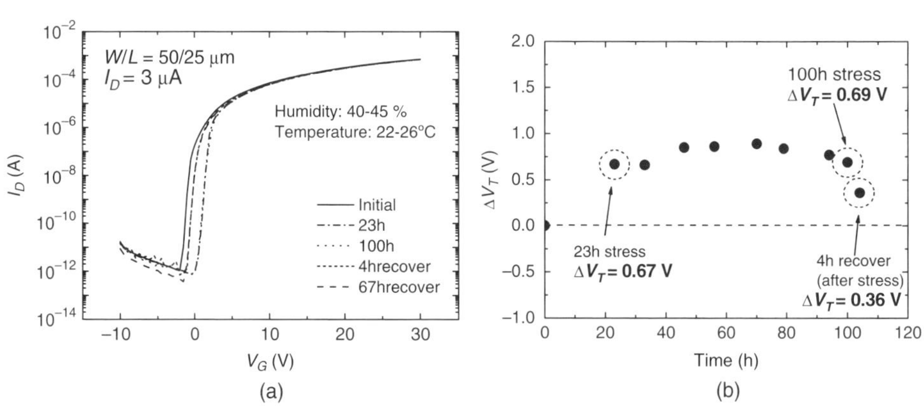

Figure 6.34 Current stress results obtained for 1:2:1 GIZO based TFTs: (a) evolution of transfer characteristics; (b) ΔV_T evolution. Devices annealed at 300 °C

(a) (b)

Figure 6.35 *(a) Complete pixel with gate and source/drain electrodes produced using Ti/IZO and lift-off and (b) comparison between fully transparent (right IZO) and semitransparent (Ti/IZO) matrices*

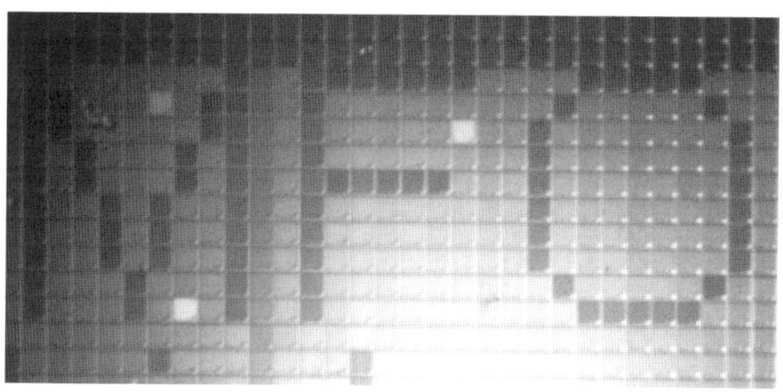

Figure 6.36 *Prototype of a LCD display with a resolution of 128 × 128 pixels*

Jahinuzzaman *et al.* [100], which is a good indication of the excellent stability of amorphous oxide semiconductors.

Figure 6.35 shows some prototypes of a-GIZO matrices produced in our laboratory, while Figure 6.36 shows a working display with a resolution of 128 × 128 pixels.

6.4 Emerging Devices Based on Cellulose Paper: Paper FETs

There is currently a strong interest in the use of biopolymers for electronic applications, mainly driven by low-cost applications. Cellulose is the Earth's major biopolymer and is of tremendous global economic importance. This natural polymer represents about one-third of plant tissues and can be restocked by photosynthesis. The biosynthesis of this polymer is about 1000 tonnes per year worldwide. Cellulose is the structural component of the primary cell wall of green plants. About 33% of all plant matter is cellulose (the cellulose content of cotton is 90% and that of wood is 50%). For industrial use, cellulose is mainly obtained from wood pulp and cotton. It is mainly used to produce cardboard and paper; to a smaller extent it is converted into a wide variety of derivative products such as cellophane and rayon. The possibility to integrate electronic

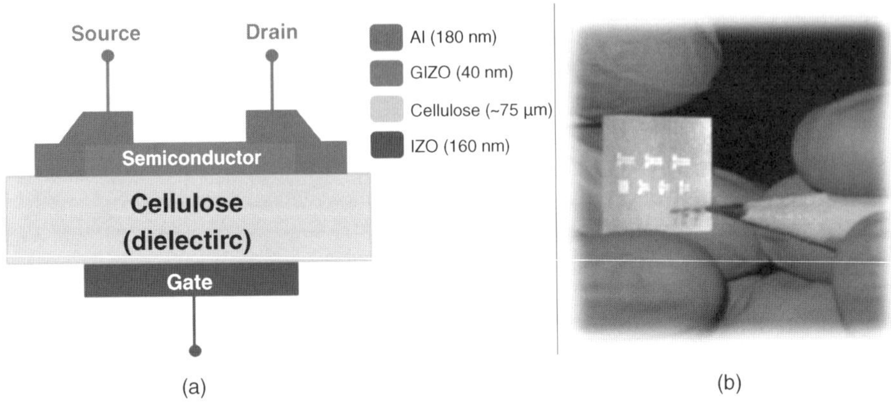

Figure 6.37 *(a) Schematic diagram of the FET structure using the cellulose sheet as the gate dielectric and (b) a typical photograph of several paper transistors*

and optoelectronic functions within the production methods of the paper industry is therefore of current interest, in order to enhance and to add new functionalities to conventional cellulose fibre based paper. To fulfil these demands, materials and methods should be developed for cheap and mass production. An example is printed electronics using ink-jet technology [101–103] or even classical printing methods like reel-to-reel processing [104]. Printed electronics, being lower in cost and biodegradable, is attracting increased interest among product developers. Printed electronics and electronic paper (e-paper) will certainly transform areas such as labels, packaging and publishing.

Some reports have been presented recently using cellulose based paper as either a substrate for physical support of the processing devices, like organic TFTs [105], logic circuits [106] and electrochromic displays [107], or as an active media in thin film flexible Li batteries [108], leading to what we call printed electronics as opposed to e-paper [109]. Recently the present authors demonstrated the possibility of producing FETs [110] as well as RAMs [111] using a conventional sheet of paper.

In this new device approach we are using the cellulose fibre based paper as an 'interstate' structure since the device is built on both sides of the cellulose sheet: the gate electrode, based on a transparent conductive oxide (TCO), is deposited on one side and on the other the active semiconductor, to be used as the channel layer, and the highly conductive drain and source regions [85, 97], as can be seen in Figure 6.37.

Figure 6.38 shows the transfer characteristics of two typical GIZO FETs, with $W/L = 10.6$, in the saturation region ($V_D = 15$ V) using two types of cellulose based fibers (different finishing and chemical composition surfaces) and the dielectric layer. Saturation mobility (μ_{sat}) and threshold voltage (V_T) were calculated from the derivative and the x-axis intercept of the $\sqrt{I_D(V_G)}$ plot, respectively. The subthreshold gate swing value (S) was obtained at the maximum slope of $dV_G/d(\log I_D)$. Table 6.6 presents a comparison of the electrical parameters of the two series of paper based GIZO FETs and of a typical GIZO FET fabricated on crystalline silicon, using thermal SiO_2 as the dielectric layer (100 nm thick).

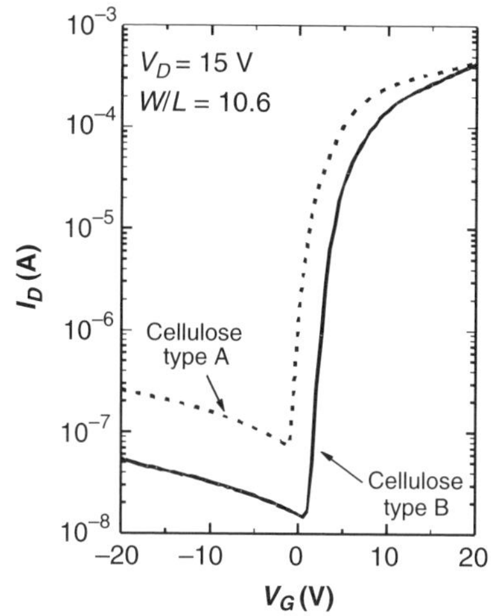

Figure 6.38 *I_D–V_G transfer characteristics obtained at $V_D = 15$ V (saturation region) for GIZO FETs using two types of cellulose fibres as dielectric layers, fully produced at room temperature*

The fabrication of high performance hybrid flexible FETs using cellulose fibre (without any treatment) as the dielectric layer and a semiconductor oxide (GIZO), deposited by RF magnetron sputtering at room temperature, as the channel layer was demostrated. The transistors processed in this way have an enhanced *n*-type operation mode and exhibit an almost zero threshold voltage, a channel saturation mobility exceeding 30 $cm^2 V^{-1} s^{-1}$, a drain-source current I_{on}/I_{off} modulation ratio above 10^4 and a subthreshold gate voltage swing of about 0.8 V $decade^{-1}$. Even 2 months after processing, the device performances were unchanged showing that they are environmentally stable (stored in air ambient conditions).

The results obtained outstrip those of amorphous Si TFTs and rival those of the actual state of art concerning oxide based TFTs produced on either glass or crystalline silicon substrates, even those processed or annealed at temperatures as high as 200–300 °C.

The compatibility of these devices with large-scale/large-area deposition techniques and low cost substrates as well as their very low operating bias delineates this as a promising approach to attain high-performance disposable electronics like paper displays, smart labels, smart packaging, radio frequency identification (RFID) and point-of-care systems for self analysis in bio-applications.

Table 6.6 *Comparison of the electrical properties of GIZO based FETs with different dielectric materials*

Substrate (dielectric)	μ_{SAT}($cm^2 V^{-1} s^{-1}$)	V_T (V)	I_{on}/I_{off}	S (V $decade^{-1}$)
Si/SiO_2	30	+1.8	8.0×10^8	0.4
Cellulose type A	30	–0.6	5.9×10^3	0.8
Cellulose type B	34	+1.9	2.9×10^4	0.8

6.5 Conclusions and Outlook

The technology to display video imagery when required on windows, automobile windshields or eyeglass lenses is at last near to becoming a reality. Transparent electronics, which comprises transparent electrodes and semiconductors, along with various transistors, circuits and other devices, is expected to become commercially possible in the near future.

It has been impossible to make transparent electronic circuits until now, because most circuits, electrodes and other elements are made of materials like metal or silicon. If transparent electronic integrated circuits can be made, it will become possible to embed them into large areas like windows. While a pane of glass might look like an ordinary window, it could actually incorporate a new type of electronics with a range of functions as *value-added glass*. Also, from an economic point of view, the world market for flat glass in 2006 was around 42 million tonnes, equivalent to 4.2 billion square metres of glass with a thickness of 4 mm (http://www.agc-flatglass.com/). This represents a monetary value at the level of primary manufacture of about US$ 20 billion. A proportion of the flat glass is further processed by laminating, tempering, coating and silvering for use in insulating glazing (for building applications) or automotive glazing. At this level, the market has a value of approximately US$ 60 billion. In terms of volume of glass used, the construction industry is the largest user (36 million tonnes), followed by the automotive industry (about 5 million tonnes). Even places like walls, desktops and other locations could be equipped with 'stick-on' electronics. Such development seems likely to bring electronic circuits into every aspect of our lives.

These transparent electronic circuits will probably not be complete products to start with but rather will appear as transparent electronic circuit components for items such as TFTs. They will make possible a range of improvements in LCD panels, OLED panels and other displays, including larger screens, finer definition, high aperture ratio and simpler manufacturing.

After the industry has built up sufficient experience in using these components, the first transparent electronic products will finally appear (Figure 6.39).

Besides the good performance, transparency is a property that does not exist in conventional electronics. As Dr M. Kawasaki from Tohoku University in Japan said, 'Transparency makes it fully possible for these materials to compete with existing silicon materials'. By merging high performances with transparency new products and new functionalities will be created which are not possible with conventional technology (Figure 6.40).

As a summary, Table 6.7 presents a comparison between transparent TFTs and the other commercially available technologies (amorphous Si and poly-Si).

Let us remember the example of Jules Verne (8 February 1828–24 March 1905), the French author who pioneered the science fiction genre. He is best known for novels such as *Twenty Thousand Leagues under the Sea* (1870). Verne wrote about space, air and underwater travel before air travel and practical submarines were invented, and before practical means of space travel had been devised. He is often popularly referred to as the 'father of science fiction'. Maybe with transparent electronics we are facing a similar situation. Hollywood has already shown the potential utility and appeal of transparent

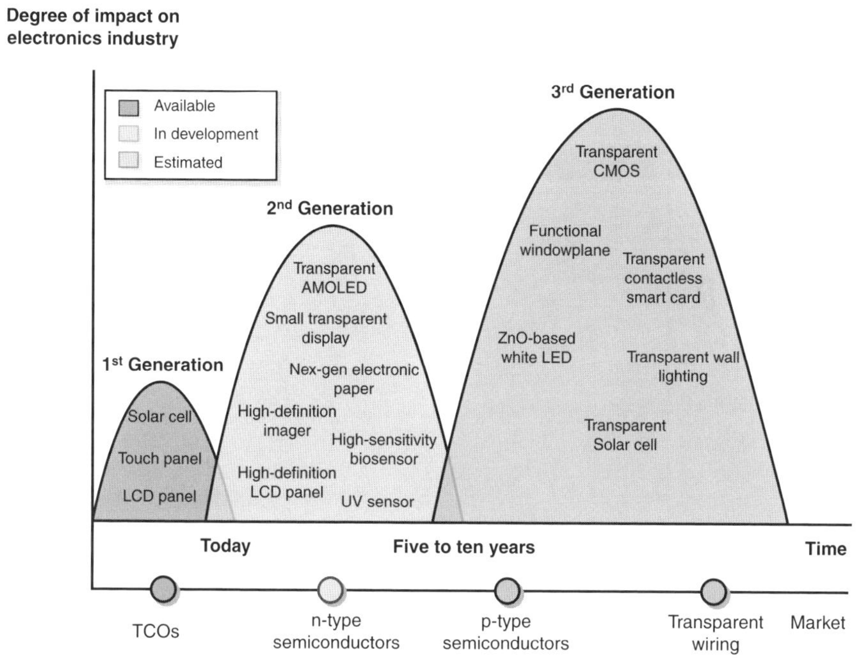

Figure 6.39 *The 1st generation in transparent electronics is based on transparent electrodes using mainly ITO for LCD and touch panels. The 2nd generation is related to the development of n-type semiconductors for TFTs and integrated circuits while the 3rd generation will envisage low cost white LEDs and low dissipation CMOS integrated circuits (adapted from [112]). See color plate section*

electronics. In a scene from the film *Minority Report*, Tom Cruise portrays a policeman in the future who manipulates images of criminals on a transparent display. Sometimes 'science fiction' is the driving force for some of the major scientific developments becoming 'science fact'.

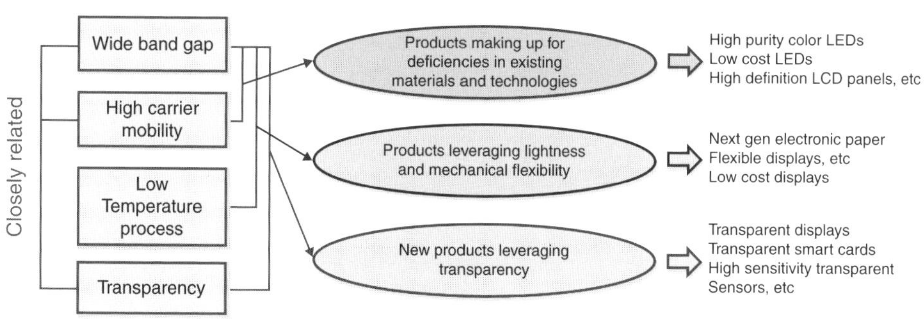

Figure 6.40 *Characteristics besides transparency. Oxide semiconductors have a number of useful characteristics, including wide band gap, relatively high carrier mobility, low temperature processing and low manufacturing costs, due to the low temperature deposition and abundant raw materials*

Table 6.7 *Comparison between the properties exhibited by TFTs produced using different technologies*

TFT semiconductor material	Transparent oxide	Amorphous Si	Low-temperature poly-Si	Characteristics of transparent oxide TFTs
Carrier mobility ($cm^2 V^{-1} s^{-1}$)	1–100	1 max	50–100	Adequate performance as TFT
Switching characteristics ($V decade^{-1}$)	0.009–0.6	0.4–0.5	0.2–0.3	
Source-drain leakage current (A)	$<10^{-13}$	$\sim10^{-12}$	$\sim10^{-12}$	
TFT characteristic variation	Low	Low	High	No circuit needed to correct characteristic variation; cost is cheap and yield may be as high as amorphous Si
TFT manufacturing for AMOLEDs	4–5 masks	4–5 masks	5–9 masks	
Manufacturing cost	Low	Low	High	
Long term TFT reliability	High (forecast)	Low	High	
Yield	High (forecast)	High	Low	
Process temperature (°C)	Room temperature to 350	About 250	250	Plastic substrates possible, expanding possible, expanding range of applications, ex paper
Applicable displays	LCD, OLED, e-paper, etc.	LDC, etc.	LCD, small organic EL	

Figure 6.41 *The Microelectronic and Optoelectronic Materials team (October 2008)*

Acknowledgements

The authors are indebted to many people for the opportunity to perform this work, in particular the Microelectronic and Optoelectronic Materials group (Figure 6.41), without whom this work would not have been possible. We would also like to acknowledge: the Multiflexioxides Project Consortium (NMP3-CT-2006-032231), the first dedicatedproject on Transparent Electronics in Europe; the project with SAMSUNG (SAIT) 'STABOXI' related to the passivation of a-GIZO TFTS; and the Electronic and Telecommunications Research Institute (ETRI) for the project IT R&D programme of MKE-2006-S079-03, 'Smart window with transparent electronic devices'.

The authors also wish to express their gratitude to CENIMAT (FCT-UNL) and CEMOP (UNINOVA) for the excellent working conditions that made the present work possible.

This work was partially funded by the Portuguese Science Foundation (FCT-MCTES) through projects PTDC/CTM/73943/2006 and PTDC/EEA-ELC/64975/2006.

References

[1] J. E. Lilienfeld, Method and apparatus for controlling electric currents, 1930. United States Patent 1745175.

[2] J. E. Lilienfeld, Device for controlling electric current, 1933. United States Patent 1900018.

[3] O. Heil, Improvements in or relating amplifiers or other control arrangements, *UK patent 439,457*, 1935.

[4] J. B. W. Shockley and W. Brattain, Three-electrode circuit element utilizing semiconductive materials, *US Patent 2,524,035*, 1947.

[5] P. K. Weimer, Tft - New Thin-Film Transistor, *Proceedings of the Institute of Radio Engineers*, vol. 50, pp. 1462, 1469, 1962.

[6] D. Kahng and M. M. Atalla, Silicon-Silicon Dioxide Field Induced Devices, *IRE Solid State Device Conference*, 1960.
[7] T. P. Brody, J. A. Asars and G. D. Dixon, 6×6 inch 20 lines-per-inch liquid-crystal display panel, *IEEE Transactions on Electron Devices*, vol. ED20, pp. 995–1001, 1973.
[8] T. P. Brody, Large-scale integration for display screens, *IEEE Transactions on Consumer Electronics*, vol. 21, pp. 260–289, 1975.
[9] W. E. Spear and P. G. Lecomber, Substitutional doping of amorphous silicon, *Solid State Communications*, vol. 17, pp. 1193-1196, 1975.
[10] W. E. Spear and P. G. Lecomber, Electronic properties of substituted doped amorphous Si and Ge, *Philosophical Magazine*, vol. 33, pp. 935–949, 1976.
[11] P. G. Lecomber, W. E. Spear and A. Ghaith, Amorphous-silicon field-effect device and possible application, *Electronics Letters*, vol. 15, pp. 179–181, 1979.
[12] W. d. Boer, *Active Matrix Liquid Crystal Displays, Fundamentals and Applications*, Newnes, 2005.
[13] DisplaySearch. http://www.displaysearch.com/cps/rde/xchg/displaysearch/hs.xsl/091019_displaysearch_forecasts_flat_panel_display_industry_recovery_in_2010.asp
[14] Y. Kuo, *Thin Film Transistors – Materials and Processes Volume 2: Amorphous Silicon Thin Film Transistors*, Kluwer Academic, 2004.
[15] S. Y. Yoon, J. Y. Oh, C. O. Kim and J. Jang, Low temperature solid phase crystallization of amorphous silicon at 380 degrees C, *Journal of Applied Physics*, vol. 84, pp. 6463–6465, 1998.
[16] S. Y. Yoon, S. J. Park, K. H. Kim and J. Jang, Metal-induced crystallization of amorphous silicon, *Thin Solid Films*, vol. 383, pp. 34–38, 2001.
[17] P. Lengsfeld, N. H. Nickel and W. Fuhs, Step-by-step excimer laser induced crystallization of a-Si:H, *Applied Physics Letters*, vol. 76, pp. 1680–1682, 2000.
[18] A. T. Voutsas, A new era of crystallization: advances in polysilicon crystallization and crystal engineering, *Applied Surface Science*, vol. 208, pp. 250–262, 2003.
[19] F. Garnier, G. Horowitz, X. H. Peng and D. Fichou, An all-organic soft thin-film transistor with very high carrier mobility, *Advanced Materials*, vol. 2, pp. 592–594, 1990.
[20] C. L. Cheng, S. R. Forrest, M. L. Kaplan, P. H. Schmidt and B. Tell, Novel Organic-on-inp field-effect transistor, *Applied Physics Letters*, vol. 47, pp. 1217–1219, 1985.
[21] C. D. Dimitrakopoulos and P. R. L. Malenfant, Organic thin film transistors for large area electronics, *Advanced Materials*, vol. 14, p. 99–117, 2002.
[22] C. D. Sheraw, L. Zhou, J. R. Huang, D. J. Gundlach, T. N. Jackson, M. G. Kane, I. G. Hill, M. S. Hammond, J. Campi, B. K. Greening, J. Francl and J. West, Organic thin-film transistor-driven polymer-dispersed liquid crystal displays on flexible polymeric substrates, *Applied Physics Letters*, vol. 80, pp. 1088–1090, 2002.
[23] H. Sirringhaus, T. Kawase, R. H. Friend, T. Shimoda, M. Inbasekaran, W. Wu and E. P. Woo, High-resolution inkjet printing of all-polymer transistor circuits, *Science*, vol. 290, pp. 2123–2126, 2000.
[24] I. Kymissis, *Organic Field Effect Transistors Theory, Fabrication and Characterization*, Springer, 2009.
[25] M. Kitamura and Y. Arakawa, Pentacene-based organic field-effect transistors, *Journal of Physics-Condensed Matter*, vol. 20, pp. 1–16, 2008.
[26] K. Nomura, H. Ohta, K. Ueda, T. Kamiya, M. Hirano and H. Hosono, Thin-film transistor fabricated in single-crystalline transparent oxide semiconductor, *Science*, vol. 300, pp. 1269–1272, 2003.
[27] J. F. Wager, Transparent electronics, *Science*, vol. 300, pp. 1245–1246, 2003.
[28] E. M. C. Fortunato, P. M. C. Barquinha, A. Pimentel, A. M. F. Goncalves, A. J. S. Marques, R. F. P. Martins and L. M. N. Pereira, Wide-bandgap high-mobility ZnO thin-film transistors produced at room temperature, *Applied Physics Letters*, vol. 85, pp. 2541–2543, 2004.
[29] S. Masuda, K. Kitamura, Y. Okumura, S. Miyatake, H. Tabata and T. Kawai, Transparent thin film transistors using ZnO as an active channel layer and their electrical properties, *Journal of Applied Physics*, vol. 93, pp. 1624–1630, 2003.

[30] B. J. Norris, J. Anderson, J. F. Wager and D. A. Keszler, Spin-coated zinc oxide transparent transistors, *Journal of Physics D: Applied Physics*, vol. 36, pp. L105–L107, 2003.
[31] R. L. Hoffman, B. J. Norris and J. F. Wager, ZnO-based transparent thin-film transistors, *Applied Physics Letters*, vol. 82, pp. 733–735, 2003.
[32] P. F. Carcia, R. S. McLean, M. H. Reilly and G. Nunes, Transparent ZnO thin-film transistor fabricated by rf magnetron sputtering, *Applied Physics Letters*, vol. 82, pp. 1117–1119, 2003.
[33] F. M. Hossain, J. Nishii, S. Takagi, A. Ohtomo, T. Fukumura, H. Fujioka, H. Ohno, H. Koinuma and M. Kawasaki, Modeling and simulation of polycrystalline ZnO thin-film transistors, *Journal of Applied Physics*, vol. 94, pp. 7768–7777, 2003.
[34] H. S. Bae, M. H. Yoon, J. H. Kim and S. Im, Photodetecting properties of ZnO-based thin-film transistors, *Applied Physics Letters*, vol. 83, pp. 5313–5315, 2003.
[35] K. Nomura, H. Ohta, A. Takagi, T. Kamiya, M. Hirano and H. Hosono, Room-temperature fabrication of transparent flexible thin-film transistors using amorphous oxide semiconductors, *Nature*, vol. 432, pp. 488–492, 2004.
[36] B. Sun and H. Sirringhaus, Solution-processed zinc oxide field-effect transistors based on self-assembly of colloidal nanorods, *Nano Letters*, vol. 5, pp. 2408–2413, 2005.
[37] R. E. Presley, C. L. Munsee, C. H. Park, D. Hong, J. F. Wager and D. A. Keszler, Tin oxide transparent thin-film transistors, *Journal of Physics D: Applied Physics*, vol. 37, pp. 2810–2813, 2004.
[38] H. A. Klasens and H. Koelmans, A tin oxide field-effect transistor, *Solid-State Electronics*, vol. 7, pp. 701–702, 1964.
[39] E. N. Dattoli, Q. Wan, W. Guo, Y. B. Chen, X. Q. Pan and W. Lu, Fully transparent thin-film transistor devices based on SnO_2 nanowires, *Nano Letters*, vol. 7, pp. 2463–2469, 2007.
[40] J. H. Noh, S. Y. Ryu, C. S. Kim, S. J. Jo, H. S. Hwang and H. K. Baik, Indium oxide thin-film transistors fabricated by room temperature rf-magnetron sputtering, *Idw '07: Proceedings of the 14th International Display Workshops,* Vols 1–3, pp. 1999–2000, 2007.
[41] J. W. Park, S. W. Han, N. Jeon, J. Jang and S. Yoo, Improved electrical characteristics of amorphous oxide TFTs based on TiO_x channel layer grown by low-temperature MOCVD, *IEEE Electron Device Letters*, vol. 29, pp. 1319–1321, 2008.
[42] J. W. Park and S. Yoo, New n-type TiO_2 transparent active channel TFTs fabricated with a solution process, *IEEE Electron Device Letters*, vol. 29, pp. 724–727, 2008.
[43] S. J. Seo, C. G. Choi, Y. H. Hwang and B. S. Bae, High performance solution-processed amorphous zinc tin oxide thin film transistor, *Journal of Physics D:- Applied Physics*, vol. 42, pp. 1–5, 2009.
[44] Y. J. Chang, D. H. Lee, G. S. Herman and C. H. Chang, High-performance, spin-coated zinc tin oxide thin-film transistors, *Electrochemical and Solid State Letters*, vol. 10, pp. H135–H138, 2007.
[45] M. K. Jayaraj, K. J. Saji, K. Nomura, T. Kamiya and H. Hosono, Optical and electrical properties of amorphous zinc tin oxide thin films examined for thin film transistor application, *Journal of Vacuum Science and Technology B*, vol. 26, pp. 495–501, 2008.
[46] H. Q. Chiang, J. F. Wager, R. L. Hoffman, J. Jeong and D. A. Keszler, High mobility transparent thin-film transistors with amorphous zinc tin oxide channel layer, *Applied Physics Letters*, vol. 86, DOI: 013503-1 to 013503-3, 2005.
[47] N. Itagaki, T. Iwasaki, H. Kumomi, T. Den, K. Nomura, T. Kamiya and H. Hosono, Zn-ln-O based thin-film transistors: compositional dependence, *Physica Status Solidi A*, vol. 205, pp. 1915–1919, 2008.
[48] M. G. McDowell and I. G. Hill, Influence of channel stoichiometry on zinc indium oxide thin-film transistor performance, *IEEE Transactions on Electron Devices*, vol. 56, pp. 343–347, 2009.
[49] D. H. Cho, S. Yang, C. Byun, M. K. Ryu, S. H. K. Park, C. S. Hwang, S. M. Yoon and H. Y. Chu, Transparent oxide thin-film transistors composed of Al and Sn-doped Zinc indium oxide, *IEEE Electron Device Letters*, vol. 30, pp. 48–50, 2009.

[50] N. L. Dehuff, E. S. Kettenring, D. Hong, H. Q. Chiang, J. F. Wager, R. L. Hoffman, C. H. Park and D. A. Keszler, Transparent thin-film transistors with zinc indium oxide channel layer, *Journal of Applied Physics*, vol. 97, DOI: 064505-1 to 064505-5, 2005.
[51] R. Martins, P. Barquinha, I. Ferreira, L. Pereira, G. Goncalves and E. Fortunato, Role of order and disorder on the electronic performances of oxide semiconductor thin film transistors, *Journal of Applied Physics*, vol. 101, DOI: 044505-1 to 044505-7, 2007.
[52] D. C. Paine, B. Yaglioglu, Z. Beiley and S. Lee, Amorphous IZO-based transparent thin film transistors, *Thin Solid Films*, vol. 516, pp. 5894–5898, 2008.
[53] C. G. Choi, S. J. Seo and B. S. Bae, Solution-processed indium-zinc oxide transparent thin-film transistors, *Electrochemical and Solid State Letters*, vol. 11, pp. H7–H9, 2008.
[54] R. E. Presley, D. Hong, H. Q. Chiang, C. M. Hung, R. L. Hoffman and J. F. Wager, Transparent ring oscillator based on indium gallium oxide thin-film transistors, *Solid-State Electronics*, vol. 50, pp. 500–503, 2006.
[55] J. Y. Kwon, K. S. Son, J. S. Jung, T. S. Kim, M. K. Ryu, K. B. Park, B. W. Yoo, J. W. Kim, Y. G. Lee, K. C. Park, S. Y. Lee and J. M. Kim, Bottom-Gate gallium indium zinc oxide thin-film transistor array for high-resolution AMOLED display, *IEEE Electron Device Letters*, vol. 29, pp. 1309–1311, 2008.
[56] P. Barquinha, A. Vila, G. Goncalves, L. Pereira, R. Martins, J. Morante and E. Fortunato, The role of source and drain material in the performance of GIZO based thin-film transistors, *Physica Status Solidi A*, vol. 205, pp. 1905–1909, 2008.
[57] H. Lim, H. Yin, J. S. Park, I. Song, C. Kim, J. Park, S. Kim, S. W. Kim, C. B. Lee, Y. C. Kim, Y. S. Park and D. Kang, Double gate GaInZnO thin film transistors, *Applied Physics Letters*, vol. 93, DOI: 063505-1 to 063505-3, 2008.
[58] Y. Park, C. J. Kim, S. I. Kim, I. Song, J. C. Park, H. Lim, S. W. Kim and E. Lee, Highly stable Ga_2O_3-In_2O_3-ZnO thin-film transistors for AM-OLED application, *Idw '07: Proceedings of the 14th International Display Workshops*, Vols 1–3, pp. 1775–1778, 2007.
[59] C. Y. Koo, D. Kim, S. Jeong, J. Moon, C. Park, M. Jeon, W. C. Sin, J. Jung, H. J. Woo, S. H. Kim and J. Ha, Sol-gel derived Ga-In-Zn-O semiconductor layers for solution-processed thin-film transistors, *Journal of the Korean Physical Society*, vol. 53, pp. 218–222, 2008.
[60] E. M. C. Fortunato, L. M. N. Pereira, P. M. C. Barquinha, A. M. B. do Rego, G. Goncalves, A. Vila, J. R. Morante and R. F. P. Martins, High mobility indium free amorphous oxide thin film transistors, *Applied Physics Letters*, vol. 92, DOI: 222103-1 to 222103-3, 2008.
[61] H. Yabuta, M. Sano, K. Abe, T. Aiba, T. Den, H. Kumomi, K. Nomura, T. Kamiya and H. Hosono, High-mobility thin-film transistor with amorphous $InGaZnO_4$ channel fabricated by room temperature rf-magnetron sputtering, *Applied Physics Letters*, vol. 89, DOI: 112123-1 to 112123-3, 2006.
[62] C. J. Kim, D. Kang, H. Song, J. C. Park, H. Lim, S. Kim, E. Lee, R. J. Chung, J. C. Lee and Y. Park, Highly stable Ga_2O_3-In_2O_3-ZnO TFT for active-matrix organic light-emitting diode display application, *2006 International Electron Devices Meeting*, Vols 1 and 2, pp. 41–44, 2006.
[63] H. N. Lee, J. W. Kyung, S. K. Kang, D. Y. Kim, M. C. Sung, S. J. Kim, C. N. Kim, H. G. Kim and S. T. Kim, Current status of, challenges to, and perspective view of AM-OLED, Idw '06: Proceedings of the 13th International Display Workshops, Vols 1–3, pp. 663–666, 2006.
[64] H. N. Lee, J. Kyung, S. K. Kang, D. Y. Kim, M. C. Sung, S. J. Kim, C. N. Kim, H. G. Kim and S. T. Kim, 3.5 inch QCIF+ AM-OLED panel based on oxide TFT backplane, *SID International Symposium, Digest of Technical Papers*, vol. 38, pp. 1826–1829, 2007.
[65] http://www.printedelectronicsworld.com/articles/samsung_expected_to_slash_manufacturing_costs_of_oled_displays_00001188.asp?sc=1).
[66] H. Yin, S. Kim, C. J. Kim, I. Song, J. Park, S. Kim and Y. Park, Fully transparent nonvolatile memory employing amorphous oxides as charge trap and transistor's channel layer, *Applied Physics Letters*, vol. 93, DOI: 172109-1 to 172109-3, 2008.
[67] S. Das, Cheap, flexible transistor looks good on paper, *New Scientist*, vol. 199, pp. 26–26, 2008.

[68] M. Ito, M. Kon, M. Ishizaki, C. Miyazaki, K. Imayoshi, M. Tamakoshi, Y. Ugajin and N. Sekine, A novel display structure for color electronic paper driven with fully transparent amorphous oxide TFT array, *Idw '06: Proceedings of the 13th International Display Workshops,* Vols 1–3, pp. 585–586, 2006.
[69] K. S. Son, T. S. Kim, J. S. Jung, M. K. Ryu, K. B. Park, B. W. Yoo, J. W. Kim, Y. G. Lee, J. Y. Kwon, S. Y. Lee and J. M. Kim, Late-news paper: 4 inch QVGA AMOLED driven by the threshold voltage controlled amorphous GIZO (Ga2O3-In2O3-ZnO) TFT, *SID International Symposium, Digest of Technical Papers*, vol. 39, pp. 633–636, 2008.
[70] J. K. Jeong, J. H. Jeong, J. H. Choi, J. S. Im, S. H. Kim, H. W. Yang, K. N. Kang, K. S. Kim, T. K. Ahn, H. J. Chung, M. Kim, B. S. Gu, J. S. Park, Y. G. Mo, H. D. Kim and H. K. Chung, Distinguished paper: 12.1-inch WXGA AMOLED display driven by indium-gallium-zinc oxide TFTs array, *SID International Symposium, Digest of Technical Papers*, vol. 39, pp. 1–4, 2008.
[71] J. H. Lee, D. H. Kim, D. J. Yang, S. Y. Hong, K. S. Yoon, P. S. Hong, C. O. Jeong, H. S. Park, S. Y. Kim, S. K. Lim, S. S. Kim, K. S. Son, T. S. Kim, J. Y. Kwon, and S. Y. Lee, World's largest (15-inch) XGA AMLCD panel using IGZO oxide TFT, *SID International Symposium, Digest of Technical Papers*, vol. 39, pp. 625–628, 2008.
[72] D. K. John Wager and R. Presley, *Transparent Electronics*, Springer, 2008.
[73] Jae Kyeong Jeong, Hyun-Joong Chung, Yeon-Gon Mo and Hye Dong Kim, A New Era of Oxide Thin-Film Transistors for Large-Sized AMOLED Displays, *Information Display*, vol. 24, pp. 20–23, 2008.
[74] T. Minami, Transparent conducting oxide semiconductors for transparent electrodes, *Semiconductor Science and Technology*, vol. 20, pp. S35–S44, 2005.
[75] A. L. D. H. L. Hartnagel, A.K. Jain and C. Jagadish, *Semiconducting Transparent Thin Films*, Institute of Physics, 1995.
[76] K. Ellmer, Resistivity of polycrystalline zinc oxide films: current status and physical limit, *Journal of Physics D: Applied Physics*, vol. 34, pp. 3097–3108, 2001.
[77] R. G. Gordon, Criteria for choosing transparent conductors, *MRS Bulletin*, vol. 25, pp. 52–57, 2000.
[78] H. J. Ko, Y. F. Chen, S. K. Hong, H. Wenisch, T. Yao and D. C. Look, Ga-doped ZnO films grown on GaN templates by plasma-assisted molecular-beam epitaxy, *Applied Physics Letters*, vol. 77, pp. 3761–3763, 2000.
[79] E. Fortunato, D. Ginley, H. Hosono and D. C. Paine, Transparent conducting oxides for photovoltaics, *MRS Bulletin*, vol. 32, pp. 242–247, 2007.
[80] V. Assuncao, E. Fortunato, A. Marques, H. Aguas, I. Ferreira, M. E. V. Costa and R. Martins, Influence of the deposition pressure on the properties of transparent and conductive ZnO: Ga thin-film produced by r.f. sputtering at room temperature, *Thin Solid Films*, vol. 427, pp. 401–405, 2003.
[81] E. Fortunato, A. Goncalves, A. Marques, V. Assuncao, I. Ferreira, H. Aguas, L. Pereira and R. Martins, Gallium zinc oxide coated polymeric substrates for optoelectronic applications, *Flexible Electronics-Materials and Device Technology*, vol. 769, pp. 291–296, 2003.
[82] V. Assuncao, E. Fortunato, A. Marques, A. Goncalves, I. Ferreira, H. Aguas and R. Martins, New challenges on gallium-doped zinc oxide films prepared by r.f. magnetron sputtering, *Thin Solid Films*, vol. 442, pp. 102–106, 2003.
[83] E. Fortunato, A. Goncalves, V. Assuncao, A. Marques, H. Aguas, L. Pereira, I. Ferreira and R. Martins, Growth of ZnO: Ga thin films at room temperature on polymeric substrates: thickness dependence, *Thin Solid Films*, vol. 442, pp. 121–126, 2003.
[84] A. Klaus Ellmer,Bernd Rech *Transparent Conductive Zinc Oxide - Basics and Applications in Thin Film Solar Cells*, Springer, 2008.
[85] E. Fortunato, V. Assuncao, A. Goncalves, A. Marques, H. Aguas, L. Pereira, I. Ferreira, P. Vilarinho, and R. Martins, High quality conductive gallium-doped zinc oxide films deposited at room temperature, *Thin Solid Films*, vol. 451, pp. 443–447, 2004.
[86] E. P. Denton, H. Rawson and J. E. Stanworth, Vanadate glasses, *Nature*, vol. 173, pp. 1030–1032, 1954.

[87] E. Fortunato, A. Pimentel, A. Goncalves, A. Marques and R. Martins, High mobility amorphous/nanocrystalline indium zinc oxide deposited at room temperature, *Thin Solid Films*, vol. 502, pp. 104–107, 2006.
[88] R. Martins, P. Barquinha, A. Pimentel, L. Pereira and E. Fortunato, Transport in high mobility amorphous wide band gap indium zinc oxide films, *Physica Status Solidi A*, vol. 202, pp. R95–R97, 2005.
[89] R. Martins, P. Almeida, P. Barquinha, L. Pereira, A. Pimentel, I. Ferreira and E. Fortunato, Electron transport and optical characteristics in amorphous indium zinc oxide films, *Journal of Non-Crystalline Solids*, vol. 352, pp. 1471–1474, 2006.
[90] R. Martins, P. Barquinha, A. Pimentel, L. Pereira, E. Fortunato, D. Kang, I. Song, C. Kim, J. Park and Y. Park, Electron transport in single and multicomponent n-type oxide semiconductors, *Thin Solid Films*, vol. 516, pp. 1322–1325, 2008.
[91] N. Ito, Y. Sato, P. K. Song, A. Kaijio, K. Inoue and Y. Shigesato, Electrical and optical properties of amorphous indium zinc oxide films, *Thin Solid Films*, vol. 496, pp. 99–103, 2006.
[92] B. Yaglioglu, H. Y. Yeom and D. C. Paine, Crystallization of amorphous In_2O_3-10 wt % ZnO thin films annealed in air, *Applied Physics Letters*, vol. 86, DOI: 261908-1 to 261908-3, 2005.
[93] P. Barquinha, G. Goncalves, L. Pereira, R. Martins and E. Fortunato, Effect of annealing temperature on the properties of IZO films and IZO based transparent TFTs, *Thin Solid Films*, vol. 515, pp. 8450–8454, 2007.
[94] European Project: Multiflexioxides (NMP3-CT-2006-032231), Second year technical report, 2008.
[95] E. Fortunato, P. Barquinha, A. Pimentel, L. Pereira, G. Goncalves and R. Martins, Amorphous IZO TTFTs with saturation mobilities exceeding 100 cm(2)/Vs, *Physica Status Solidi-Rapid Research Letters*, vol. 1, pp. R34–R36, 2007.
[96] B. Yaglioglu, H. Y. Yeom, R. Beresford and D. C. Paine, High-mobility amorphous In_2O_3-10 wt %ZnO thin film transistors, *Applied Physics Letters*, vol. 89, DOI: 062103-1 to 023502-3, 2006.
[97] E. Fortunato, P. Barquinha, G. Goncalves, L. Pereira and R. Martins, High mobility and low threshold voltage transparent thin film transistors based on amorphous indium zinc oxide semiconductors, *Solid-State Electronics*, vol. 52, pp. 443–448, 2008.
[98] P. Barquinha, L. Pereira, G. Goncalves, R. Martins and E. Fortunato, The effect of deposition conditions and annealing on the performance of high-mobility GIZO TFTs, *Electrochemical and Solid State Letters*, vol. 11, pp. H248–H251, 2008.
[99] P. Barquinha, L. Pereira, G. Gonçalves, R. Martins and E. Fortunato, Toward high-performance amorphous GIZO TFTs, *Journal of the. Electrochemical Society*, vol. 156, pp. H161–H168, 2009.
[100] S. M. Jahinuzzaman, A. Sultana, K. Sakariya, P. Servati and A. Nathan, Threshold voltage instability of amorphous silicon thin-film transistors under constant current stress, *Applied Physics Letters*, vol. 87, 2005.
[101] B. J. de Gans, P. C. Duineveld and U. S. Schubert, Inkjet printing of polymers: state of the art and future developments, *Advanced Materials*, vol. 16, pp. 203–213, 2004.
[102] M. L. Chabinyc, W. S. Wong, A. C. Arias, S. Ready, R. A. Lujan, J. H. Daniel, B. Krusor, R. B. Apte, A. Salleo and R. A. Street, Printing methods and materials for large-area electronic devices, *Proceedings of the IEEE*, vol. 93, pp. 1491–1499, 2005.
[103] L. Yang, A. Rida, R. Vyas and M. M. Tentzeris, RFID tag and RF structures on a paper substrate using inkjet-printing technology, *IEEE Transactions on Microwave Theory and Techniques*, vol. 55, pp. 2894–2901, 2007.
[104] B. Zimmermann, M. Glatthaar, M. Niggemann, M. K. Riede, A. Hinsch and A. S Gombert, ITO-free wrap through organic solar cells - a module concept for cost-efficient reel-to-reel production, *Solar Energy Materials and Solar Cells*, vol. 91, pp. 374–378, 2007.
[105] Y. H. Kim, D. G. Moon, and J. I. Han, Organic TFT array on a paper substrate, *IEEE Electron Device Letters*, vol. 25, pp. 702–704, 2004.
[106] D. Nilsson, N. Robinson, M. Berggren and R. Forchheimer, Electrochemical logic circuits, *Advanced Materials*, vol. 17, pp. 353-+, 2005.

[107] P. Andersson, D. Nilsson, P. O. Svensson, M. X. Chen, A. Malmstrom, T. Remonen, T. Kugler and M. Berggren, Active matrix displays based on all-organic electrochemical smart pixels printed on paper, *Advanced Materials*, vol. 14, pp. 1460–1464, 2002.
[108] V. L. Pushparaj, M. M. Shaijumon, A. Kumar, S. Murugesan, L. Ci, R. Vajtai, R. J. Linhardt, O. Nalamasu and P. M. Ajayan, Flexible energy storage devices based on nanocomposite paper, *Proceedings of the National Academy of Sciences of the United States of America*, vol. 104, pp. 13 574–13 577, 2007.
[109] A. Henzen and J. van de Kamer, The present and future of electronic paper, *Journal of the Society for Information Display*, vol. 14, pp. 437–442, 2006.
[110] E. Fortunato, N. Correia, P. Barquinha, L. Pereira, G. Goncalves and R. Martins, High-performance flexible hybrid field-effect transistors based on cellulose fiber paper, *IEEE Electron Device Letters*, vol. 29, pp. 988–990, 2008.
[111] R. Martins, P. Barquinha, L. Pereira, N. Correia, G. Goncalves, I. Ferreira and E. Fortunato, Write-erase and read paper memory transistor, *Applied Physics Letters*, vol. 93, DOI: 203501-1 to 203501-3, 2008.
[112] Tech-On! http://techon.nikkeibp.co.jp.

7

Carbon Nanotube Transparent Electrodes

Teresa M. Barnes and Jeffrey L. Blackburn
National Renewable Energy Laboratory, Golden, USA

7.1 Introduction

Thin networks of intertwined single-wall carbon nanotubes (SWCNTs) form optically transparent and electrically conducting coatings, which have been suggested as a potential replacement for transparent conducting oxides (TCOs) in a variety of applications [1–4]. SWCNT networks can be deposited readily on flexible substrates using low-cost, low-temperature deposition methods. These networks can have a wide range of conductivities and optical transparencies depending on their thickness [5]. One major difference between SWCNT films and TCOs such as In_2O_3:Sn (ITO), ZnO and SnO_2 is that their fabrication process readily produces a preferentially hole-conducting contact [1]. This is particularly advantageous in organic photovoltaic (OPV) devices, where the SWCNT network can replace both ITO and PEDOT:PSS in a typical bulk-heterojunction device [6]. Other researchers have also had success replacing the ITO layer in a bulk-heterojunction device on a flexible substrate [7]. SWCNT transparent films can successfully replace the ZnO bi-layer in an efficient $Cu(In_{0.7}Ga_{0.3})Se_2$/CdS device [8] and the back contact in a transparent CdTe device [9]. Studies on *p*-GaN indicates that SWCNT network electrodes can produce an ohmic contact to a *p*-type semiconductor [10]. Transparent SWCNT networks have also been incorporated into flexible, transparent transistors [11], organic light emitting diodes (OLEDs) [12] and chemical sensors [13]. Although the conductivity and transparency of SWCNT networks is not yet equal to the best TCOs, the array of functional devices produced

Transparent Electronics: From Synthesis to Applications Edited by Antonio Facchetti and Tobin J. Marks

with SWCNT electrodes clearly demonstrates their potential as transparent contacts. In this chapter, we will present data on the optoelectronic properties of SWCNT networks and several strategies to improve their performance. Although we focus on their applications as transparent electrical contacts, particularly in photovoltaics (PVs), SWCNT networks have the potential to play many roles in transparent electronics.

7.2 Chirality and Band Structure of SWCNTs

SWCNTs can be thought of as rolled graphene sheets. A section of the graphene lattice is shown in Figure 7.1 [14, 15]. Every other carbon atom along the 'zigzag' edge of the lattice is given a pair of (n,m) indices to indicate its position on the lattice. Conceptually, single-walled nanotubes are constructed by folding this graphene sheet into a tube along a 'roll-up vector' that connects the point (0,0) to another point (n,m) on the graphene lattice. The angle of the roll-up vector with respect to the zigzag edge is defined as the chiral angle. As Figure 7.1 shows, unique SWCNTs are created from roll-up vectors spanning from a chiral angle of 0° (zigzag SWCNTs) to 30° (armchair SWCNTs). The electronic structure of each of these SWCNTs is unique, and it can either be metallic or semiconducting, as determined by the zone-folding approximation for a graphene sheet. The electronic structure tends to follow trends based on 'families' of SWCNTs that share mathematical commonalities in their (n,m) assignments [16]. The most fundamental family relationship relates nanotubes that have the same remainder r for the equation $2n+m=3p+r$. For example, when the quantity $(2n+m)$ is evenly divisible by 3 and $r=0$, the SWCNTs have metallic band structures. These electronic structure relationships dictate that one-third of all possible SWCNTs have a metallic band structure, and most current synthetic methods produce a

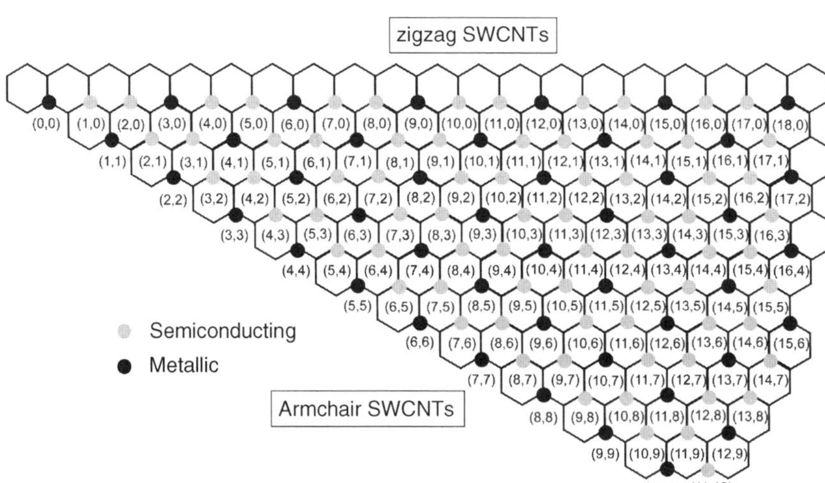

Figure 7.1 *Section of the graphene lattice with indices for unique SWCNTs that are created by a roll-up vector from the point (0,0) to another point (n,m) on the lattice*

mixture of semiconducting and metallic SWCNTs (s-SWCNTs and m-SWCNTs) with an abundance ratio of ~2:1, respectively.

7.3 Synthesis, Purification, and Dispersion of SWCNTs

SWCNTs can be produced by a number of different synthesis techniques including pulsed-laser vaporization (PLV) [17, 18], electric-arc breakdown (Arc), high-pressure carbon monoxide conversion (HiPCO) [19], and several chemical vapor deposition (CVD) processes [20]. Each growth technique yields SWCNTs with unique properties including length distributions, defect densities, and diameter distributions. No technique is yet capable of producing a single type or size of tubes, but they each tend to produce tubes with a slightly different chirality distribution and size range. PLV, Arc, HiPCO, and CVD have all been used to produce nanotubes for transparent conducting (TC) films. As-prepared, a 'soot' of single-walled nanotube material contains some amounts of the following: single-walled carbon nanotubes, metal catalyst particles, amorphous carbon, graphitic impurities, and, in some cases, other fullerene-based nanostructures such as bucky balls and nanohorns. The elimination of residual metal catalysts and non-nanotube carbon is typically necessary before using such a soot in electronic applications, and several purification strategies have been developed to meet these needs. Most purification strategies involve soaking, extracting, sonicating, or refluxing the nanotube soot in a solution of a strong acid, such as HNO_3 or HCl, to remove residual metal catalyst particles [17, 21–23]. Often, the catalyst particles are coated with graphitic shells that must be chemically damaged to allow penetration of acid for metal dissolution. Such carbon coatings have been removed using an oxidation step in air [22, 24] aqueous peroxide chemistry [25], or a high-temperature bake in CO_2. Following this acid treatment, amorphous carbon impurities are removed by techniques such as cross-flow filtration with a basic solution [26] or high-temperature hydrogen treatment [27]. The resulting purified mat is a thin (5–20 μm), flat, free-standing black paper typically called a 'bucky paper.' Unless annealed, these bucky papers usually contain residual acid from the reflux step that is actually quite beneficial for conductivity enhancement when the end product is a transparent conducting film [28]. Other purification strategies involve high-speed centrifugation of a nitric-acid-treated SWCNT soot to separate SWCNTs from non-nanotube carbon based on their different electrokinetic potentials and buoyant densities [29, 30].

Raw or purified SWCNT samples contain large bundles of nanotubes due to van der Waals interactions between individual tubes; therefore, they must first be dispersed in solution to cast a transparent film. This requires a micelle-forming surfactant in the solution to yield high concentrations of isolated tubes in aqueous solutions. A variety of surfactant molecules may be employed, including: anionic surfactants such as sodium dodecyl sulfate (SDS), sodium dodecylbenzyl sulfonate (SDBS), or sodium cholate hydrate (cholate); cationic surfactants such as dodecyltrimethylammonium bromide (DTAB); and nonionic surfactants such as Triton X-100, TWEEN, and a host of other polymers [31, 32]. Once surfactant is added to the solution, sonication is used to debundle and disperse the tubes, and varying degrees of ultracentrifugation may be used to eliminate non-nanotube carbon, large SWCNT bundles, and other impurities.

7.4 Deposition of SWCNT Networks

Several methods exist in the literature for SWCNT network deposition after the nanotubes are dispersed in solution. The most commonly used method is vacuum filtration followed by transfer of the SWCNT film to the desired substrate, as first described by Wu *et al.* in 2004 [1]. In this process, a SWCNT dispersion is vacuum filtered through a mixed cellulose ester membrane until dry, then rinsed with additional water to remove the surfactant. The network can then be wet pressed onto the substrate and vacuum dried or transferred by means of an acetone vapor bath [1]. Another variation for film transfer involves using an alumina filter to form the network, and then transferring it using a pre-patterned polydimethylsiloxane (PDMS) stamp [33].

Ultrasonic or air-brush spray deposition has been used successfully by several groups to deposit SWCNT networks. Unlike vacuum filtration/membrane transfer processes, spray deposition requires a significant effort in ink development to produce high-quality films. Ink-dispersion parameters, such as sonication power and duration, and surfactant choice can strongly affect SWCNT network properties [34, 35], and the effects of ultrasonic spray conditions are not generally well understood. However, empirical optimization of spray conditions and ink formulations has led to films that are competitive with the best membrane-transferred material [5, 35]. Spray deposition is readily scalable to large area and highly amenable to producing patterned films.

Other deposition techniques such as spin-coating and dip-coating of SWCNT networks have also been used. These methods often involve more than a hundred coating cycles, and they are quite labor and time intensive [36, 37]. Direct synthesis by floating-catalyst CVD (FC-CVD) is a method by which the nanotubes are grown *in situ* as an interconnected thin transparent film. This method shows some promise for the deposition of networks without forming a SWCNT dispersion or ink [38]. However, it is difficult to assess how residual impurities from the catalyst and amorphous carbon would affect the films.

7.5 Effects of Chemical Doping

As discussed above, raw SWCNTs are typically exposed to strong acids when the tubes are purified, and this is known to produce a preferentially hole-conducting material [1, 39]. Like most nanomaterials, SWCNT networks have large surface area, so they are very effective absorbers of atmospheric impurities such as O_2 and water, which are known to dope the networks [40]. In short, it is nearly impossible to produce a truly undoped SWCNT network.

Many molecular dopants have been used to increase the conductivity of SWCNTs following film deposition. *p*-type doping, which is most relevant for TC applications, is typically achieved through simple molecular adsorption by soaking the SWCNT film in a solution of the desired, typically acidic, molecule. This strategy has been used to dope SWCNTs with thionyl chloride [41–44], nitric acid [28, 44, 45], sulfuric acid [45], and hydrochloric acid. Molecular vapors of highly electronegative species may also be used to *p*-dope SWCNTs, as has been done using halogens such as Br_2 and I_2 [46]. Treatment with highly reducing agents such as hydrazine has been shown to chemically dedope the SWCNTs by adding excess electrons to the doped films and restoring the Fermi level to mid-gap [44, 47]. As all of these strategies rely on molecular adsorption, the potential well for the adsorbed dopant is fairly shallow, and dopant desorption is a significant problem for the retention of conductivity. Other groups have used approaches that rely on the formation of covalent bonds in an attempt to increase the stability of conductivity [48, 49].

It has been shown that a post-deposition acid treatment performs two crucial functions in the networks. First, it helps remove residual surfactant and densifies the networks. For example, in sprayed films, the thickness decreases by a factor of nearly 10 after acid treatment, and the sheet resistance drops several orders of magnitude from its 'as-sprayed' value ($\sim 10^6\,\Omega\,\mathrm{sq}^{-1}$). Effective removal of residual surfactant or dispersing agents by acid treatment, evaporation, or rinsing is essential for obtaining reliable electrical data. If surfactants are not removed after the networks are deposited, they can lead to very high values of sheet resistance for the films. SWCNT networks do not exhibit similar conductivity to single tubes, and it is believed that inter-tube or inter-bundle tunneling processes limit their conductivity [44, 50]. Even trace amounts of residual insulators between tubes may severely limit network conductivity by increasing the effective barrier height at the interfaces. Other impurities, such as amorphous carbon or residual catalyst metals, may also affect electrical data, but their impact is less clear. Some applications (i.e. OPV) are sensitive to high levels of residual acids, the acid-treated films are rinsed several times for those applications. After rinsing, the sheet resistance increases modestly, typically from $60\,\Omega\,\mathrm{sq}^{-1}$ after acid treatment to about $110\,\Omega\,\mathrm{sq}^{-1}$ after rinsing.

Secondly, as discussed above, acidic species directly affect the fundamental transport properties of the individual SWCNTs and SWCNT networks. Charge transfer to adsorbed protons injects significant hole density into the SWCNTs, and it has been shown to lower the Fermi level by as much as 0.6 V [45]. This is well below the valence-band maximum for semiconducting SWCNTs, creating a degenerately doped hole-conducting semiconductor with properties very much like that of a one-dimensional metal. Additionally, acids have been shown in the literature to dramatically affect the properties of transport barriers between SWCNTs, increasing the transmission probability through these barriers and consequently increasing film conductivity [44, 50].

7.6 Optical Properties of SWCNTs and SWCNT Networks

Figure 7.2 compares the density of states (DOS) for two ~1.4-nm-diameter SWCNTs. The DOS for both the s-SWCNT (10,5) and the m-SWCNT (8,8) contains sharp singularities at various energies that arise from quantum confinement of the electron wavefunctions in two dimensions [44]. In such a single-particle picture, electronic excitations between these so-called van Hove singularities (vHS) dominate the photo-physics of a particular SWCNT, and they give rise to discrete peaks in the absorbance spectrum. A large body of work has demonstrated that the SWCNT optical resonances arise from excitons, with a diameter-dependent binding energy of up to ~500 meV [51–53]. Thus, the unique DOS for each chirality of SWCNTs leads to distributions of excitonic absorption peaks characteristic of s-SWCNTs and m-SWCNTs in each sample.

Figure 7.3 shows the absorbance spectra of aqueous dispersions of SWCNTs produced by several different synthetic methods. The average energy ranges for semiconducting and metallic transitions observed in the visible and near-infrared (NIR) region are labeled at the top of the figure. The S_{xx} and M_{xx} transitions appear as envelopes of several different

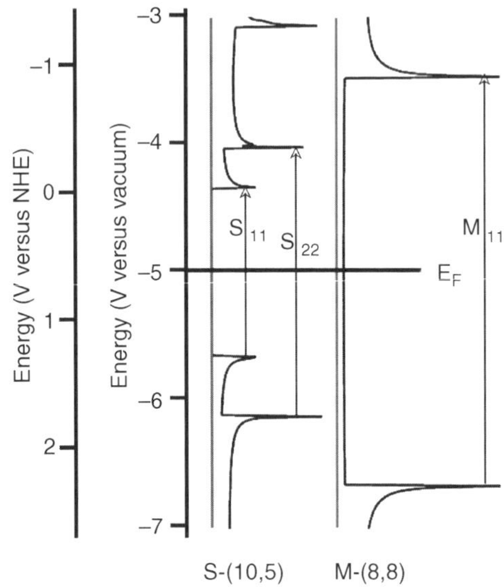

Figure 7.2 *Calculated density of states for (10,5) s-SWCNTs and (8,8) m-SWCNTs, representative of the 1–1.4-nm-diameter distribution produced by laser vaporization, plotted on an absolute energy scale, versus the normal hydrogen electrode (NHE) and versus vacuum*

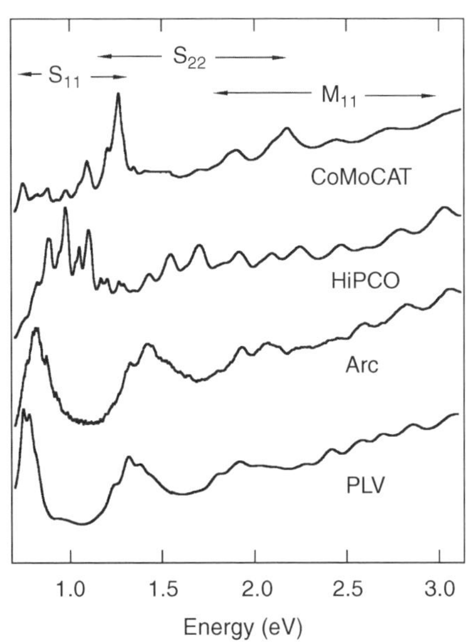

Figure 7.3 *Absorbance spectra as a function of energy for aqueous dispersions of SWCNTs produced by several different synthetic methods*

individual narrow peaks representing the transitions of individual SWCNT chiralities. Any given production method produces a distribution of SWCNT diameters centered around some average diameter, so there is an average energy that directly correlates to the average diameter and an energetic spread for these transition envelopes that depends on how broad or narrow is the distribution of diameters. For example, although the CoMoCAT and HiPCO processes are both CVD processes, the CoMoCAT process produces a very narrow diameter distribution centered around the (6,5) SWCNT. Additionally, the transitions shift to longer wavelengths for production methods such as laser vaporization, which produce larger-diameter SWCNTs.

7.6.1 Optical Transparency

Figure 7.4 shows the optical transmission, reflection, and absorption of a typical 50 $\Omega\,sq^{-1}$ SWCNT network deposited using ultrasonic spray by Eikos, Inc. [54] on quartz. There is a large peak around 280 nm that corresponds to the π-plasmon resonance of the SWCNT network that severely impacts transmission in the visible range. The π-plasmon arises from a collective excitation of the π-electron system of the SWCNTs [55–58], and the relative contributions of metallic tubes, non-nanotube impurities, and bundling to this absorption are still being explored. Optical transmission is considerably higher at wavelengths above 1100 nm. More resistive networks (200–1000 $\Omega\,sq^{-1}$) exhibit extremely high transparency ($>80\%$) throughout the visible and NIR [9]. The transmission and reflection data shown in Figure 7.4 were taken using an integrating sphere to account for diffuse reflectance (haze). The SWCNT networks have very low reflectivity, which stays constant at about 7%, and no detectable haze. Looking more carefully at the absorption spectrum in Figure 7.4, there is a

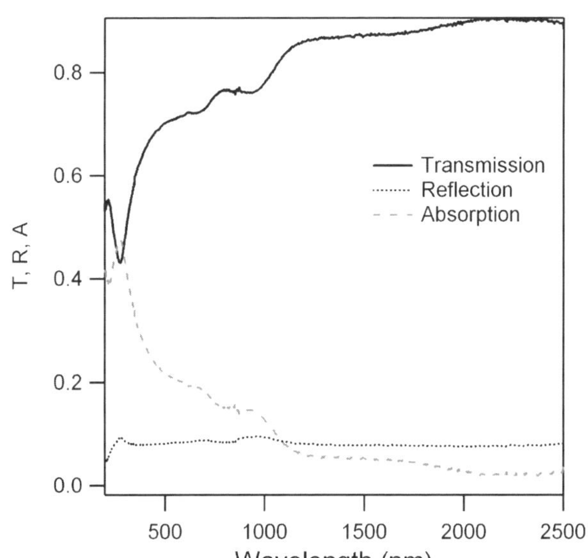

Figure 7.4 *Measured transmission and reflection spectra for a SWCNT film on quartz. The absorption spectrum was calculated and shows the strong π-plasmon absorbance that severely reduces visible transmittance*

small peak in this spectrum at around 1700 nm (0.73 eV) and a distinct peak at 950 nm (1.30 eV), which are due to the S_{11} and S_{22} transitions of the SWCNTs [56, 59]. These values are well within the expected ranges of peak locations for arc-produced SWCNT given in Figure 7.3. The M_{11} transition at 690 nm (1.80 eV) is difficult to resolve in this spectrum. All of the fundamental absorption peaks are weak, indicating that the SWCNTs are doped (discussed below) [28].

7.6.2 Optical Constants

Optical constants for SWCNT networks, made by Eikos, Inc., have been obtained by spectroscopic ellipsometry (SE) [9]. SE measurements were performed at 50°, 60° and 70° angle of incidence, along with plane-polarized transmission measurements. The optical constants were fit using the WVASE32 software package [60]. A Maxwell-Garnett effective medium approximation was used to model the void fraction in the films. Film thicknesses and void fractions determined from the optical model were found to agree with atomic force microscopy (AFM) and scanning electron microscopy (SEM) data [5]. This model was developed phenomenologically based on fits to the ellipsometry data [5]. Standard models for transparent conductors, such as the Drude [61] model, could not fit the optical data satisfactorily. This is further evidence for conduction dominated by something other than free carriers in the networks [62]. SWCNTs are known to exhibit optical anisotropy [63–65], but the degree of anisotropy in bundles and networks is not well understood. The model presented by Barnes *et al.* represents an isotropic approximation for the optical properties of the SWCNT networks.

Figure 7.5 shows the optical constants as a function of wavelength for a 50 $\Omega\,sq^{-1}$ film. The values for the optical constants are identical for SWCNT network films up to 200 $\Omega\,sq^{-1}$,

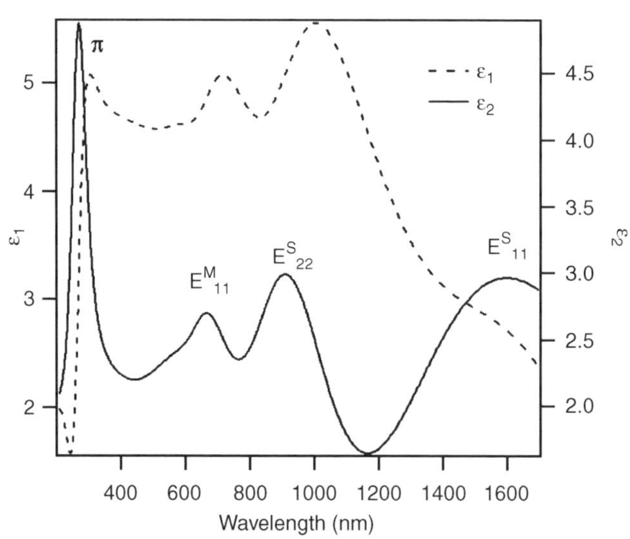

Figure 7.5 *Optical constants for a 50 $\Omega\,sq^{-1}$ SWCNT network as a function of energy. The optical constants clearly show strong absorbance features corresponding to the three fundamental transitions (E^s_{11}, E^s_{22}, E^m_{11}) and the π-plasmon oscillation. Reprinted with permission from [5] Copyright (2007) Institute of Physics*

which is different from previous results reporting a change in the magnitude of the optical constants with thickness and conductivity [2]. Optical constants could not be determined for the higher sheet-resistance layers due to the low reflectivity of the extremely thin SWCNT layers. The optical constants show a strong spectral dependence, with intense peaks corresponding to the fundamental transitions of the SWCNT. The strong peak in ε_2 at 250 nm can be attributed to the π-plasmon resonance [55].

There are strong correlations between the ellipsometry and transmission data for these films. The S_{11} and S_{22} peaks have nearly identical energies when measured with either technique. The location of the M_{11} peak is similar, but the peak is more easily seen with ellipsometry. The magnitude and spectral shape of both ε_1 and ε_2 agree well with the low-energy portions of the spectrum derived from electron energy loss spectroscopy (EELS) data [58]. Additionally, ε_1 remains positive and non-zero throughout this range, which suggests that the networks do not behave purely as metallic conductors at room temperature. This is in agreement with the results from the temperature-dependent resistivity study, which does not show a 'metallic' temperature dependence for the networks (discussed below) [44, 50].

7.7 Electrical Properties of SWCNT Networks

Individual carbon nanotubes are known to be extremely conductive due to their nearly ballistic mobility [66, 67]. Several groups have described field effect transistors (FETs) using individual nanotubes [47, 66, 68], and these devices have been successful tools for studying nanotube physics and illustrating the possibilities of nanotube-based electronics. However, single-tube devices are impractical for commercial applications due to their cost, low yield, requirement for very small contacts, and other processing challenges [69]. Commercially viable nanotube-based electronic devices need to rely on high-yield and readily produced ensembles of tubes such as bundles or networks of multiple tubes that can be deposited over large areas.

High conductivities have been measured in bundles, ropes, and networks of tubes, but these show different electrical properties from individual tubes and each other [70]. The behavior of relatively sparse transparent networks is most relevant for transparent contacts [70–72]. These works have identified several factors including network density [70, 73], interactions between tubes or bundles [70, 73], bundle length [71], and doping [41] as having a strong influence on network conductivity.

7.8 Sheet Resistance and Transport Measurements

Sheet resistance can be measured with a four-point probe, and values between 50 $\Omega\,sq^{-1}$ and 1000 $\Omega\,sq^{-1}$ are typical of SWCNT films with appreciable transparency in the visible region of the spectrum (e.g. 60–95%T). Obtaining bulk resistivity or conductivity values is more difficult because it is somewhat challenging to assign a precise thickness to the networks. Evaluating transport properties (i.e. carrier density and mobility) of the films was found to be nontrivial due to low carrier mobility in the networks. Low carrier mobility could be caused by several factors, including high resistivity between SWCNT bundles due to

Schottky barriers between individual nanotubes [74]. Raman spectroscopy and Fourier transform infrared (FTIR) spectroscopy data suggest that the carrier density in doped films is of the order 10^{20}–10^{21} cm^{-3} [44].

Seebeck effect measurements on SWCNT networks show a positive slope with very little scatter, indicating that the networks are hole-conducting [5]. The measurement yields a Seebeck coefficient of 16.5 ± 1.4 mV K^{-1}, which is in agreement with previous results for SWCNT samples [41, 75]. Seebeck measurements are less sensitive to low carrier mobility than Hall measurements; therefore, they are more reliable than Hall measurements for determining carrier sign in the networks. Previous work has shown that the Seebeck coefficient of SWCNTs is influenced by ambient gases, depending on their molecular mass [75]. The results presented by Barnes *et al* [5]. were measured in vacuum, and they would be expected to change in magnitude, but not sign, depending on the measurement ambient. Further evidence for the hole-selective conductivity of SWCNT networks is abundant in the literature. The standard oxidative purification process is known to induce *p*-type charge-transfer doping of the nanotubes [28, 45], and several groups have observed that SWCNT networks behave like *p*-type semiconductors in FETs [76, 77]. Photovoltaic [6], organic light emitting diode (LED) [12], and inorganic LED [10] devices have successfully used SWCNT networks as hole-selective contacts. Additionally, SWCNT networks often form rectifying contacts in applications where n^+-doped semiconductors are normally employed [78].

Measuring temperature-dependent resistivity (or conductivity) can provide useful insights into the conductivity mechanism in conductive films. Semiconductors generally show a decreasing resistivity with increasing temperature, whereas transport in metals is dominated by phonon scattering, so that resistivity increases with increasing temperatures. SWCNT networks display both behaviors, leading to a characteristic U-shaped $R(T)$ curve. This behavior evokes the comparison with conducting polymers, which show similar behavior, and have traditionally been modeled by considering metallic regions of conductivity separated by thin tunneling barriers [70, 79, 80]. Equation (7.1) has been used to describe this 'hybrid' transport behavior between semiconducting and metallic conductivity:

$$R_S = \alpha T + \beta \exp\left(\frac{T_b}{T_s + T}\right). \tag{7.1}$$

In Equation (7.1), the linear term represents metal-like conductivity (dR/dT is positive), and the second term represents the thermally assisted tunneling contribution that has a semiconductor-like temperature dependence (dR/dT is negative). T_s can be described as the temperature above which thermal-assisted fluctuation effects become important. T_b is a considerably more complex function of the tunneling barrier height and shape as affected by the image force and local electric field. Smaller values of T_b are indicative of a lower effective barrier height. β is a weak function of temperature that also accounts for the barrier shape and network properties, and it can be considered constant compared with the exponential portion of Equation (7.1) [81].

Figure 7.6(a) shows the temperature dependence of the resistivity for a typical acid-purified SWCNT network produced without additional doping and for the same film after soaking overnight in $SOCl_2$. The temperature dependence of resistivity was determined by

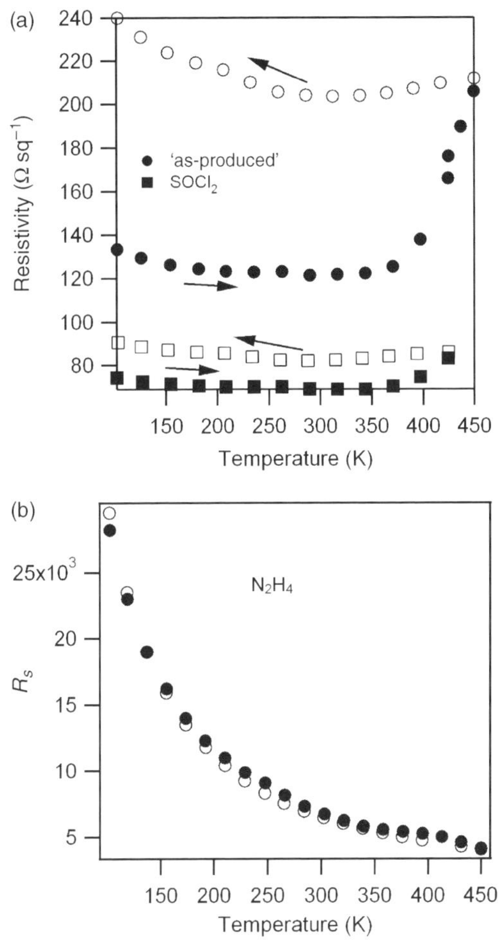

Figure 7.6 *(a) Temperature-dependent resistivity of an 'as-produced' or unintentionally doped bulk SWCNT film and one that has been soaked overnight in $SOCl_2$. The arrows indicate the direction of temperature change. (b) Reversible temperature-dependent resistivity for a film after treatment in N_2H_4. Closed symbols indicate increasing temperature and open symbols indicate decreasing temperature on both graphs*

sweeping up in temperature from 100 K to a maximum of 450 K, while measuring the sheet resistance (R_s) about every 20 K. The temperature sweep was then reversed to check for hysteresis and reversibility. The resistance data have the expected 'U' shape, with R_s initially decreasing with increasing temperature until a critical temperature ($T^* \sim 325$ K) is reached, at which point R_s increases. Both the intentionally doped and as-produced film show considerable hysteresis in their resistivity and the $SOCl_2$-doped film actually shows less hysteresis than the as-produced film. It is well known that SWCNT networks are readily doped unintentionally through materials processing and atmospheric impurities [39, 40], and Figure 7.6(a) indicates that this doping is highly unstable. The hysteresis shown here is not presented in most reports of temperature-dependent resistivity; however, it has been consistently observed, including films in which the SWCNTs were never exposed to acids [44, 50]. Additionally, the temperature dependence becomes reversible (at higher R_s values) after multiple measurement cycles. Note that hysteresis would not be observed if the increase in resistivity with temperature was due solely to scattering processes associated with metallic conductivity. Furthermore, transmission spectra of both films show evidence of bleached excitonic transitions before heating, and a restoration of oscillator strength after heating. These highly reproducible optical and electrical data demonstrate that the film conductivity decreases primarily due to dedoping at temperatures above T^*.

If the increase in R_s above T^* is indeed caused primarily by dopant desorption, then a completely 'dedoped' SWCNT network should not exhibit this behavior. Figure 7.6(b) shows $R_s(T)$ for a hydrazine-treated, dedoped SWCNT film. Samples treated in hydrazine undergo a completely reversible change in R_s as the temperature increases and then decreases. Optical transmission measurements demonstrate that the hydrazine-treated films undergo no change in E_{11} intensities upon heating, in dramatic contrast to intentionally redox-doped films or films left in ambient conditions, which are oxygen-doped. Additionally, unlike doped samples, the data in Figure 7.6(b) show no evidence of an increase in resistivity with temperature, further supporting the notion that this behavior is related to dopant desorption. This behavior is markedly different from what has been observed in the literature for nominally 'undoped' films and ascribed to metallic conductivity [23, 70, 80, 82]. These data suggest that increases in R_s with temperature can be explained by dopant desorption, rather than the onset of metallic conductivity.

Theory, Raman spectroscopy and terahertz spectroscopy suggest a possible hole density as high as 10^{21} cm^{-3} with thionyl chloride doping [44]. This doping level would easily be considered degenerate in most materials, but the electrical properties of these networks suggest that many of these carriers do not contribute toward the conductivity. It is likely that interactions associated with tube-tube junctions primarily control the resistivity in these films. The FET mobility in single SWCNTs (intrinsic m-SWCNTs or s-SWCNTs in the 'on' state) is extremely high, on the order of 1000–10 000 cm^2 V^{-1} s^{-1}, with a mean free path of 0.5 to several μm [83]. Because this distance is much longer than the typical distance between tube-tube junctions, it is reasonable to conclude that the relatively high resistance of SWCNT *films* is due to the large density of tube-tube junctions. Each junction creates a tunnel barrier through which electrons must propagate with some finite transmission probability. Therefore, carriers localized on one SWCNT may either tunnel into an adjacent SWCNT with some probability that depends on tube-tube barriers, or remain localized on the SWCNT. Those carries that remain localized do not contribute to DC transport, whereas the delocalized carriers do. The effect of these tube-tube barriers is captured in the second (tunneling) term of Equation (7.1), and it is discussed in more detail below in the context of temperature-dependent measurements of chirality-separated SWCNT films [44, 50].

7.9 Morphology of SWCNT Networks

SWCNT networks exhibit a unique morphology for transparent contacts. Figure 7.7 illustrates the morphology of a typical SWCNT network with a R_s of 50 Ω sq^{-1} and visible transmittance greater than 75%. The fibrillar nature of the film shown in Figure 7.7 allows a

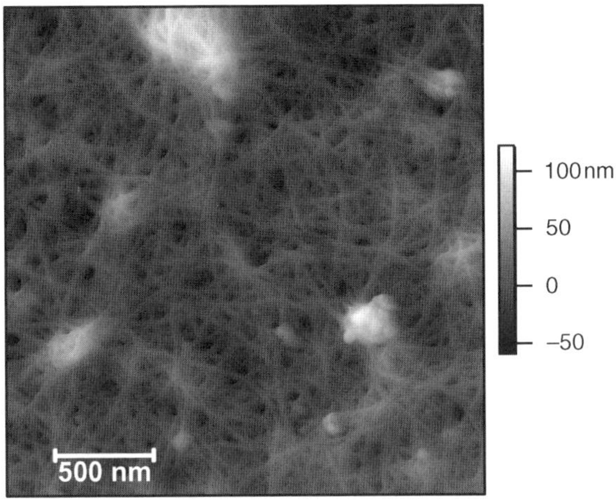

Figure 7.7 *AFM image of a typical 50 Ω sq^{-1} SWCNT network on quartz. This film is 40 nm thick with an RMS roughness of 31 nm. Reprinted with permission from [5] Copyright (2007) Institute of Physics*

large void fraction (~30%), which enables the networks to be transparent [5]. Note that the ropes shown in Figure 7.7 are bundles rather than individual tubes. Figure 7.7 illustrates a number of key points about SWCNT network morphology that determine the optoelectronic performance of the film. First, the film is very thin. Secondly, the films can have roughness approximately equal to their thickness. Thickness measurements were made on this and other films by imaging over an area with a scratch from a razor blade. Line averaging over 2 μm of a 50 Ω sq^{-1} layer yielded a thickness of 40 nm with a root-mean-square (RMS) roughness of ±31 nm. The 100 Ω sq^{-1} film was 30 nm thick, with an RMS roughness of ±26 nm.

Importantly, the peak feature height may be a more useful metric than roughness for assessing the electrode's potential utility in applications requiring a smooth film. The peak feature size in Figure 7.7 is at least 100 nm, which means that portions of the electrode could readily penetrate a thin layer. This is a serious problem in OPV because the active layers are only ~200–300 nm thick, and the SWCNT electrodes can short the devices if the peak feature size is too high. More recently, sprayed electrodes having peak feature sizes below 50 nm have enabled reproducible and moderately efficient OPV devices [84].

Tube or bundle length is known to have a strong effect on conductivity, but it can be difficult to assess quantitatively. It is not possible to find any tube 'ends' in Figure 7.7. They may be obscured by the larger impurities or bundles, but no ends are readily apparent. This suggests that the bundles in the network shown here are quite long. The image is about 2 μm on each side, and the bundles appear significantly longer than that. SWCNT length is believed to strongly affect conductivity in the networks. The junctions between tubes limit conductivity, and limiting the number of junctions required to traverse the film should increase the conductivity [71].

The networks are thin and sparse compared with solid TCO films that conduct like plane-parallel slabs. Consequently, their volume conductivities are quite high despite their relatively high sheet resistance. Based on the measured sheet resistances thickness, and void fraction, a coarse volume conductivity on the order of 3300–5000 S cm^{-1} is typical of the networks. This volume conductivity is on the order of good-quality TCOs [85], and it is orders of magnitude higher than that of PEDOT:PSS, a commercially available transparent conducting polymer [86].

7.10 Literature Results on Transparent SWCNT Networks

7.10.1 Optical and Electrical Properties of SWCNT Networks

An early report on transparent SWCNT network films showed the tremendous promise of a material with a sheet resistance around 30 $\Omega\ sq^{-1}$ and extremely high optical transparency in the infrared (IR) [1]. Subsequent work has shown a wide disparity in sheet resistance and transparency data, revealing that the film properties depend strongly on the tube growth method, how they are processed, and if any post-treatment is employed. A single figure of merit is often useful to compare a wide array of transparent contact data [85], and one reasonable figure of merit is given as ϕ in Equation (7.2).

$$\phi = \frac{100}{-R_s{*}\log(T)} \tag{7.2}$$

ϕ is defined here as the inverse of R_s multiplied by the logarithm of the transmittance at 550 nm. This is not the ideal definition because it does not provide any information on the transmission spectrum through the visible range, but it is useful for making comparisons between reported data. A high value of ϕ indicates a transparent contact with the required combination of low R_s and high visible transmission.

Table 7.1 includes a selection of data from the literature with information on the film with the lowest R_s from each work in order of decreasing ϕ. The transparency data are reported simply as the measured value at 550 nm. The tube synthesis method describes how the SWCNTs were grown. The deposition method is the technique used to transfer the SWCNT film to the substrate, and the post-treatment column describes any acid soaks or water rinses

Table 7.1 *Comparison of literature data for transparent SWCNT networks showing R_s and %T at 500 nm for the film with the lowest R_s in each work. Tube-synthesis method, film deposition technique and post-treatment are given for each film*

R_s	%T at 550 nm	ϕ	Tube synthesis	Film deposition	Post-treatment	Year	Ref.
30	0.78	30.9	PLV	Vacuum filtration	None	2004	28
39	0.74	19.6	PLV	Spray	HNO_3	2008	35
50	0.71	13.5	Arc	Spray	None	2007	5
50	0.70	12.9	CVD	Direct CVD growth	None	2008	38
85	0.80	12.1	Arc	Spin-coat	HNO_3	2008	37
30	0.45	9.61	HiPCO	Vacuum filtration	$HNO_3/SOCl_2$	2007	100
80	0.60	5.63	Arc	Vacuum filtration/PDMS	None	2006	33
340	0.80	3.03	Arc	Dip-coat/spray	None	2008	36
1000	0.90	2.19	PLV	Spray	Water rinse	2005	101
1000	0.85	1.42	HiPCO	Vacuum filtration	None	2004	73
900	0.70	0.72	HiPCO	Electrophoretic	None	2008	102

after the film is deposited. The two films with the highest values of ϕ were produced by PLV, and HiPCO-grown SWCNT yielded films with the two lowest (least desirable) values of ϕ according to the data in Table 7.1. Arc produced SWCNT films are consistently between the two extremes of HiPCO and PLV in quality. However, there is considerable variation in ϕ values for each growth method and deposition technique depending on sample quality and whether or not the films are doped/post-treated. Vacuum filtration followed by membrane transfer to either a stamp or a flat substrate is the most commonly used film deposition technique from the papers in Table 7.1. This technique is easy to adopt in the laboratory and requires relatively little optimization of SWCNT solutions used to make the films. Very few papers provide details of their post-treatment processes, and these steps can be crucial for surfactant removal, doping, or dedoping of the SWCNT networks, all of which strongly affect conductivity.

7.10.2 SWCNT Network Properties Compared with Common TCOs

Although they are continually improving, SWCNT networks have not yet matched the optoelectronic performance of commonly used TCOs, such as indium-doped tin oxide (ITO). This may be acceptable in many applications (i.e. touch screens) that do not need the simultaneously high conductivity and transparency of the best TCOs, but more demanding applications such as PVs may not be able to use SWCNT networks on a large scale until their properties improve. Beyond conductivity and transparency, other material properties, such as high-temperature stability, flexibility, durability, roughness, cost, and manufacturing compatibility may be more critical factors in electrode choice than simply picking the material with the best figure-of-merit performance. Several examples of this exist in PVs. SnO_2:F is used in CdTe devices because of its high-temperature stability, durability, and cost–despite its relatively high absorbance in the visible. $CuIn_xGa_{1-x}Se_2$ (CIGS) devices often rely on a ZnO:Al top contact because it exhibits the required combination of high transmission and conductivity with low-temperature deposition at low costs; however, the stability and durability of this contact are poor [85]. Recent work has shown that contact morphology may matter more than conductivity in OPV devices due to their extremely thin active layers [84].

Figure 7.8 compares optical transmission at various sheet resistances for several different TCOs and SWCNT networks. Data from SnO_2:F and ZnO:Al were taken on films produced at the National Renewable Energy Laboratory (NREL) specifically for CdTe and CIGS devices. The ITO film is a commercial product (Colorado Concept Coating) used in NREL's OPV devices. The first SWCNT film is a commercially available product (Eikos, Inc.), and the second two were made by membrane transfer and spray deposition at NREL. The transmission spectra for the TCOs and SWCNT films differ greatly. SWCNT networks tend to be thinner (30–50 nm) and rougher than most TCO films, so there are no interference fringes apparent in the SWCNT spectra. Also, SWCNT networks have far higher transmission at long wavelengths than TCOs due to the absence of free-carrier absorption in this range for the SWCNT networks. Free-carrier absorption generally starts at about 1100 nm for TCOs, and long-wavelength absorption losses in the TCO can hinder the ultimate conversion efficiency of a PV device. Tandem solar cells, IR detectors, and other applications require long-wavelength transmission, and SWCNT networks have definite advantages over TCOs for these types of applications [87].

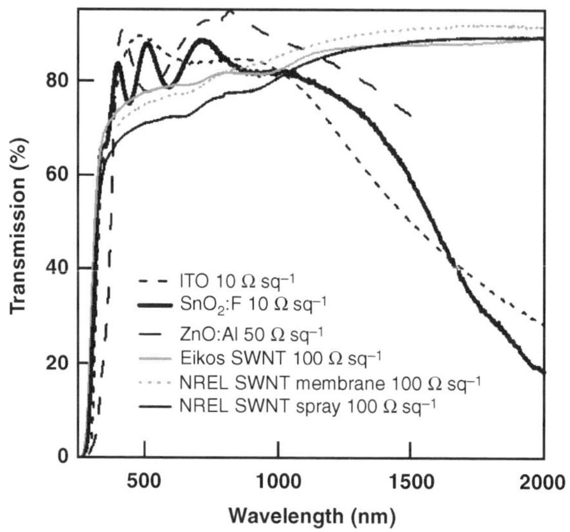

Figure 7.8 *Optical transmission as a function of wavelength for several common TCOs and SWCNT networks. The SnO_2:F and ZnO:Al were grown at NREL and are representative of the TCOs used in NREL's typical CdTe and CIGS solar cells. The ITO is the commercial product used here for OPV devices. The first SWCNT film is a commercially available product from Eikos; the second was made at NREL by the membrane transfer process, and the third was deposited by ultrasonic spray*

SWCNT networks exhibit lower transmission in the visible range than the TCOs shown here, but their performance is continually improving.

7.10.3 Networks Containing Separated SWCNTs

Conductivity in SWCNT networks is complicated due to their fibrillar nature, morphology, affinity for impurities, and the fundamental electronic structure of the tubes themselves. As produced, two-thirds of the tubes in a SWCNT soot have a semiconducting band structure, and one-third have a metallic band structure. Numerous theories exist in the literature about the effects of different conductivity types on electrical properties, and efficient separation of SWCNTs by conductivity type is required to answer many of the outlying questions in the literature.

Many strategies have been attempted, with varying degrees of success, to separate SWCNTs by electronic structure. Several strategies rely on the differing chemical reactivities of m- and s-SWCNTs to species such as diazonium salts [88–90] or amines [91, 92]. In many cases, the selective interaction of surfactants or the selective covalent functionalization by, for example, diazonium moieties affects the surface conductance of s- and m-SWCNTs to different degrees, allowing separation via dielectrophoresis [89, 90, 93]. Most recently, the preparation of highly pure (~99%) fractions of s- and m-SWCNTs has been achieved using density-gradient centrifugation [94].

Reports are just beginning to emerge on the use of separated SWCNT fractions to produce transparent conducting thin films. In one report comparing an enriched m-SWCNT film to a bulk, mixed film, the conductivity of the m-SWCNT film was found to be more stable upon soaking in nitric acid [95]. In another report, films of varying colors were produced by m-SWCNT fractions from different production methods (e.g. laser, HiPCO), along with conductivity enhancements for films enriched in m-SWCNTs compared with bulk films [96]. Finally, a recent report has shown that multiple fractions, with varying degrees of m-SWCNT enrichment, can be combined to span the entire range of m-SWCNT content, from ~5% m-SWCNTs to ~99% m-SWCNTs [44]. Such films provide a means to understand how tubes of differing electronic structures contribute to the properties observed for films that have been studied to date.

Figure 7.9 shows background-subtracted [97] UV-Vis-IR spectra for a series of transparent SWCNT films with varying m- and s-SWCNT contents. The peak envelopes corresponding to intrinsic excitonic transitions of semiconducting (S_{11}, S_{22}, S_{33}) and metallic (M_{11}, M_{22}) SWCNTs can be clearly seen in these spectra, and their relative weighting varies as the fraction of each type of nanotube is varied. The fraction of m-SWCNTs in each film is used as a relative metric by which we may evaluate type-dependent trends. The calculation of m-SWCNT content is based on the integrated areas underneath the S_{22} and M_{11} peak envelopes.

Chemical doping affects metal- and semiconductor-enriched films very differently, as shown in Figure 7.10(a). $SOCl_2$ and HNO_3 remove essentially all oscillator strength from the S_{11} and S_{22} peak envelopes in the semiconductor-enriched films, while dramatically decreasing the sheet resistance to ~180 Ω sq^{-1}. A free-carrier-induced plasma absorption

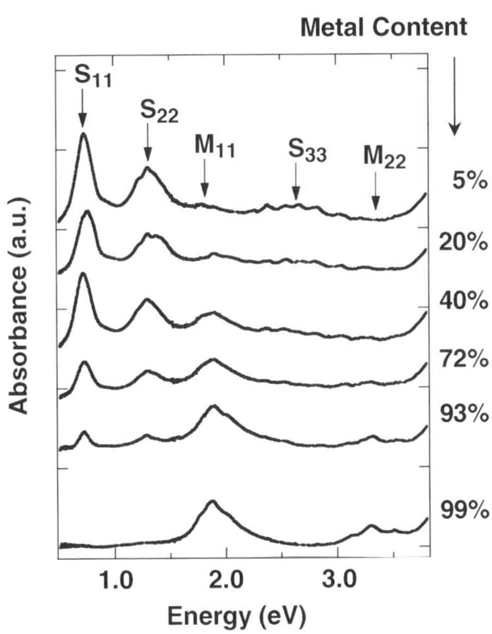

Figure 7.9 *Corrected absorbance spectra of a range of films with varying m- and s-SWCNT contents. For clarity, the spectra have had the rising ultraviolet background subtracted to produce a relatively flat background in the range of 0.5–4 eV, based on the procedure of Nair et al. [97]*

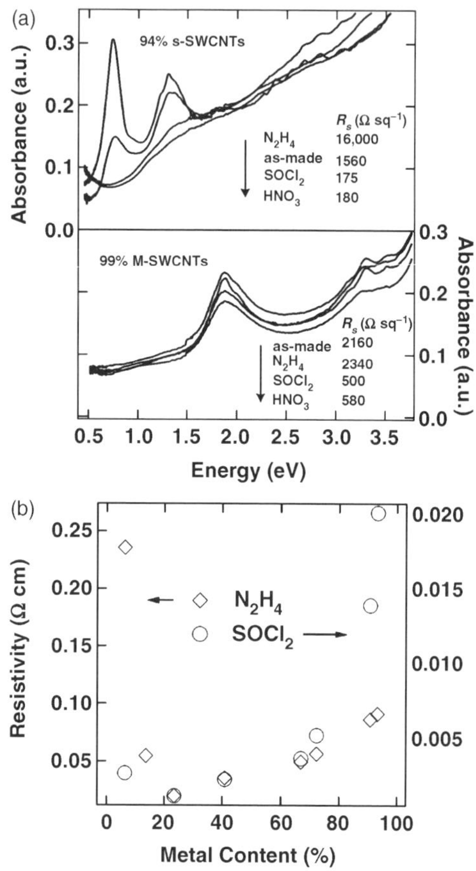

Figure 7.10 *(a) Absorbance spectra of s-SWCNT (top) and m-SWCNT (bottom) enriched films after different chemical treatments (or as-prepared). The chemical treatment for each film is shown, as well as the associated sheet resistance measured after each chemical treatment. Note the y-axis for the m-SWCNT film is on the right axis. (b) Resistivity as a function of metal content for networks doped with $SOCl_2$ and N_2H_4*

feature appears in the highly doped films in the IR as well. These optical and electrical changes can be reversed by dedoping the film with hydrazine. In contrast, the optical and electrical changes observed for the metal-enriched films are much less pronounced. Upon doping with $SOCl_2$ or HNO_3, the M_{11} and M_{22} peak envelopes are broadened and lose some of the fine structure arising from individual m-SWCNT transitions, but do not suffer a significant loss (<5%) of integrated optical density [44].

Figure 7.10(b) shows the variation of film resistivity as a function of the m-SWCNT content for films treated with hydrazine and thionyl chloride. For hydrazine-treated films, semiconductor-enriched films are much more resistive than metal-enriched films, consistent with a low density of carriers at the Fermi level for the intrinsic s-SWCNTs and significant carrier density for the intrinsic m-SWCNTs. In the thionyl chloride-treated films, the opposite trend is seen, and semiconductor-enriched films out-perform the metal-enriched films. For samples

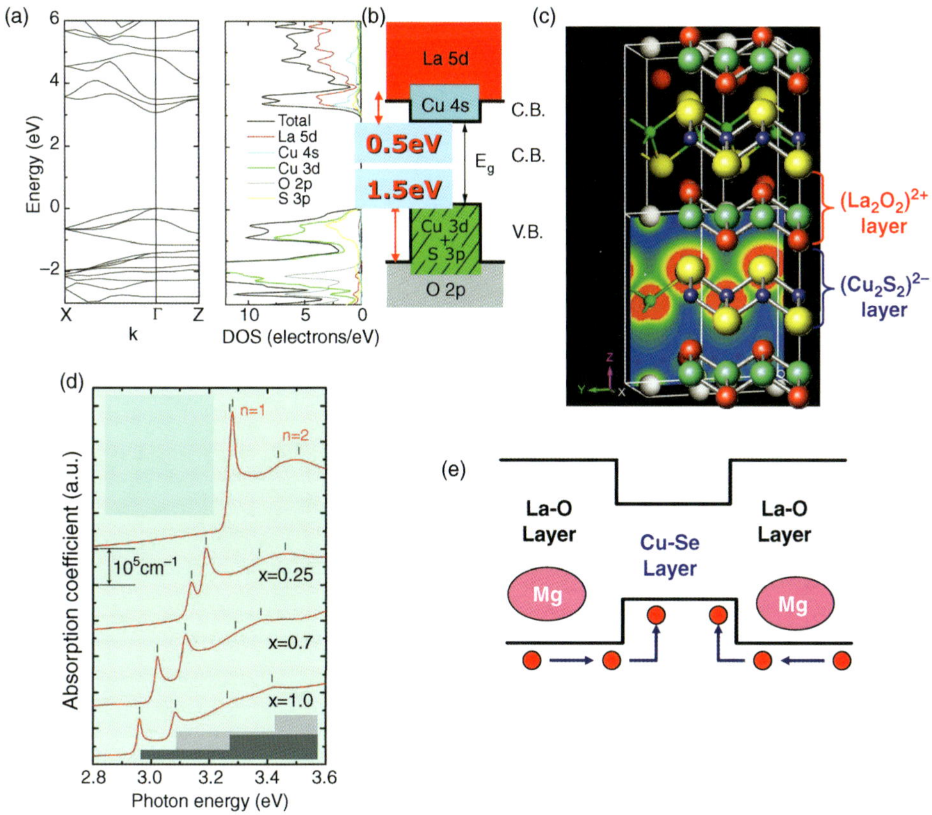

Figure 2.9 *Two-dimensional electronic structure in LnCuOCh. (a) Band structure of LaCuOS, (b) schematic illustration of electronic structure near the band gap, (c) hole density map, (d) two-dimensional optical absorption spectra obtained from LaCuO($S_{1-x}Se_x$) and (e) schematic illustration of local electronic structure to explain natural modulation doping of wide band gap semiconductors*

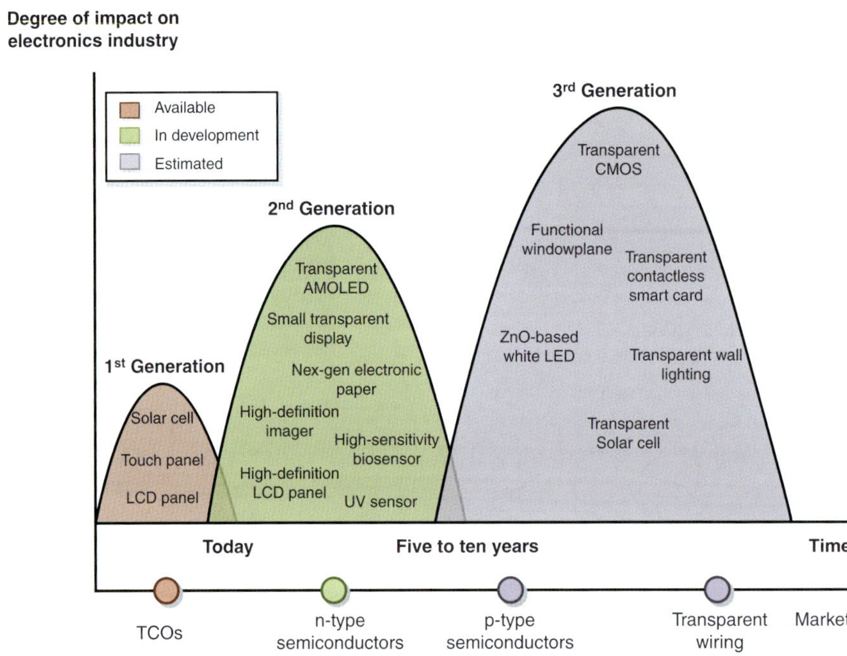

Figure 6.39 *The 1st generation in transparent electronics is based on transparent electrodes using mainly ITO for LCD and touch panels. The 2nd generation is related to the development of n-type semiconductors for TFTs and integrated circuits while the 3rd generation will envisage low cost white LEDs and low dissipation CMOS integrated circuits (adapted from [112])*

Figure 8.8 *Photographs of a reversible display seen from the TFT array side (a) and the frontplane side (b). Reversible images are displayed on the opposite side (c)*

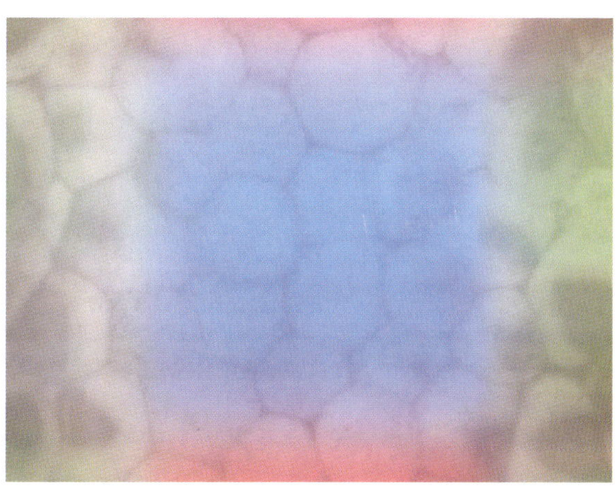

Figure 8.12 *Micrograph of microcapsules seen through a color filter array and transparent TFT array*

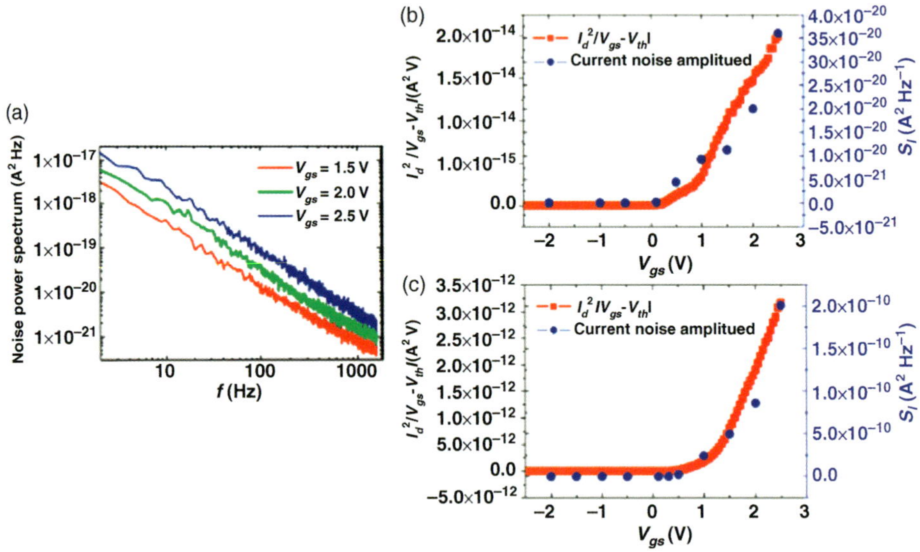

Figure 10.5 (a) Current noise power spectrum of a SnO_2 NWT as a function of frequency at a constant drain bias of 0.1 V (from top to bottom: $V_{gs}=2.5$, 2.0 and 1.5 V). (b) Measured $I_d^2(V_{gs}-V_{th})$ and the amplitude of current noise spectrum at 100 Hz versus gate bias at drain bias of 0.1 V. (c) Measured $I_d^2(V_{gs}-V_{th})$ and current noise spectrum amplitude at 100 Hz versus gate bias at drain bias of 1 V. Reprinted with permission from ref. 28. Copyright 2008, American Institute of Physics

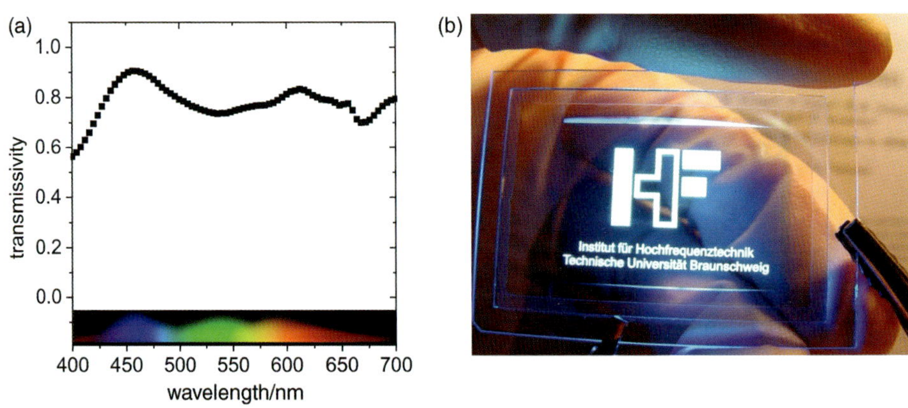

Figure 12.4 (a) Overall transmissivity of a device with a 60 nm thick WO_3 buffer layer and (b) photograph of a transparent white light OLED

with very low m-SWCNT content, the resistivity of the hydrazine-treated films is 100 times that of the highly conductive redox-doped films, whereas only a factor of 4 difference is observed for the high m-SWCNT contents. In fact, a doped s-SWCNT film (6% m-SWCNT) has significantly lower sheet resistance than a similarly doped 94% m-SWCNT film [44].

7.10.4 Temperature-Dependent Effects and Transport

Temperature-dependent transport measurements can also be used to learn more about optoelectronic trends in the separated SWCNT films. Figure 7.11 shows temperature-dependent transport measurements on a highly metal-enriched and highly semiconductor-enriched film. For both films, as the temperature is raised from 100 K to room temperature, the sheet resistance decreases with an exponential dependence. The temperature-dependent resistance data strongly support the hypothesis that inter-tube junctions limit the conductivity of these transparent SWCNT films, because a progressive decrease in resistance with increasing temperature is not expected for a film with purely metallic conduction (e.g. the m-SWCNT film). The tunneling term in Equation (7.1) was used to fit the temperature-dependent data shown in Figure 7.11, and the fits are shown as solid lines. The T_b parameter is a function of the tunnel barrier height and shape as affected by the image force and local field [81]. Upon *p*-type doping, T_b is significantly lowered for both semiconductor- and metal-enriched films. If barrier energies are lowered via chemical modification, conductance

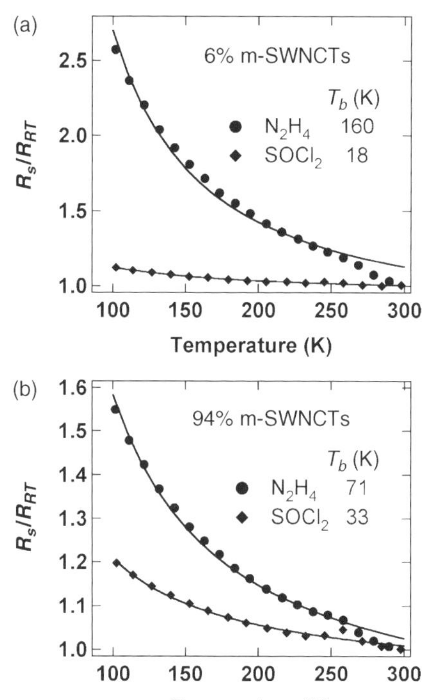

Figure 7.11 *Resistance as a function of temperature for a semiconductor-enriched (a) and metal-enriched (b) SWCNT film. Solid lines show fits to the tunneling term of Equation (7.1)*

through the barriers increases and the exponential temperature dependence becomes shallower [45]. This suggests that $SOCl_2$, HNO_3, or O_2 molecules, adsorbed at or near tube-tube junctions, create local electric fields that distort the barriers and reduce the mean barrier height. Figure 7.11 demonstrates that T_b is reduced for the semiconductor-enriched films to a greater degree than for the metal-enriched films. This suggests that the *p*-type dopants increase conductance through tunnel junctions between s-SWCNTs more effectively than junctions between m-SWCNTs, helping to explain the trends seen in Figure 7.9(b).

Finally, for separated SWCNT films, it is particularly interesting to note that evidence of apparent metallic behavior at $T > T^*$ (positive dR_s/dT) decreases as the metallic fraction increases, as shown in Figure 7.12(a). That is, R_s at $T > T^*$ increases dramatically for

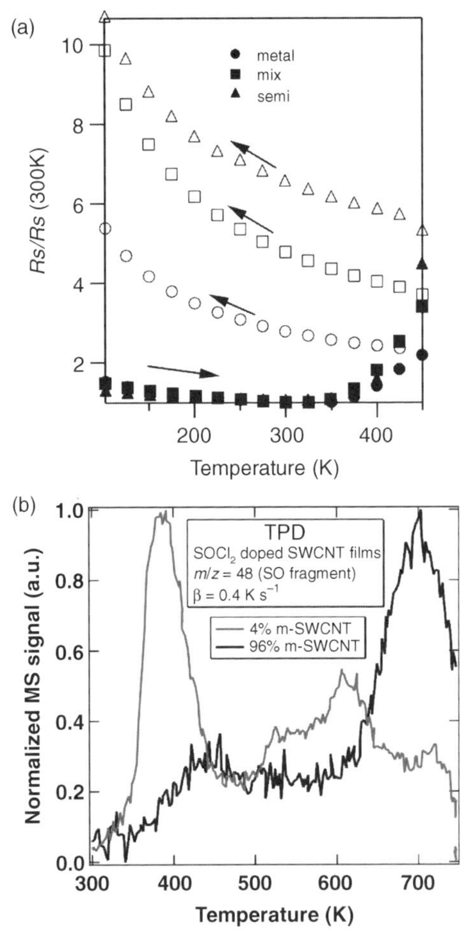

Figure 7.12 *(a) R_s of nitric acid treated samples as a function of temperature. Samples were first cooled, then R_s was measured during one heating and cooling cycle. The arrows indicate the direction of temperature change. (b) TPD data showing thermal desorption characteristics of $SOCl_2$. Note that the $SOCl_2$ desorbs from the semiconducting sample at ~380 K compared with at ~700 K from the metal sample*

samples with the lowest metallic tube contents, and it either decreases or stays stable for the nearly pure metallic tubes. Above T^*, there is a consistent decrease in the slope of R_s as a function of temperature with increasing metallic content. Also, as shown in Figure 7.11, the metal-enriched SWCNT films show the least hysteresis with temperature cycling, whereas the semiconducting samples show the most. These trends suggest a difference in binding energy for redox dopants on metallic and semiconducting SWCNTs. Dopant desorption from the networks can be readily confirmed by temperature-programmed desorption (TPD) measurements on $SOCl_2$-treated films. Figure 7.12 shows that $SOCl_2$ desorbs rapidly at about 400 K from the semiconducting network, whereas the desorption peak in the metal-enriched network is at almost 700 K. This is consistent with the temperature-dependent resistivity measurements that show that metal-enriched films undergo less change after heating to 450 K. The TPD results demonstrate that significant amounts of dopant desorb from the semiconductor-enriched samples at temperatures well below 450 K.

The use of separated SWCNTs to optimize transparency and conductivity in SWCNT thin films is still in the very early stages. Inevitably, many new reports will be published on this rapidly growing field of research before the publication of this book.

7.11 Conclusions

SWCNTs are particularly suited to applications where the advantages of their inherent flexibility, high hole-conductivity, and amenability to solution processing outweigh any potential losses due to absorbance. Amorphous metal oxides, such as $In_xZn_{1-x}O$, appear to meet the requirement of flexibility [98] but these materials still contain costly indium. Satisfactory *p*-type or hole conducting has not been yet demonstrated in a readily available TCO material [99]. Demanding applications such as flexible PV place increasingly stringent requirements on transparent contacts such as stability, flexibility, conductivity, cost, and transparency, and researchers have begun to explore beyond the traditional TCO realm to meet them. SWCNT or other nanostructured networks show considerable advantages for many applications, and their use will increase as their optoelectronic performance improves.

Acknowledgements

This work was supported under DOE contract no. DE-AC36-99GO10337. We thank the High-Efficiency Concepts Program and the US Department of Energy's Laboratory Directed Research and Development Fund for support. We also thank Tim Coutts, Mike Heben, Robert Tenent and Jao van de Lagemaat for their insights on this work.

References

[1] Wu, Z.; Zhihong, C.; Xu, D.; Logan, J. M.; Sippel, J.; Nikolou, M.; Kamaras, K.; Reynolds, J. R.; Tanner, D. B.; Hebard, A. F.; Rinzler, A. G., Transparent, conductive carbon nanotube films. *Science*, **305**, 1273–1276 (2004).

[2] Fanchini, G.; Unalan, H. E.; Chhowalla, M., Optoelectronic properties of transparent and conducting single-wall carbon nanotube thin films. *Appl. Phys. Lett.*, **88**, 191 919 (2006).

[3] Yu, X.; Rajamani, R.; Stelson, K. A.; Cui, T., Carbon nanotube based transparent conductive thin films. *J. Nanosci. Nanotech.*, **6**, 1939–1944 (2006).

[4] Gruner, G., Carbon nanotube films for transparent and plastic electronics. *J. Mater. Chem.*, **16**, 3533–3539 (2006).

[5] Barnes, T. M.; van de Lagemaat, J.; Levi, D.; Rumbles, G.; Coutts, T. J.; Weeks, C. L.; Britz, D. A.; Levitsky, I.; Peltola, J.; Glatkowski, P., Optical characterization of highly conductive single-wall carbon-nanotube transparent electrodes. *Phy. Rev. B*, **75**, 235 410 (2007).

[6] van de Lagemaat, J.; Barnes, T. M.; Rumbles, G.; Shaheen, S. E.; Coutts, T. J.; Weeks, C.; Levitsky, I.; Peltola, J.; Glatkowski, P., Organic solar cells with carbon nanotubes replacing In_2O_3:Sn as the transparent electrode. *Appl. Phys. Lett.*, **88**, 233 503 (2006).

[7] Rowell, M. W.; Topinka, M. A.; McGehee, M. D.; Prall, H. J.; Dennler, G.; Sariciftci, N. S.; Hu, L. B.; Gruner, G., Organic solar cells with carbon nanotube network electrodes. *Appl. Phys. Lett.*, **88**, 233 506 (2006).

[8] Contreras, M. A.; Barnes, T.; vandeLagemaat, J.; Rumbles, G.; Coutts, T. J.; Weeks, C.; Glatkowski, P.; Levitsky, I.; Peltola, J.; Britz, D. A., Replacement of transparent conductive oxides by single-wall carbon nanotubes in Cu(In,Ga)Se_2-based solar cells. *J. Phys. Chem. C*, **111**, 14 045–14 048 (2007).

[9] Barnes, T. M.; Wu, X.; Zhou, J.; Duda, A.; van de Lagemaat, J.; Coutts, T. J.; Weeks, C. L.; Britz, D. A.; Glatkowski, P., Single-wall carbon nanotube networks as a transparent back contact in CdTe solar cells. *Appl. Phys. Lett.*, **90**, 243 503 (2007).

[10] Lee, K.; Wu, Z.; Chen, Z.; Ren, F.; Pearton, S. J.; Rinzler, A. G., Single wall carbon nanotubes for p-type ohmic contacts to GaN light-emitting diodes. *Nano Lett.*, **4**, 911–914 (2004).

[11] Artukovic, E.; Kaempgen, M.; Hecht, D. S.; Roth, S.; GrUner, G., Transparent and flexible carbon nanotube transistors. *Nano Lett.*, **5**, 757–760 (2005).

[12] Aguirre, C. M.; Auvray, S.; Pigeon, S.; Izquierdo, R.; Desjardins, P.; Martel, R., Carbon nanotube sheets as electrodes in organic light-emitting diodes. *Appl. Phys. Lett.*, **88**, 183 104 (2006).

[13] Snow, E. S.; Perkins, F. K.; Houser, E. J.; Badescu, S. C.; Reinecke, T. L., Chemical detection with a single-walled carbon nanotube capacitor. *Science*, **307**, 1942–1945 (2005).

[14] Dresselhaus, M. S.; Dresselhaus, G.; Saito, R., C_{60}-related tubules. *Solid State Commun.*, **84**, 201–205 (1992).

[15] Dillon, A. C.; Yudasaka, M.; Dresselhaus, M. S., Employing Raman spectroscopy to qualitatively evaluate the purity of carbon single-wall nanotube materials. *J. Nanosci. Nanotech.*, **4**, 691–703 (2004).

[16] Samsonidze, G. G.; Saito, R.; Kobayashi, N.; Gruneis, A.; Jiang, J.; Jorio, A.; Chou, S. G.; Dresselhaus, G.; Dresselhaus, M. S., Family behavior of the optical transition energies in single-wall carbon nanotubes of smaller diameters. *Appl. Phys. Lett.*, **85**, 5703–5705 (2004).

[17] Dillon, A. C.; Gennett, T.; Jones, K. M.; Alleman, J. L.; Parilla, P. A.; Heben, M. J., A simple and complete purification of single-walled carbon nanotube materials. *Adv. Mater. (Weinheim, Ger.)*, **11**, 1354–1358 (1999).

[18] Dillon, A. C.; Parilla, P. A.; Alleman, J. L.; Perkins, J. D.; Heben, M. J., Controlling single-wall nanotube diameters with variation in laser pulse power. *Chem. Phys. Lett.*, **316**, 13–18 (2000).

[19] Bronikowski, M. J.; Willis, P. A.; Colbert, D. T.; Smith, K. A.; Smalley, R. E., Gas-phase production of carbon single-walled nanotubes from carbon monoxide via the HipCO process: a parametric study. *NAN06*, AVS, Boston, MA, pp 1800–1805 (2001).

[20] Dillon, A. C.; Mahan, A. H.; Alleman, J. L.; Heben, M. J.; Parilla, P. A.; Jones, K. M., Hot-wire chemical vapor deposition of carbon nanotubes. *Thin Solid Films*, **430**, 292–295 (2003).

[21] Chiang, I. W.; Brinson, B. E.; Huang, A. Y.; Willis, P. A.; Bronikowski, M. J.; Margrave, J. L.; Smalley, R. E.; Hauge, R. H., Purification and characterization of single-wall carbon nanotubes (SWNTs) obtained from the gas-phase decomposition of CO (HipCO process). *J. Phys. Chem. B*, **105**, 8297–8301 (2001).

[22] Xu, Y. Q.; Peng, H.; Hauge, R. H.; Smalley, R. E., Controlled multistep purification of single-walled carbon nanotubes. *Nano Lett.*, **5**, 163–168 (2005).

[23] Rinzler, A.; Liu, J.; Dai, H.; Nikolaev, P.; Huffman, C. B.; Rodriguez-Macias, F. J.; Boul, P. J.; Lu, A. H.; Heymann, D.; Colbert, D. T.; Lee, R. S.; Fischer, J. E.; Rao, A. M.; Eklund, P. C.; Smalley, R. E., Large-scale purification of single-wall carbon nanotubes: proces, product, and characterization. *Appl. Phys. A*, **67**, 29–37 (1998).

[24] Chiang, I. W.; Brinson, B. E.; Smalley, R. E.; Margrave, J. L.; Hauge, R. H., Purification and characterization of single-wall carbon nanotubes. *J. Phys. Chem. B*, **105**, 1157 (2001).

[25] Wang, Y. H.; Shan, H.; Hauge, R. H.; Pasquali, M.; Smalley, R. E., A highly selective, one-pot purification method for single-walled carbon nanotubes. *J. Phys. Chem. B*, **111**, 1249–1252 (2007).

[26] Rinzler, A. G.; Liu, J.; Dai, H.; Nikolaev, P.; Huffman, C. B.; Rodriguez-Macias, F. J.; Boul, P. J.; Lu, A. H.; Heymann, D.; Colbert, D. T.; Lee, R. S.; Fischer, J. E.; Rao, A. M.; Eklund, P. C.; Smalley, R. E., Large-scale purification of single-wall carbon nanotubes: process, product, and characterization. *Appl. Phys. A*, **67**, 29–37 (1998).

[27] Vivekchand, S. R. C.; Govindaraj, A.; Seikh, M. M.; Rao, C. N. R., New method of purification of carbon nanotubes based on hydrogen treatment. *J. Phys. Chem. B*, **108**, 6935–6937 (2004).

[28] Wu, Z.; Chen, Z.; Du, X.; Logan, J. M.; Sippel, J.; Nikolou, M.; Kamaras, K.; Reynolds, J. R.; Tanner, D. B.; Hebard, A. F.; Rinzler, A. G., Transparent, conductive carbon nanotube films. *Science*, **305**, 1273–1276 (2004).

[29] Yu, A.; Bekyarova, E.; Itkis, M. E.; Fakhrutdinov, D.; Webster, R.; Haddon, R. C., Application of centrifugation to the large-scale purification of electric arc-produced single-walled carbon nanotubes. *J. Am. Chem. Soc.*, **128**, 9902–9908 (2006).

[30] Hu, H.; Yu, A.; Kim, E.; Zhao, B.; Itkis, M. E.; Bekyarova, E.; Haddon, R. C., Influence of the zeta potential on the dispersability and purification of single-walled carbon nanotubes. *J. Phys. Chem. B*, **109**, 11520–11524 (2005).

[31] Moore, V. C.; Strano, M. S.; Haroz, E. H.; Hauge, R. H.; Smalley, R. E., Individually suspended single-walled carbon nanotubes in various surfactants. *Nano Lett.*, **3**, 1379 (2003).

[32] Islam, M. F.; Rojas, E.; Bergey, D. M.; Johnson, A. T.; Yodh, A. G., High weight fraction surfactant solubilization of single-wall carbon nanotubes in water. *Nano Lett.*, **3**, 269–273 (2003).

[33] Zhou, Y.; Hu, L.; Gruner, G., A method of printing carbon nanotube thin films. *Appl. Phys. Lett.*, **88**, 123109–3 (2006).

[34] Kaempgen, M.; Lebert, M.; Haluska, M.; Nicoloso, N.; Roth, S., Sonochemical optimization of the conductivity of single wall carbon nanotube networks. *Adv. Mater.*, **20**, 616–620 (2008).

[35] Tenent, R. C.; Barnes, T. M.; Bergeson, J. D.; Ferguson, A. J.; To, B.; Gedvilas, L. M.; Heben, M. J.; Blackburn, J. L., Ultrasmooth, large-area, high-uniformity, conductive transparent single-walled-carbon-nanotube films for photovoltaics produced by ultrasonic spraying. *Adv. Mater.*, **21**, 3210–3216 (2009).

[36] Song, Y. I.; Yang, C. M.; Kim, D. Y.; Kanoh, H.; Kaneko, K., Flexible transparent conducting single-wall carbon nanotube film with network bridging method. *J. Colloid Interface Sci.*, **318**, 365–371 (2008).

[37] Yim, J. H.; Kim, Y. S.; Koh, K. H.; Lee, S., Fabrication of transparent single wall carbon nanotube films with low sheet resistance. *J. Vac. Sci. Tech. B*, **26**, 851–855 (2008).

[38] Ma, W. J.; Song, L.; Yang, R.; Zhang, T. H.; Zhao, Y. C.; Sun, L. F.; Ren, Y.; Liu, D. F.; Liu, L. F.; Shen, J.; Zhang, Z. X.; Xiang, Y. J.; Zhou, W. Y.; Xie, S. S., Directly synthesized strong, highly conducting, transparent single-walled carbon nanotube films. *Nano Lett.*, **7**, 2307–2311 (2007).

[39] Skakalova, V.; Kaiser, A. B.; Dettlaff-Weglikowska, U.; Hrncarikova, K.; Roth, S., Effect of chemical treatment on electrical conductivity, infrared absorption, and raman spectra of single-walled carbon nanotubes. *J. Phys. Chem. B*, **109**, 7174–7181 (2005).

[40] Collins, P. G.; Bradley, K.; Ishigami, M.; Zettl, A., Extreme oxygen sensitivity of electronic properties of carbon nanotubes. *Science*, **287**, 1801 (2000).

[41] Dettlaff-Weglikowska, U.; Skakalova, V.; Graupner, R.; Jhang, S. H.; Kim, B. H.; Lee, H. J.; Ley, L.; Park, Y. W.; Berber, S.; Tomanek, D.; Roth, S., Effect of $SOCl_2$ treatment on electrical and mechanical properties of single-wall carbon nanotube networks. *J. Am. Chem. Soc.*, **127**, 5125–5131 (2005).

[42] Skakalova, V.; Kaiser, A. B.; Dettlaff-Weglikowska, U.; Hrncarikova, K.; Roth, S., Effect of chemical treatment on electrical conductivity, infrared absorption, and raman spectra of SWCNT. *J. Phys. Chem. B*, **109**, 7174–7181 (2005).

[43] Zhang, D.; Ryu, K.; Liu, X.; Polikarpov, E.; Ly, J.; Tompson, M. E.; Zhou, C., Transparent, conductive, and flexible carbon nanotube films and their application in organic light-emitting diodes. *Nano Lett.*, **6**, 1880–1886 (2006).

[44] Blackburn, J. L.; Barnes, T. M.; Beard, M. C.; Kim, Y.-H.; Tenent, R. C.; McDonald, T. J.; Coutts, T. J.; To, B.; Heben, M. J., Transparent conductive single-walled carbon nanotube networks with precisely tunable ratios of semiconducting and metallic nanotubes. *ACS Nano*, **2**, 1266–1274 (2008).

[45] Zhou, W.; Vavro, J.; Nemes, N. M.; Fischer, J. E.; Borondics, F.; Kamaras, K.; Tanner, D. B., Charge transfer and fermi level shift in p-doped single-walled carbon nanotubes. *Phys. Rev. B*, **71**, 205423 (2005).

[46] Lee, R. S.; Kim, H. J.; Fischer, J. E.; Thess, A.; Smalley, R. E., Conductivity enhancement in single-walled carbon nanotube bundles doped with K and Br. *Nature*, **388**, 255–257 (1997).

[47] Klinke, C.; Chen, J.; Afzali, A.; Avouris, P., Charge transfer induced polarity switching in carbon nanotube transistors. *Nano Lett.*, **5**, 555–558 (2005).

[48] Bhavin, B. P.; Giovanni, F.; Goki, E.; Manish, C., Improved conductivity of transparent single-wall carbon nanotube thin films via stable postdeposition functionalization. *Appl. Phys. Lett.*, **90**, 121913 (2007).

[49] Giovanni, F.; Husnu Emrah, U.; Manish, C., Modification of transparent and conducting single wall carbon nanotube thin films via bromine functionalization. *Appl. Phys. Lett.*, **90**, 092 114 (2007).

[50] Barnes, T. M.; Blackburn, J. L.; van de Lagemaat, J.; Coutts, T. J.; Heben, M. J., Reversibility, dopant desorption, and tunneling in the temperature-dependent conductivity of type-separated, conductive carbon nanotube networks. *ACS Nano*, **2**, 1968–1976 (2008).

[51] Wang, F.; Dukovic, G.; Brus, L. E.; Heinz, T. F., The optical resonances in carbon nanotubes arise from excitons. *Science*, **308**, 838 (2005).

[52] Ma, Y.-Z.; Valkunas, L.; Bachilo, S. M.; Fleming, G. R., Exciton binding energy in semiconducting single-walled carbon nanotubes. *J. Phys. Chem. B*, **109**, 15671–15674 (2005).

[53] Wang, F.; Cho, D. J.; Kessler, B.; Deslippe, J.; Schuck, P. J.; Louie, S. G.; Zettl, A.; Heinz, T. F.; Shen, Y. R., Observation of excitons in one-dimensional metallic single-walled carbon nanotubes. *Phys. Rev. Lett.*, **99**, 227 401 (2007).

[54] Trottier, C. M.; Glatkowski, P.; Wallis, P.; Luo, J., Properties and characterization of carbon-nanotube-based transparent conductive coating. *J. Soc. Inform. Disp.*, **13**, 759–763 (2005).

[55] Landi, B. J.; Ruf, H. J.; Evans, C. M.; Cress, C. D.; Raffaelle, R. P., Purity assessment of single-wall carbon nanotubes, using optical absorption spectroscopy. *J. Phys. Chem. B*, **109**, 9952–9965 (2005).

[56] Kataura, H.; Kumazawa, Y.; Maniwa, Y.; Umezu, I.; Suzuki, S.; Ohtsuka, Y.; Achiba, Y., Optical properties of single-wall carbon nanotubes. *Synth. Met.*, **103**, 2555–2558 (1999).

[57] Liu, X.; Pichler, T.; Knupfer, M.; Golden, M. S.; Fink, J.; Walters, D. A.; Casavant, M. J.; Schmidt, J.; Smalley, R. E., An electron energy-loss study of the structural and electronic

properties of magnetically aligned single wall carbon nanotubes. *Synth. Met.*, **121**, 1183–1186 (2001).
[58] Pichler, T.; Knupfer, M.; Golden, M. S.; Fink, J.; Rinzler, A.; Smalley, R. E., Localized and delocalized electronic states in single-wall carbon nanotubes. *Phys. Rev. Lett.*, **80**, 4729–4732 (1998).
[59] Liu, X.; Pichler, T.; Knupfer, M.; Golden, M. S.; Fink, J.; Kataura, H.; Achiba, Y., Detailed analysis of the mean diameter and diameter distribution of single-wall carbon nanotubes from their optical response. *Phys. Rev. B*, **66**, 045411 (2002).
[60] Johs, B.; Herzinger, C.; Guenther, B., *V.A.S.E. For windows,* 3.511. J. A. Woolam Co., Lincoln, NE (2005).
[61] Borondics, F.; Kamaras, K.; Nikolou, M.; Tanner, D. B.; Chen, Z. H.; Rinzler, A. G., Charge dynamics in transparent single-walled carbon nanotube films from optical transmission measurements. *Phys. Rev. B*, **74**, 045 431 (2006).
[62] Coutts, T. J.; Young, D. L.; Li, X., Characterization of transparent conducting oxides. *MRS Bull.*, **25**, 58 (2000).
[63] Gruneis, A.; Saito, R.; Samsonidze, G. G.; Kimura, T.; Pimenta, M. A.; Jorio, A.; Souza, A. G.; Dresselhaus, G.; Dresselhaus, M. S., Inhomogeneous optical absorption around the k point in graphite and carbon nanotubes. *Phys. Rev. B*, **67**, 165 402 (2003).
[64] Hwang, J.; Gommans, H. H.; Ugawa, A.; Tashiro, H.; Haggenmueller, R.; Winey, K. I.; Fischer, J. E.; Tanner, D. B.; Rinzler, A. G., Polarized spectroscopy of aligned single-wall carbon nanotubes. *Phys. Rev. B*, **62**, R13310 (2000).
[65] Murakami, Y.; Einarsson, E.; Edamura, T.; Maruyama, S., Polarization dependance of the optical absorption of single-walled carbon nanotubes. *Phys. Rev. Lett.*, **94**, 087402 (2005).
[66] Javey, A.; Guo, J.; Wang, Q.; Lundstrom, M.; Dai, H., Ballistic carbon nanotube field-effect transistors. *Nature*, **424**, 654–657 (2003).
[67] Kong, J.; Yenilmez, E.; Tombler, T. W.; Kim, W.; Dai, H.; Laughlin, R. B.; Liu, L.; Jayanthi, C. S.; Wu, S. Y., Quantum interference and ballistic transmission in nanotube electron waveguides. *Phys. Rev. Lett.*, **87**, 106801/1–106801/4 (2001).
[68] Misewich, J. A.; Martel, R.; Avouris, P.; Tsang, J. C.; Heinze, S.; Tersoff, J., Electrically induced optical emission from a carbon nanotube FET. *Science*, **300**, 783–786 (2003).
[69] Baughman, R. H.; Zakhidov, A. A.; de Heer, W. A., Carbon nanotubes – the route towards applications. *Science*, **297**, 787 (2002).
[70] Skakalova, V.; Kaiser, A. B.; Woo, Y. S.; Roth, S., Electronic transport in carbon nanotubes: from individual nanotubes to thin and thick networks. *Phys. Rev. B*, **74**, 085403/1–085403/10 (2006).
[71] Hecht, D. S.; Hu, L.; Gruner, G., Conductivity scaling with bundle length and diameter in single walled carbon nanotube networks. *Appl. Phys. Lett.*, **89**, 133112 (2006).
[72] Kaiser, A. B.; Park, Y. W.; Kim, G. T.; Choi, E. S.; Dusberg, G.; Roth, S., Electronic transport in carbon nanotube ropes and mats. *Synth. Met.*, **103**, 2547–2550 (1999).
[73] Hu, L.; Hecht, D. S.; Gruner, G., Percolation in transparent and conducting carbon nanotube networks. *Nano Lett.*, **4**, 2513–2517 (2004).
[74] Stadermann, M.; Papadakis, S. J.; Falvo, M. R.; Novak, J.; Snow, E.; Fu, Q.; Liu, J.; Fridman, Y.; Boland, J. J.; Superfine, R.; Washburn, S., Nanoscale study of conduction through carbon nanotube networks. *Phys. Rev. B*, **69**, 201402(R) (2004).
[75] Romero, H. E.; Bolton, K.; Rosen, A.; Eklund, P. C., Atom collision-induced resistivity of carbon nanotubes. *Science*, **307**, 89–93 (2005).
[76] Bradley, K.; Gabriel, J.-C. P.; GrUner, G., Flexible nanotube electronics. *Nano Lett.*, **3**, 1353 (2003).
[77] Snow, E. S.; Novak, J. P.; Campbell, P. M.; Park, D., Random networks of carbon nanotubes as an electronic material. *Appl. Phys. Lett.*, **82**, 2145 (2003).
[78] Contreras, M.; Barnes, T. M.; van de Lagemaat, J.; Rumbles, G.; Coutts, T. J.; Weeks, C.; Glatkowski, P.; Levitsky, I.; Peltola, J., Application of single-wall carbon nanotubes as transparent electrodes in Cu(In,Ga)Se_2-based solar cells. *IEEE 4th WCPEC, IEEE,* Waikoloa, HI (2006).

[79] Fischer, J. E.; Dai, H.; Thess, A.; Lee, R.; Hanjani, N. M.; Dehaas, D. L.; Smalley, R. E., Metallic resistivity in crystalline ropes of single-wall carbon nanotubes. *Phys. Rev. B*, **55**, R4921 (1997).
[80] Kaiser, A. B.; Dusberg, G.; Roth, S., Heterogeneous model for conduction in carbon nanotubes. *Phys. Rev. B*, **57**, 1418 (1998).
[81] Sheng, P., Fluctuation-induced tunneling conduction in disordered materials. *Phys. Rev. B*, **21**, 2180 (1980).
[82] Shiraishi, M.; Ata, M., Conduction mechanisms in single-walled carbon nanotubes. *Synth. Met.*, **128**, 235–239 (2002).
[83] Durkop, T.; Kim, B. M.; Fuhrer, M. S., Properties and applications of high-mobility semiconducting nanotubes. *J. Phys.: Condens. Matter*, **16**, R553–R580 (2004).
[84] Barnes, T.; Blackburn, J. L.; Tenent, R. C.; Morfa, A. J.; Heben, M. J.; Coutts, T. J., Optimizing carbon nanotube contacts for use in organic photovoltaics. *IEEE PVSC 33*, IEEE, San Diego, CA (2008).
[85] Gordon, R. G., Criteria for choosing transparent conductors. *MRS Bull.*, **25**, 52–57 (2000).
[86] Jonsson, S. K. M.; Birgerson, J.; Crispin, X.; Greczynski, G.; Osikowicz, W.; van der Gon, A. W. D.; Salaneck, W. R.; Fahlman, M., The effects of solvents on the morphology and sheet resistance in poly (3,4-ethylenedioxythiophene)-polystyrenesulfonic acid (PEDOT-PSS) films. *Synth. Met.*, **139**, 1–10 (2003).
[87] Coutts, T. J.; Ward, J. S.; Young, D. L.; Emery, K. A.; Gessert, T. A.; Noufi, R., Critical issues in the design of polycrystalline, thin-film tandem solar cells. *Prog. Photovol.*, **11**, 359–375 (2003).
[88] Doyle, C. D.; Rocha, J.-D. R.; Weisman, R. B.; Tour, J. M., Structure-dependent reactivity of semiconducting single-walled carbon nanotubes with benzenediazonium salts. *J. Am. Chem. Soc.*, **130**, 6795–6800 (2008).
[89] Kim, W. J.; Usrey, M. L.; Strano, M. S., Selective functionalization and free solution electrophoresis of single-walled carbon nanotubes: Separate enrichment of metallic and semiconducting SWNT. *Chem. Mater.*, **19**, 1571–1576 (2007).
[90] Baik, S.; Usrey, M.; Rotkina, L.; Strano, M., Using the selective functionalization of metallic single-walled carbon nanotubes to control dielectrophoretic mobility. *J. Phys. Chem. B*, **108**, 15560–15564 (2004).
[91] Chattopadhyay, D.; Galeska, I.; Papadimitrakopoulos, F., A route for bulk separation of semiconducting from metallic single-wall carbon nanotubes. *J. Am. Chem. Soc.*, **125**, 3370 (2003).
[92] Maeda, Y.; Kimura, S. i.; Kanda, M.; Hirashima, Y.; Hasegawa, T.; Wakahara, T.; Lian, Y.; Nakahodo, T.; Tsuchiya, T.; Akasaka, T.; Lu, J.; Zhang, X.; Gao, Y.; Yu, Y.; Nagase, S.; Kazaoui, S.; Minami, N.; Shimizu, T.; Tokumoto, H.; Saito, R., Large-scale separation of metallic and semiconducting single-walled carbon nanotubes. *J. Am. Chem. Soc.*, **127**, 10287–10290 (2005).
[93] Kim, Y.; Hong, S.; Jung, S.; Strano, M. S.; Choi, J.; Baik, S., Dielectrophoresis of surface conductance modulated single-walled carbon nanotubes using catanionic surfactants. *J. Phys. Chem. B*, **110**, 1541–1545 (2006).
[94] Arnold, M. S.; Green, A. A.; Hulvat, J. F.; Stupp, S. I.; Hersam, M. C., Sorting carbon nanotubes by electronic structure using density differentiation. *Nature Nanotech.*, **1**, 60–65 (2006).
[95] Miyata, Y.; Yanagi, K.; Maniwa, Y.; Kataura, H., Highly stabilized conductivity of metallic single wall carbon nanotube thin films. *J. Phys. Chem. C*, **110**, 3591–3596 (2008).
[96] Green, A. A.; Hersam, M. C., Colored semitransparent conductive coatings consisting of monodisperse metallic single-walled carbon nanotubes. *Nano Lett.*, **8**, 1417–1422 (2008).
[97] Nair, N.; Usrey, M. L.; Kim, W.-J.; Braatz, R. D.; Strano, M. S., Estimation of the (n,m) concentration distribution of single-walled carbon nanotubes from photoabsorption spectra. *Anal. Chem.*, **78**, 7689–7696 (2006).
[98] Kang, J.-W.; Jeong, W.-I.; Kim, J.-J.; Kim, H.-K.; Kim, D.-G.; Lee, G.-H., High-performance flexible organic light-emitting diodes using amorphous indium zinc oxide anode. *Electrochem. Solid State Lett.*, **10**, J75–J78 (2007).

[99] Granqvist, C. G., Transparent conductors as solar energy materials: a panoramic review. *Sol. Energy Mater. Sol. Cells*, **91**, 1529–1598 (2007).
[100] Parekh, B. B.; Fanchini, G.; Eda, G.; Chhowalla, M., Improved conductivity of transparent single-wall carbon nanotube thin films via stable postdeposition functionalization. *Appl. Phys. Lett.*, **90**, 121913-1–121913-3 (2007).
[101] Kaempgen, M.; Duesberg, G. S.; Roth, S., Transparent carbon nanotube coatings. *Appl. Surf. Sci.*, **252**, 425 (2005).
[102] Lima, M. D.; de Andrade, M. J.; Bergmann, C. P.; Roth, S., Thin, conductive, carbon nanotube networks over transparent substrates by electrophoretic deposition. *J. Mater. Chem.*, **18**, 776–779 (2008).

8

Application of Transparent Amorphous Oxide Thin Film Transistors to Electronic Paper

Manabu Ito

Technical Research Institute, Toppan Printing Co., Ltd., Japan

8.1 Introduction

Paper has been used as the most dominant medium for more than 2000 years. It is thin, light and flexible, has a high contrast of excellent readability and offers convenience for storage and portability. Moreover, it consumes no power when it is read. However, ink on paper cannot be updated, so an enormous amount of paper is printed and thrown out. If paper can be replaced by an electronically rewritable form, a considerable number of forests would be saved.

Electronic display can be dynamically rewritable and compatible with digital contents. Information can be stored in the external device: therefore, there is no need to print out lots of documents. However, electronic display consumes a considerable amount of electricity. Low power consumption is preferable in a world, where hand-held devices are used anywhere and at any time. Moreover, present displays, for example, liquid crystal displays (LCDs), plasma display panels (PDPs) and organic light emitting diodes (OLEDs), are not suitable for 'intensive' or 'immersive' reading, compared with paper. Few people would imagine that a person would be willing to read a whole novel by electronic display.

Recently, electronic paper emerged as a next generation media. It has the advantages of both paper and electronic display. Electronic paper is electronically rewritable and

Transparent Electronics: From Synthesis to Applications Edited by Antonio Facchetti and Tobin J. Marks

consumes ultra low power with high brightness and contrast with full viewing angle like real paper. Applications of electronic paper include electronic book readers, electronic newspapers, electronic pricing labels in retail shops, watches, cellular phones, electronic billboards, time tables at bus stations and destination boards of subway trains. The market of electronic paper is growing rapidly. In 2004, the first electronic book, Σ Book, was commercialized by Matsushita, followed by LIBRIé of SONY. In 2006, Motorola brought an electronic paper cellular phone into the market. This cellular phone is designed for low-end and developing countries, using an electronic paper display instead of a conventional LCD display. Its function is limited but it has the advantages of low price, low power consumption and light weight (68 g). The Hamburg Hochbahn train installed the world's first onboard destination displays using electronic paper in 2006. Electronic paper display is regarded as an important innovation for rail transport and air traffic due its excellent readability, extremely low power consumption and simplicity of integration into existing passenger information systems given its thin profile and light weight. In 2007, Amazon Kindle, an electronic paper book, was launched by Amazon. With its outstanding business model, Amazon's first offering of the Kindle sold out in just five and a half hours. In 2008, a French newspaper company, Les Echos, began to distribute an electronic version of its newspaper to its dedicated reading devices on a subscription basis. In addition, Polymer Vision, a spin-off of Philips, announced that they would introduce the world's first rollable electronic paper display, Readius, to the market by mid 2008. In Readius, the display extends up to 5 in. and is stored away after use by being rolled up. The time has come for electronic paper to be available in our daily life.

In Figure 8.1 the required demands for electronic paper are summarized from high to low priority. Rewritable, reflective and image stable characteristics are indispensable for electronic paper. The rewritable aspect is the most fundamental one in electronic paper. The latter two characteristics are requisite for realizing ultra low power consumption. In reflective display, outside light is inevitable to view the displayed image: however, reflective display consumes far less power than emissive display, which requires considerable energy to generate its own light. Image stability offers further power saving because energy is consumed only when the image is changed, not to maintain it.

Additional benefits of electronic paper are that it should be thin and light, have a high contrast and display color image with high resolution and flexibility. In order to realize such an ideal electronic paper, not only the frontplane but also the backplane of the electronic

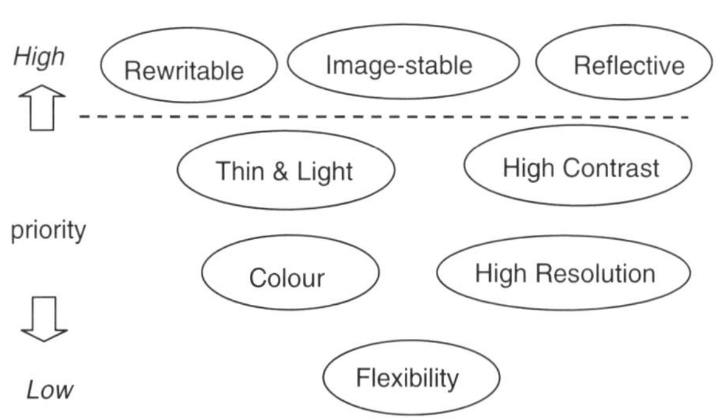

Figure 8.1 *Requirements for electronic paper*

paper play a critical role. A light, thin, flexible, low cost and high resolution thin film transistor (TFT) is needed.

Lately, oxide semiconductors have attracted considerable attention as novel candidates for channel materials of TFTs [1]. Various kinds of oxide semiconductors have been reported in the last few years, such as ZnO [2], $Mg_xZn_{1-x}O$ [3], Zn-Sn-O (ZTO) [4], In-Zn-O (IZO) [5], SnO_2 [6], Ga_2O_3 [7], In-Ga-O (IGO) [8], In_2O_3 [9], In-Sn-O (ITO) [10], single-crystalline $InGaO_3(ZnO)_5$ superlattice [11] and amorphous In-Ga-Zn-O (a-InGaZnO) [12].

One of the most important criteria for choosing an appropriate oxide semiconductor was proposed by Hosono *et al.* in 1996 [13]. They predicted that amorphous double oxides composed of heavy metal cations with an electronic configuration $(n\text{-}1)d^{10}ns^0$ $(n \geq 4)$ are promising candidates for semiconductor. These ns orbitals have large radii, so there is a large overlap between the adjacent orbitals, which leads to insensitivity to the distorted metal-oxygen-metal chemical bonds. Based upon the abovementioned criteria, they successfully fabricated a-InGaZnO TFT in 2004 [11]. Surprisingly, a high mobility of $7\,cm^2\,V^{-1}\,s^{-1}$ and high on/off ratio of more than 5 orders of magnitude were achieved even at room temperature. There are other oxide semiconductors; however, a-IGZO is superior in its high mobility, low process temperature, controllability of carrier density and long-term chemical stability.

Compared with other types of semiconductors, for example, amorphous Si (a-Si), low temperature poly-Si (LTPS) and organic semiconductors, transparent amorphous oxide semiconductors (TAOSs) can be characterized by the following features:

- high mobility
- processability for large area
- low process temperature
- transparency.

No other semiconductors can meet the abovementioned four characteristics simultaneously. The high mobility of TAOS TFTs is attractive for current-demanding applications, such as OLED display. Oxide semiconductors can be deposited by standard sputtering technique [14]; therefore, TAOS TFTs can be processed for large area such as mass production process of a-Si TFTs. An amorphous structure is also desirable for uniformity in a large area, grain-boundary-free structure, smooth surface and low process temperature. Presence of grain boundary tends to cause serious problems, such as segregation or degradation by grain boundary diffusion. As a TAOS is grain-boundary-free, it is expected that a TAOS TFT is stable for a longer period. A smooth surface is also preferable for realizing excellent device properties.

In this chapter, we focus our attention on the low process temperature and transparency of the TAOS and demonstrate the feasibility of combining a microencapsulated electrophoretic frontplane and a-IGZO TFT backplane [15].

8.2 Microencapsulated Electrophoretic Display

The first electronic paper, proposed by Sheridon and Berkovitz in 1977, was called Gyricon display. The Gyricon display consists of spherical beads, with optically

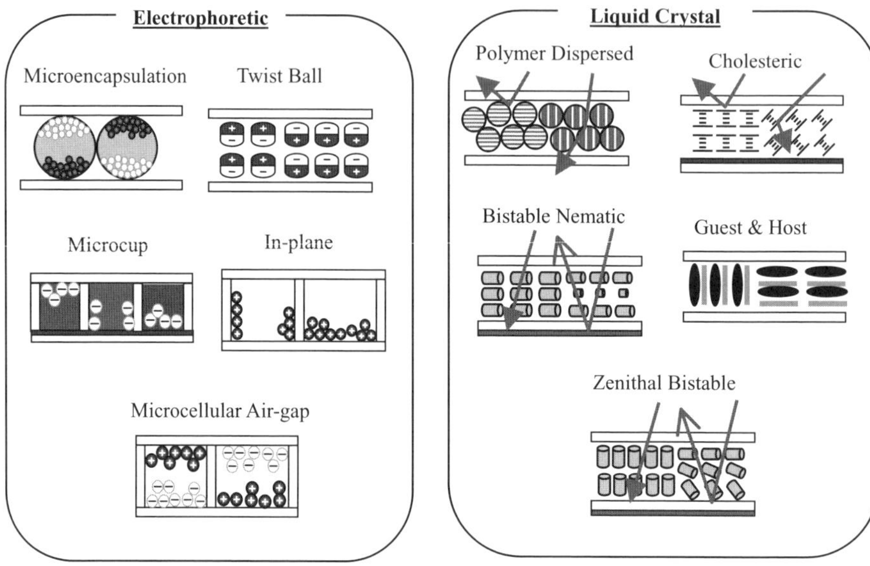

Figure 8.2 *Representative examples of electronic paper technology*

contrasting hemispheres, dispersed in a transparent rubber sheet. A spherical cavity around each bead is filled with oil, allowing it to rotate in response to an electric field.

Since then, various types of electronic paper technologies have been proposed, as shown in Figure 8.2. The electronic paper display technologies are roughly classified into two categories, liquid crystal type and electrophoretic type. In the liquid crystal type, liquid crystals changes their states by applying an electric filed and, in that manner, light can be managed. In the electrophoretic type, charged objects in a fluid translate in response to an electric field. Optical contrast is obtained by pulling the charged pigment particles to the front or back of the display. Electrophoretic display is also categorized into two groups from the fluid of the electrophoretic medium; liquid or air. Microencapsulation, twist ball, Microcup and in-plane displays belong to the liquid fluid type and the microcellular air-gap type uses air as a fluid. The switching speed of the electrophoretic display is inversely proportional the viscosity of the fluid. As the viscosity of the air is approximately 0.02 cP and that of fluid is typically on the order of 1 cP, switching speed of microcellular air-gap type display is expected to be two orders of magnitude higher than liquid fluid electrophoretic display. The switching time of the microcellular air-gap type display, named the Quick-Response Liquid Powder Display (QR-LPD), which is developed by Bridgestone, is quite fast at about 0.2 ms. However, the lack of long-term stability and high driving voltage of QR-LPD are still problems.

Among a number of electronic paper technologies, it is widely recognized that the microencapsulated electrophoretic display, which is developed by E Ink Corporation, is far superior to other types of technology from the technological and manufacturing point of view [16]. What distinguishes microencapsulated electrophoretic display is that the electrophoretic fluid is microencapsulated, instead of filling pre-existing microcellular structures [17]. Microencapsulated electrophoretic ink solved a longstanding problem of

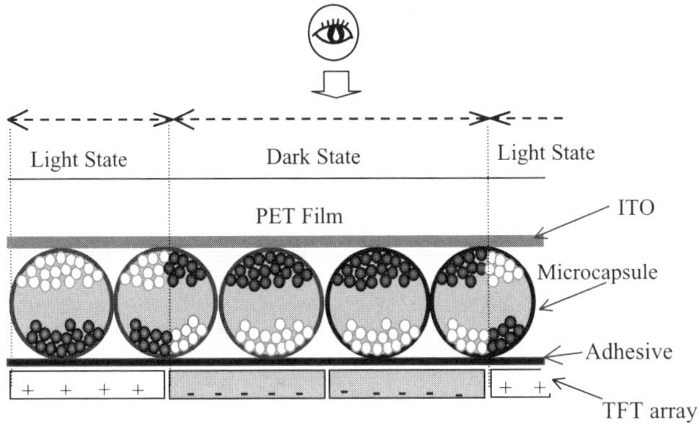

Figure 8.3 *Schematic cross-section view of the E Ink display*

electrophoretic display, which is the rather short lifetime due to particle clustering and agglomeration and lateral migration. Electrophoretic imaging film, which is called Vizplex imaging film, is composed of poly(ethylene terephthalate) (PET), an ITO layer, an ink layer, adhesive and release sheet, as shown in Figure 8.3. The adhesive layer and release sheet is coated over the microcapsule layer to facilitate the attachment to a backplane. The ink layer is comprised of millions of tiny microcapsules, which contain positively charged white pigment particles and negatively charged black particles suspended in a clear liquid fluid. When a negative electric field is applied to the top of the microcapsules, the white particles move to the top of the microcapsule where they become visible to the user. This makes the surface appear white on that spot. By reversing this process, a black appearance can be seen. The grayscale in Vizplex imaging film is achieved by partially addressing between the black and white states. Current Vizplex imaging film supports eight levels (3-bit) of grayscale.

The microencapsulated electrophoretic display, which is called the E Ink display, can be fabricated by removing the release sheet and laminating the Vizplex imaging film onto a TFT array. One of the most attractive features of the E Ink display, compared with Microcup or QR-LPD, is that the complicated alignment between imaging film and TFT array is not necessary. It should be noted that the resolution of the display is determined not by the size of the microcapsules but by the size of the pixel electrodes on the backplane because subcapsule addressing is possible, as shown in Figure 8.3.

E Ink display offers three key benefits. First, it has a paper like readability enjoying high reflectance, good contrast ratio and wide viewing angle because the mechanism of displaying the image in E Ink display is reflective, which is similar to that of ink on paper. Therefore, E Ink display is suitable for 'immersive reading', like reading a novel in display for hours as one does with a book. Secondly, E Ink display consumes ultra low power due to its image stability. Power is needed only when the image is changed. Power consumption of E Ink display is estimated to be less than a hundredth of that of an LCD. Thirdly, E Ink display is thin and light, as Vizplex imaging film is fabricated onto a PET substrate. A flexible display can be easily realized by combination with a flexible TFT array.

The driving voltage of E Ink display is 15 V, which is relatively low compared with other electronic paper displays: such as 80 V for QR-LPD. E Ink display can be driven by an active matrix or passive matrix backplane depending on the application. Obviously, active matrix drive is preferable in high resolution displays, such as an electronic book reader or electronic newspaper.

In E Ink display, charged pigment particles move up and down in liquid fluid. Therefore, switching speed of E Ink display is rather slow compared with a conventional flat panel display, for example, an LCD, PDP or OLED. Early microencapsulated electrophoretic films had a response time of 1 s. However, significant improvements have been made and a 30 ms black-to-white update time at 15 V drive voltage has been reported without a large sacrifice in other display characteristics, including image stability [18].

8.3 Flexible Electronic Paper

8.3.1 Flexible Display

Flexible display has been a topic of interest for a long time, and frontplane and backplane technology has now reached the point where reasonable flexible displays can be demonstrated [19]. Televisions on the wall or as handheld rollable displays will reach the market very soon. iSuppli Corporation estimated that the flexible market will grow to approximately US$ 350 million by 2013. Most of that market is estimated to be from the electrophoretic type of display, as exemplified by E Ink display in their market forecast. Therefore, until the early part of the next decade, E Ink and other types of electrophoretic display will be the main market for the flexible display. In this market, what kind of 'flexibility' is required? The definition of flexible displays can vary. Omodani proposed four states in flexible display [20]:

1. elastic but used in a flat rigid form;
2. curved or conformed;
3. bendable, like paper or cloth;
4. foldable, folding into a small area.

The first category, elastic, offers quite important benefits in the actual applications, even though it is not truly flexible. Display fabricated onto plastic substrate is nonfragile, thin and lightweight. Such kind of display can be readily used in mobile devices, where light weight and space-saving is strongly desired and rough handling can occur. For the second category, curved or conformed, displays are shaped during manufacturing but not flexed during use. A possible application is for an automotive application, such as front panels of cars. A curved display is appealing from the aesthetic point of view. The third category, bendable, offers a roll-up display. Roll-up display will drastically improve the portability of the handheld display. However, durability of such a display is still being questioned at present. The fourth category, foldable, is the most challenging one. A foldable display can be folded into a small area and can be stored in, say, pockets or bags. Such a type of display is, however, very ambitious, although the prospective benefits are high. One should choose the category carefully, taking into consideration of the applications.

An additional advantage of flexible display is the applicability of using a roll-to-roll manufacturing process. A continuous roll-to-roll production process will reduce the manufacturing cost to a great extent. Low cost with high volume products has been achieved by the roll-to-roll process, such as printing magazines and wallpapers, depositing low cost films. For manufacturing displays, especially the backplane, precise patterning is necessary, where a thin film is deposited and then patterned with conventional lithography. These processes have been well established for Si wafers or glass substrates; however, there are still significant unsolved problems to apply these processes to continuous roll-to-roll manufacturing. For example, in a roll-to-roll process, tension is applied to the substrate materials. As the substrate elongates under applied tension, alignment would be extremely difficult. Further technological development is necessary, both in the process and in the manufacturing equipment.

With regard to E Ink display, Vizplex imaging film is already flexible. Therefore, if flexible backplanes, that is flexible TFT arrays, are available, flexible display can be easily realized.

A considerable number of attempts have been made to fabricate flexible TFTs, such as low temperature deposition of a-Si [21], using a transfer technique named SUFTLA (Surface Free Technology by Laser Annealing/Ablation) [22] and organic semiconductor deposition [23]. As for a-Si, a high deposition temperature above 250 °C is preferable to extract a better performance of a-Si TFT. Low temperature deposited a-Si tends to show degradation of the performance. SUFTLA technology was developed by Seiko Epson. In this method, low temperature poly-Si (LTPS) TFT is fabricated on a quartz or glass substrate. Then, the TFT device is lifted off from the original substrate and attached on to the plastic film. The mobility of TFT made by SUFTLA exceeds 100 $cm^2\,V^{-1}\,s^{-1}$ [22] and, moreover, peripheral circuits can be integrated onto flexible substrate. However, the manufacturing cost of SUFTLA TFT is extremely high, hence the application of SUFTLA technology is limited. Organic semiconductors can be deposited at low process temperature. Organic semiconductors can be fabricated by a vacuum-free printing process, where drastic cost reduction is expected. Nonetheless, low performance and lack of long-term stability in organic TFTs are still problems to be solved.

Under these circumstances, a TAOS seems to be the most plausible candidate as a channel layer for a flexible TFT. The paramount important feature of a TAOS in flexible display is its low process temperature. A TAOS shows high performance even in a room temperature process where conventional low-cost plastics films can be used as a substrate. The combination of Vizplex imaging film and a flexible TAOS TFT array is expected to be the best solution for flexible electronic paper.

8.3.2 Flexible Electronic Paper Driven by an a-IGZO TFT Array

The fabrication process of flexible a-IGZO TFT on poly(ethylene naphthalate) PEN film is described in detail. For a flexible TFT array, 125 μm thick PEN Q65FA (supplied by Teijin-Dupont) was used as a substrate [24]. Bottom gate and top contact structure was employed. All the TFT layers were vacuum-coated by RF sputtering or electron beam (EB) evaporation technique. A 35 nm thick a-InGaZnO layer was deposited by RF magnetron sputtering technique using polycrystalline $InGaZnO_4$ target in Ar and O_2 gas ambient. The deposition conditions of the a-IGZO layer was optimized by carefully choosing the oxygen flow rate during deposition. It should be pointed out that the optimum deposition condition of a-IGZO is different depending on the substrate materials, more specifically, from PEN to glass. The cause of this difference is not clear, however, outgassing from polymer substrate during deposition may be one of the reasons. A 300 nm thick SiON layer was also deposited by RF magnetron sputtering for the gate insulator using Si_3N_4 as a target in Ar and O_2 gas

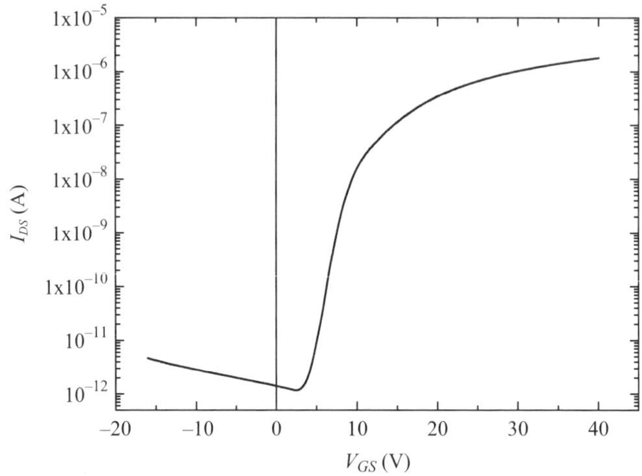

Figure 8.4 *Transfer characteristic of a pixel TFT on PEN substrate at $V_{DS} = 15$ V*

ambient. As source, drain and gate electrodes, 50 nm Al was coated by EB evaporation. All the layers were deposited at room temperature. Neither post-deposition annealing nor treatment was carried out. Source, drain, gate and channel layers were defined by standard photolithography and lift-off techniques. We fabricated 2 in. displays with a pixel number of 80×60 and a pixel size of 500 μm × 500 μm. The channel length (L) and channel width (W) of a pixel TFT are both 50 μm. After fabrication of a flexible TFT array, Vizplex imaging film was subsequently laminated onto a TFT array. Display performance was investigated before and after bending. In Figure 8.4 the transfer characteristic of a pixel TFT of an a-IGZO TFT array on PEN at $V_{DS} = 15$ V is shown. The device exhibits a sharp on-off ratio of more than six orders of magnitude. The a-IGZO TFT on PEN exhibits normally off characteristics with mobility (μ) of 5.1 $cm^2\,V^{-1}\,s^{-1}$ and threshold voltage (V_{th}) of 5.8 V. A flexible TFT fabricated onto PEN substrate is shown in Figure 8.5(a). After fabrication of the TFT array, it is

(a) (b)

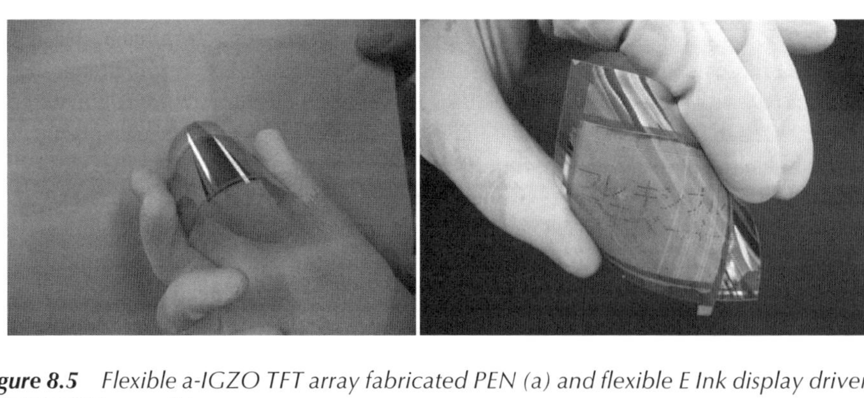

Figure 8.5 *Flexible a-IGZO TFT array fabricated PEN (a) and flexible E Ink display driven by a-IGZO TFT array (b)*

laminated onto Vizplex imaging film. Figure 8.5(b) shows a 2 in. flexible electronic paper driven by an a-IGZO TFT array. The display image was not affected by bending. The flexible electronic paper is 320 μm thick, which is one-sixth of a LCD display, and the 2 in. display weighs just 1.3 g. This was the first demonstration of a flexible display driven by an oxide-based TFT. The combination of Vizplex imaging film and TAOS TFT array offers an ideal solution for flexible electronic paper.

8.4 Application of Transparent Electronics

As the oxide semiconductors are wide band gap materials, transparent TFTs can be easily realized by the combination of transparent electrodes and insulators. Transparency is one of the most significant features of TAOS TFTs. As the band gap of a-Si is 1.7 eV and that of crystalline-Si is 1.1 eV, 'transparent electronics' cannot be realized in Si technology. In TAOS TFTs, features of high mobility or low process temperature have attracted a lot of attention. However, transparency has been underestimated or even neglected in the research and development of TAOSs. Few examples of actual applications have been reported exploiting the transparency of TAOSs until now [25, 26]. Transparent circuits will have unprecedented applications in flat panel displays and other electronic devices, such as see-through display or novel display structures. Here, two practical examples taking advantage of the transparency of TAOS TFTs in electronic paper are summarized.

8.4.1 Reversible Display

Reversible electronic paper display can be demonstrated using fully transparent TFTs. A schematic cross-section view of a reversible display is shown in Figure 8.6. With the

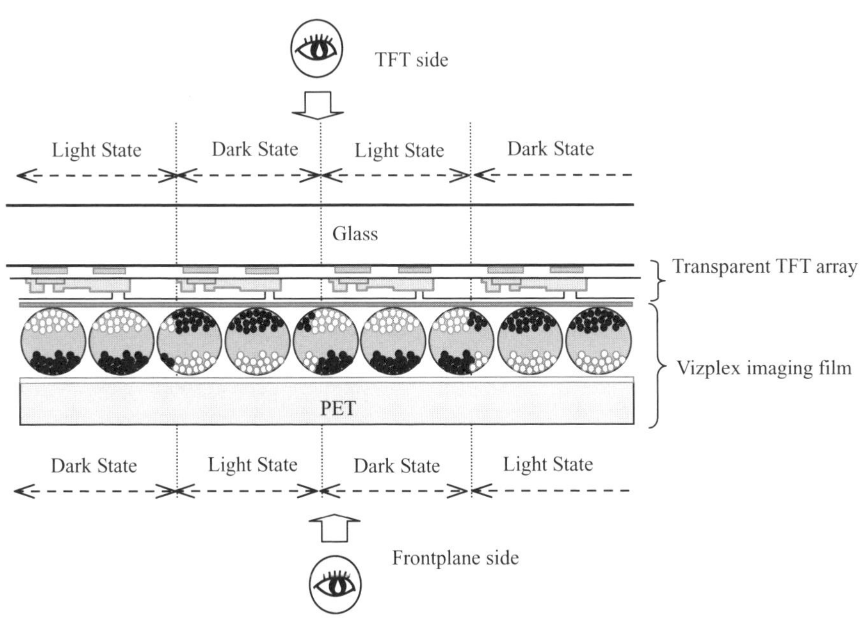

Figure 8.6 *Schematic cross-section view of a reversible display*

Table 8.1 Specifications for each layer in a reversible display

Layer	Material	Thickness	Deposition process
Gate	ITO	100 nm	DC sputtering
Gate insulator	SiON	350 nm	RF sputtering
Channel	a-IGZO	30 nm	RF sputtering
Source and drain	ITO	100 nm	DC sputtering
Passivation	SiON	50 nm	RF sputtering
Interlayer insulator	Polymer	2.5 μm	Spin-coating
Pixel electrode	ITO	100 nm	DC sputtering

combination of transparent TFT array and the Vizplex imaging film, a display can be seen not only from the frontplane side but also from the TFT array side. In order to demonstrate the reversible display, a fully transparent TFT array was fabricated onto a glass substrate. The specifications for the TFT layers are shown in Table 8.1. It goes without saying that all the layers, which are employed in this TFT, are transparent. The display is 5.35 in. with a resolution of 150 ppi (640 × 480: VGA). The transmittance spectrum and a photograph of the transparent a-IGZO TFT array are shown in Figure 8.7(a) and (b), respectively. Transmittance in the visible wavelength region is well over 70%. As evident from Figure 8.7(b), a fully transparent a-IGZO TFT array can be seen through.

After fabrication of the TFT array, Vizplex imaging film was laminated on to it. The displayed image of the reversible display is shown in Figure 8.8. As the E Ink display is a white-and-black dual-particle system, a completely reversible image is shown on the opposite side of the display, that is, for the frontplane side and the TFT array side. It seems that the transparent TFT array does not affect the visibility of the display. In this manner, reversible images can be displayed with one frontplane and one TAOS TFT array.

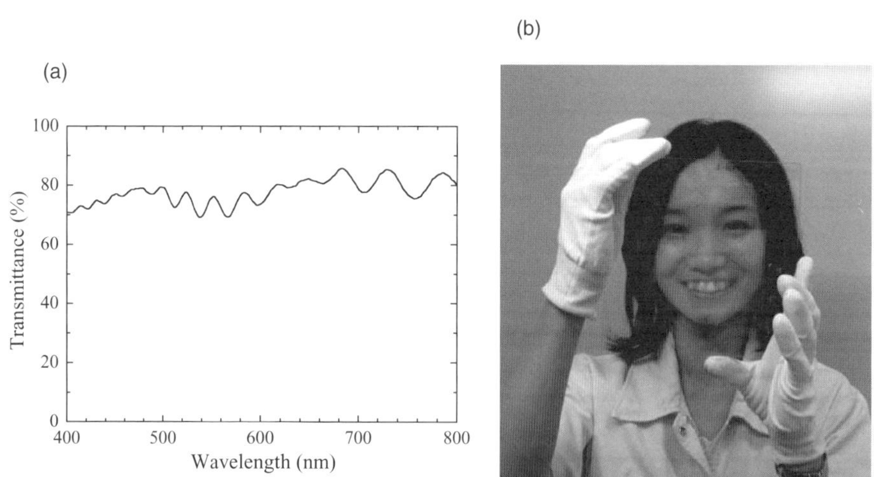

Figure 8.7 *Transmittance spectrum (a) and photograph (b) of a transparent TFT array*

Figure 8.8 Photographs of a reversible display seen from the TFT array side (a) and the frontplane side (b). Reversible images are displayed on the opposite side (c). See color plate section

8.4.2 'Front Drive' Structure for Color Electronic Paper

Here we investigate the further possibility of the transparent property of TAOS TFTsin a color electronic paper display.

8.4.2.1 Color Microencapsulated Electrophoretic Display

In principle, electrophoretic displays can display two colors by choosing appropriate pairs of pigments or a pigment and dye. For example, with a combination of red pigment with a blue dye or red and blue pigments with the opposite charge, red and blue colors can be displayed. There are several ways to realize full color. One way is to use three combinations of two-color electrophoretic fluids to express red, green and blue [27]. However, in this way, a complicated patterning of subpixels and each color fluid is required. Another way is to use a color filter array over a black and white electrophoretic film, as in a LCD. One can use the infrastructure of a color filter array of an LCD and this method is much more low risk with a simplified process. Here, the structure and the fabrication process of a full color E Ink display are described in detail.

In this structure, the color filter array is processed onto glass substrate in the first instance. Subsequently, microencapsulated electrophoretic inks are coated on it. The TFT array is

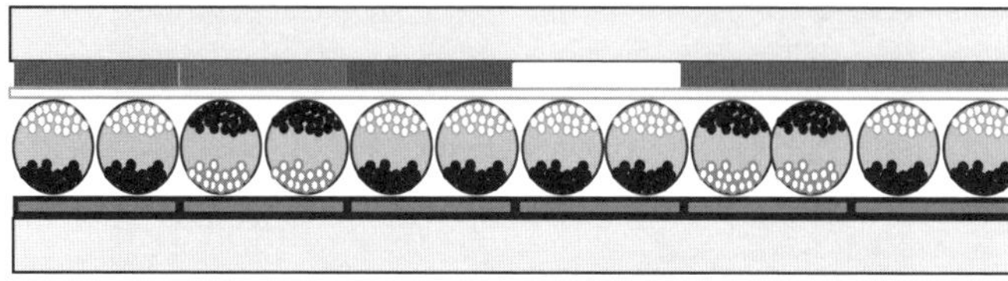

Figure 8.9 Conventional structure of the color E Ink display

fabricated onto another substrate and then combined with the color filter array. (The structure of a conventional color E Ink display is shown in Figure 8.9.) In this process, however, the integration of the display involved precise alignment of the color filter array and TFT array in both the horizontal and vertical axis. In the color E Ink display, the alignment between the upper color filter array and lower TFT array is more challenging than that of an LCD due to the following two reasons:

1. thickness of microcapsules;
2. presence of adhesive.

First, in the LCD case, the color filter array and TFT array can be aligned through a 4–6 μm height spacer. As for the E Ink display, however, the upper and lower substrate should be aligned through a 40–50 μm thick ink layer. The thickness of the inner layer is different by one order of magnitude between the LCD and E Ink display. This difference makes the manufacturing process of the color E Ink display extremely difficult.

Secondly, the color filter array and TFT array are laminated through adhesive in the E Ink display. Once two substrates are adhered, position adjustment is practically impossible. These problems of alignment lead to the low process yield and high cost of the color E Ink display.

8.4.2.2 Novel Display Structure – Front Drive Structure

In order to solve abovementioned problems, we propose a novel display structure taking advantage of the transparent property of the TAOS TFT. In this structure, a transparent TFT array is directly deposited onto the color filter array. Furthermore, this TFT array on the color filter array is positioned at the viewing side of the display, as shown in Figure 8.10. As the TFT array is fabricated directly onto the color filter array, alignment between the color filter array and the TFT array can be easily realized. It is expected that a fully transparent TFT array does not affect the visibility of the display, as in the case of a reversible display. Moreover, amorphous oxide TFT can be deposited at low temperature. Therefore, heat damage to the color filter during deposition of the TFT can be avoided. This display structure is referred to as a 'Front Drive' structure, because the display is driven from the front side of the display.

We investigated the feasibility of this 'Front Drive' structure for E Ink display and fabricated a 4 in. full color display. The color filter array was prepared by the standard photolithography process using pigment dispersion color resists with a staggered red, green, blue and white (RGBW) square subpixel layout. White subpixel was added in order to enhance the brightness and dynamic range of the display [28]. The four RGBW subpixels constitute one pixel, whose size is 250 μm × 250 μm. After fabrication of the RGBW subpixels, the transparent overcoat layer was formed in order to smooth the surface

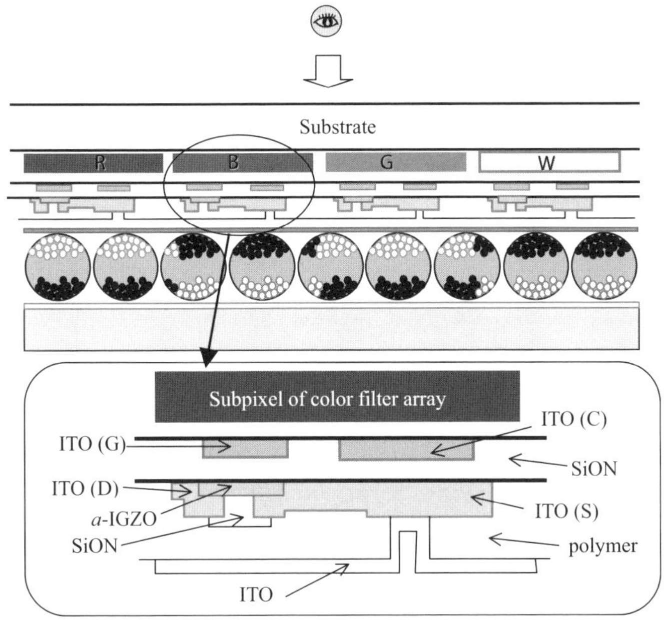

Figure 8.10 Schematic cross-section view of the 'Front Drive' structure for the E Ink display

roughness of the color filter array. The surface roughness of the color filter array is remarkably smoothed by carefully optimizing the coating condition of the overcoat layer [29]. A smooth surface is crucial in this structure, because the TFT array is fabricated directly onto the color filter array, otherwise disconnection or peeling of the TFT array would occur. After fabrication of the color filter array, the TFT array was processed onto the color filter array. In order to realize a fully transparent TFT, transparent materials are chosen as components of the TFT layers, as in the case of a reversible display. We employed bottom gate and top contact structure for the 'Front Drive' structure display. As for the gate, capacitor, source and drain electrodes, 100 nm thick indium tin oxide (ITO) was used. SiON layers of 300 nm and 40 nm thickness serve as the gate insulator and passivation layer, respectively. As a channel layer, 30 nm thick a-IGZO was used. After fabrication of the source and drain electrodes, 2.6 μm thick transparent polymer film was spin-coated as an interlayer insulator and annealed at 230 °C for 1 h. Finally, ITO film was deposited and patterned as a pixel electrode.

The transfer characteristics of a fully transparent a-IGZO TFT fabricated on a color filter array are shown in Figure 8.11. The a-IGZO TFT on color filter exhibits normally off characteristics with μ of $6.0\,cm^2\,V^{-1}\,s^{-1}$, V_{th} of 1.6 V and on/off ratio of more than 10^4. The TFT array was successfully fabricated onto a color filter array without any disconnection or peeling. After fabrication of the color filter and transparent TFT array, Vizplex imaging film was laminated onto the TFT array. A micrograph of Vizplex imaging film laminated onto a color filter and TFT array is shown in Figure 8.12. As can be seen from Figure 8.12, microcapsules of Vizplex imaging film can be seen through the color filter and transparent TFT array.

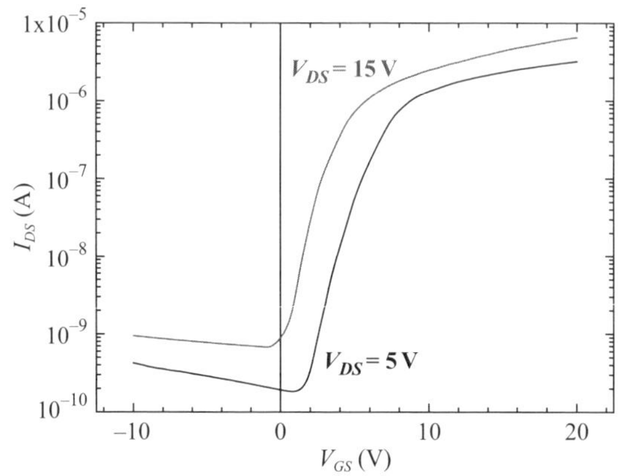

Figure 8.11 *Transfer characteristic of a pixel TFT on a color filter array*

A 4 in. E Ink display driven by a 'Front Drive' structure is shown in Figure 8.13. Specifications for the display are summarized in Table 8.2. We have successfully driven the 4 in. E Ink display with our novel 'Front Drive' structure. Our 'Front Drive' structure will increase the yield and reduce the cost of the display dramatically, due to its simplicity and the easy-to-manufacture process. Moreover, the 'Front Drive' structure can be applied not only to electronic paper but also to LCDs and OLEDs.

Finally, we would like to comment on the photoresponse of oxide TFTs in the 'Front Drive' structure. In the 'Front Drive' structure illustrated here, transmitted light to the TFT was not shielded either by the black matrix of the color filter or by the metal gate electrode as in the case of the LCD. Therefore, the influence of the transmitted light to the TFT should be carefully investigated. The influence of the light irradiation to the oxide TFT has been

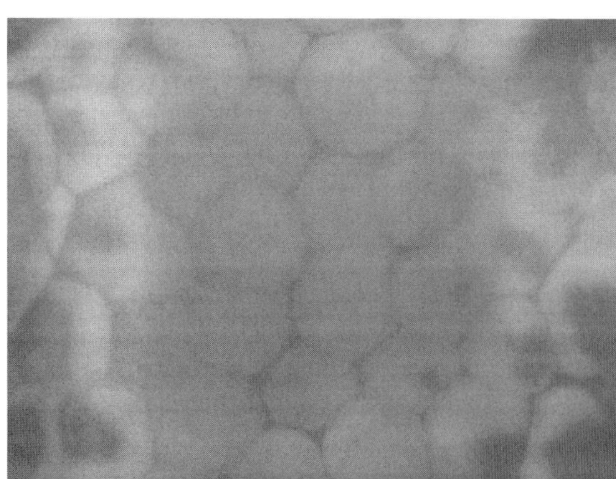

Figure 8.12 *Micrograph of microcapsules seen through a color filter array and transparent TFT array. See color plate section*

Table 8.2 *Specifications for the 'Front Drive' color E Ink display*

Display size	4 in. diagonal
Display type	QVGA [320 (×2) × 240 (×2)]
Resolution	100 ppi
Color filter layout	Staggered RGBW square
Size of TFT	$W/L = 5\,\mu m/20\,\mu m$
Pixel pitch	$250\,\mu m \times 250\,\mu m$

Figure 8.13 *4 in. QVGA color electronic paper driven by a 'Front Drive' structure*

already reported [30, 31]. As the optical band gap of a-IGZO is approximately 3.2 eV, illumination of near ultraviolet light to the channel layer has an adverse effect on the operation of the TFT. In our case, any detectable photoresponse was not observed under indoor lighting conditions. The photoresponse of the TFT was strongly influenced not only by the optical band gap of the semiconductor material but also by the thickness of the channel layer and the intensity of the illuminated light. We employed a rather thin channel layer thickness of 30 nm to avoid the effect of photoresponse. The presence of a color filter may absorb near ultraviolet light to some extent. It should be also pointed out that the driving of the display was performed in 'not-so-bright' indoor lighting conditions. Further systematic study is necessary to understand the effect of light illumination on fully transparent oxide TFTs. Without overcoming the photoresponse problem of transparent TFTs, 'transparent electronics' cannot be realized.

8.5 Conclusion

This chapter has presented the applications of TAOS TFTs to electronic paper display. TAOS TFTs have the characteristics of high mobility, processability for large area, low temperature processing, and transparency. However, as is often said, 'new material' requires new applications. Recently, electronic paper emerged as a next generation media. Electronic paper inherits the advantages of paper and electronic displays. A number of electronic paper technologies have been proposed; however, the microencapsulated electrophoretic type, which is called the E Ink display, is regarded as the leading-edge technology in electronic paper.

We focused our attention on the low-temperature processing and transparency of a-IGZO TFTs and demonstrated the applicability of a-IGZO TFTs to E Ink display. While the low process temperature of a-IGZO is promising in realizing 'flexible' displays, a novel display structure, the 'Front Drive' Structure, was successfully realized taking advantage of the transparent property of TAOSs. We believe that the combination of Vizplex imaging film and a-IGZO TFT backplane increase the possibilities of electronic paper.

Acknowledgements

The author would like to express his deepest appreciation to Professor Hosono of Tokyo Institute of Technology for providing valuable suggestions. The author is grateful to E Ink Corp. for supplying the Vizplex film for this study. The author would like to thank to Ms C. Miyazaki, Mr N. Ikeda, Dr M. Ishizaki, Mr Y. Kokubo, Mr T. Okubo, Mr R. Matsubara, Mr K. Hatta, Ms Y. Ugajin and Mr N. Sekine of Toppan Printing Co., Ltd for offering beneficial suggestions.

References

[1] J. F. Wager, Transparent electronics, *Science*, **300**, 1245–1246 (2003).

[2] S. Masuda, K. Kitamura, Y. Okumura, S. Miyatake, H. Tabata and T. Kawai, Transparent oxide thin film transistors using ZnO as an active channel layer and their electrical properties, *J. Appl. Phys.*, **93**, 1624–1630 (2003).

[3] Y. Kwon, Y. Li, Y. Heo, M. Jones, P. Holloway, D. Norton, Z. Park and S. Li, Enhancement-mode thin-film field-effect transistor using phosphorus-doped (Zn,Mg)O channel, *Appl Phys. Lett.*, **84**, 2685–2687 (2004).

[4] D. Hong and J. F. Wager, Passivation of zinc–tin–oxide thin-film transistors, *J. Vac. Sci. Technol.*, **B23**, L25–L27 (2005).

[5] N. L. Dehuff, E. S. Kettenring, D. Hong, H. Q. Chiang, J. F. Wager, R. L. Hoffman, C.-H. Park and D. A. Keszler, Transparent thin-film transistors with zinc indium oxide channel layer, *J. Appl. Phys.*, **97**, 64505 (2005).

[6] M. W. J. Prins, S. E. Zinnemers, J. F. Cillessen and J. B. Giesbers, Depletion-type thin-film transistors with a ferroelectric insulator, *Appl. Phys. Lett.*, **70**, 458–460 (1997).

[7] K. Matsuzaki, H. Yanagi, T. Kamiya, H. Hiramatsu, K. Nomura, M. Hirano and H. Hosono, Field-induced current modulation in epitaxial film of deep-ultraviolet transparent oxide semiconductor Ga_2O_3, *Appl. Phys. Lett.*, **88**, 92106 (2006).

[8] R. E. Presley, D. Hong, H. Q. Chiang, C. M. Hung, R. L. Hoffman and J. F. Wager, Transparent ring oscillator based on indium gallium oxide thin-film transistors, *Solid-State Electron.*, **50**, 500–503 (2006).

[9] L. Wang, M. Yoon, G. Lu, Y. Yang, A. Facchetti and T. Marks, High-performance transparent inorganic–organic hybrid thin-film-type transistors, *Nature Mater.*, **5**, 893–900 (2006).

[10] T. Miyasako, M. Senoo and E. Tokumitsu, Ferroelectric-gate thin-film transistors using indium-tin-oxide channel with large charge controllability, *Appl. Phys. Lett.*, **86**, 162902 (2005).

[11] K. Nomura, H. Ohta, K. Ueda, T. Kamiya, M. Hirano and H. Hosono, Thin-film transistor fabricated in single-crystalline transparent oxide semiconductor, *Science*, **300**, 1269–1272 (2003).
[12] K. Nomura, H. Ohta, A. Takagi, T. Kamiya, M. Hirano and H. Hosono, Room-temperature fabrication of transparent flexible thin-film transistors using amorphous oxide semiconductors, *Nature*, **432**, 488–492 (2004).
[13] H. Hosono, N. Kikuchi, N. Ueda and H. Kawazoe, Working hypothesis to explore novel wide band gap electrically conducting amorphous oxides and examples, *J. Non-Cryst. Solids*, **198-200**, 165–169 (1996).
[14] H. Yabuta, M. Sano, K. Abe, T. Aiba, T. Den, H. Kumomi, K. Nomura, T. Kamiya and H. Hosono, High-mobility thin-film transistor with amorphous $InGaZnO_4$ channel fabricated by room temperature rf-magnetron sputtering, *Appl. Phys. Lett.*, **89**, 112123 (2006).
[15] M. Ito, C. Miyazaki, M. Ishizaki, M. Kon, N. Ikeda, T. Okubo, R. Matsubara, K. Hatta, Y. Ugajin and N. Sekine, Application of amorphous oxide TFT to electrophoretic display, *J. Non-Cryst. Solids*, **354**, 2777–2782 (2008).
[16] A. Henzen, The present and future of electronic paper, *Proc. EuroDisplay 2005*, 174–175 (2005).
[17] B. Comiskey, J. D. Alber, H. Yoshizawa and J. Jacobson, An electrophoretic ink for all-printed reflective electronic displays, *Nature*, **394**, 253–255 (1998).
[18] T. Whitesides, M. Walls, R. Paolini, S. Sohn, H. Gates, M. McCreary and J. Jacobson, Towards video-rate microencapsulated dual-particle electrophoretic displays, *Proc. SID 2004*, 133–135 (2004).
[19] G. Crawford (Ed.), *Flexible Flat Panel Displays*, John Wiley & Sons, Ltd, Chichester. (2005).
[20] M. Omodani, What is electronic paper? The expectations, *Proc. SID 2004*, 128–131 (2004).
[21] T. H. Hwang, W. Lee, W. S. Hong, S. J. Kim, S. I. Kim, N. S. Roh, I. Nikulin, J. Y. Choi, H. I. Jeon, S. J. Hong, J. K. Lee, M. J. Han, S. J. Baek, M. Kim, S. U. Lee and S. S. Shin, 14.3 inch active matrix-based plastic electrophoretic display using low temperature processes, *Proc. SID 2007*, 1684–1685 (2007).
[22] M. Miyasaka, Suftla flexible microelectronics on their way to business, *Proc. SID 2007*, 1673–1676 (2007).
[23] E. Huitema, E. van Veenendaal, N. van Aerle, F. Touwslager, J. Hamers and P. van Lieshout, Rollable displays – a technology development enabling breakthrough mobile devices, *Proc. SID 2008*, 927–930 (2008).
[24] M. Ito, M. Kon, M. Ishizaki and N. Sekine, A flexible active-matrix TFT array with amorphous oxide semiconductors for electronic paper, *Proc. IDW/AD 05*, 845–846 (2005).
[25] P. Görrn, M. Sander, J. Meyer, M. Kröger, E. Becker, H. Johannes, W. Kowalsky and T. Riedl, Towards see-through displays: fully transparent thin-film transistors driving transparent organic light-emitting diodes, *Adv. Mater.*, **18**, 738–741 (2006).
[26] C.-W. Byun, C.-S Hwang, S.-H.K. Park, J. H. Shin, M. Ryu, S. Yang, J.-I. Lee, D.-H. Cho, W.-S. Cheong, S.-M. Yoon, H. Y. Chu and K. I. Cho, Transparent and high-aperture-ratio AMOLED panel using very stable ZnO TFT, *Proc. IDW 2007*, 1787–1788 (2007).
[27] S. Wang, H. M. Zang and P. Li, Roll-to-roll manufacturing process for full color electrophoretic film, *Proc. SID 2006*, 1587–1589 (2006).
[28] A. Bouchard, H. Doshi, B. Kalhori and A. Oleson, Advances in active-matrix color displays using electrophoretic ink and color filters, *Proc. SID 2006*, 1934–1937 (2006).
[29] M. Ito, M. Kon, C. Miyazaki, N. Ikeda, M. Ishizaki, Y. Ugajin and N. Sekine, 'Front Drive' display structure for color electronic paper using fully transparent amorphous oxide TFT array, *IEICE Trans. Electron*, **E90-C**, 2105–2111 (2007).
[30] H. Bae and S. Im, Ultraviolet detecting properties of ZnO-based thin film transistors, *Thin Solid Films*, **469-470**, 75–79 (2004).
[31] P. Barquinha, A. Pimentel, A. Marques, L. Pereira, R. Martins and E. Fortunato, Influence of the semiconductor thickness on the electrical properties of transparent TFTs based on indium zinc oxide, *J. Non-Cryst. Solids*, **352**, 1749–1752 (2006).

9

Solution-Processed Electronics Based on Transparent Conductive Oxides

Vivek Subramanian
EECS, University of California, USA

9.1 Introduction

As discussed throughout this book, transparent conductive oxides have received substantial attention in recent years for their applications to a wide range of electronic systems. In particular, transparent conductive oxides are extremely attractive candidates for use in displays. Transparent conductors based on indium tin oxide (ITO) are already widely used in displays. More recently, there has been tremendous interest in the use of transparent transistors based on semiconducting transparent oxides such as zinc oxide, zinc indium oxide and zinc indium gallium oxide.

Transparent electronics is attractive for displays since transparent transistors may be used in display pixels without degrading the aperture ratio of the display. As a result, it may be possible to implement a large range of display functionality, including pixel transistors and even potentially drivers and other peripheral circuitry, using transparent transistors integrated onto the display substrate. As a result, it is clear that some of the most important applications of transparent electronics are in large area electronic systems such as displays.

Given the focus on large area electronics, it is natural, therefore, that there has been substantial interest in the use of solution-based processing techniques such as printing for the fabrication of transparent electronics. In this chapter, we examine the reasoning for solution-processed transparent electronics, and review the state of the art and technology trends of the same. We discuss issues to be solved and summarize the outlook for solution-processed electronics using transparent conductive oxides.

Transparent Electronics: From Synthesis to Applications Edited by Antonio Facchetti and Tobin J. Marks

9.1.1 The Case for Printed Electronics

Printed electronics is considered promising for large area electronic systems for several reasons. To understand the reasons, it is worth contrasting printed electronics to conventional lithographically patterned electronics, particularly with respect to large area electronics fabrication. In conventional large area electronics, subtractive processes are commonly used. For example, to create a patterned conductor, a blanket film of conductor material is deposited (typically by sputtering). The substrate is then coated with photoresist. This is then exposed using a stepper; since the field size of steppers is typically smaller than the size of modern display panels, the pattern is often stitched using multiple exposures over the entire glass surface. After exposure, the resist is then developed. A plasma etch is then used to etch the conductor, followed by a resist strip step and cleaning step. Thus, overall, the process required to form a single patterned conductor requires multiple steps, including several steps involving vacuum processing and also several steps involving expensive processing equipment, including steppers, plasma etchers, and large area sputtering tools.

In contrast, to deposit a conductor using printing, the overall process is much simpler. In the simplest case, a patterned conductor is directly printed and subsequently annealed to create the requisite thin film. In practice, the process may be somewhat more complicated to create the appropriate surfaces prior to printing, and annealing processes may be multi-step processes. However, overall, the process is expected to be substantially simpler than conventional subtractive processes. Given the additive nature of printing technology, there is a reduction in overall process complexity, as well as a reduction in the overall capital expenditure due to the reduced dependence on vacuum processing, expensive tooling, and overall tool count. As a result, it is expected that printed electronics will result in a substantial reduction in cost per unit area of substrate [1]. This makes printed electronics an attractive proposition for large area electronics.

Of course, there are some tradeoffs. First, printing technology today is unable to achieve the linewidths of current or future display steppers. As a result, while the cost per unit area of printed electronics is lower than subtractively processed conventional electronics, the cost per transistor (or equivalently, the cost per function) of printed electronics is higher. Current generation display steppers are able to pattern lines of dimension much less than 5 μm, while the best high-speed printing techniques today are unable to produce features smaller than 20 μm. This results in an increase in cost per function.

Additionally, the manufacturing models associated with printed electronics are not yet clear. Tool yields, etc., are still to be determined, and as a result, the precise cost associated with printed electronics is not yet clear, and therefore, substantial risk remains.

Finally, for most material systems that are printable today, there is degradation in performance relative to sputtered or evaporated thin films. Printed films typically retain contamination from the ink, including residual solvent, binder, etc., which degrade performance. Additionally, many printed thin films have worse morphology and grain structure than sputtered or evaporated thin films, also resulting in degraded performance. Finally, many printed thin films have substantial porosity or incorporation of gaseous species such as oxygen, etc. These effects tend to degrade performance as well. As a result, therefore, there tends to be an overall degradation in performance of printed devices relative to vacuum and lithographically processed conventional devices for large area electronics.

9.1.2 A Survey of Printed Materials for Electronics

Printed electronics has been a field of active research for almost two decades. It is predicted that products based on printed transistors and display devices will become available in the near future; certainly, numerous companies are in advanced stages of development of products that incorporate at least some printed layers. It is worthwhile, therefore, to review the state of the art of printed electronic materials as a starting point for understanding the interest in solution-processed transparent conductive oxides.

Printed conductors have been in widespread use for several decades. Printed conductors based on silver flake inks are widely used in the fabrication of printed circuit boards for, for example, automotive applications. Such inks consist of flakes mixed with polymeric anticoagulants and solvents. Upon printing and drying, a percolated network of conductive flakes interspersed with nonconductive polymer chains forms the conductor film. While the conductivity of these conductors is poor relative to metals, they are suitable for conductive traces for many printed circuit board applications.

More recently, there has been increased interest in nanoparticle-based conductors. In these conductors, nanoscale particles of metal are dispersed in a solvent. The particles themselves are typically encapsulated in an organic ligand. Upon printing and subsequent sintering, the ligand boils off and the particles sinter together to form conductive thin films with conductivity substantially higher than flake inks. Nanoparticles of numerous conductive materials have been made, including gold, silver and copper [2–4]. The advantage of these materials over silver flake inks is their higher conductivity. Additionally, since they do not retain any polymeric binder, they tend to show less moisture-induced swelling, instability, etc.

In conjunction with polymer and/or particle-based dielectrics, as well as with carbon-based resistors, the aforementioned conductors have been used to print passive components for a wide range of printed circuit board applications, and indeed, it can be argued that this has been a reasonable success story for printed electronics. Most recently, over the last decade or so, there has correspondingly evolved an interest in printed transistors as well.

Numerous demonstrations of printed transistors have been made over the last decade. The vast majority of these have made use of organic semiconductors. By using a range of organic semiconductors, including polymers [5], soluble small molecules [6], and soluble small molecule precursors for insoluble small molecular organic semiconductors [7], various groups have demonstrated organic transistors fabricated using printing techniques. In general, the vast majority of these have been *p*-type semiconductors, since high quality solution-processable, air-stable *n*-type organic semiconductors have only recently become available. The performance of these devices has been steadily increasing, with materials offering field effect mobility greater than $1\ cm^2\ V^{-1}\ s^{-1}$ having been demonstrated. In fully printed form, devices with mobility of $\sim 0.2\ cm^2\ V^{-1}\ s^{-1}$ have been demonstrated using organic materials. More recently, there have also been demonstrations of printed silicon transistors using solution-processed silanes, but the temperatures required to fabricate these have prevented their use on plastic substrates. The tremendous air sensitivity of these has also necessitated the use of inert ambient processing.

Summarizing the state of the art in printed electronics, therefore, it is apparent that substantial progress has been made on realizing printed passive components, though active components have lagged to a degree. Organic transistors on plastic have made steady progress, but the performance of these is still relatively poor, and stability is marginal. Inorganic transistors using silane have been demonstrated, but the air sensitivity and high temperature processing has limited their applicability. Indeed, these last points are addressable using transparent conductive oxides. This has thus driven interest in the use of solution-processed transparent conductive oxides in printed electronics.

9.1.3 The Case for Solution-Processed Transparent Conductive Oxides

As discussed in other chapters, transparent conductive oxides are attractive for several reasons. Briefly, from the perspective of large area electronics and displays, the transparency of these materials allows them to be incorporated into systems where light transparency is important, such as their use in conductors for liquid crystal displays and organic light emitting diodes. Additionally, as discussed in other chapters, relatively high mobilities ($>20\,cm^2\,V^{-1}\,s^{-1}$) have been achieved in transistors formed out of these materials using semiconducting transparent oxides deposited using sputtering or laser deposition techniques [8, 9]. These two factors alone have fueled interest in the use of these materials in printed electronics; since the primary application of these materials is often in displays, all the economic factors discussed above that are potentially advantageous for printing over conventional subtractive processing apply here. Additionally, since these materials are oxygen stable, they overcome a major disadvantage of other printed semiconductors. Since they are oxygen stable, they may be printed in air and processed in air, which points to very simple fabrication processes being potentially possible. Finally, since transistors with relatively high mobilities have been achieved using these materials in amorphous or fine-grain thin film material deposited using conventional techniques, there is hope that similar or perhaps only slightly lower performance may be achieved using these materials in solution-processed form.

9.2 Solution-Processed Transparent Conductive Oxides

Over the last five years, there has a surge in activity in the area of solution-processed transparent conductive oxides. Broadly speaking, this activity has focused on three main areas – nanoparticles, nanowires, and solution-deposited thin films. Each shall be reviewed in turn here.

9.2.1 Transparent Conductive Oxide Nanoparticles

Nanoparticles have received substantial interest from the printed electronics community in general. Nanoparticles have been shown to be an excellent route to realizing printed thin film conductors. The use of nanoparticles is driven by two main factors. First, nanoparticles of suitable diameter show a substantial reduction in melting point relative to their bulk counterparts. This is a well-understood phenomenon, and is due to the increased surface area to volume ratio that exists as particle diameter is reduced. As a consequence, when thin films containing contiguous small-diameter nanoparticles are heated to low-to-moderate temperatures, the particles fuse together, producing polycrystalline films with increased grain size. This in turn therefore provides a pathway for realization of high-quality thin films out of solution without necessitating the use of high-temperature annealing; the size and structure of the nanoparticle may be optimized to tune the requisite annealing temperature.

A second benefit of nanoparticles is that they may be encapsulated in an organic ligand to tune their behavior. This ligand typically serves two functions. First, from a synthetic perspective, it may be used to control particle diameter during synthesis, and may also be used to prevent agglomeration of particles. Second, it may be used to tune the solubility of the particle and its stability in an ink form. For example, by using a ligand with a nonpolar outer group, it is possible to realize particles that may form stable inks in a range of nonpolar solvents. Correspondingly, by using a ligand with a polar outer group, it is possible to realize particles that may form stable inks in polar solvents. The ligand itself also has an impact on the sintering temperature of the particle. Since the first step in the sintering process is the evolution of the ligand, the choice of ligand can determine the sintering temperature of the particle. For example, ligands that strongly bind to the nanoparticle core will typically produce films with higher sintering temperatures than ligands that are weakly bound.

Given the above discussion, a pathway for the realization of solution-processed films of transparent conductive oxides using nanoparticles is clear. First, nanoparticles of transparent conductive oxides are synthesized, with suitable organic encapsulants as appropriate. For example, zinc oxide is easily synthesized in nanoparticle form (Figure 9.1) [10–12]. This may be achieved by precipitating zinc oxide out of, for example, zinc acetate in an isopropanol bath by addition of sodium hydroxide. The addition of sodium hydroxide results in the formation of zinc oxide particular precipitates. The size of the precipitates increases with time. By adding an encapsulant such as an alkanethiol at an appropriate time, it is possible to control the size of the resulting nanoparticles. The resulting particles consist

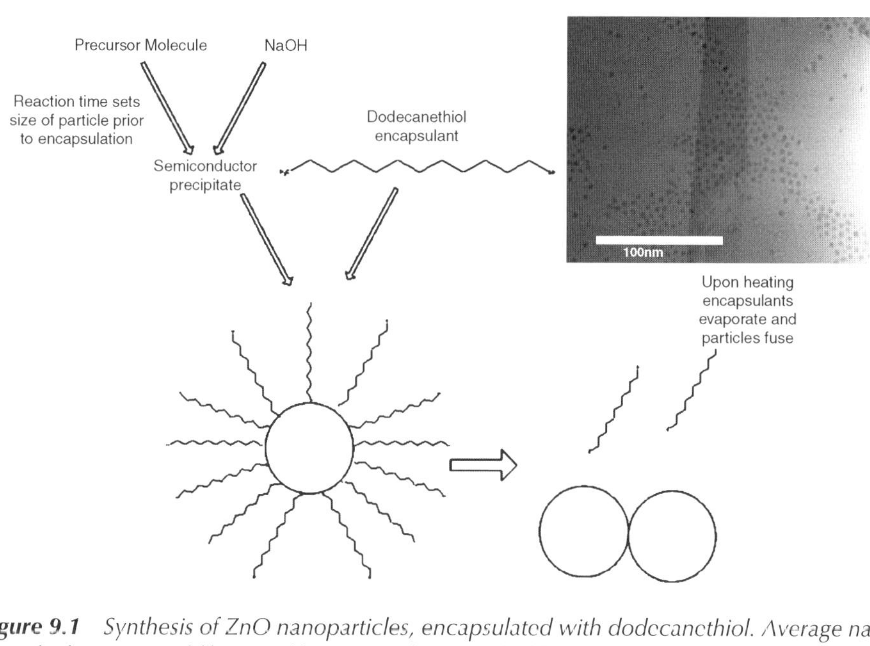

Figure 9.1 Synthesis of ZnO nanoparticles, encapsulated with dodecanethiol. Average nanoparticle diameter could be tuned by varying the time of addition of the thiol after NaOH addition. Average diameters in the range of 1–3 nm were easily realizable

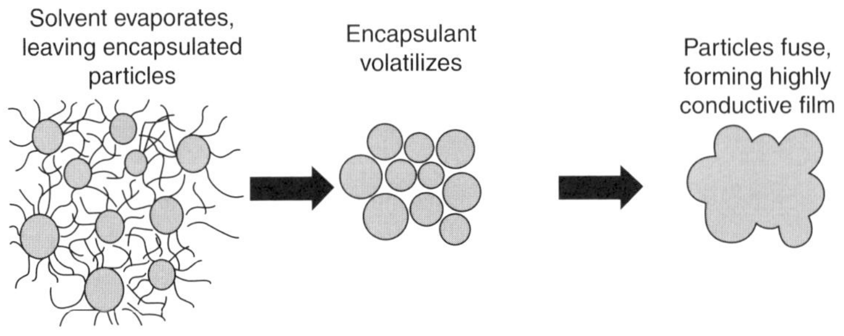

Figure 9.2 *Schematic process for thin film formation using nanoparticles*

of zinc oxide, encapsulated by an alkanethiol ligand. Suitable washing, concentration, and ink formulation can then be used to realize a zinc oxide nanoparticle ink that is suitable for use in the formation of solution-processed zinc oxide thin films. Such processes are not limited purely to the formation of zinc oxide; for example, doped zinc oxide nanoparticles have also been demonstrated [13], and such processes should therefore also be applicable to doped zinc oxide nanoparticles; indeed, applicability to other transparent conductive oxides such as zinc indium oxide, also appears likely, and as a result, this is an area of heavy research activity.

Once nanoparticle inks are formed, they may be deposited onto a relevant substrate using a variety of deposition techniques, for example, spin-coating and inkjet printing may be used. An advantage of nanoparticles is that the formulation can be tuned to suit the needs of printing by changing the choice of ligand and the resultant usable solvent system. Upon depositing onto the substrate of choice, the film is dried to drive out residual solvent. The resulting film therefore consists of a densely packed thin film of individual ligand-encapsulated nanoparticles. This film is then heated to a suitable temperature to drive off the ligand. At this point, the small diameter nanoparticles come into contact with each other, and annealing at a suitable temperature causes them to fuse, resulting in grain growth and the realization of a polycrystalline thin film (Figure 9.2). The quality of the final thin film depends strongly on temperature (Figure 9.3). For example, sintering at too low a

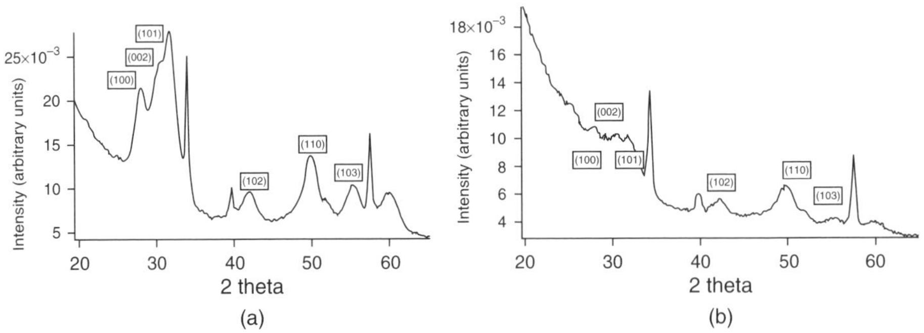

Figure 9.3 *X-ray diffraction characteristics of films formed from nanoparticles annealed at (a) 150 °C and (b) 170 °C. Note that a strong difference in crystallinity is seen between the two samples, attesting to the need for careful tuning of sintering conditions*

temperature prevents efficient removal of the ligand, resulting in poor quality, carbon-contaminated thin films. Sintering at too high a temperature results in rapid, 'explosive' ligand removal and rapid sintering, forming small grain thin films with poor electrical quality. However, at intermediate temperatures, it is possible to produce thin films that have gone through good sintering processes, resulting in the formation of films with improved thin film characteristics.

The resultant thin films may then be used to form thin film transistors [14]. For example, in conjunction with SiO_2 gate dielectrics and Au contacts, simple bottom-gated ZnO thin film transistors have been demonstrated (Figure 9.4). Despite the fact that Au is known to form a poor contact to ZnO due to the large workfunction mismatch, reasonable solution-processed ZnO thin film transistors have been realized, with performance comparable with those of many solution-processed organic semiconductor devices. The ZnO films show good transparency, and, importantly, all processing and operation may be done in air owing to the inherent oxygen stability of ZnO thin films realized in this way.

At the time of writing, ZnO nanoparticles remain an active area of research as a means of realizing solution-processed thin films of transparent conductive oxides. Further work remains to be done to clarify the relationship between film formation conditions, nanoparticle structure and sintering conditions; however, it is clear that nanoparticles are a promising means of realizing transparent conductive oxide thin films out of solution, and thus may represent a pathway to realize such systems on large area electronics on plastic, glass, etc.

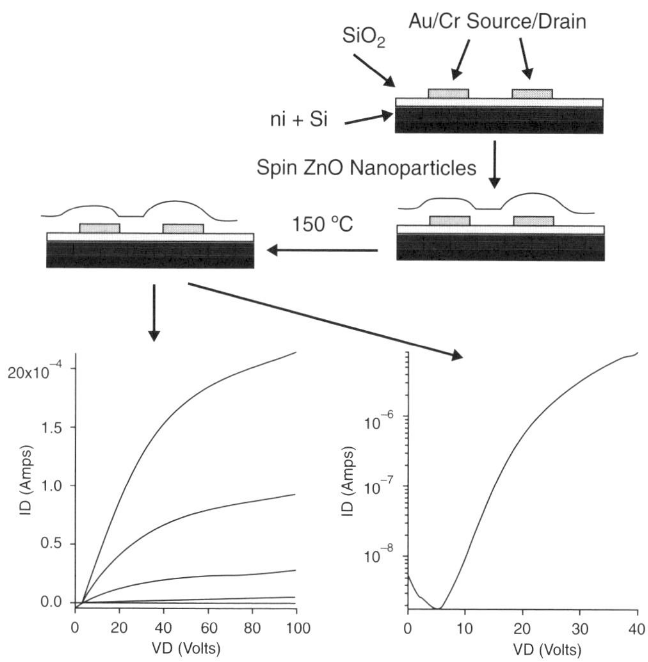

Figure 9.4 *ZnO thin film transistors realized using spin-coated ZnO nanoparticles. Field effect mobilities as high as 0.2 $cm^2 V^{-1} s^{-1}$ have been obtained*

9.2.2 Nanowire-Based Transparent Conducting Oxide Devices

Just as it is possible to synthesize zinc oxide nanoparticles, it has correspondingly been extensively shown that it is possible to synthesize zinc oxide nanowires by exploiting the strongly asymmetric crystal growth of the same during solution-based synthesis. By choosing appropriate synthetic conditions, it is possible to cause nanoparticle growth rates to be dramatically higher along certain crystal planes than on others, resulting in the formation of elongated nanowires. These nanowires in turn may be used to fabricate thin film transistors in a manner that is roughly analogous to the fabrication of thin film transistors using nanoparticles. Various demonstrations of ZnO nanowire-based thin film transistors have been shown [15, 16]. The main potential advantage of nanorods over nanoparticle-based devices is, of course, the larger average crystallite size that exists in nanorods. Assuming a tight packing of nanorods, a percolation path for strong conduction between two terminals can be achieved with the bulk of the carrier transport occurring within individual nanorods. In theory, this should result in improved mobility in ZnO nanorod thin films versus sintered nanoparticle thin films. Certainly, results have been shown that attest to an improvement in transport characteristics as rod anisotropy is increased, provided packing density of the rods is maintained.

This last point is rather important. In practice, obtaining dense arrays of nanorods is difficult. Typically, sparse meshes of rods are obtained, leading to degraded transport caused by a reduced overall current path. Therefore, while it is possible to use nanorods to realize ZnO-based transistors, the experimentally demonstrated performance has not proven to be substantially better than that achieved using sintered films of nanoparticles. This is likely due to the poor actual effective coverage of nanorods that is achieved; though transport within individual rods will certainly be better than transport in the small grain films resulting from sintered nanoparticles, the poor coverage, porous film results in degraded effective transport path, causing degradation in performance (Figure 9.5).

Another disadvantage of using nanowires is the large amount of scattering that exists between wires. As a result, films formed using these materials typically are not particularly

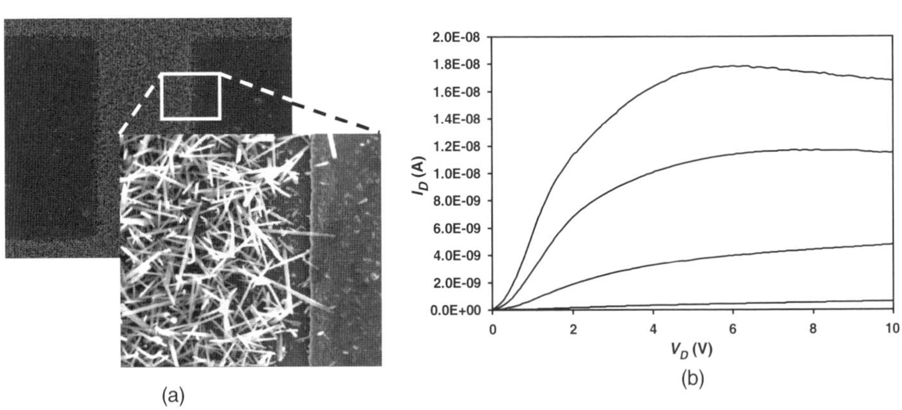

Figure 9.5 (a) ZnO nanorods dispersed in a transistor channel. The poor coverage of the rods degrades transistor characteristics (b) Additionally, the large amount of rod-to-rod light scattering degrades optical transparency of the resulting films

transparent, and show a substantial amount of light scattering. This therefore reduces the interest in these materials for transparent electronics. However, the large surface area of the nanorods potentially makes them interesting for sensor applications, since the surface of the rods may be functionalized to enhance sensitivity to specific chemical species; thus, interest in the use of nanorods for sensors has persisted.

9.2.3 Solution-Deposited Thin Films

In the aforementioned sections, processes for forming ZnO thin films based on nanostructured materials have been discussed. In these processes, nanostructured ZnO was formed in separate synthesis reactions, and the resulting ZnO nanoparticles or nanorods were deposited into a substrate of choice to form ZnO thin films. An obvious alternative would be to perform the chemical reaction directly on the target substrate, thus producing ZnO directly on the same. This is the conceptual basis that underlies numerous pathways that have been evolved to produce ZnO thin films out of solution onto a target substrate, without requiring the prior synthesis of nanoparticles or nanowires.

Conceptually, the advantage of this process is the elimination of the need for an encapsulant species. In theory, therefore, this should result in the formation of thin films of higher quality with lower carbon contamination levels. However, the disadvantage of this process is that the chemical conversion to ZnO is performed directly on the target substrate, which may potentially place temperature constraints on the same. Additionally, since many of these chemical reactions are blanket reactions, to achieve additive processing, it is necessary to develop means to target the chemical reaction. To date, several different methods of forming ZnO directly on target substrates out of solution have been developed, and a few major categories will be reviewed here.

9.2.3.1 Chemical Bath Deposition

Chemical bath deposition is conceptually a very simple process. In chemical bath deposition, a precipitation reaction is caused to occur directly on the surface of a target substrate. To prevent large scale precipitation in the bulk of the solution, the precipitation reaction is designed to be surface sensitive, such that precipitation only occurs at the growing solid-liquid interface. As a result, many such reactions often require a nucleation step to start the growth; once the growth is initiated, it proceeds as a surface precipitation reaction, directly producing, for example, ZnO, on a target substrate.

Chemical bath deposition of ZnO has been demonstrated using the following chemistry [17]:

$$NH_4OH + ZnCl_2 + H_2O \rightarrow ZnO$$

The precipitating ZnO deposits on the wafer surface, but does not precipitate in the bulk of the reaction environment, provided specific ranges of bath temperature and pH are maintained. Thus, by appropriately controlling bath pH and temperature, it is possible to produce ZnO thin films on appropriate substrates without causing large scale precipitate formation within the bath itself. As a consequence, the transport characteristics of devices fabricated using such processes depend strongly on bath conditions (Figure 9.6).

In addition to the impact of bath conditions, performance also depends on annealing conditions. After the chemical bath deposition process is complete, the substrate is washed

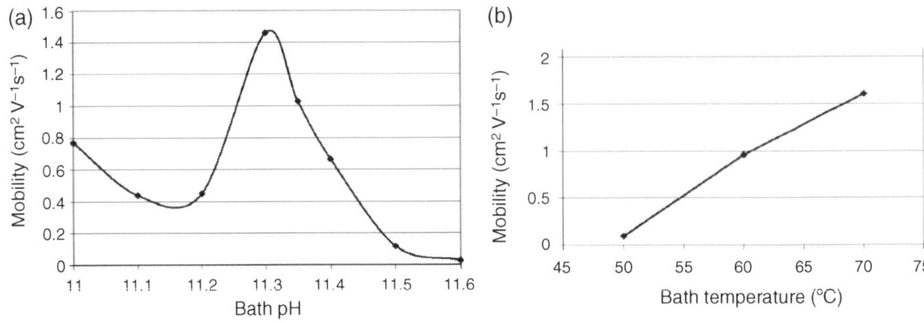

Figure 9.6 Variation in carrier mobility in thin film transistors fabricated using chemical bath deposited ZnO channels as a function of (a) bath pH and (b) bath temperature

and dried and subsequently annealed at elevated temperature to improve grain morphology. In general, performance is found to improve at elevated temperatures, as expected (Figure 9.7).

One disadvantage of chemical bath deposition versus nanoparticle-based film formation is that the chemical bath deposition process is not purely additive in nature. The chemical bath deposited film will tend to deposit on any appropriate surface, requiring a subsequent patterning step. As discussed previously, this is highly disadvantageous, since the main reason to consider solution-based processing is to exploit the cost advantages of additive processing. Fortunately, it is possible to cause chemical bath deposition to be quasi-additive by exploiting the fact that precipitation occurs only at surfaces. Therefore, by pretreating the surfaces to make some regions hydrophilic while others are hydrophobic, it is possible to control the deposition such that it only occurs in hydrophilic regions. Thus, direct patterned deposition using chemical bath deposition is achievable, albeit with slightly increased process complexity.

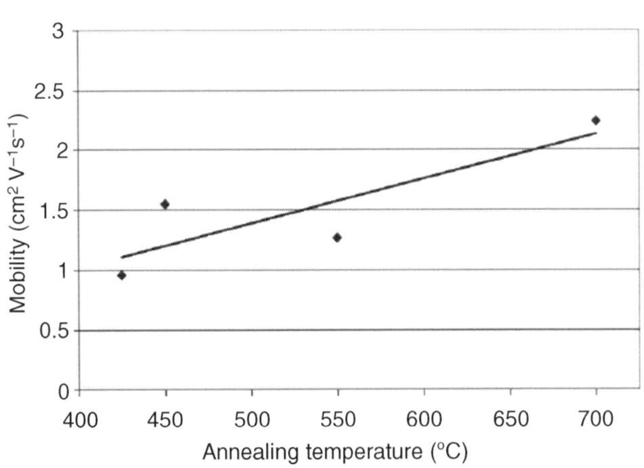

Figure 9.7 Effect of annealing temperature on carrier mobility in chemical bath deposited ZnO thin film transistors

9.2.3.2 *Direct Chemical Reaction*

The chemical bath deposition process discussed above produced ZnO by a precipitation reaction on the target substrate surface. The disadvantage of this process is twofold. First, the reaction is highly surface and reaction condition sensitive; this makes optimization of the process difficult and places constraints on repeatability and reliability of the same. Second, there is likelihood of incorporation of the reaction by-products within the chemical bath deposited thin film, resulting in decreased characteristics. Simpler routes to formation of ZnO have also been demonstrated, including sol-gel techniques [18], and, most simply, direct chemical conversion techniques [19]. In this process, a zinc salt (such as zinc nitrate, for example) is deposited onto the target substrate surface out of solution, and subsequently annealed in oxygen. The resulting film converts to zinc oxide at elevated temperatures. Since the process is extremely simple, it has the potential to be more manufacturable and also may result in the formation of films with higher quality and purity. However, given the on-surface chemical conversion, a high temperature sintering step is required, which limits substrate compatibility.

9.3 Summary

The solution-based deposition of transparent conductive oxides is an attractive technique for realization low-cost transparent electronics on large area substrates. The use of solution processing potentially enables a fully additive process flow, resulting in reduced process complexity and capital expenditure. This in turn is expected to lead to a reduction in overall cost per unit area for transparent electronics.

To realize solution-based deposition of transparent conductive oxides, numerous routes have been pursued by various groups worldwide. Nanoparticles of various transparent conductive oxides have been synthesized. These can be deposited onto target substrates and sintered to produce polycrystalline thin films. Similarly, nanowires can be used in the same way. Finally, various direct chemical deposition techniques have been developed, including chemical bath deposition, sol-gel techniques, and direct oxidation of solution-deposited precursors.

Currently, the performance of solution-processed transparent conductive oxides lags behind that achieved using vacuum processing techniques. However, the performance of these is close to that achieved by competitive solution-based semiconductor technologies. Given the stability and transparency advantages of transparent conductive oxides, it is likely, therefore, that with suitable optimization and technology improvement, solution-processed transparent conductive oxides will represent an attractive pathway for the realization of large area electronics on a wide range of substrates.

References

[1] V. Subramanian, J. M. J. Fréchet, P. C. Chang, D. Huang, J. B. Lee, S. E. Molesa, A. R. Murphy, D. R. Redinger, and S. K. Volkman, Progress towards development of all-printed RFID tags: materials, processes, and devices, *Proc. IEEE*, **93**, 1330–1338 (2005).

[2] D. Huang, F. Liao, S. Molesa, D. Redinger, and V. Subramanian, Plastic-compatible low-resistance printable gold nanoparticle conductors for flexible electronics, *J. Electrochem. Soc.*, **150**, 412 (2003).
[3] D. Redinger, R. Farshchi, and V. Subramanian, An ink-jet-deposited passive component process for RFID, *IEEE Trans. Electron Devices*, **51**, 1978 (2004).
[4] S. K. Volkman, Y. Pei, D. Redinger, S. Yin, and V. Subramanian, Ink-jetted silver/copper conductors for printed RFID applications, *Mater. Res. Soc. Symp. Proc.*, **814**, I7. 8 (2004).
[5] H. Sirringhaus, T. Kawase, R. H. Friend, T. Shimoda, M. Inbasekaran, W. Wu, and E. P. Woo, High-resolution inkjet printing of all-polymer transistor circuits, *Science*, **290**, 2123–2126 (2000).
[6] S. K. Park, T. N. Jackson, J. E. Anthony, and D. A. Mourey, High mobility solution processed 6,13-bis(triisopropyl-silylethynyl) pentacene organic thin film transistors, *Appl. Phys. Lett.*, **91**, 063514 (2007).
[7] A. Afzali, C. D. Dimitrakopoulos, and T. L. Breen, High-performance, solution-processed organic thin film transistors from a novel pentacene precursor, *J. Am. Chem. Soc.*, **124**, 8812–8813 (2002).
[8] J. Nishii, F. M. Hossain, S. Takagi, T. Aita, K. Saikusa, Y. Ohmaki, I. Ohkubo, S. Kishimoto, A. Ohtomo, T. Fukumura, F. Matsukura, Y. Ohno, H. Koinuma, H. Ohno, and M. Kawasaki, High mobility thin film transistors with transparent ZnO channels, *Jpn. J. Appl. Phys., Part 2*, **42**, L347–9 (2003).
[9] R. L. Hoffman, B. J. Norris, and J. F. Wager, ZnO-based transparent thin-film transistors, *Appl. Phys. Lett.*, **82**, 733–5 (2003).
[10] N. S. Pesika, Z. Hu, K. J. Stebe, and P. C. Searson, Quenching of growth of ZnO nanoparticles by adsorption of octanethiol, *J. Phys. Chem. B*, **106**, 6985–6990 (2002).
[11] M. Shim and P. Guyot-Sionnest, Organic-capped ZnO nanocrystals: synthesis and n-type character, *J. Am. Chem. Soc*, **123**, 11651–11654 (2001).
[12] Z. Hu, D. J. Escamilla Ramirez, H. E. Heredia Cerveva, G. Oskam, and P. C. Searson, Synthesis of ZnO nanoparticles in 2-proponol by reaction with water, *J. Phys. Chem. B.*, **109**, 11209–11214 (2005).
[13] R. Viswanatha, S. Sapra, S. S. Gupta, B. Satpati, P. V. Satyam, B. N. Dev, and D. D. Sarma, Synthesis and characterization of Mn-doped ZnO nanocrystals, *J. Phys. Chem. B*, **108**, 6303–6310 (2004).
[14] S. K. Volkman, S. E. Molesa, J. B. Lee, B. A. Mattis, A. de la Fuente Vornbrock, T. Bakhishev, and V. Subramanian, A novel transparent air-stable printable n-type semiconductor technology using ZnO nanoparticles, *IEEE International Electron Device Meeting Technical Digest*, **769** (2004).
[15] B. Sun, and H. Sirringhaus, Solution-processed zinc oxide field-effect transistors based on self-assembly of colloidal nanorods, *Nano Lett.*, **5**, 2408–2413 (2005).
[16] T. Bakhishev, S. K. Volkman, and V. Subramanian, Solution-processed ZnO nanowire-network thin film transistors for transparent electronics, *Mater. Res. Soc. Symp.*, Paper DD5.5 (2005).
[17] D. Redinger, and V. Subramanian, High-performance chemical-bath-deposited zinc oxide thin-film transistors, *IEEE Trans. Electron Devices*, **54**, 1301–1307 (2007).
[18] H.-C. Cheng, C.-F. Chen, and C.-Y. Tsay, Transparent ZnO thin film transistor fabricated by sol-gel and chemical bath deposition combination method, *Appl. Phys. Lett.*, **90**, 012113 (2007).
[19] B. J. Norris, J. Anderson, J. F. Wager, and D. A. Keszler, Spin-coated zinc oxide transparent transistors, *J. Phys. D: Appl. Phys.*, **36**, L105–L107 (2003).

10

Transparent Metal Oxide Nanowire Electronics

Rocío Ponce Ortiz, Antonio Facchetti and Tobin J. Marks
Department of Chemistry and the Materials Research Center, Northwestern University, USA

10.1 Introduction

Nanowire field effect transistors (NWTs) have gained great interest as promising candidates to sustain the progress in CMOS scaling [1–3]. Semiconducting nanowires with diameters below 100 nm and high aspect ratios exhibit unique characteristics due to their size and anisotropic geometries and are ideal models to assess fundamental materials/device properties such as charge/spin transport, light coupling, and magnetic ordering in low dimensions. The principal semiconducting nanowire attractions are: (i) high-yield scalable syntheses with reproducible electronic properties [4–6]; (ii) since the diameter of the nanowires can approach 10 nm and below, it is possible to decrease thin film transistor (TFT) channel widths to limits well beyond those possible with lithography [7]; and (iii) high carrier mobilities due to their crystalline microstructures and smooth surfaces [8, 9]. These combined properties allow gate scaling without compromising electrical properties, which is currently a major issue in conventional MOSFETs.

Several reviews have addressed recent advances in nanowire-based electronics [10, 11]. Some of the most relevant nanowire materials and their corresponding TFT performance metrics are summarized in Table 10.1. In this Chapter, we will focus on nanowire integration in optically transparent TFTs and devices since another great potential

Transparent Electronics: From Synthesis to Applications Edited by Antonio Facchetti and Tobin J. Marks

Table 10.1 *Summary of device performance for some nanowire transistors*

	Single wire			Multiple wires		
Nanowire material	μ ($cm^2 V^{-1} s^{-1}$)	$I_{on}:I_{off}$	Ref.	μ ($cm^2 V^{-1} s^{-1}$)	$I_{on}:I_{off}$	Ref.
Si	1350	10^4	14	NA	NA	15
Si	270	10^4	16	NA	NA	17
Si/Ge	730	10^5	9	NA	NA	18
Ge	2	NA	19	NA	NA	20
$Ge_{1-x}Mn_x$	340	10^4	21	—	—	
GaN	100	$>10^2$	22	—	—	
ZnO	1000	10^6	23	25	10^3–10^5	24
ZnO	4120	10^4	25	—	—	
In_2O_3	1450–300	10^6	26	258	10^5	27
SnO_2	172	10^6	28	145	10^3	29

application of semiconducting nanowires is in the transparent and flexible electronics field [12, 13]. This new technology aims to realize electronic and optoelectronic circuits on optically transparent substrates such as glass or plastics. Interesting semiconducting material candidates include amorphous inorganics (metal oxides) and transparent organic semiconductors since they can be processed at low temperatures. However, these materials present the disadvantages of low TFT carrier mobilities (~0.1–1 $cm^2 V^{-1} s^{-1}$), and the corresponding circuits operate at relatively low speeds (<1.0 kHz) [30]. Higher mobilities can be achieved using crystalline semiconductors but the higher annealing temperatures required in the fabrication process inhibit their use as thin films with inexpensive glass and plastic substrates [13]. However, in the case of nanowire-based transistors [31] the ability to decouple high-temperature material growth processes from device fabrication allows the combination of crystalline semiconductor structures and enhanced electrical performance with the utilization of glass and plastic substrates [29, 32]. In this case, the fabrication process consists of synthesizing the nanowires under optimized conditions (usually high temperatures), followed by transfer to the substrate in a separate step, using solution- or dry-based transfer/alignment methods to complete device fabrication at low temperatures. This approach has led to the realization of logic, memory, and photonic devices on glass and plastic substrates using single-crystalline Si, SnO_2, Ge/Si and GaN nanowires [32] to cite just a few examples, with TFT performance far surpassing that of amorphous Si or organic-based devices [30, 31].

One class of semiconducting nanowires widely studied and perfectly suited for transparent electronics are metal oxides, including ZnO, SnO_2, and In_2O_3. These nanowires present several advantages versus thin film Si TFTs such as optical transparency, high mobility, and mechanical flexibility. Another interesting feature of these metal oxide systems is that their electronic properties can be tuned by controlled doping during nanowire growth [33]. Therefore, to use nanowire semiconductors in microelectronics, it is essential to develop reproducible synthetic methods to enable production of high-quality samples. These synthetic methods include thermal evaporation/vapor transport [34], pulsed laser deposition [35], hydrothermal growth [36], metal-organic chemical vapor deposition [37]

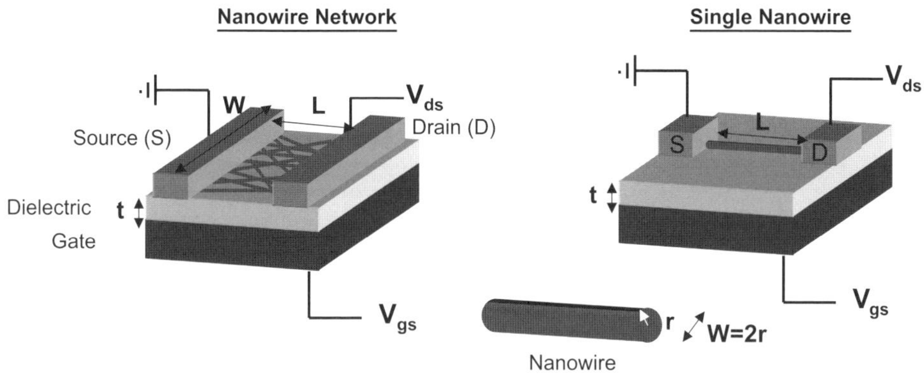

Figure 10.1 *Schematic back-gated nanowire network and single nanowire TFTs. The key geometric parameters for electrical characterization are shown*

and molecular beam epitaxy [38]. The most common technique, due to low equipment and operation cost, is thermal evaporation/vapor transport [39].

In order to fabricate electronic circuitry based on these nanowires, the reliable fabrication of nanowire-based TFTs is essential. Nanowire-based field effect transistors can be fabricated using nanowire networks or a single nanowire (Figure 10.1). Nanowire network films can be deposited on the substrate/gate/dielectric by spin-coating, drop-casting, and thermal transfer. The TFT structure is completed by physical vapor deposition or printing of the source/drain contacts. The field effect mobilities in these TFTs can be extracted by conventional equations used for amorphous silicon and estimating the capacitance of the dielectric using a parallel plate model [40]. For the single nanowire-based devices, source and drain electrodes are typically patterned onto the two ends of the nanowire channel (usually deposited by casting) by thermal evaporation, photolithography or e-beam lithography. In addition, a weakly capacitively coupled terminal is employed to provide precise control of the channel conduction via the application of a transverse electric field [41]. To examine device operation, four different gate configurations can be fabricated: back (or bottom), top, side and surrounding gates. The back gate configuration is the most common geometry (Figure 10.1) due to its fabrication simplicity, however, it makes difficult the control of individual channel segments.

In single wire NWTs, the field effect mobility can be calculated from the measured transconductance $g_m = dI_{ds}/dV_{gs}$ using the equation $\mu = dI_{ds}/dV_{gs} \times L^2/C_i \times 1/V_{ds}$, where L is the device channel length, V_{ds} is the source-drain voltage and C_i is the gate-to-channel capacitance estimated using the cylinder-on-plane model:

$$C_i = 2\pi\varepsilon_0 k_{\text{eff}} L/\cosh^{-1}(1 + t/r)$$

where ε_0 is the vacuum dielectric constant, k_{eff} is the effective dielectric constant of the gate dielectric, t is the thickness of the dielectric layer, and r is the radius of the nanowire (Figure 10.1) [42–44].

10.2 Nanowire Transistors

In this section we will discuss the fabrication and performance of representative nanowire TFTs based on ZnO, In_2O_3, and SnO_2 since these are the most investigated metal oxide semiconductors.

10.2.1 ZnO Nanowire Transistors

Among semiconducting nanowires, ZnO nanowires have been extensively studied due to their unique physical properties and versatile device applications [42, 45, 46]. In particular, ZnO is promising for optoelectronic devices due to the wide direct bandgap of 3.37 eV at room temperature and the relatively large exciton binding energy of 60 meV. ZnO usually crystallizes in a stable wurtzite crystal structure and is a natural *n*-type semiconductor [46]. The *n*-type electrical conductivity originates from lattice defects, such as oxygen vacancies and zinc interstitials, or incidental hydrogen [47], all of which are strongly dependent upon sample preparation methods. In general, ZnO nanowires can be produced by either vapor- [39, 48] or solution-phase growth methods [49]. Using physical and thermal evaporation methods, ZnO nanowires have been successfully prepared from ZnO powders (available from Alfa Aesar or Sigma-Aldrich) using either catalysts [50, 51] or graphites [48a]. The formation of ZnO nanoribbons has also been reported by physical evaporation at high temperatures [52]. In the solution synthetic approach, by using porous templates, polycrystalline ZnO nanowires have been prepared via sol-gel [53] and electrodeposition techniques [54]. In the latter, Zn is electrodeposited in alumina templates, followed by oxidation of the Zn nanowire arrays.

The typical field effect mobility of back-gated single ZnO nanowire devices based on a thermally grown SiO_2 dielectric layer on doped Si gate substrates ranges between $3\,cm^2\,V^{-1}\,s^{-1}$ and $80\,cm^2\,V^{-1}\,s^{-1}$ with an I_{on}/I_{off} ratio of $\sim 10^4$–10^6 [55]. Top, side, and surrounding gate TFT structures involve a more complicated fabrication process but afford enhanced device performance [56]. Field effect mobilities approaching $1000\,cm^2\,V^{-1}\,s^{-1}$ have been reported for side-gated TFTs, due to excellent gate modulation efficiency [23]. However, these side-gated devices are not transparent since Si/SiO_2 substrates and niobium (Nb) were used as dielectric and contact materials.

ZnO NWT performance can be further enhanced by passivation of the ZnO nanowire surface [25, 57] to compensate for surface defects acting as charge carrier traps [58], translating in higher mobilities and more uniform threshold voltages. Using ZnO nanowires grown on different substrates (Au-coated sapphire substrates or Au-catalyst-free ZnO films) both enhancement and depletion mode operation were demonstrated by Hong *et al.* for single-wire devices on Si/SiO_2 gate/dielectric substrates [59], and the noise spectra have a classic $1/f$ dependence at room temperature [60].

ZnO nanowire networks have also been used in TFT [61, 62] and complementary inverter devices [24]. For example, Sun *et al.* fabricated devices made from 65-nm-long and 10-nm-wide nanorods deposited by spin coating onto Si/SiO_2 substrates at moderate temperatures (230 °C). These TFTs exhibit mobilities of $0.61\,cm^2\,V^{-1}\,s^{-1}$ and I_{on}/I_{off} of 3×10^5 [61]. Also using a low-temperature process (under 140 °C), a nanowire network transistor fabrication on a polymer substrate [poly(vinyl phenol)/Al/polyimide] was demonstrated by Ko *et al.* using direct nanoimprinting of Au nanoparticles to define

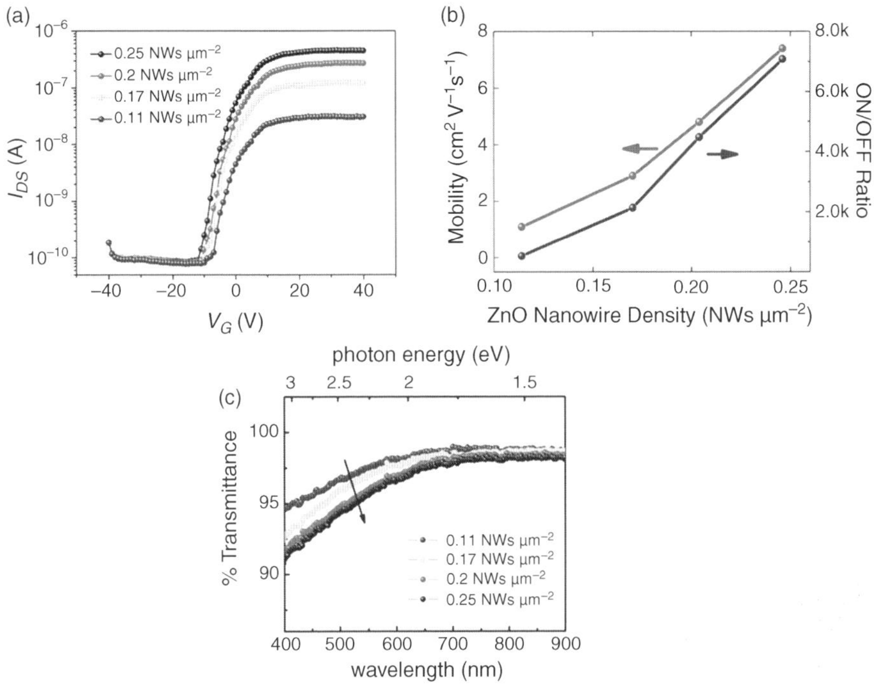

Figure 10.2 *(a) Effect of nanowire density (from top to bottom: 0.25, 0.2, 0.17 and 0.11 NWs μm^{-2}) on the transfer characteristics of 20 μm channel devices ($V_{ds} = 4$ V). (b) Mobility and I_{on}/I_{off} ratio of TFT devices with respect to the ZnO nanowire density. (c) Transmittance as a function of wavelength for the different ZnO nanowire densities (from top to bottom: 0.11, 0.17, 0.2 and 0.25 NWs μm^{-2}). Reprinted with permission from ref. 24. Copyright 2009, American Institute of Physics*

source and drain electrodes [62]. Unalan *et al.* fabricated semiconducting random networks by depositing high temperature-grown ZnO nanowires onto Si/SiO_2 receiver substrates at room temperature [25]. This approach is a less lithographically intense alternative to individual nanowire devices and yields TFTs exhibiting *n*-channel depletion mode behavior with mobilities up to 25 $cm^2\,V^{-1}\,s^{-1}$ and I_{on}/I_{off} ranging from 10^3 to 10^5. The effect of nanowire density on the performance of the corresponding TFTs were analyzed by Unalan *et al.*, and the results indicate that both mobility and I_{on}/I_{off} ratio increase with nanowire density while optical transmittance values (with the substrate absorption removed) range between 95% and 90%, suggesting that the losses due to overlapping nanowires are negligible (Figure 10.2). This example indicates the possibility of fabricating high performance TFTs and inverters based on transparent nanowires using low-cost and low-temperature manufacturing [24].

10.2.2 In_2O_3 Nanowire Transistors

Another promising metal oxide nanowire material is In_2O_3 due to its easy preparative access (usually synthesized by pulsed laser ablation using an InAs target, available from

Alfa Aesar) chemical stability, and wide band gap (3.6 eV) [63–66]. One-dimensional transport of these metal oxide nanowires at different temperatures have been studied by Liu *et al.* [67], finding Luttinger liquid behavior close to that of single-wall carbon nanotubes (SWCNTs) and multi-wall carbon nanotubes (MWCNTs) [68, 69]. Furthermore, differential conductance gaps and current gaps found at relatively low gate bias can be explained using conventional low-temperature TFT theories.

Single nanowire TFTs fabricated with this In_2O_3 nanowire on oxide dielectrics, such as on SiO_2/Si substrates with Ti/Au contacts, typically exhibit mobilities ranging from 7 to 280 $cm^2\,V^{-1}\,s^{-1}$ [70–73] vs. ~160 $cm^2\,V^{-1}\,s^{-1}$ for the corresponding single-crystal based TFTs [74]. Enhanced device performance has been demonstrated by proper selection of the gate dielectric. In this regard, the group at Northwestern University developed an unconventional gate dielectric, namely a self-assembled nanodielectric (SAND) [75–77], fabricated with alternating σ and π constituent molecular layers, strongly linked by crosslinking siloxane bonds/layers. As shown in Figure 10.3, the fabrication of SAND involves an iterative combination of: (i) self-limiting chemisorption of σ-π siloxane building blocks, such as α,ω-difunctionalized hydrocarbon chains (Alk), or highly polarizable, siloxy-protected stilbazolium layers (Stb); and (ii) *in situ* siloxy group removal, concurrent with 'capping' using an octachlorotrisiloxane-derived layer (Cap), which is essential to stabilize/planarize the molecular layers and regenerate a reactive hydroxyl surface for further monolayer depositions. Different types of multilayers can thus be fabricated by the combination of these building blocks (Alk + Cap (TYPE I), Stb + Cap (TYPE II), and Alk + Cap + Stb + Cap (TYPE III)) (Figure 10.3).

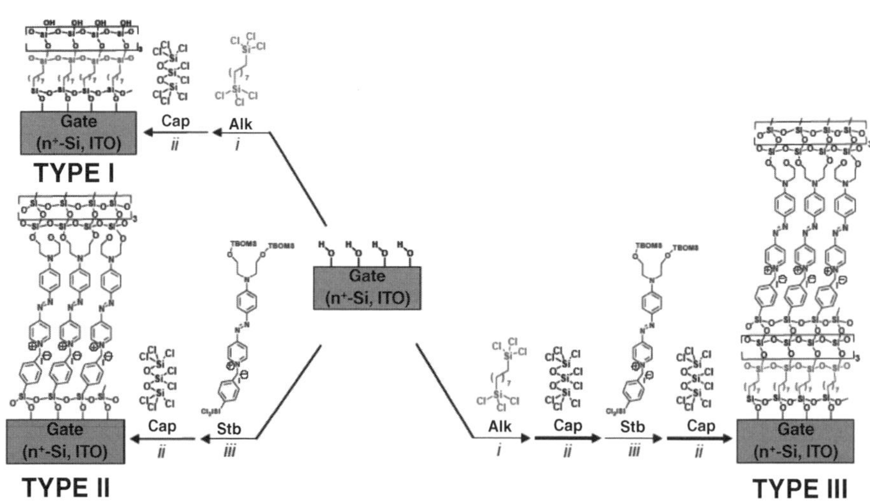

Figure 10.3 Scheme for the fabrication of types I–III self-assembled nanodielectrics (SANDs). The SAND typically used in combination with nanowire semiconductors is a trilayer of type III, denoted $(TYPE\ III)_3$, having a thickness of ~16 nm and a accumulation capacitance of ~180 $nF\,cm^{-2}$. Reproduced with permission from ref. 75. Copyright 2005 National Academy of Sciences, USA

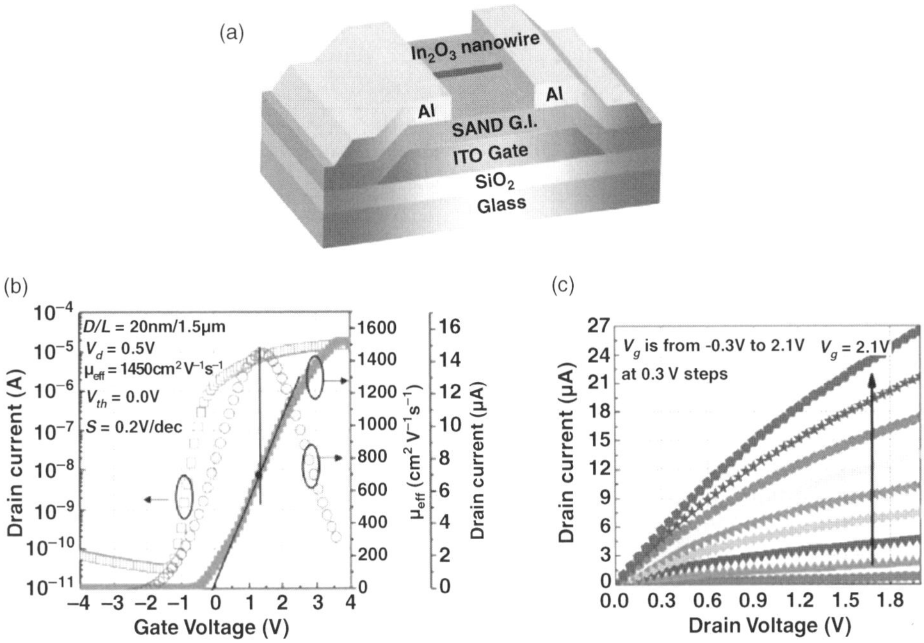

Figure 10.4 *(a) Structure of SAND-based* In_2O_3 *nanowire TFT. (b) Drain current vs. gate-source voltage* ($I_{ds}-V_{gs}$) *characteristics at* $V_d = 0.5$ *V. From left to right the linear-scale* $I_{ds}-V_{gs}$*, mobility* μ *and log-scale* $I_{ds}-V_{gs}$*, respectively, are represented. (c) Drain current vs. drain-source voltage* ($I_{ds}-V_{ds}$) *characteristics for various values of* V_g*. Reprinted with permission from ref. 26. Copyright 2008, American Institute of Physics*

Using (TYPE III)$_3$ SAND gate dielectrics (capacitance $\sim$180 nF cm^{-2}), we have fabricated nanowire-based TFTs [26] consisting of individual In_2O_3 nanowires as semiconductor channels, demonstrating significant advances in performance over previously reported nanowire TFTs employing In_2O_3 or other mid/wide band gap metal oxide nanowires. The corresponding SAND-based devices exhibit a subthreshold slope (S) of 0.2 V decade^{-1}, and I_{on}/I_{off} ratio of 10^6 and a threshold voltage of 0.0 V. The leakage current through the SAND layer is only 30–40 pA at 4 V, indicating negligible leakage current through the gate dielectric. The field effect mobilities varied from $\sim$1450 to $\sim$300 cm^2 V^{-1} s^{-1} over a gate bias range from –0.3 to 2.1 V (Figure 10.4). These values are much higher than the maximum value of $\sim$280 cm^2 V^{-1} s^{-1} obtained using In_2O_3 nanowires on SiO_2/Si dielectrics [71] and the value of $\sim$75 cm^2 V^{-1} s^{-1} reported for a high-k lead zirconate titanate (PZT) gate dielectric [78]. This excellent mobility can be attributed to the single-crystal nature of the nanowire along with the quasi one-dimensional electronic structure [79], which inhibits low-angle carrier scattering, combined with the low interfacial charge trapping aptitude of SAND dielectrics.

The fabrication of high-performance, fully transparent and flexible NWTs consisting of single Zn-doped In_2O_3 nanowires has been reported by Zhang *et al.* using a high-k SiN_x layer as the gate dielectric on poly(ethylene terephthalate) (PET) substrates [80].

By Zn doping, the resistivity decreased from 1.83×10^{-1} to $2.15 \times 10^{-2}\,\Omega\,cm$, without observable structural or orientation changes in the nanowires. The electron mobility of these TFTs is $\sim$630 $cm^2\,V^{-1}\,s^{-1}$ in the linear operation regime, a value significantly higher than the corresponding ones for In_2O_3/SAND TFTs [81] ($\sim$160 $cm^2\,V^{-1}\,s^{-1}$), and comparable with undoped In_2O_3 nanowire transistors on glass substrates [82]. These high performance devices show over 90% optical transmittance in the visible range and good mechanical flexibility and negligible performance degradation under bending conditions.

10.2.3 SnO_2 Nanowire Transistors

SnO_2 nanowires are attractive for low-cost transparent electronics due to their high transparency, ease of obtaining Ohmic contacts with conventional transparent conductive electrodes, and low growth cost compared with other metal oxide materials [81] as well as silicon nanowires and carbon nanotubes [33]. SnO_2 with a tetragonal crystal structure is an excellent candidate for integrating into high performance TFTs, flexible and transparent NWTs and gas sensors [83] due to the wide band gap of 3.6 eV and high surface-to-volume ratio [84].

SnO_2 nanowires can be used as undoped materials due to the presence of impurities in the form of oxygen vacancies but usually they exhibit low carrier concentrations and are very sensitive to ambient conditions [85, 86]. An alternative is to dope SnO_2 nanowires [33]. In particular, lightly doping with Ta yields single nanowire TFTs with mobilities exceeding 100 $cm^2\,V^{-1}\,s^{-1}$ on Si/SiO_2 substrates [29]. The fabrication of transparent devices has been possible by using glass substrates and ITO electrodes, achieving a field effect mobility of 179 $cm^2\,V^{-1}\,s^{-1}$ in the linear regime for a single Ta-doped SnO_2 nanowire transistor. Fully transparent low temperature processed TFTs were also fabricated using arrays of parallel Ta-doped SnO_2 nanowires as the transistor channel with mobilities of 145 $cm^2\,V^{-1}\,s^{-1}$ in the linear region and 112 $cm^2\,V^{-1}\,s^{-1}$ in the saturation regime, in a process compatible with flexible substrates [29].

We exploited the combination of undoped SnO_2 nanowires with SAND dielectrics and demonstrated TFTs with a field effect mobility of $\sim$172 $cm^2\,V^{-1}\,s^{-1}$ (for a single wire transistor), subthreshold slope of $\sim$0.3 V $decade^{-1}$, I_{on}/I_{off} of 10^6, and threshold voltage of –1.9 V [28]. For these NWTs, low frequency noise measurements were carried out to analyze current fluctuation and understand the origin of the enhanced TFT performance with SAND vs. SiO_2 as gate dielectric. The current noise power spectrum (S_I), as a function of gate bias (at a constant drain bias of 1.0 V), is shown in Figure 10.5(a). In the low frequency regime, S_I varies as $f^{-\beta}$, with β ranging between 1 and 1.15.

In Figure 10.5(b) and (c), S_I at 100 Hz and the normalized square of the drain current $I_d^2(V_{gs}-V_{th})$ are plotted as a function of gate bias at drain biases of 0.1 and 1.0 V, respectively, resulting in Hooge's constant (α_H) values of 4.5×10^{-2} and 5.1×10^{-2}, respectively. These results indicate that SAND nanodielectric and various SnO_2 nanowire surface treatments and annealing steps can be used successfully to enhance the quality of the interfaces, thus reducing dielectric/semiconductor interface state densities and improving the $1/f$ noise with respect to untreated devices [28]. Furthermore, the interface density was found much lower for SAND ($\alpha_H = 3.3$–5.1×10^{-2}) than in SiO_2-based SnO_2 TFTs ($\alpha_H = 3.5 \times 10^{-1}$) [87].

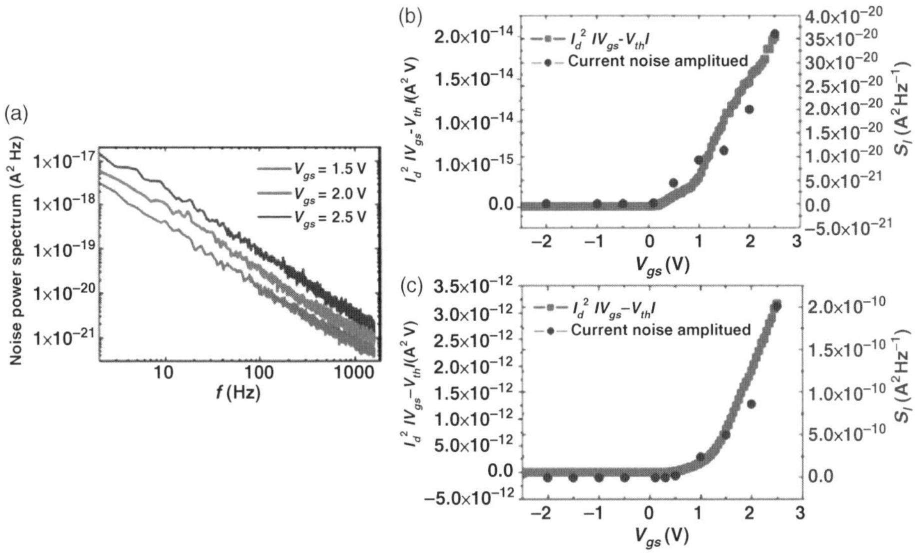

Figure 10.5 *(a) Current noise power spectrum of a SnO_2 NWT as a function of frequency at a constant drain bias of 0.1 V (from top to bottom: $V_{gs} = 2.5$, 2.0 and 1.5 V). (b) Measured $I_d^2(V_{gs} - V_{th})$ and the amplitude of current noise spectrum at 100 Hz versus gate bias at drain bias of 0.1 V. (c) Measured $I_d^2(V_{gs} - V_{th})$ and current noise spectrum amplitude at 100 Hz versus gate bias at drain bias of 1 V. Reprinted with permission from ref. 28. Copyright 2008, American Institute of Physics. See color plate section*

10.3 Transparent Nanowire Circuits and Displays

The development of optically transparent and mechanically flexible electronic circuitry is essential for next-generation visual technologies and portable electronics. NWTs are of particular interest for future display devices because of their several unique and interesting features: (i) enhanced field effect mobility compared with the bulk mobility for the same semiconductor; (ii) amenability to low-temperature processing, and (iii) optical transparency and mechanical flexibility.

As shown in the previous sections, In_2O_3 and ZnO nanowires are particularly promising candidates for transistor active channels since these materials are both transparent and mechanically robust/flexible. In order to be used in commercially viable logic circuits and display devices, TFTs based on these semiconductors must exhibit high on-current (I_{on}), high on/off current ratio (I_{on}/I_{off}), high field effect mobility (μ), steep subthreshold slope (S), and small threshold voltage (V_{th}) variation during device operation. In this sense, we recently demonstrated fully transparent NWTs using all-transparent In_2O_3 and ZnO nanowires (Figure 10.6) [82]. The structure of these TFTs consists of a SiO_2 buffer layer, followed by a patterned IZO gate electrode, and an atomic layer deposition (ALD)-derived high-k Al_2O_3 gate insulator, a single crystal semiconducting In_2O_3 or ZnO nanowire (with a diameter of 20 nm and length of 1.80 μm) for the active channel and ITO for the source/drain electrodes. In_2O_3 NWTs on glass substrates exhibit ~82% visible transparency and

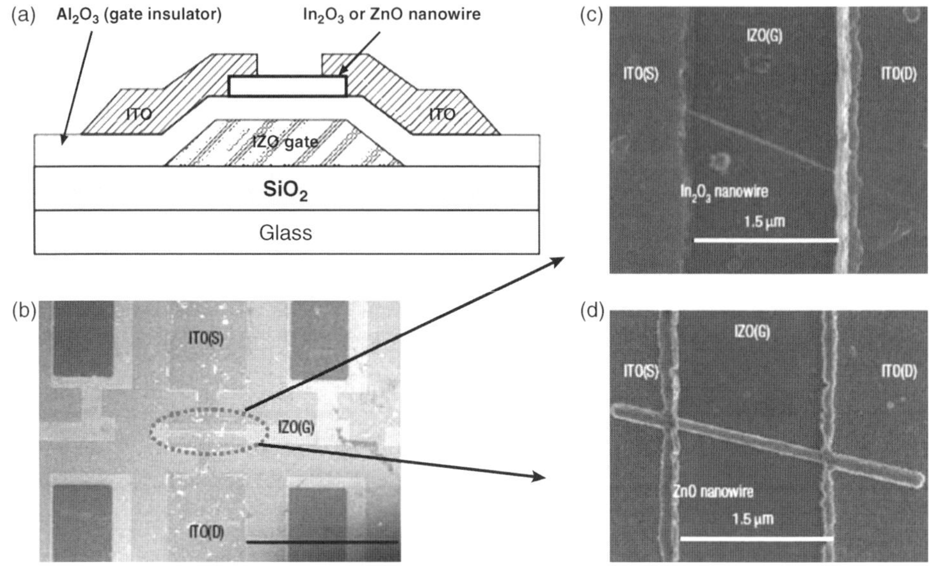

Figure 10.6 *(a) Cross-sectional view of fully transparent NWT device structures consisting of a 500 nm thickness SiO_2 buffer layer, a 120 nm patterned IZO gate electrode, 18 nm ALD-deposited Al_2O_3 gate insulator, single In_2O_3 or ZnO nanowires as the active channel, and ITO source and drain electrodes. (b) Top-view field emission scanning electron microscopy (FE-SEM) images of the devices. (c) Top-view SEM of a single In_2O_3 nanowire. (d) Top-view SEM of a single ZnO nanowire. Reprinted by permission from Macmillan Publishers Ltd: from ref. 82, copyright 2007*

n-type transistor characteristics with a S of 0.16 V decade^{-1}, an I_{on}/I_{off} of 10^6, a V_{th} of –0.27 V, and a μ varying from 514 to 300 cm^2 V^{-1} s^{-1}. ZnO NWTs on glass substrates exhibit ~83% visible transparency with S of 0.3 V decade^{-1}, I_{on}/I_{off} ~10^6, V_{th} of –0.07 V and μ varying from 96 to 70 cm^2 V^{-1} s^{-1}. Fully transparent and flexible In_2O_3 NWTs with optical transmission of ~81% have been also fabricated on PET plastic substrates with S of 0.9 V decade^{-1}, I_{on}/I_{off} of ~10^5, V_{th} of –0.6 V and μ of ~120–167 cm^2 V^{-1} s^{-1} (Figure 10.7). These results were obtained by applying an ozone treatment to the nanowire surface, which removes defects and contamination, and enhances charge injection by modifying the work function [88, 89]. Furthermore, ozone treatment enhances oxygen vacancy density thus increasing nanowire conductivity [90, 91]. These high-performance NWTs are attractive as pixel switching and driving transistors in active matrix organic light emitting diode (AMOLED) displays. Having an optically transparent circuit driving the pixel significantly enhances aperture ratio efficiency in active matrix arrays and thus decreases power consumption.

In 2008 we demonstrated the first optically transparent AMOLED display driven exclusively by nanowire electronics [27]. The circuitry required for each AM display pixel must contain at least one switching transistor, one driver transistor, and a storage capacitor, with appropriate scan and data lines to allow selective pixel addressing (Figure 10.8).

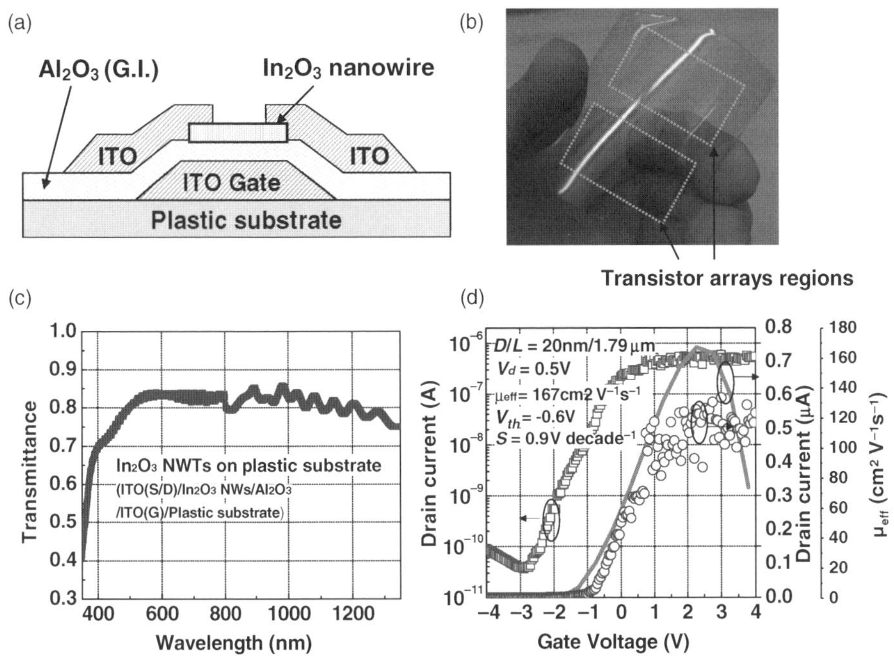

Figure 10.7 (a) Cross-sectional view of a fully transparent and flexible NWT using a single In_2O_3 nanowire for the active channel, ITO as contact electrodes, and ALD-deposited Al_2O_3 as the gate insulator. (b) Image of In_2O_3 NWTs on a plastic substrate showing the optical clarity and mechanical flexibility. (c) Optical transmission of the aforementioned In_2O_3 NWTs. (d) I_{ds}–V_{gs} characteristics at $V_d = 0.5$ V. Circles, squares and solid line correspond to linear-scale I_{ds}–V_{gs}, log-scale I_{ds}–V_{gs}, and mobility μ_{eff}, respectively. Reprinted by permission from Macmillan Publishers Ltd: from ref. 82, copyright 2007

These AMOLED display elements were fabricated at room temperature via a simple and scalable process, requiring only 4 photomasks versus ~8–10 photomasks for a conventional poly-Si TFT AMOLED display. In_2O_3 nanowires were used as active channel materials, the SAND nanodielectric as the gate insulator, and ITO as the transparent conducting gate and source-drain electrodes [Figure 10.8(a)]. Green-emitting polymer LEDs with high-efficiency interfacial charge-blocking and electroluminescent materials were integrated with a transparent bottom contact electrode for efficient optical emission through the glass substrate, and with an opaque or semitransparent top contact cathode. This allows the emission through both the top and bottom sides of the structure. In this structure, the NWTs are the control and drive circuitry for $54 \times 176\,\mu m$ OLED pixels. The SAND gate insulator provides a high breakdown voltage and low interface trap density, while yielding high mobility, a steep subthreshold slope, low operating voltage, high I_{on}/I_{off} ratios, and modest hysteresis [79, 81, 92, 93]. The In_2O_3 NWTs used for this study exhibit an impressive I_{on} ~1 μA (at $V_{gs} = 3.0$ V, $V_{ds} = 0.1$ V), I_{on}/I_{off} of 10^5, V_{th} of 0.1 V, S of 0.25 V decade^{-1} and a calculated mobility of ~258 cm^2 V^{-1} s^{-1}. The NWT circuit yields are ~ 90% and are also able to fulfil AMOLED transistor requirements for rapid switching and high speeds.

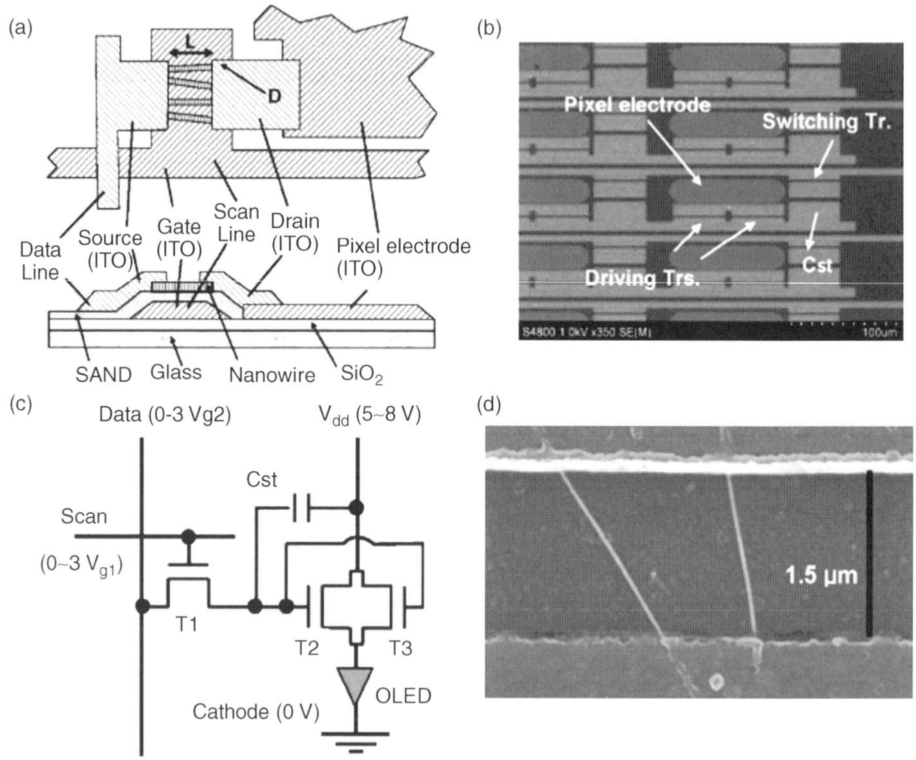

Figure 10.8 *(a) Top and cross-sectional views of the NWT structure consisting of a SiO_2 buffer layer (200 nm), patterned ITO gate electrode (100 nm), SAND gate dielectric (24 nm), multiple In_2O_3 nanowires for the active channel, ITO for the source-drain electrodes (100 nm), and a SiO_2 passivation layer (200 nm). The ITO layer on the right served as the OLED cathode. (b) Top-view FE-SEM image of several $54 \times 176\ \mu m$ pixels within a 2×2 mm NWT array layout. (c) Schematic for the circuit of a single pixel, consisting of one switching transistor (T1), two driving transistors (T2 and T3), and one storage capacitor. (d) FE-SEM image of a region within a NWT transistor channel showing multiple In_2O_3 nanowires connected between source and drain electrodes. Reprinted with permission from ref. 27. Copyright 2008 American Chemical Society*

The cross-section of the full NW-AMOLED device structure is shown in Figure 10.9(a); it consists of: (i) an ITO anode, (ii) a PEDOT-PSS (poly(3,4-ethylenedioxythiophene) poly (styrenesulfonate)) hole injection layer, (iii) a TFB (poly(9,9-dioctylfluorene-*co*-*N*-(4-(3-methylpropylphenyl)diphenylamine))) + $TPDSi_2$ (4,4′-bis(*p*-trichlorosilylpropylphenyl) phenylamino)biphenyl) hole-transport/electron-blocking layer (HTL/EBL), (iv) a TFB + F8BT (poly(9,9-dioctylfluorene-*co*-benzothiadiazole)) emissive layer (EML), and (v) a LiF/Al electron-injection layer/cathode [94].

The efficiency of the NW-AMOLED device was estimated to be $\sim 11\ cd\ A^{-1}$ and the luminance $\sim 1630\ cd\ m^{-2}$ within the active pixel area, corresponding to an average

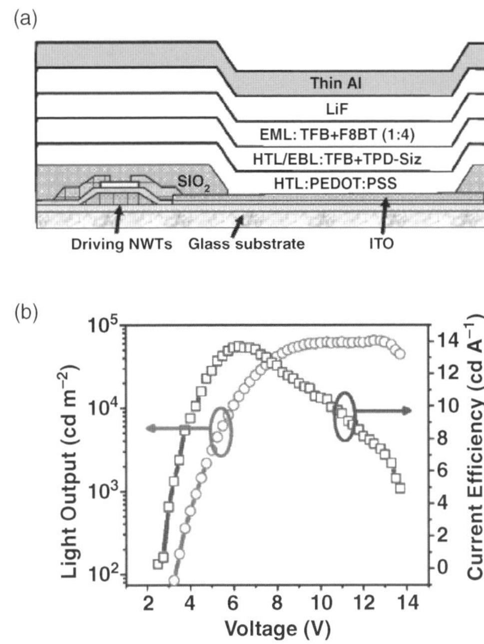

Figure 10.9 *(a) Cross-sectional view of the nanowire-driven OLED device structure. (b) Luminescence and current efficiency versus voltage characteristics of the polymer OLED structure measured in a large area device. Reprinted with permission from ref. 27. Copyright 2008 American Chemical Society*

luminance of ~300 cd m^{-2} for the overall array [Figure 10.9(b)]. In Figure 10.10, optical images of the display devices are shown.

Transparent displays were also fabricated using a very thin ~15 nm aluminium cathode, showing optical transmission of ~72% before and ~35 % after OLED deposition in the 350–1350 nm wavelength range.

Recently, we have also demonstrated the fabrication of optically transparent NWTs and AMOLED displays using high-performance arsenic-doped In_2O_3 nanowires [95]. The doping process here enhances the electrical performance of the In_2O_3 nanowires, leading to n-type field-effect mobilities ranging from 1080 to 1490 cm^2 V^{-1} s^{-1} and I_{on}/I_{off} ~5×10^{-6} for single wire transistors, when using a 50 nm atomic layer deposited Al_2O_3 dielectric, ITO source and drain electrodes, and an ITO glass substrate. The electrical properties of these NWTs were further enhanced by replacing the high-k Al_2O_3 gate dielectric with a $(TYPE\text{-}III)_3$ SAND dielectric, achieving mobilities up to 2560 cm^2 V^{-1} s^{-1} for a single wire TFT. Furthermore, the unity-gain frequency of this NWT can reach 18.8 GHz, which is the highest operation frequency ever achieved for a nanowire transparent TFT.

The fabrication of an AMOLED display was possible using As-doped In_2O_3 multiple wire TFTs. This process begins with the patterning of the OLED ITO anode to define gate electrodes, followed by deposition of a 50 nm ALD Al_2O_3 dielectric layer and subsequent wet etching to create contact holes as bottom gate electrode contacts. After that, a

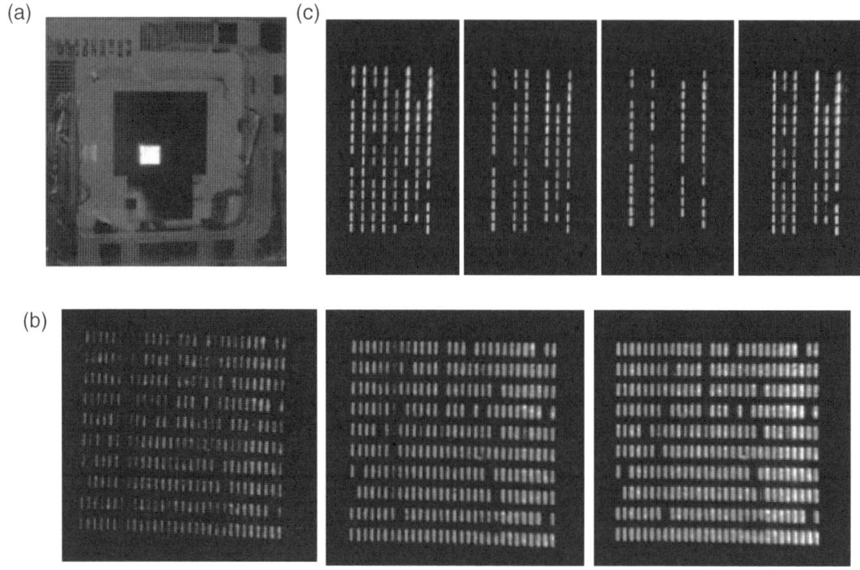

Figure 10.10 (a) Optical images of a NW-AMOLED substrate consisting of three 2 × 2 mm transistor arrays, plus unit pixels and test devices, showing emission from a 2 × 2 mm AMOLED array. (b) Magnified microscope image of a 2 × 2 mm AMOLED pixel array for various bias conditions (drive circuit: V_{data} are 1, 2, and 3 V with fixed $V_{scan} = 3$ V and $V_{dd} = 5$ V). (c) Optical images of 15 pixel × 8 pixel array, with selected pixel lines addressed independently by the control of V_{data} lines. Reprinted with permission from ref. 27. Copyright 2008 American Chemical Society

suspension of As-doped In_2O_3 nanowires in 2-propanol is dispersed on the device substrate followed by e-beam evaporation and patterning of source and drain Al electrodes. Finally, a 200 nm thick SiO_2 layer is deposited by e-beam evaporation to passivate and planarize the nanowire transistor prior to OLED fabrication. Figure 10.11(a) shows the equivalent circuit diagram of the seven-segment AMOLED display circuitry, while Figure 10.11(c) shows optical images of these displays with different numerical digits illuminated at different data line voltages. These devices exhibit an optical transmittance of ~83 and ~35% in the visible region before [Figure 10.11(b)] and after OLED layer deposition, respectively. To the best of our knowledge, this is the first demonstration of a numerical AMOLED display driven exclusively by transparent TFT circuits.

Finally, the fabrication of fully transparent TFTs has been also demonstrated by Kim *et al.* using aligned SWCNT arrays as the active channel, an ALD HfO_2 gate dielectric, and ITO electrodes [96]. In order to improve the on/off ratio of the TFTs, electrical 'burning' was performed to remove the metallic SWCNTs contained in the SWCNT mixture. These TFTs exhibit μ of ~285 $cm^2\,V^{-1}\,s^{-1}$ and I_{on}/I_{off} of ~10^4 for single wire transistors and μ of ~756 $cm^2\,V^{-1}\,s^{-1}$ for aligned-array SWCNT-TFTs. The optical transmission of these devices is ~83%, with the predominant absorption attributed to the ITO contacts. These results indicate the potential of NWTs and SWCNT-TFTs for next-generation display technologies, including 'see-through' products.

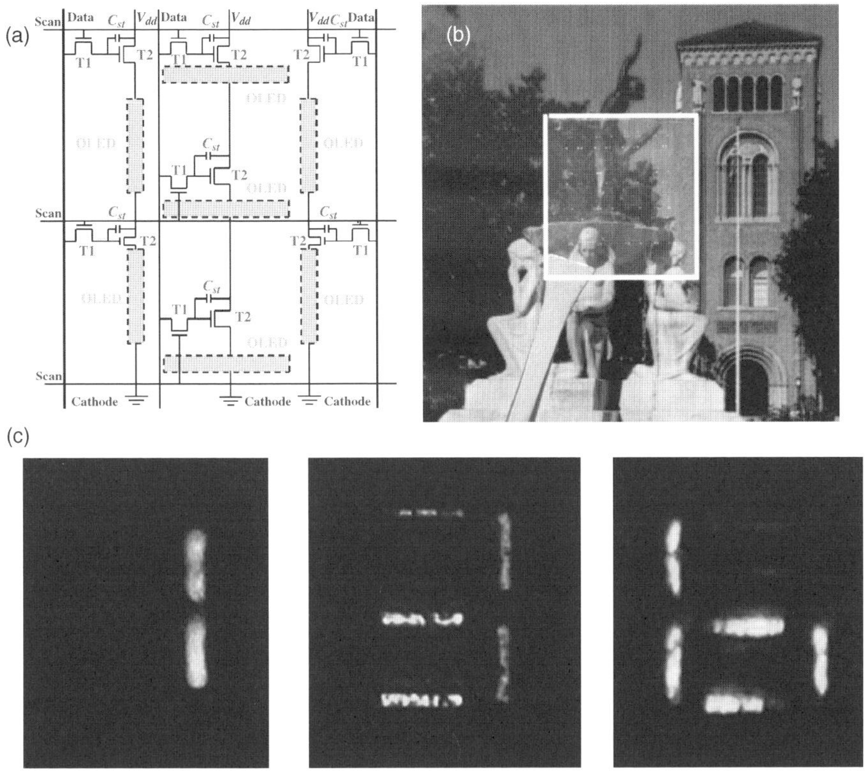

Figure 10.11 (a) Schematic diagram of a seven-segment digit of AMOLED pixel, which consists of one switching transistor (T1), one driving transistor (T2), and one storage capacitor (C_{st}). (b) Optical photograph of fully transparent AMOLED display before OLED layer deposition with the substrate area highlighted for clarity. (c) Optical photograph of AMOLED animation. The seven-segment digit displays the numbers 1, 3 and 6, respectively. Reprinted with permission from ref. 95. Copyright 2009 American Chemical Society

10.4 Conclusions

In the development of optically transparent and mechanically flexible electronics and in the achievement of the miniaturization levels required for new technologies, metal oxide nanowires have attracted great interest. Using these nanowires, transparent and flexible displays have been already demonstrated. However, it is still desirable to optimize transistor performance even further by complete understanding of the physics of nanowire transport characteristics and by controlling/maximizing the semiconductor-dielectric interface quality. Several approaches have been successfully used to enhance device performance, such as passivation of the devices, the use of organic self-assembled nanodielectrics or ozone treatments. The results are encouraging since achieving high-performance NWTs will enable low-power consumption as well as optical transparency in future display technologies.

References

[1] ITRS Organization, *International Technology Roadmap for Semiconductors Report 2007 edition*, 2007.
[2] E. Y. Wu, J. H. Stathis, L.-K. Han, Ultra-thin oxide reliability for USLI applications, *Semiconductor Sci. Technol.* **15**, 425–435 (2000).
[3] (a) Y. Qin, X. D. Wang, Z. L. Wang, Microfibre-nanowire hybrid structure for energy scavenging, *Nature* **451**, 809–813 (2008); (b) A. I. Hochbaum, R. K. Chen, R. D. Delgado, W. J. Liang, E. C. Garnett, N. Najarian, A. Majumdar, P. D. Yang, Enhanced thermoelectric performance of rough silicon nanowires, *Nature* **451**, 163–167 (2008); (c) C. K. Chan, H. L. Peng, G. Liu, K. McIlwrath, X. F. Zhang, R. A. Huggins, Y. Cui, High-performance lithium battery anodes using silicon nanowires, *Nat. Nanotechnol.* **3**, 31–35 (2008).
[4] C. M. Lieber, Z. L. Wang, Functional nanowires, *MRS Bull.* **32**, 99–108 (2007).
[5] W. Lu, C. M. Lieber, Semiconductor nanowires, *J. Phys. D: Appl. Phys.* **39**, R387–R406 (2006).
[6] A. M. Morales, C. M. Lieber, A laser ablation method for the synthesis of crystalline semiconductor nanowires, *Science* **279**, 208–211 (1998).
[7] Y. Wu, Y. Cui, L. Huynh, C. J. Barrelet, D. C. Bell, C. M. Lieber, Controlled growth and structures of molecular-scale silicon nanowires, *Nano Lett.* **4**, 433–436 (2004).
[8] W. Lu, J. Xiang, B. P. Timko, Y. Wu, C. M. Lieber, One dimensional hole gas in germanium/silicon nanowire heterostructures, *Proc. Natl Acad. Sci. USA* **102**, 10 046–10 051 (2005).
[9] J. Xiang, W. Lu, Y. Hu, Y. Wu, H. Yan, C. M. Lieber, Ge/Si nanowire heterostructures as high-performance field-effect transistors, *Nature* **441**, 489–493 (2006).
[10] W. Lu, P. Xie, C. M. Lieber, Nanowire transistor performance limits and applications, *IEEE Trans. Elect. Dev.* **55**, 2859–2876 (2008).
[11] X. Wu, J. S. Kulkarni, G. Collins, N. Petkov, D. Almécija, J. J. Boland, D. Erts, J. D. Holmes, Synthesis and electrical and mechanical properties of silicon and germanium nanowires, *Chem. Mater.* **20**, 5954–5967 (2008).
[12] (a) R. F. Service, New age semiconductors pick up the pace, *Science* **287**, 415–417 (2000); (b) J. M. Shaw, P. F. Seidler, Organic electronics: introduction, *IBM J. Res. Dev.* **45**, 3–9 (2001); (c) E. Lueder, *Liquid Crystal Displays: Addressing Schemes and Electro-Optical Effects* John Wiley & Sons, Ltd Chichester (2001).
[13] S. Uchikoga, Low-temperature polycrystalline silicon thin-film transistor technologies for system-on-glass displays, *MRS Bull.* **27**, 881–886 (2002).
[14] Y. Cui, Z. Zhong, D. Wang, W. U. Wang, C. M. Lieber, High performance silicon nanowire field effect transistors, *Nano Lett.* **3**, 149–152 (2003).
[15] B.-R. Huang, J.-F. Hsu, C.-S. Huang, Y.-T. Shih, K.-S. Lu, Silicon nanowire networks for the application of field effect phototransistor, *Mater. Sci. Eng. C* **27**, 1197–1200 (2007).
[16] G. Zheng, W. Lu, S. Jin, C. M. Lieber, Synthesis and fabrication of high-performance n-type silicon nanowire transistors, *Adv. Mater.* **16**, 1890–1893 (2004).
[17] K. Heo, E. Cho, J.-E. Yang, M.-H. Kim, M. Lee, B. Y. Lee, S. G. Kwon, M.-S. Lee, M.-H. Jo, H.-J. Choi, T. Hyeon, S. Hong, Large-scale assembly of silicon nanowire network-based devices using conventional microfabrication facilities, *Nano Lett.* **8**, 4523–4527 (2008).
[18] A. Javey, S. Nam, R. S. Friedman, H. Yan, C. M. Lieber, Layer-by-layer assembly of nanowires for three-dimensional, multifunctional electronics, *Nano Lett.* **7**, 773–777 (2007).
[19] T. Hanrath, B. A. Korgel, Influence of surface states on electron transport through intrinsic Ge nanowires, *J. Phys. Chem. B* **109**, 5518–5524 (2005).
[20] (a) D. Erts, B. Polyakov, B. Daly, M. A. Morris, S. Ellingboe, J. D. Boland, J. D. Holmes, High density germanium nanowire assemblies: contact challenges and electrical characterization, *J. Phys. Chem. B* **110**, 820–826 (2006); (b) K. J. Ziegler, B. Polyakov, J. S. Kulkarni, T. A. Crowley, K. M. Ryan, M. A. Morris, D. Erts, J. D. Holmes, Conductive films of ordered nanowire arrays, *J. Mater. Chem.* **14**, 585–589 (2004).
[21] M. I. van der Meulen, N. Petkov, M. A. Morris, O. Kazakova, X. Han, K. L. Wang, A. P. Jacob, J. D. Holmes, *Nano Lett.* **9**, 50–56 (2009).
[22] J.-W. Lee, M.-H. Ham, J.-M. Myoung, Gate-controlled transport in GaN nanowire devices with high-k Si_3N_4 gate dielectrics, *Solid State Commun.* **145**, 327–331 (2008).

[23] S. N. Cha, J. E. Jang, Y. Choi, G. A. J. Amaratunga, G. W. Ho, M. E. Welland, D. G. Hasko, D. J. Kang, J. M. Kim, High performance ZnO nanowire field effect transistor using self-aligned nanogap gate electrodes, *Appl. Phys. Lett.* **89**, 263102/1–263102/3 (2006).

[24] H. E. Unalan, Y. Zhang, P. Hiralal, S. Dalal, D. Chu, G. Eda, K. B. K. Teo, M. Chhowalla, W. I. Milne, G. A. J. Amaratunga, Zinc oxide nanowire networks for macroelectronic devices, *Appl. Phys. Lett.* **94**, 163501/1–163501/3 (2009).

[25] P. C. Chang, Z. Fan, C. J. Chien, D. Stichtenoth, C. Ronning, J. G. Lu, High-performance ZnO nanowire field effect transistors, *Appl. Phys. Lett.* **89**, 133113/1–133113/3 (2006).

[26] S. Ju, F. Ishikawa, P. Chen, H.-K. Chang, C. Zhou, Y.-G. Ha, J. Liu, A. Facchetti, T. J. Marks, D. B. Janes, High performance In_2O_3 nanowire transistors using organic gate nanodielectrics, *Appl. Phys. Lett.* **92**, 222105/1–222105/3 (2008).

[27] S. Ju, J. Li, J. Liu, P.-C. Chen, Y.-G. Ha, F. Ishikawa, H. Chang, C. Zhou, A. Facchetti, D. B. Janes, T. J. Marks, Transparent active matrix organic light-emitting diode displays driven by nanowire transistor circuitry, *Nano Lett.* **8**, 997–1004 (2008).

[28] S. Ju, P. Chen, C. Zhou, Y.-G. Ha, A. Facchetti, T. J. Marks, S. K. Kim, S. Mohammadi, D. B. Janes, 1/f noise of SnO_2 nanowire transistors, *Appl. Phys. Lett.* **92**, 243120/1–243120/3 (2008).

[29] (a) E. N. Dattoli, Q. Wan, W. Guo, Y. Chen, X. Pan, W. Lu, Fully transparent thin-film transistor devices based on SnO_2 nanowires, *Nano Lett.* **7**, 2463–2469 (2007).

[30] (a) C. D. Dimitrakopoulos, D. J. Mascaro, Organic thin-film transistors: a review of recent advances, *IBM J. Res. Dev.* **45**, 11–27 (2001); (b) Y. Chen, J. An, P. Kazlas, A. Ritenour, H. Gates, M. McMreary, Electronic paper: flexible active-matrix electronic ink display, *Nature* **423**, 136 (2003).

[31] X. Duan, C. Niu, V. Sahi, J. Chen, J. W. Parce, S. Empedocles, J. L. Goldman, High-performance thin-film transistors using semiconductor nanowires and nanoribbons, *Nature* **425**, 274–278 (2003).

[32] (a) M. C. McAlpine, R. S. Friedman, S. Jin, K. H. Lin, W. U. Wang, C. M. Lieber, High-performance nanowire electronics and photonics on glass and plastic substrates, *Nano Lett.* **3**, 1531–1535 (2003); (b) R. S. Friedman, M. C. McAlpine, D. S. Ricketts, D. Ham, C. M. Lieber, High-speed integrated nanowire circuits, *Nature* **434**, 1085 (2005).

[33] (a) Q. Wan, E. Dattoli, W. Lu, Doping-dependent electrical characteristics of SnO_2 nanowires, *Small* **4**, 451–454 (2008); (b) Q. Wan, E. N. Dattoli, W. Y. Fung, W. Guo, Y. Chen, X. Pan, W. Lu, High-performance transparent conducting oxide nanowires, *Nano Lett.* **6**, 2909–2915 (2006); (c) Q. Wan, E. N. Dattoli, W. Lu, Transparent metallic Sb-doped SnO_2 nanowires, *Appl. Phys. Lett.* **90**, 222107 (2007).

[34] (a) Y. C. Kong, D. P. Yu, B. Zhang, W. Fang, S. Q. Feng, Ultraviolet-emitting ZnO nanowires synthesized by a physical vapor deposition approach, *Appl. Phys. Lett.* **78**, 407–409 (2001); (b) Z. W. Pan, S. Dai, C. M. Rouleau, D. H. Lowndes, Germanium-catalyzed growth of zinc oxide nanowires: a semiconductor catalyst for nanowire synthesis, *Angew. Chem. Int. Ed. Engl.* **44**, 274–278 (2005).

[35] Z. W. Liu, C. K. Ong, T. Yu, Z. X. Shen, Catalyst-free pulsed-laser-deposited ZnO nanorods and their room-temperature photoluminescence properties, *Appl. Phys. Lett.* **88**, 053110/1–053110/2 (2006).

[36] M. Yang, G. S. Pang, L. F. Jiang, S. H. Feng, Hydrothermal synthesis of one-dimensional zinc oxide with different precursors, *Nanotechnology* **17**, 206–212 (2006).

[37] (a) W. I. Park, D. H. Kim, S. W. Jung, G. C. Yi, Metalorganic vapor-phase epitaxial growth of vertically well-aligned ZnO nanorods, *Appl. Phys. Lett.* **80**, 4232–4234 (2002); (b) S. W. Kim, S. Fujita, S. Fujita, ZnO nanowires with high aspect ratios grown by metalorganic chemical vapor deposition using gold nanoparticles, *Appl. Phys. Lett.* **86**, 153119 (2005).

[38] Y. W. Heo, V. Varadarajan, M. Kaufman, K. Kim, D. P. Norton, F. Ren, P. H. Fleming, Site-specific growth of ZnO nanorods using catalysis-driven molecular-beam epitaxy, *Appl. Phys. Lett.* **81**, 3046–3048 (2002).

[39] Z. R. Dai, Z. W. Pan, Z. L. Wang, Novel nanostructures of functional oxides synthesized by thermal evaporation, *Adv. Funct. Mater.* **13**, 9–24 (2003).

[40] (a) S. Sze, *Semiconductor Devices: Physics and Technology* 2nd edn, John Wiley & Sons, Ltd New York (1985); (b) B. G. Streetman, S. K. Banerjee, *Solid State Electronic Devices*, Pearson

Prentice Hall New Jersey (2006); (c) R. A. Martin, V. M. Da Costa, M. Hack, J. G. Shaw, High-voltage amorphous silicon thin-film transistors, *IEEE Trans. Electr. Dev.* **4**, 634–644 (1993).

[41] (a) J. Guo, J. Wang, E. Polizzi, S. Datta, M. Lundstrom, Electrostatics of nanowire transistors, *IEEE Trans. Nanotechnol.* **2**, 329–334 (2003); (b) Y. C. Yeo, Metal gate technology for nanoscale transistors - material selection and process integration issues, *Thin Solid Films* **462/463**, 34–41 (2004).

[42] (a) P.-C. Chang, J. G. Lu, ZnO nanowire field-effect transistors, *IEEE Trans. Electron Devices* **55**, 2977–2978 (2008); (b) S. J. Pearton, D. P. Norton, L.-C. Tien, J. Guo, Modeling and fabrication of ZnO nanowire transistors, *IEEE Trans. Electron Devices* **55**, 3012–3019 (2008).

[43] R. Martel, T. Schmidt, H. R. Shea, T. Hertel, P. Avouris, Single- and multi-wall carbon nanotube field-effect transistors, *Appl. Phys. Lett.* **73**, 2247–2449 (1998).

[44] D. Wang, Q. Wang, A. Javey, R. Tu, H. Dai, H. Kim, P. C. McIntyre, T. Krishnamohan, K. C. Saraswat, Germanium nanowire field-effect transistors with SiO_2 and high-k HfO_2 gate dielectrics, *Appl. Phys. Lett.* **83**, 2432–2434 (2003).

[45] (a) A. Tsukazaki, A. Ohtomo, T. Onuma, M. Ohtani, T. Makino, M. Suyima, K. Ohtani, S. Chichibu, S. Fuke, Y. Segawa, H. Ohno, H. Koinuma, M. Kawasaki, Repeated temperature modulation epitaxy for p-type doping and light-emitting diode based on ZnO, *Nat. Mater.* **4**, 42–46 (2005); (b) S. J. Pearton, D. P. Norton, K. Ip, Y. W. Heo, T. Steiner, Recent progress in processing and properties of ZnO, *Prog. Mater. Sci.* **50**, 293–340 (2005).

[46] C. Jagadish,S. J. Pearton, *Zinc Oxide Bulk, Thin Films and Nanostructures*Elsevier Oxford (2006).

[47] D. Seghier, H. P. Gislason, Characterization of donor states in ZnO, *Phys. Rev. B: Condens. Matter* **401/402**, 404–407 (2007).

[48] (a) B. D. Yao, Y. F. Chan, N. I. Wang, Formation of ZnO nanostructures by a simple way of thermal evaporation, *Appl. Phys. Lett.* **81**, 757–759 (2002); (b) H. W. Kim, N. H. Kim, J. H. Shim, N. H. Cho, C. Lee, Catalyst-free MOCVD growth of ZnO nanorods and their structural characterization, *J. Mater. Sci., Mater. Electron.* **16**, 13–15 (2005); (c) H. J. Kim, K. H. Sung, K. S. An, Y. K. Lee, C. G. Kim, Y. H. Lee, Y. Kim, ZnO nanowhiskers on ZnO nanoparticle-deposited Si(111) by MOCVD, *J. Mater. Chem.* **14**, 3396–3397 (2004); (d) Y. Sun, G. M. Fuge, M. N. R. Ashfold, Growth mechanisms for ZnO nanorods formed by pulsed laser deposition, *Superlattices Microstruct.* **39**, 33–40 (2006).

[49] K. S. Shankar, A. K. Raychaudhuri, Fabrication of nanowires of multicomponent oxides: Review of recent advances, *Mater. Sci. Eng., C* **25**, 738–751 (2005).

[50] M. H. Huang, Y. W. Henning Feick, N. Tran, E. Weber, P. Yang,Catalytic growth of zinc oxide by vapor transport, *Adv. Mater.* **13**, 113–116 (2001).

[51] M. H. Huang, S. Mao, H. Feick, H. Yan, Y. Wu, H. Kind, E. Weber, R. Russo, P. Yang, Room-temperature ultraviolet nanowire nanolasers, *Science* **292**, 1897–1899 (2001).

[52] Z. W. Pan, Z. R. Dai, Z. L. Wang, Nanobelts of semiconducting oxides, *Science* **291**, 1947–1949 (2001).

[53] B. B. Lakshmi, P. K. Dorhout, C. R. Martin, Sol-gel template synthesis of semiconductor nanostructures, *Chem. Mater.* **9**, 857–862 (1997).

[54] Y. Li, G. W. Meng, L. D. Zhang, F. Phillipp, Ordered semiconductor ZnO nanowire arrays and their photoluminescence properties, *Appl. Phys. Lett.* **76**, 2011–2013 (2000).

[55] (a) J. R. LaRoche, Y. W. Heo, B. S. Kang, L. C. Tien, Y. Kwon, D. P. Norton, B. P. Gila, F. Ren, S. J. Pearton, Fabrication approaches to ZnO nanowire devices, *J. Electron. Mater.* **34**, 404–408 (2005); (b) Y. W. Heo, D. P. Norton, L. C. Tien, Y. Kwon, B. S. Kang, F. Ren, S. J. Pearson, J. R. LaRoche, ZnO nanowire growth and devices, *Mater. Sci. Eng., R Rep.* **47**, 1–47 (2004).

[56] (a) Y. Huang, X. F. Duan, Y. Cui, L. J. Lauhon, K. H. Kim, C. M. Lieber, Logic gates and computation from assembled nanowire building blocks, *Science* **294**, 1313–1317 (2001); (b) Y. W. Heo, L. C. Tien, Y. Kwon, D. P. Norton, S. J. Pearton, B. S. Kang, F. Ren, Depletion-mode ZnO nanowire field-effect transistor, *Appl. Phys. Lett.* **85**, 2274–2276 (2004); (c) S. Kim, K. Keem, J. Kang, C. Yoon, D. Yeom, D.-Y. Jeong, B.-M. Moon, A fabrication technique for top-gate ZnO nanowire field-effect transistors by a photolithography process, *Microelectron. Eng.* **84**, 1622–1626 (2007); (d) H. T. Ng, J. Han, T. Yamada, P. Nguyen, Y. P. Chen, M. Meyyappan,

Single cristal nanowire vertical surround-gate field-effect transistor, *Nano Lett.* **4**, 1247–1252 (2004).

[57] (a) S. H. Jo, J. Y. Lao, Z. F. Ren, R. A. Farrer, T. Baldacchini, J. T. Fourkas, Field-emission studies on thin films of zinc oxide nanowires, *Appl. Phys. Lett.* **83**, 4821–4823 (2003); (b) S. Ju, K. Lee, M. H. Yoon, A. Facchetti, T. J. Marks, D. B. Janes, High performance ZnO nanowire field effect transistors with organic gate nanodielectrics: Effects of metal contacts and ozone treatment, *Nanotechnology* **18**, 155201/1–155201/7 (2007); (c) W. I. Park, J. S. Kim, G. C. Yi, M. H. Bae, H. J. Lee, Fabrication and electrical characteristics of high-performance ZnO nanorod field-effect transistors, *Appl. Phys. Lett.* **85**, 5052–5054 (2004); (d) C. Bekeny, T. Voss, B. Hilker, J. Gutowski, R. Hauschild, H. Kalt, B. Postels, A. Bakin, A. Waag, Influence of ZnO seed crystals and annealing on the optical quality of low-temperature grown ZnO nanorods, *J. Appl. Phys.* **102**, 044908/1–044908/3 (2007); (e) L. Shi, Y. M. Xu, S. K. Hark, Y. Liu, S. Wang, L. M. Peng, K. W. Wong, Q. Li, Optical and electrical performance of SnO_2 capped ZnO nanowire arrays, *Nano Lett.* **7**, 3559–3563 (2007).

[58] G. Eranna, B. C. Joshi, D. P. Runthala, R. P. Gupta, Oxide materials for development of integrated gas sensors - a comprehensive review, *Crit. Rev. Solid State Mater. Sci.* **29**, 111–188 (2004).

[59] W.-K. Hong, D.-K. Hwang, I.-K. Park, G. Jo, S. Song, S.-J. Park, T. Lee, B.-J. Kim, E. A. Stach, Realization of highly reproducible ZnO nanowire field effect transistors with n-channel depletion and enhancement modes, *Appl. Phys. Lett.* **90**, 243103/1–243103/3 (2007).

[60] H. D. Xiong, W. Wang, Q. Li, C. A. Richter, J. S. Suehle, W.-K. Hong, T. Lee, D. M. Fleetwood, Random telegraph signals in n-type ZnO nanowire field effect transistors at low temperature, *Appl. Phys. Lett.* **91**, 053107/1–053107/3 (2007).

[61] B. Sun, H. Sirringhaus, Solution-processed zinc oxide field-effect transistors based on self-assembly of colloidal nanorods, *Nano Lett.* **5**, 2408–2413 (2005).

[62] S. H. Ko, I. Park, M. N. Heng, M. S. Rogers, C. P. Grigoropoulos, A. P. Pisano, ZnO nanowire network transistor fabrication on a polymer substrate by low-temperature, all-inorganic nano-particle solution process, *Appl. Phys. Lett.* **92**, 154102/1–154102/3 (2008).

[63] K. Sreenivas, T. S. Rao, A. Mansingh, Preparation and characterization of rf sputtered indium tin oxide films, *J. Appl. Phys.* **57**, 384–392 (1985).

[64] Y. Shigesato, S. Takaki, T. Haranoh, Electrical and structural properties of low resistivity tin-doped indium oxide films, *J. Appl. Phys.* **71**, 3356–3364 (1992).

[65] J. Tamaki, C. Naruo, Y. Yamamoto, M. Mastuoka, Sensing properties to dilute chlorine gas of indium oxide based thin film sensors prepared by electron beam evaporation, *Sens. Actuators, B* **83**, 190–194 (2002).

[66] M. Liess, Electric-field-induced migration of chemisorbed gas molecules on a sensitive film - a new chemical sensor, *Thin Solid Films* **410**, 183–187 (2002).

[67] F. Liu, M. Bao, K. L. Wang, C. Li, B. Lei, C. Zhou, One-dimensional transport of In_2O_3 nanowires, *Appl. Phys. Lett.* **86**, 213101/1–213101/3 (2005).

[68] M. Bockrath, D. H. Cobden, J. Lu, A. G. Rinzler, R. E. Smalley, L. Balents, P. L. McEuen, Luttinger-liquid behaviour in carbon nanotubes, *Nature* **397**, 598–600 (1999).

[69] A. Bachtold, M. De Jonge, K. Grove-Rasmussen, P. L. McEuen, Suppression of tunneling into multiwall carbon nanotubes, *Phys. Rev. Lett.* **87**, 166801/1–166801/4 (2001).

[70] C. Li, D. Zhang, S. Han, X. Liu, T. Tang, C. Zhou, Diameter-controlled growth of single-crystalline In_2O_3 nanowires and their electronic properties, *Adv. Mater.* **15**, 143–146 (2003).

[71] B. Lei, C. Li, D. Zhang, T. Tang, C. Zhou, Tuning electronic properties of In_2O_3 nanowires by doping control, *Appl. Phys. A* **79**, 439–442 (2004).

[72] D. Zhang, C. Li, S. Han, X. Liu, T. Tang, W. Jin, C. Zhou, Electronic transport Studies of single-crystalline In_2O_3 nanowires, *Appl. Phys. Lett.* **82**, 112–114 (2003).

[73] N. Pho, H. T. Ng, T. Yamada, M. K. Smith, L. Jun, H. Jie, M. Meyyappan, Direct integration of metal oxide nanowire in vertical field-effect transistor, *Nano Lett.* **4**, 651–657 (2004).

[74] R. L. Weiher, Electrical properties of single crystal of indium oxide, *J. Appl. Phys.* **33**, 2834–2839 (1962).

[75] M.-H. Yoon, A. Facchetti, T. J. Marks, σ-π molecular dielectric multilayers for low-voltage organic thin-film transistors, *Proc. Natl Acad. Sci. USA* **102**, 4678–4682 (2005).

[76] A. Facchetti, M.-H. Yoon, T. J. Marks, Gate dielectrics for organic field-effect transistors: new opportunities for organic electronics, *Adv. Mater.* **17**, 1705–1725 (2005).
[77] S. A. DiBenedetto, A. Facchetti, M. A. Ratner, T. J. Marks, Molecular self-assembled monolayers and multilayers for organic and unconventional inorganic thin-film transistor applications, *Adv. Mater.* **21**, 1407–1433 (2009).
[78] B. Lei, C. Li, D. Zhang, Q. F. Zhou, K. K. Shung, C. Zhou, Nanowire transistors with ferroelectric gate dielectric: enhanced performance and memory effects, *Appl. Phys. Lett.* **84**, 4553–4555 (2004).
[79] S. Ju, K. Lee, D. B. Janes, M.-H. Yoon, A. Facchetti, T. J. Marks, Low operating voltage single ZnO nanowire field-effect transistors enabled by self-assembled organic gate nanodielectrics, *Nano Lett.* **5**, 2281–2286 (2005).
[80] W. F. Zhang, Z. B. He, G. D. Yuan, J. S. Jie, L. B. Luo, X. J. Zhang, Z. H. Chen, C. S. Lee, W. J. Zhang, S. T. Lee, High-performance, fully transparent, and flexible zinc-doped indium oxide nanowire transistors, *Appl. Phys. Lett.* **94**, 123103/1–123103/3 (2009).
[81] L. Wang, M. H. Yoon, G. Lu, Y. Yang, A. Facchetti, T. J. Marks, High-performance transparent inorganic-organic hybrid thin-film n-type transistors, *Nature Mater.* **5**, 893–900 (2006).
[82] S. Ju, A. Facchetti, Y. Xuan, J. Liu, F. Ishikawa, P. Ye, C. Zhou, T. J. Marks, D. B. Janes, Fabrication of fully transparent nanowire transistors for transparent and flexible electronics, *Nat. Nanotechnol.* **2**, 378–384 (2007).
[83] (a) A. Kolmakov, Y. Zhang, G. Cheng, M. Moskovits, Detection of CO and O_2 using tin oxide nanowire sensors, *Adv. Mater.* **15**, 997–1000 (2003); (b) R. E. Presley, C. L. Munsee, C.-H. Park, D. Hong, J. F. Wager, D. A. Keszler, Tin oxide transparent thin-film transistors, *J. Phys. D: Appl. Phys.* **37**, 2810–2813 (2004); (c) S. Kalinin, J. Shin, S. Jesse, D. Geohegan, A. Baddorf, Y. Lilach, M. Moskovits, A. Kolmakov, Electronic transport imaging in a multiwire SnO_2 chemical field-effect transistor device, *J. Appl. Phys.* **98**, 044503/1–044503/8 (2005); (d) Y. Zhang, A. Kolmakov, Y. Lilach, M. Moskovits, Electronic control of chemistry and catalysis at the surface of an individual tin oxide nanowire, *J. Phys. Chem. B* **109**, 1923–1929 (2005).
[84] F. Hernandez-Ramirez, A. Tarancon, O. Casals, J. Rodriguez, A. Romano-Rodriguez, J. R. Morante, S. Barth, S. Mathur, T. Y. Choi, D. Poulikakos, V. Callegari, P. M. Nellen, Fabrication and electrical characterization of circuits based on individual tin oxide nanowires, *Nanotechnology* **17**, 5577–5583 (2006).
[85] Z. Liu, D. Zhang, S. Han, C. Li, T. Tang, W. Jin, X. Liu, B. Lei, C. Zhou, Laser ablation synthesis and electron transport studies of tin oxide nanowires, *Adv. Mater.* **15**, 1754–1757 (2003).
[86] M. S. Arnold, P. Avouris, Z. W. Pan, Z. L. Wang, Field-effect transistors based on single semiconducting oxide nanobelts, *J. Phys. Chem. B* **107**, 659–663 (2003).
[87] S. Ju, S. Kim, S. Mohammadi, D. B. Janes, Y.-G. Ha, A. Facchetti, T. J. Marks, Interface studies of ZnO nanowire transistors and temperature-dependent I-V measurements, *Appl. Phys. Lett.* **92**, 022104/1–022104/3 (2008).
[88] O. Lang, C. Pettenkofer, J. F. Sánchez-Royo, A. Segura, A. Klein, W. Jaegermann, Thin film growth and band lineup of In_2O_3 on the layered semiconductor InSe, *J. Appl. Phys.* **86**, 5687–5691 (1999).
[89] Y. Gassenbauer, R. Schafranek, A. Klein, S. Zafeiratos, M. Hävecker, A. Knop-Gericke, R. Schlögl, Surface states, surface potentials, and segregation at surfaces of tin-doped In_2O_3, *Phys. Rev. B* **73**, 245312 (2006).
[90] J. Goldberger, D. Sirbuly, M. Law, P. Yang, ZnO nanowire transistors, *J. Phys. Chem. B* **109**, 9–14 (2005).
[91] R. Martins, E. Fortunato, P. Nunes, I. Ferreira, A. Marques, M. Bender, N. Katsarakis, V. Cimalla, G. Kiriakidis, Zinc oxide as an ozone sensor, *J. Appl. Phys.* **96**, 1398–1408 (2004).
[92] S.-H. Hur, M.-H. Yoon, A. Gaur, A. Facchetti, T. J. Marks, J. A. Rogers, Organic nanodielectrics for low voltage carbon nanotube thin film transistors and complementary logic gates, *J. Am. Chem. Soc.* **127**, 13808–13809 (2005).
[93] S. Ju, D. B. Janes, G. Lu, A. Facchetti, T. J. Marks, Effect of bias stress on ZnO nanowire field-effect transistors fabricated with organic gate nanodielectrics, *Appl. Phys. Lett.* **89**, 193506/1–193506/3 (2006).

[94] H. Yan, P. Lee, N. Armstrong, A. Graham, G. Evmenenko, P. Dutta, T. J. Marks, High-performance hole-transport layers for polymer light-emitting diodes. Implementation of organosiloxane cross-linking chemistry in polymeric electroluminescent devices, *J. Am. Chem. Soc.* **127**, 3172–3183 (2005).

[95] P.-C. Chen, G. Shen, H. Chen, Y.-G. Ha, C. Wu, S. Sukcharoenchoke, Y. Fu, J. Liu, A. Facchetti, T. J. Marks, M. Thompson, C. Zhou, High-performance single-crystalline arsenic-doped indium oxide nanowires and self-assembled gate dielectrics for transparent thin film transistors and active matrix organic light-emitting diode displays, *ACS Nano* **3**, 3383–3390 (2009).

[96] S. Kim, S. Ju, J. H. Back, Y. Xuan, P. D. Ye, M. Shim, D. B. Janes, S. Mohammadi, Fully transparent thin-film transistors based on aligned carbon nanotube arrays and indium tin oxide electrodes, *Adv. Mater.* **21**, 564–568 (2009).

11

Application of Transparent Oxide Semiconductors for Flexible Electronics

Peter F. Carcia

DuPont CR&D Experimental Station, Wilmington, USA

11.1 Introduction

Electronics on flexible plastic substrates [1, 2] are attractive for portable applications, because plastic is lighter weight, thinner, bendable, and more rugged than glass. Flexible plastic substrates can also be processed reel-to-reel, which should reduce manufacturing cost. Some applications envisioned for flexible electronics include (rollable) displays [3], low cost radio frequency identification (RFID) tags [4], and large area or macroelectronics [5], in general.

Plastic substrates, however, are temperature sensitive. Common, inexpensive, plastic substrates, such as poly(ethylene terephthalate) (e.g. PET-Mylar®), can withstand processing only up to ~100 °C, while poly(ethylene naphthalate) (PEN), in the same polyester family, can be used up to ~200 °C. (More detail about plastic substrates will be given later in this chapter). However, low mobility (~1 $cm^2 V^{-1} s^{-1}$), amorphous silicon (a-Si) device technology [6, 7], used in liquid crystal displays (LCDs) [8] for laptop computers and large screen flat-panel TVs, requires processing up to 350 °C. Moreover, higher performance electronics, based on high mobility (~100 $cm^2 V^{-1} s^{-1}$) polycrystalline Si technology [6, 9], which drives newer organic light emitting displays (OLEDs) [10], require still higher process temperatures, ~600 °C. (A high mobility translates into a higher output current, and lower voltage and higher frequency operation). However, post-annealing processes used in

Transparent Electronics: From Synthesis to Applications Edited by Antonio Facchetti and Tobin J. Marks

polycrystalline Si technology [9] can be the source of nonuniform device behavior, which impedes cost-effective application to large area electronics and displays.

The challenge then for flexible electronics, especially for high performance applications, is either to develop a new high temperature plastic, or find new semiconductor materials and processes that work at lower temperature. This, in part, has fueled a broad interest in *organic* semiconductors [11, 12] as an inexpensive, low-temperature class of alternative materials, especially for thin film transistors (TFTs), although the mobility of most organic semiconductors is no better ($<1\,cm^2\,V^{-1}\,s^{-1}$) than in a-Si. Further, many organic materials degrade [13, 14] in normal atmospheric conditions, requiring protection strategies. Nonetheless, there are applications, such as electrophoretic displays, for which organic TFTs function adequately as the electronic backplane [15].

While progress has been made toward developing a new high temperature plastic at DuPont, even if such a substrate technology existed today, it would only be compatible with a-Si and organic semiconductor technologies. For this reason, oxide semiconductors, such as ZnO [16–18], are gaining increasing interest, because devices can be fabricated at low temperature with moderate mobility [19–21] ($\sim 20\,cm^2\,V^{-1}\,s^{-1}$). Oxide semiconductors thus have the potential to enable large area, flexible electronics with high performance, and this has motivated research in our laboratory on oxides. A parallel and arguably more popular attraction of oxide semiconductors is their optical transparency.

Transparent electronics [22], with as yet unspecified applications, captures the imagination. However, transparency could, in fact, prove to be a practical advantage in active matrix schemes to drive displays, because Si-based transistors are light sensitive [7], requiring shielding from ambient light, and this reduces the pixel 'aperture', in proportion to the area the TFT occupies. For this reason an all transparent oxide TFT could increase the effective aperture and be insensitive to visible light. This is especially relevant to OLEDs, in which multiple transistors supply and control the current to each pixel in order to achieve stable operation [23, 24].

Transparent oxide semiconductors (TOSs), comprising ZnO, In_2O_3, SnO_2, CdO and Ga_2O_3, are already ubiquitous in electronics as transparent conducting electrodes [25]. While research on oxide semiconductor devices, such as ZnO for thin film transistors (TFTs), lags the development of organic devices, transparent conducting oxide electrodes (e.g. indium-tin oxide) in electronics, and in particular on plastic, have been used reliably for decades in commercial products [25].

Electrical conduction in undoped films of TOSs is attributable to intrinsic defects, such as interstitial metal ions or oxygen vacancies, which donate free electrons [26]. These oxide semiconductors are normally *n*-type, or electron conductors, and oxide semiconducting films can be made with electron carrier concentrations more than $10^{20}\,cm^{-3}$ and a mobility greater than $50\,cm^2\,V^{-1}\,s^{-1}$, even though their structure is nanocrystalline or amorphous [25, 27, 28]! High electron concentrations occur in TOSs, because self-compensating defects do not form [29]. The mobility of polycrystalline thin film TOSs depends on their electron carrier concentration [30]. At high concentration, scattering by ionized impurities (doped) or native defects (undoped) reduces mobility. At lower carrier concentration, grain boundary scattering dominates the mobility, which increases with the number of carriers because of Coulombic screening and trap filling. Consequently the dependence of mobility on carrier concentration frequently exhibits a maximum, on account of the competing effects of defect and grain boundary scattering. (Of course an oxide semiconductor channel of a TFT should have low carrier concentration for a low device off-current and high on/off ratio).

The high mobility of TOSs, even when their structure is amorphous or nanocrystalline, is associated with s-band conduction [31, 32]. Since the conduction band is formed by overlap of *spherically* symmetric *s orbitals* of the heavy metals (In, Sn, Zn, Cd and Ga), structural disorder changes that overlap much less than if the bonding was directional. Consequently charge transport is less sensitive to bond angle distortion, as might occur in a disordered structure, especially for low temperature growth that is needed for temperature-sensitive plastic substrates.

The success of a future oxide semiconductor device technology depends not only on a high mobility, but as importantly on reliable, stable operation. In a device, this is characterized by negligible operational hysteresis, small shift in threshold or turn-on voltage, particularly with electrical stress, and low off-state current for low power consumption. A critical element in a stable oxide transistor is identifying a gate dielectric material and process, which form a low defect interface with the oxide semiconductor [33, 34].

In contrast to Si-based device technologies [6], where processing and materials are well established, oxide semiconductor device technology is much less mature. Choices for the oxide semiconductor and gate dielectric are the subject of contemporary research. A number of groups are investigating ternary [35, 36] and quaternary oxides [37], because they form stable amorphous films without grain boundaries, where charge trapping can occur. Of course amorphous semiconductors are not free of bulk traps, which can originate from imperfect (dangling) bonds. In a-Si, adding hydrogen during synthesis passivates the defects and reduces the number of traps [38]. Related strategies may also be applicable to binary oxides, such as ZnO. It is also reasonable to expect that a simple binary oxide would have advantages for uniformly coating large area electronics. In this chapter we will emphasize device development with binary oxides, in particular ZnO.

The organization of the material in this chapter includes a discussion of thin film binary oxide properties, devices on rigid substrates, studies of devices on different gate dielectrics, a summary of devices made on plastic substrates, and finally discussion of strategy for patterning devices on plastic substrates.

11.2 Zinc Oxide

11.2.1 ZnO Thin Film Properties

ZnO is a wide band gap (3.37 eV), stable, *n*-type semiconductor possessing hexagonal wurtzite structure and a high melting temperature (1975 °C) [39]. Sputtering is a common technique used to fabricate ZnO thin films for electrode applications [30]. Sputtering is also attractive because it can be easily scaled up for manufacturing. Specifically, in our laboratory, we synthesize ZnO thin films by radio frequency (RF) magnetron sputtering in a stainless steel chamber which is a cryo-pumped to $\sim 2 \times 10^{-7}$ Torr prior to deposition. The target is 6.5 in. diameter, dense ZnO (SCM, Tallman, NY, USA), and the substrate-to-target distance is 76 mm. Substrates are placed on a water-cooled table directly beneath the sputtering target. This avoids any unintentional heating of substrates during sputtering of

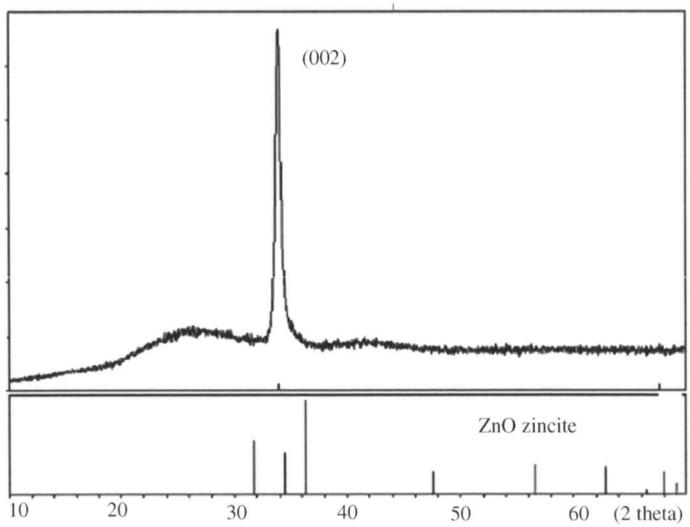

Figure 11.1 X-ray diffraction pattern of ZnO thin film with (002) texture and reference line pattern for ZnO zincite. Reprinted with permission from [19]. Copyright (2005) SID

ZnO. Sputtering is in Ar with a metered partial pressure of O_2 at low power (100 W) and low cathode voltage (–100 V). These less energetic sputtering conditions mitigate potentially damaging bombardment of the growing ZnO film by energetic oxygen ions [40], which can cause high stress. For the same reasons we also sputter at a relatively high total gas pressure (~20 mTorr), for which more energy dissipating collisions occur. At 100 W (~0.5 W cm^{-2}) and 20 mTorr the ZnO deposition rate is ~0.2 nm s^{-1}.

By X-ray diffraction sputtered films are polycrystalline with (002) c-axis texture, as shown in Figure 11.1. (Notably, Hsieh and Wu [41] report that ZnO thinner than 40 nm becomes amorphous.) Usually there is only a single prominent diffraction peak in polycrystalline ZnO with inter-planar spacing expanded relative to the reference spacing for zincite, $d(002) = 0.2603$ nm. We interpreted this lattice expansion along the c-axis to be related to a biaxial compressive stress [42] in the plane of the film, that is

$$\sigma = -\frac{E}{2\nu} \cdot \frac{\delta d}{d} \tag{11.1}$$

where the film stress (σ) is related to the fractional change ($\delta d/d$) in inter-planar lattice spacing in the c-axis direction and the ZnO elastic properties, i.e. E is Young's modulus and ν is Poisson's ratio. The negative sign in Equation (11.1) corresponds to compressive stress for lattice expansion. The stress in our ZnO films, which we determined from the measured difference in substrate curvature [43] before and after film deposition, is compressive. For the conditions used to make ZnO TFT devices: $pO_2 \sim 10^{-5}$ Torr and a total Ar + O_2 pressure of 20 mTorr, the compressive stress was <0.5 GPa. We did find that devices made at higher total sputtering pressure (20 mTorr) have lower stress and better transistor properties.

From surface imaging with atomic force microscopy (AFM), films sputtered at 20 mTorr and $pO_2 = 1 \times 10^{-5}$ Torr possess granular morphology with average granule size in the range

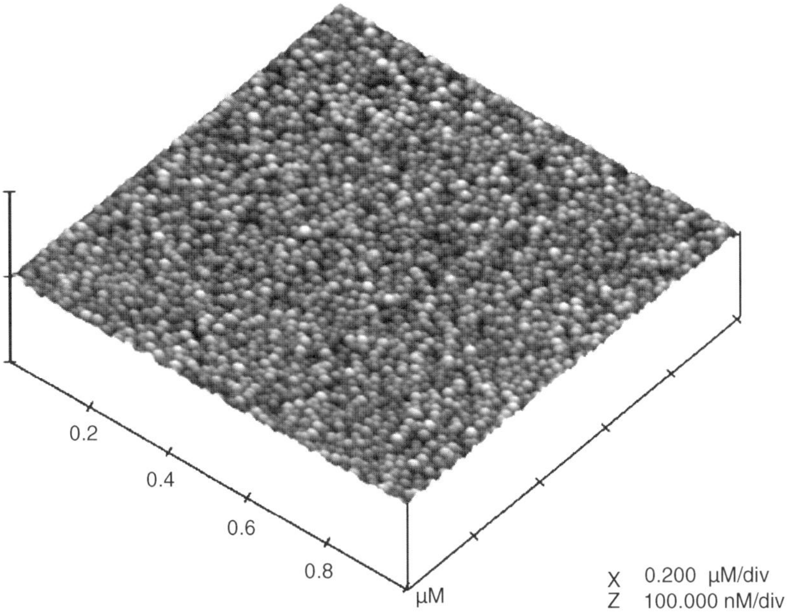

Figure 11.2 Atomic force microscope image of ZnO surface topography for sputtering conditions of 20 mTorr (Ar + O_2) and $pO_2 = 1 \times 10^{-5}$ Torr. The ZnO film thickness was 50 nm. The image height or z-range is 30 nm and the scan size is 1 µm × 1 µm. Reprinted with permission from [19]. Copyright (2005) SID

32–38 nm and an average surface roughness of ~1.9 nm, as shown in Figure 11.2. Changing the oxygen partial pressure during sputtering in the range 10^{-6}–10^{-3} Torr, has little influence on the morphological structure of ZnO films. Also, independent of pO_2, ZnO films are transparent in the visible with a measured optical transmission typically above 80% for an optical wavelength >400 nm, as shown in Figure 11.3.

In contrast, the electrical resistivity of ZnO films does depend strongly on pO_2 during sputtering. Undoped ZnO is an *n*-type conductor, because native defects such as interstitial

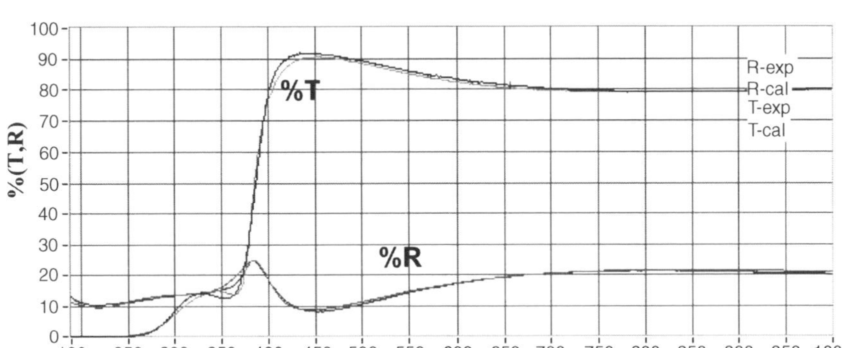

Figure 11.3 Optical transmission and reflectance for ~100 nm thick ZnO thin film RF magnetron sputtered at 20 mTorr (Ar) with $pO_2 = 1 \times 10^{-5}$ Torr. Substrate is Corning 7059 glass. Reprinted with permission from [19]. Copyright (2005) SID

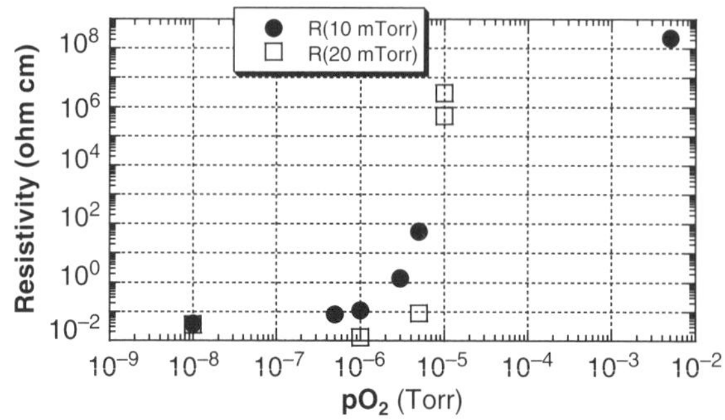

Figure 11.4 Dependence of electrical resistivity of magnetron sputtered ZnO thin films on partial pressure of oxygen (pO_2) used in sputtering at total $Ar + O_2$ pressure of 10 mTorr (●) and 20 mTorr (□). Reprinted with permission from [18]. Copyright (2003) American Institute of Physics

Zn ions or oxygen vacancies contribute free electrons for electrical conduction [26]. Figure 11.4 shows that the resistivity undergoes an abrupt transition from semiconducting ($\rho \sim 0.03$ ohm cm) at low oxygen partial pressure to semi-insulating ($\rho = 10^6$–10^8 ohm cm) at higher pO_2. In the transition region the dependence of resistivity is exponential-like, and can be fitted approximately by the expression $\rho = \rho_o \exp (1.5 \times pO_2)$, where ρ_o is about 0.03 ohm cm and pO_2 is in units of μTorr (1×10^{-6} Torr). Films with a resistivity less than 1 ohm cm have Hall mobility in the range 12–25 $cm^2 V^{-1} s^{-1}$ (*n*-type). For comparison, the mobility of single crystalline ZnO is ~200 $cm^2 V^{-1} s^{-1}$ [39]. For growth above a critical pO_2 (~10^{-5} Torr), we believe that films become close to stoichiometric, with fewer structural defects and consequently much higher resistivity, which is preferred for ZnO TFTs with low off-state current.

11.2.2 ZnO Thin Film Transistors

TFTs of ZnO have been evaluated using a simple test structure with inverted coplanar geometry on a Si substrate, as shown in Figure 11.5. TFT fabrication begins with a 100 nm thick SiO_2 layer thermally grown on a heavily doped *n*-type Si substrate, which serves as the

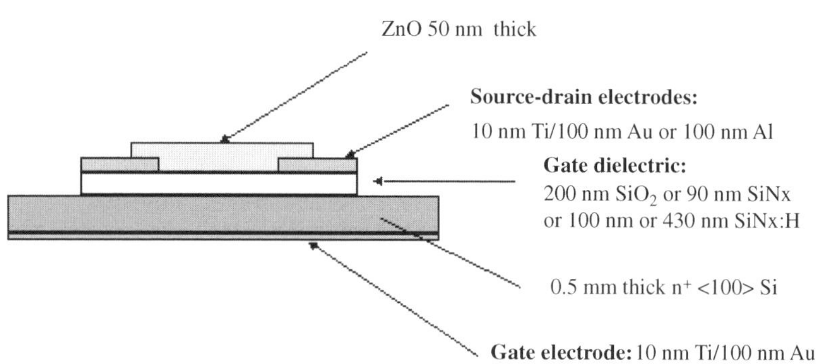

Figure 11.5 Cross-section of the ZnO thin film transistor test structure with the inverted coplanar geometry on conducting Si substrate. Reprinted with permission from [34]. Copyright (2007) American Institute of Physics

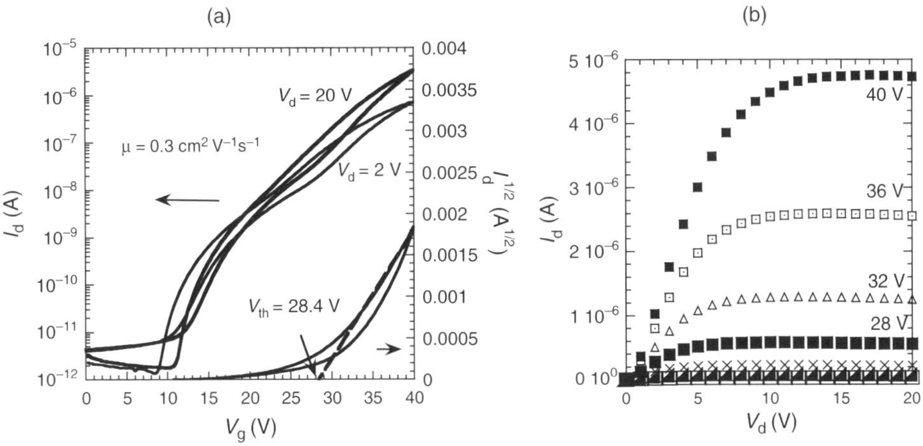

Figure 11.6 *Transfer (a) and output (b) curves for ZnO thin film transistors on 200 nm thick, thermally grown SiO_2 dielectric. Reprinted with permission from [19]. Copyright (2005) SID*

common gate electrode. Ti-Au source and drain electrodes (10 nm Ti followed by 100 nm Au), 200 μm wide with a 20 μm gap are then deposited and patterned directly on the thermal silicon oxide layer by traditional photolithography. Ti-Au is also deposited on the back of the Si as a common gate electrode. ZnO about 50 nm thick is then sputtered between source and drain electrodes using a shadow mask.

Figure 11.6 illustrates transfer and output characteristics for a typical ZnO transistor sputtered on thermally grown SiO_2 at 20 mTorr total pressure and $pO_2 = 2 \times 10^{-5}$ Torr. The output curves show well-behaved linearity at low source drain voltage, followed by distinct saturation at higher voltage, and systematic current modulation with increasing gate voltage. From the transfer curves this device operates in the enhancement mode, corresponding to absence of a conducting ZnO channel at zero gate voltage. All ZnO transistors we have fabricated operate in the accumulation mode, i.e. application of a positive gate voltage increases the concentration of channel electrons, consistent with n-type conduction in undoped ZnO. For device analysis we used the standard thin transistors equations [44]. Below saturation, the drain current is given by Equation (11.2), and in saturation by the square law dependence of Equation (11.3):

$$I_d = (W/L)\mu C_g[(V_g - V_{th})V_d - V_d^2/2], \quad V_d < (V_g - V_{th}) \tag{11.2}$$

$$I_d = (W/2L)\mu C_g(V_g - V_{th})^2, \quad V_d > (V_g - V_{th})(\text{saturation}) \tag{11.3}$$

where I_d and V_d are, respectively, the current and voltage between source and drain electrodes, V_g is the voltage between the gate and source reference electrode, V_{th} is the saturation threshold voltage, and μ is the device field effect mobility. C_g ($F\,cm^{-2}$) is the capacitance of the gate dielectric, W is the device width and L is the channel length. In the saturation regime, device mobility and the threshold voltage can be determined from a plot of $\sqrt{I_d}$ against gate voltage V_g, as displayed on the right-hand axis of Figure 11.6(a). From analysis of the transfer curves using Equation (11.3), this device has a mobility

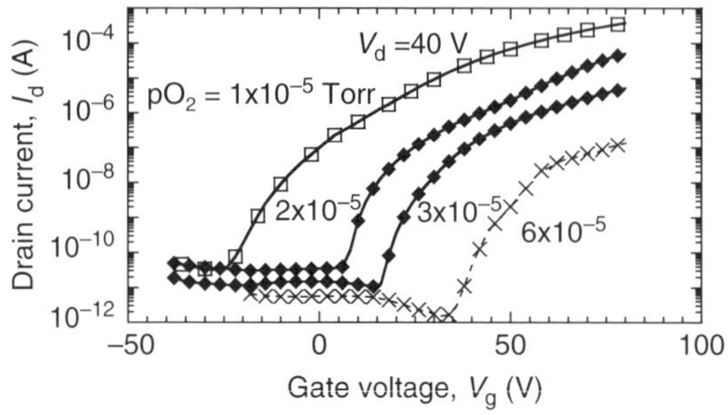

Figure 11.7 Transfer curves, as a function of the oxygen partial pressure pO_2 during ZnO sputtering, for ZnO transistors on 200 nm thick SiO_2 thermally grown on Si. Reprinted with permission from [34]. Copyright (2007) American Institute of Physics

$\mu = 0.3\ cm^2\ V^{-1}\ s^{-1}$, $V_{th} = 28.4$ V, an on/off ratio $>10^6$ with an inverse subthreshold slope, $S = (d\log_{10} I_d/dV_g)^{-1} \sim 4$ V decade^{-1}. The high V_{th} and S and hysteresis in the transfer curve are characteristic of a nonideal device. In this case trapping of charge carriers in the ZnO channel is suspected to cause this nonideal behavior.

To further probe this trapping phenomenon, we synthesized ZnO TFTs over a range of $pO_2 = 1 \times 10^{-5}$, 2×10^{-5}, 3×10^{-5} and 6×10^{-5} Torr. The transfer curves for TFTs made at these different oxygen partial pressures are shown in Figure 11.7, and Table 11.1 is a summary of the corresponding μ, V_{th} and on/off ratio. While the device fabricated at the lowest pO_2 operates in the depletion mode (a conducting ZnO channel exists without application of a positive gate voltage), devices made at higher pO_2 operate in the enhancement mode. Depletion mode behavior at low pO_2 is consistent with a more electrically conducting ZnO, as shown in Figure 11.4. Application of a negative gate voltage is needed to deplete the ZnO channel of conduction electrons. As pO_2 increases, the turn-on voltage, corresponding to where the drain current perceptibly increases, shifts to more positive values, V_{th} is above 40 V, and mobility falls dramatically by more than 1000 times.

Since ZnO thin films are polycrystalline, we expect their properties to be described by a model that accounts for grain boundary trapping of mobile carriers. In Levinson's model [45] of polycrystalline TFTs, the mobility of charge carriers is given by

$$\mu^{-1} = \mu_o^{-1} + [\mu_{gb}\exp(-E_b/kT)]^{-1} \tag{11.4}$$

Table 11.1 *Dependence of field effect mobility (μ), threshold voltage (V_{th}), and on/off ratio in ZnO for TFTs on pO_2, used in sputtering ZnO channel*

pO_2 (Torr)	μ ($cm^2\ V^{-1}\ s^{-1}$)	V_{th} (V)	On/off
1×10^{-5}	1.6	28.5	10^7
2×10^{-5}	0.6	48.8	1.7×10^6
3×10^{-5}	0.04	40.7	5×10^5
6×10^{-5}	0.001	43.0	$\sim 1 \times 10^5$

where μ_o is the mobility in a crystalline ZnO grain, ~200 $cm^2 V^{-1} s^{-1}$, and $\mu_{gb}\exp(-E_b/kT)$ is the contribution accounting for grain boundary scattering. In this model the barrier height or energy, $E_b \approx q^2 N_t^2/8\varepsilon n$, where q is the electronic charge, N_t is equal to the density of traps in the grain boundary, n is the density of mobile carriers in a grain and ε is the dielectric constant for ZnO. Applying Levinson plots of $\ln(I_d/V_g)$ versus $(1/V_g)$ at a drain voltage of 1 V, we calculate $N_t \sim 2 \times 10^{12}$ $cm^{-2} eV^{-1}$ for ZnO TFTs with $pO_2 = 3 \times 10^{-5}$ Torr. Assuming that $\mu_b \sim 10$ $cm^2 V^{-1} s^{-1}$, which is in the middle of the range of Hall mobility measured in ZnO thin films, $\varepsilon = 10$ for ZnO, Equation (11.4) with $n \approx 7 \times 10^{17}$ cm^{-3} approximately fits the measured mobility of 0.04 $cm^2 V^{-1} s^{-1}$. This corresponds to E_b ~130 meV. For comparison, in their grain-boundary model of electrical conductivity for sputtered ZnO, Weissenrieder and Muller [46] report a trap density of ~2×10^{12} cm^{-2} and a barrier energy of ~140 meV.

The trap density N_t can also be used to predict V_{th}. Using the expression applicable to polycrystalline ZnO channel, where its thickness is assumed equal to the average grain size [44],

$$V_{th} \approx qN_t E_G / 2C_g \tag{11.5}$$

Here N_t is a uniform trap density ($cm^{-2} eV^{-1}$), E_G (eV) is the ZnO band gap, C_g ($F\,cm^{-2}$) is the gate capacitance and q the electronic charge. For $N_t \sim 2 \times 10^{12}$ $cm^{-2} eV^{-1}$, we calculate $V_{th} = 31.5$ V, a little lower than the measured 40.7 V. Clearly, though, traps in ZnO can explain high V_{th} for ZnO TFTs on SiO_2.

Previously we found that N_t increases at higher oxygen process pressure, and this may also explain the strong dependence of field effect mobility with pO_2 in ZnO TFTs. A reduction in (N_t^2/n) by 50% (relative to the value calculated for $pO_2 = 3 \times 10^{-5}$ Torr) gives $\mu \sim 0.8$ $cm^2 V^{-1} s^{-1}$, whereas a corresponding 50% increase reduces μ to ~0.004 $cm^2 V^{-1} s^{-1}$. This is roughly the range (1.6–0.001 $cm^2 V^{-1} s^{-1}$) of mobility change we observe with pO_2. Therefore we attribute the large reduction in mobility with pO_2 to an increase in trap density. From photoluminescence of ZnO sputtered on thermally grown SiO_2 on Si, we conclude that likely trapping sites are electron acceptor defects, such as oxygen adsorbed in grain boundaries or interstitially in ZnO [34]. A higher oxygen process pressure could increase the concentration of acceptor defects, which then degrade the field effect mobility of ZnO TFTs with a SiO_2 gate dielectric.

11.3 Indium Oxide

11.3.1 In_2O_3 Thin Film Properties

Indium oxide (IO) is a transparent conducting oxide, which crystallizes in the bixbyite or C-type rare-earth structure [47]. The electrical conduction is n-type, which arises from ionized oxygen vacancies that donate free electrons, i.e. $O_O \rightarrow V_O + 2e^- + 1/2\,O_2$ (g), where V_O is a vacancy at an oxygen site [48]. When doped with tin oxide (~9 at%), indium-tin oxide (ITO) is a widely used transparent conducting electrode in electronics [25]. Applications include transparent electrodes for displays and solar cells, and demand for ITO coatings is currently so great it is threatening a worldwide shortage of indium metal.

Sputtering from an oxide target is the common method to deposit ITO thin films, and we also sputter IO thin films from an In_2O_3 target. For films deposited on unheated substrates,

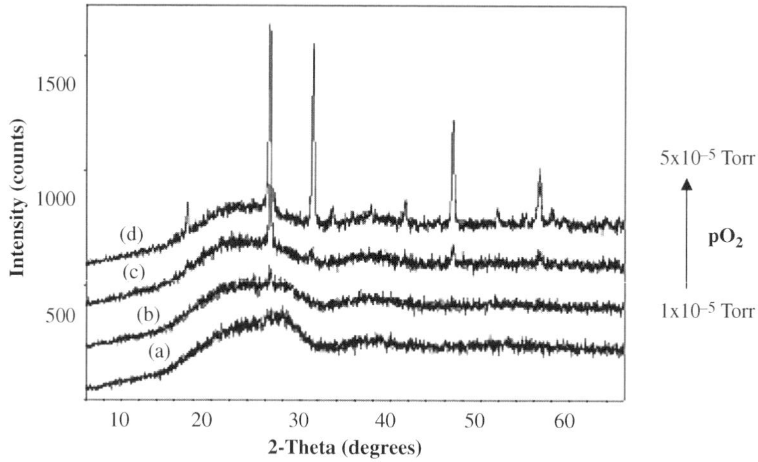

Figure 11.8 *X-ray diffraction (Cu Kα) patterns for indium oxide films grown at 100 W in 12 mTorr Ar and (a) $pO_2 = 5 \times 10^{-6}$, (b) $pO_2 = 1 \times 10^{-4}$ Torr, (c) $pO_2 = 5 \times 10^{-4}$ and (d) $pO_2 = 1 \times 10^{-3}$ Torr. Reprinted with permission from [28]. Copyright (2003) Society of Vacuum Coaters*

such as polymers, we find little to no differentiation in structural and electrical properties for IO and ITO films. The film structural properties, as determined by X-ray diffraction and atomic force microscopy (AFM), are summarized in Figures 11.8 and 11.9. For IO films the surface morphology consists of very fine granular features (~10 nm) with small surface

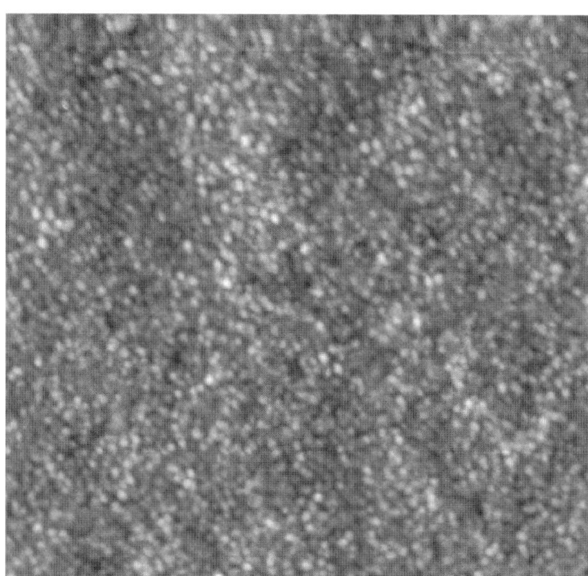

Figure 11.9 *AFM topography image of IO deposited on PET substrate at 12 mTorr Ar and $pO_2 = 5 \times 10^{-6}$ Torr. Image corresponds to 1 μm × 1 μm area. Reprinted with permission from [28]. Copyright (2003) Society of Vacuum Coaters*

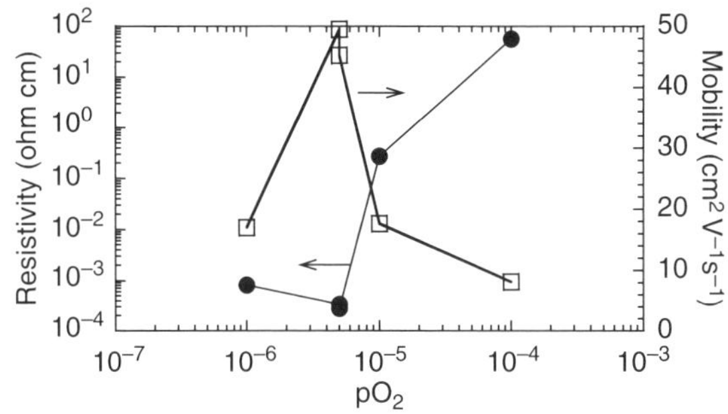

Figure 11.10 *Dependence of electrical resistivity and Hall mobility on partial pressure of oxygen for indium oxide thin films grown at 100 W in 12 mTorr Ar. Reprinted with permission from [28]. Copyright (2003) Society of Vacuum Coaters*

roughness (<1 nm) for low resistivity films. By X-ray diffraction we find that crystallinity of IO films increases with pO_2, and at lower pO_2 they have (222)-texture. At higher pO_2 films are polycrystalline with more random orientation, while films deposited at the lowest pO_2 are amorphous. We have shown that bombardment by energetic oxygen ions, as pO_2 increases, causes films to crystallize [27, 28]. Consistent with increasing crystallinity, the surface roughness also increases.

The electrical resistivity and electron mobility for IO films grown at 12 mTorr and 100 W as a function of pO_2 on PET substrates are shown in Figure 11.10. The minimum in IO resistivity is $\sim 3 \times 10^{-4}$ ohm cm with a peak Hall mobility, $\mu_H \sim 50\,cm^2\,V^{-1}$. Remarkably, even amorphous films of IO have a respectably high Hall mobility. From the electrical data we calculated the dependence of electron concentration n on pO_2. For IO thin films, the electron concentration falls rapidly at higher pO_2, consistent with a reduction in the electron concentration from oxygen vacancies. Films sputtered at higher pO_2 are closer to stoichiometric and therefore are less conducting because of fewer oxygen vacancies. In a TFT, low off-state device current requires low conductivity films. Unfortunately it is difficult to consistently fabricate and maintain In_2O_3 films with insulating properties, which may be related to reactivity of IO with moisture to create oxygen vacancies or simply ambient annealing [49]. In the case of IO the magnitude of n at higher pO_2 was at least 100 times smaller than in ITO [27, 28]. In ITO, Sn^{4+} ions donate electrons, and this could account for a higher electron concentration at high pO_2. Near the resistivity minimum for IO and ITO, their electrical properties are essentially the same, which suggests that native defects dominate electron donation in both IO and ITO grown at room temperature and low pO_2.

11.3.2 In_2O_3 Thin Film Transistors

We investigated IO TFTs that were less than ~50 nm thick and sputtered at 4 mTorr total pressure with pO_2 between 0.1 mTorr and 1.0 mTorr. We find that IO films have better TFT performance when their initial sheet resistance is >500 Mohm square^{-1}, corresponding to conductivity less than $\sim 10^{-4}\,ohm^{-1}\,cm^{-1}$. Compared with IO films sputtered at higher total pressure (>12 mTorr), IO films sputtered at 4 mTorr are better crystallized, which we concluded from sharper and more intense X-ray diffraction peaks. In a lower pressure sputtering environment, less frequent collisions of energetic ions with sputtering gas atoms occur, and consequently the growing film experiences more bombardment, causing crystallization.

The TFT device test structure we used was similar to that shown in Figure 11.5. The gate oxide was either 200 nm thick SiO_2 thermally grown on Si, or a 500 nm thick layer of Al_2O_3 deposited by electron beam evaporation on Si. On SiO_2 source-drain electrodes were photolithographically patterned Ti/Au electrodes, where the channel length was 5, 10, or 20 μm with device width-to-length ratio of 10. On Al_2O_3 we thermally evaporated rectangular shaped Al source-drain electrodes with nominal channel length of ~100 μm and width ~1 mm through a shadow mask. The In_2O_3 semiconductor was then sputtered, also through a shadow mask, between source and drain electrodes.

Figures 11.11 and 11.12 summarize the output and transfer characteristics of IO TFTs sputtered at 4 mTorr on thermal oxide of SiO_2 and on evaporated Al_2O_3 [50]. On SiO_2 dielectric, we measure an initial saturation mobility, $\mu_{sat} = 0.028\ cm^2\ V^{-1}\ s^{-1}$, and an on/off ratio $\sim 10^4$ with a threshold voltage $V_{th} = 8.2$ V. When this same device was remeasured, after storing in the laboratory ambient for 16 days, the mobility increased to $\mu_{sat} = 0.36\ cm^2\ V^{-1}\ s^{-1}$, V_{th} decreased to 1.6 V, and the on/off ratio increased to $\sim 10^5$. Better properties are achieved for IO TFTs on SiO_2, when IO, ~50 nm thick, is sputtered at a higher total pressure of 20 mTorr ($pO_2 = 1 \times 10^{-4}$ Torr). For these conditions IO is

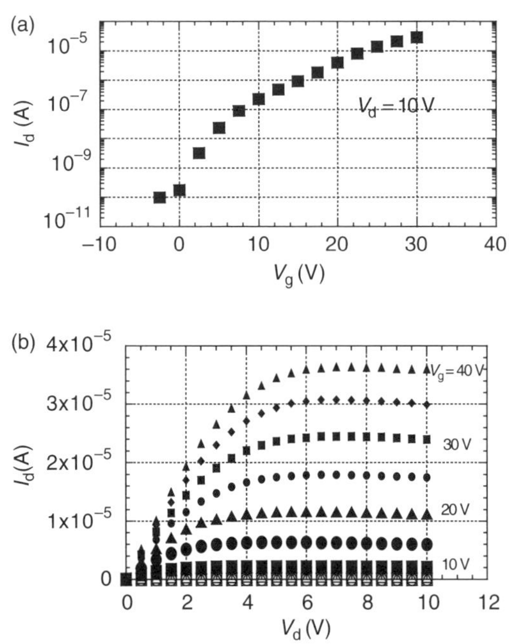

Figure 11.11 Transfer (a) and output (b) curves for indium oxide thin film transistors on 200 nm thick thermally grown SiO_2 dielectric on Si substrate. Indium oxide was sputtered at 4 mTorr in an Ar/O_2 gas mixture. Reprinted with permission from [50]. Copyright (2005) Electrochemical Society

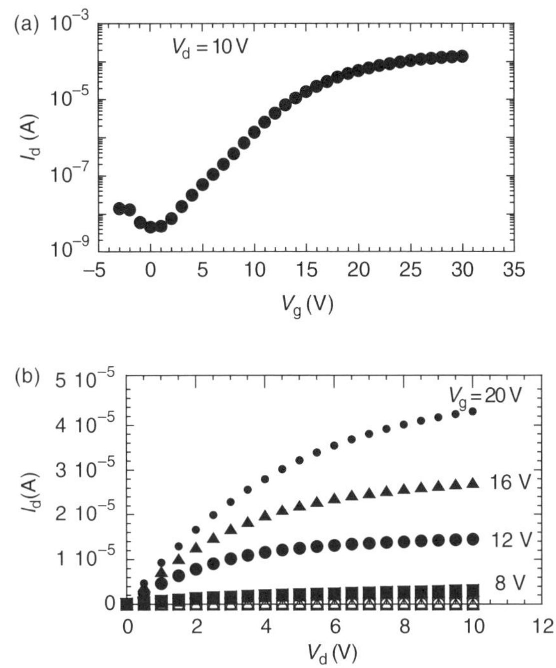

Figure 11.12 *Transfer (a) and output (b) curves for indium oxide thin film transistors on e-beam evaporated 500 nm thick Al_2O_3 dielectric on Si substrate. Indium oxide was sputtered at 4 mTorr in an Ar/O_2 gas mixture. Reprinted with permission from [50]. Copyright (2005) Electrochemical Society*

amorphous, as determined by X-ray diffraction. The transistor transfer and output characteristics for this device, which operates in the depletion mode, are shown in Figure 11.13. The mobility is $3\,cm^2\,V^{-1}\,s^{-1}$, $V_{th} = 7.8$ V, on/off ratio $>10^8$, and the inverse subthreshold slope $S = 1.2$ V decade^{-1}. Notably V_{th} is consistently lower for IO

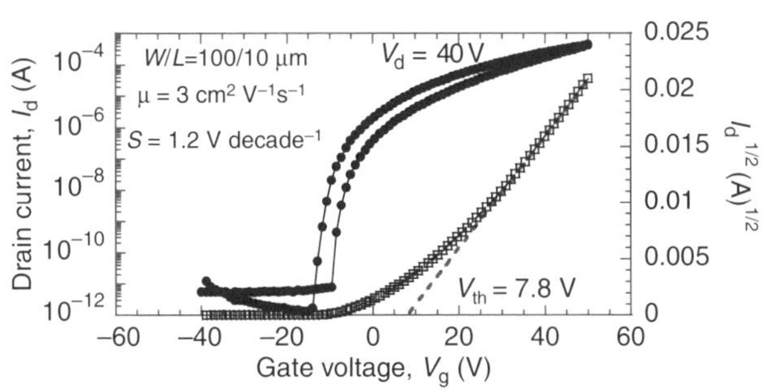

Figure 11.13 *Transfer characteristic for indium oxide thin film transistor on 200 nm thick thermally grown SiO_2 dielectric on Si substrate. Indium oxide was sputtered at 20 mTorr in an Ar/O_2 gas mixture*

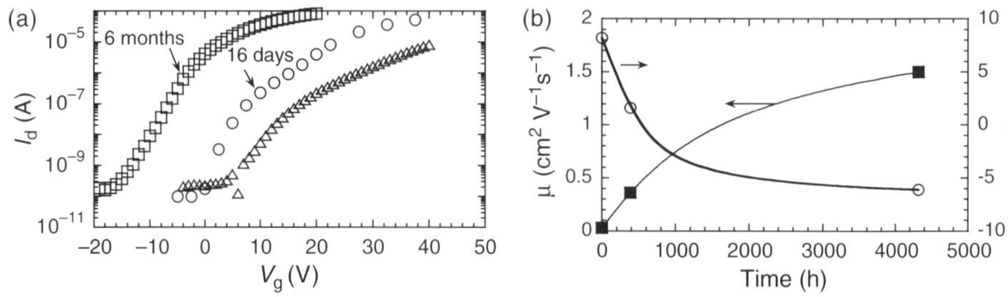

Figure 11.14 *Transfer characteristic (a) for indium oxide thin film transistor initially and after ambient ageing for 16 days and 6 months. Corresponding changes (b) in mobility (μ) and threshold voltage (V_{th}) with agwing time*

TFTs on a SiO_2 gate dielectric than ZnO TFTs on SiO_2. A smaller trap density in IO films may be the reason, and this could be associated with a noncrystalline IO film structure.

We also examined the effect of pO_2 on In_2O_3 TFTs performance for devices simultaneously sputtered on SiO_2 and Al_2O_3 dielectrics. As with ZnO TFTs on SiO_2, we observe an exponential decrease of field effect mobility with increasing pO_2. This is qualitatively consistent with the grain-boundary model of mobility, which we discussed for ZnO TFTs on SiO_2.

Finally we observe changes in IO transistor characteristics after storage in an open laboratory environment. After ~6 months in ambient, without light shielding, the mobility of IO devices fabricated on SiO_2 dielectric increased from 0.03 to 1.5 $cm^2 V^{-1} s^{-1}$, and the threshold voltage shifted from +8.2 to–6.1 V. These changes in transfer characteristic and device parameters are shown in Figure 11.14. With no gate voltage applied, the drain current measured at $V_d = 10$ V increased from ~10^{-10} to ~5×10^{-6} A, requiring –20 V to deplete electrons in the channel. (Even larger and more rapid changes in channel conductivity, not shown, occurred on the evaporated Al_2O_3 dielectric.) The increase in mobility and source-drain current, without an applied gate voltage, is consistent with an increase in IO channel conductance with time. This may be related to the assertion by Graham *et al.* [49] that the conductivity of insulating (amorphous) IO films is unstable, increasing with time due to a self annealing effect, or reaction with moisture, as we speculate. In comparison, the transfer curve for ZnO TFTs did not change after storing in ambient conditions for 6 months. Clearly this ageing phenomenon is problematic for TFTs with an IO semiconductor channel.

As a further comparison, Zhang *et al.* [51] observed air sensitivity in single crystalline, In_2O_3 nanowire TFTs. In that work, the channel conductivity increased when the device was measured in vacuum – oxygen reducing conditions – whereas the channel conductivity decreased in air. The authors attribute lower conductivity and lower device off-state in air to trapping of electrons by surface-adsorbed oxygen. This is different compared to what we and others observe for IO films, where the conductivity in air increases with time. Although Zhang *et al.* do not report about longer term effects for their device, it is conceivable that single crystalline IO may be more stable to changes in bulk oxygen stoichiometry.

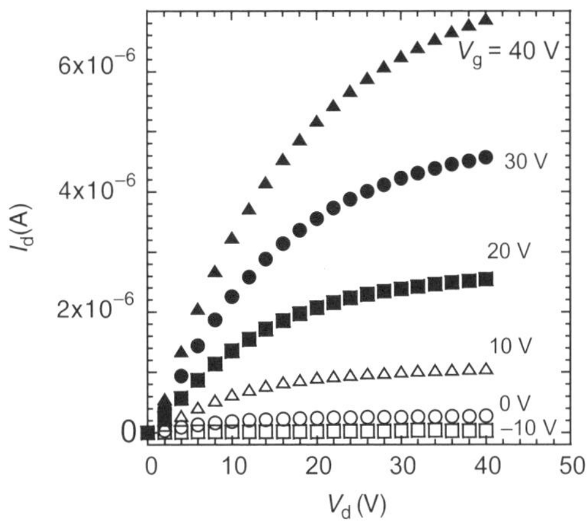

Figure 11.15 Output current for sputtered SnO_2 thin film transistor on 200 nm thick thermal SiO_2 gate dielectric on Si. Reprinted with permission from [50]. Copyright (2005) Electrochemical Society

11.4 SnO_2 Thin Film Transistors

In our laboratory we have also demonstrated the feasibility of TFTs with a sputtered SnO_2 semiconductor [50]. Figure 11.15 shows the transistor output characteristic for the sputtered SnO_2 device, fabricated on a Si wafer with 100 nm thick thermal oxide on one side and a Ti/Au common gate electrode on the back side. Al source-drain electrodes with channel length 100 μm × 1 mm width were deposited through a metal shadow mask, followed by sputtering of the SnO_2 semiconductor in 10 mTorr Ar and O_2 (5%), 87.5 nm thick, also through a shadow mask. These SnO_2 films are transparent in the visible with >80% transmission above a wavelength of 375 nm and a high index of refraction, n = 1.86 (at 633 nm). However, because the resistivity of this film is relatively low, ~50 Ω cm, the transistor off-state ($V_g = 0$) current is high, ~10^{-7} A. From a plot of $I_d^{1/2}$ versus V_g near saturation, we deduce a mobility $\mu = 0.015\,cm^2\,V^{-1}\,s^{-1}$, an on/off ratio ~300, and a threshold voltage of –11.4 V. Although these properties are not as good as for In_2O_3 or ZnO, there is wide latitude for improving the SnO_2 semiconductor properties.

11.5 Gate Dielectrics

11.5.1 Overview

While many research efforts on oxide TFTs focus on semiconductor chemistry, exploring ternary and quaternary oxide mixtures to optimize field effect mobility and stability, the development of a high performance gate dielectric is as critical and may be more so. This is especially challenging for plastic substrates, because a higher processing temperature is generally needed to produce gate dielectrics with low leakage current and high breakdown voltage [52]. Higher dielectric deposition temperature promotes a denser film growth with incorporation of fewer defects. Therefore a challenge, complementary to oxide semiconductor research, especially on plastic, is to identify a gate dielectric material and process that enable low voltage operation, reliability and stability *with* low device off-state current. Active electronic identification tags, for example, need sub-5 V operation [4], and OLEDs require high drive currents at typically less than 10 V [10]. For active-matrix driven LCDs, the static off-state current should be 0.1–1.0 pA [8].

The most important gate dielectric in modern day electronics is SiO_2 [53], thermally grown at high temperature (~1000 °C) on Si. The success and growth of modern electronics is due largely to the nearly perfect, pinhole-free dielectric properties of thermally grown SiO_2. With its exceptionally low leakage current, high breakdown voltage, and low defect density, SiO_2 provides stable, reliable performance for crystalline Si MOS transistors. While SiO_2 on Si affords a convenient vehicle for evaluating new oxide and organic thin film semiconductors, it is not a practical option for large area electronics or electronics on plastic.

A surprising finding for ZnO, sputtered at room temperature as the semiconductor channel in a TFT, is that the gate dielectric influences the type of deep level defect state formed in ZnO [34]. We measured photoluminescence spectroscopy of ZnO films [34], ~50 nm thick, sputtered at 20 mTorr Ar and $pO_2 = 2 \times 10^{-5}$ Torr *simultaneously* on Si with 90 nm thick sputtered SiN_x, 430 nm thick SiN_x:H grown by plasma-enhanced chemical vapor deposition (PECVD), thermally grown 200 nm thick SiO_2, and uncoated Si as a control. The ZnO spectra grown on these substrates are shown in Figure 11.16.

On bare Si there was a broad emission peak at 640 nm, which shifted to ~ 680 nm on sputtered and PECVD nitrides. However a unique green emission peak occurred at ~510 nm (2.43 eV) for ZnO on PECVD SiN_x:H. This emission peak was absent on sputtered SiN_x. Green emission [54] in ZnO has been attributed to oxygen vacancies (V_O) in ZnO. Ionized oxygen vacancies are donors that contribute electrons. On thermally grown SiO_2, the dominant feature was a broad yellow emission band from ZnO at 580 nm (2.13 eV). Yellow emission [55] from ZnO is associated with electron acceptor defects such O interstitials (O_i) or Zn vacancy (V_{Zn}) in ZnO. A high concentration of electron acceptor defects in ZnO, grown on SiO_2, is consistent with high V_{th} in TFTs. For these spectra we verified that significant luminescence only occurred for wavelengths less than 370 nm or 3.34 eV, corresponding to the optical absorption edge for ZnO, affirming that luminescence is from the ZnO thin film.

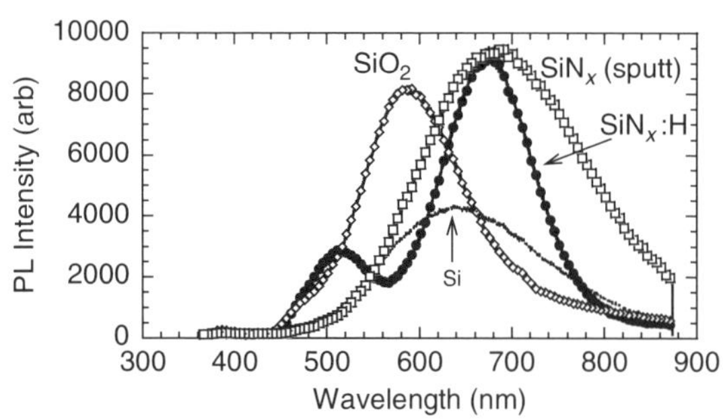

Figure 11.16 *Photoluminescence intensity versus wavelength for ZnO, grown simultaneously on sputtered SiN_x, PECVD SiN_x:H and thermally grown SiO_2 dielectric thin films on Si, including uncoated Si. Excitation source is 325 nm line of HeCd laser. Reprinted with permission from [34]. Copyright (2007) American Institute of Physics*

11.5.2 ZnO Thin Film Transistors on SiNx:H/Si Grown by Plasma-Enhanced Chemical Vapor Deposition

SiN_x:H [38], grown by PECVD, is an important dielectric in electronics. Its principal application is as the gate dielectric for *amorphous* Si TFTs that switch pixels in large area, liquid crystal, flat panel displays used in laptop computers and TVs. SiN_x:H with low leakage current and high breakdown voltage can be grown at 150 °C by RF PECVD [56]. It is noteworthy that this SiN_x:H contains a significant concentration of hydrogen, ~30 at%, (~2×10^{22} H atoms cm^{-3}). We evaluated ZnO TFTs on SiN_x:H grown by PECVD at 150 °C, which is compatible with inexpensive polyester (PET and PEN) substrates.

Figures 11.17 and 11.18 show output and transfer curves for ZnO TFTs fabricated at $pO_2 = 5 \times 10^{-5}$ Torr on 430 nm thick and 100 nm thick PECVD SiN_x:H, grown on heavily doped Si, which also serves as the common gate electrode. The ZnO was sputtered in 20 mTorr Ar/O_2 mixture between Al source-drain electrodes. On the 100 nm thick SiN_x:H, V_{on} is close to 0 V, and the device operates at lower voltage, consistent with the higher charge capacity for high relative dielectric constant ($\varepsilon_r = 7.5$), thinner SiN_x:H. For both devices the field effect mobility is ~$2\,cm^2\,V^{-1}\,s^{-1}$, the threshold voltage ~1 V, and the gate leakage current was respectably low, $<10^{-10}$ A, even for a gate voltage up to 40 V, and there is only small hysteresis for gate voltage sweeps in the positive and negative directions. On 100 nm SiN_x:H, the best TFTs had an on/off ratio ~10^6 and subthreshold slope S ~0.6 V $decade^{-1}$.

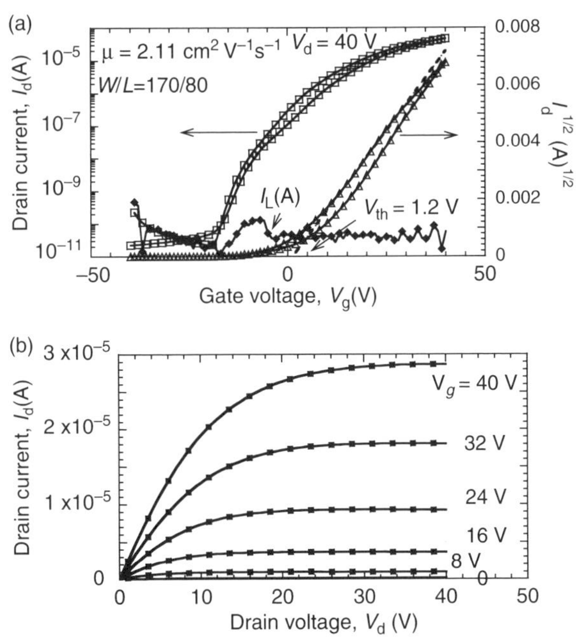

Figure 11.17 *Transfer (a) and output (b) characteristics for ZnO transistor on 430 nm thick SiN_x:H dielectric, grown by PECVD at 150 °C on Si. The clockwise hysteresis in transfer curve corresponds to drain current response to gate voltage sweep from −40 to 40 V and back to −40 V. I_L(A) is the gate leakage current. Reprinted with permission from [34]. Copyright (2007) American Institute of Physics*

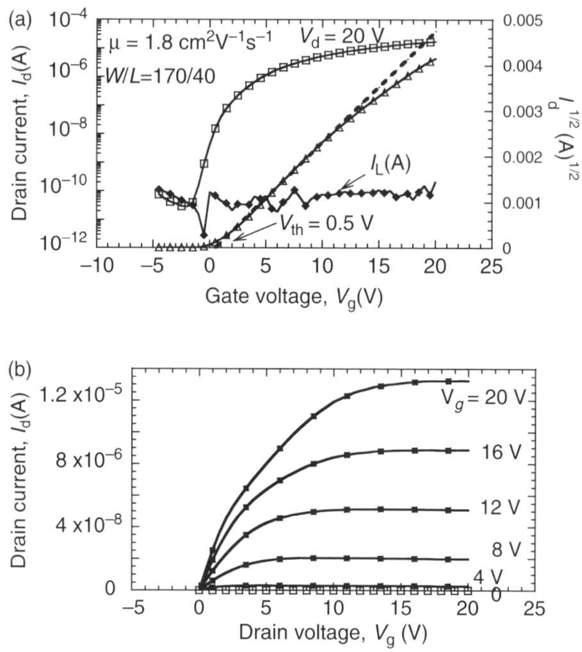

Figure 11.18 *Transfer (a) and output (b) characteristics for ZnO transistor on 100 nm thick SiN_x:H dielectric, grown by PECVD at 150 °C on Si. Gate voltage is a single sweep from −5 to 20 V. Reprinted with permission from [34]. Copyright (2007) American Institute of Physics*

ZnO TFTs on a SiN_x:H dielectric operate in the depletion mode, even when ZnO is sputtered in higher pO_2.

From photoluminescence spectroscopy of ZnO sputtered at room temperature on SiN_x:H, the green emission we observe and attribute to formation of oxygen vacancies in ZnO could account for depletion mode behavior of ZnO TFTs. The low threshold voltage is consistent with low interfacial and bulk trapping for ZnO on SiN_x:H. One speculation is that hydrogen diffuses from SiN_x:H into ZnO during sputtering and passivates acceptor trap sites. The weak dependence of field effect mobility with pO_2 [34] for ZnO TFTs on SiN_x:H supports this speculation for passivation of traps with hydrogen. For comparison, the field effect mobility in ZnO TFTs on thermally grown SiO_2 gate dielectric falls dramatically with increasing pO_2, while the measured trap density increases concurrently. From V_{th} (1–5 V) for devices on SiN_x:H, we estimate a trap density, $\sim$1–3 $\times$ 10^{11} cm^{-2} eV^{-1}, about 10 times less than for SiO_2.

11.5.3 Gate Dielectrics Grown by Atomic Layer Deposition

Atomic layer deposition (ALD) [57, 58] is a process based on self-limiting, sequential, surface reactions, which are thermally driven. The film growth mode is ideally a layer-by-layer process, which favors high density, pinhole-free films with a featureless microstructure, making ALD an ideal process for growing gate dielectrics. The superiority of ALD for growing ultra-thin gate dielectrics has already been demonstrated in the integrated circuits industry, where high-k ALD HfO_2 is a candidate to replace thermally grown SiO_2 in the

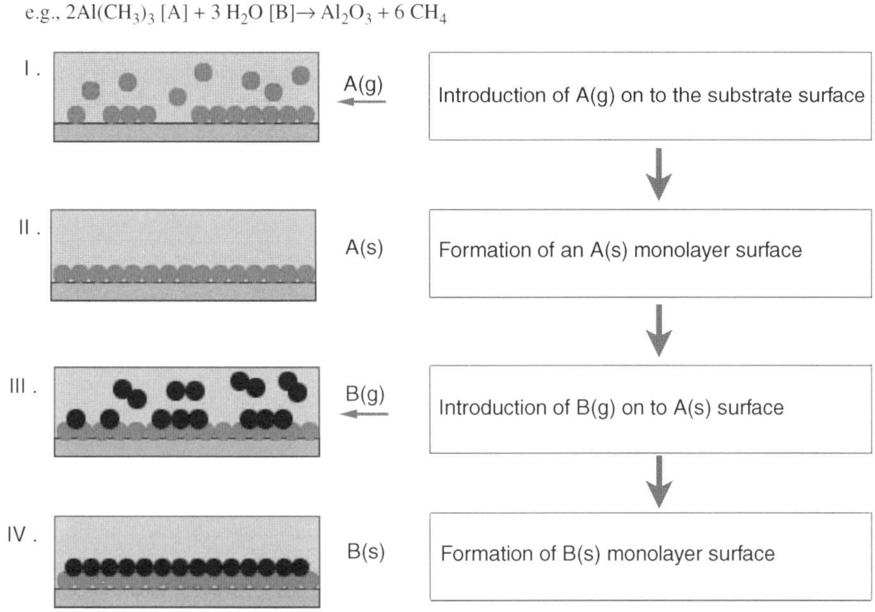

Figure 11.19 *Representation of the stepwise process of atomic layer deposition to produce a compound from precursors A and B, e.g., formation of* Al_2O_3 *is from trimethylaluminum (A) with water (B) as the oxidant*

highest performance MOS transistors. While ALD films are commonly grown at moderate substrate temperature (>300 °C), introduction of new precursor chemistry [59] and plasma-assist [60] have reduced the need for high substrate temperature, making ALD relevant to plastic substrates. Figure 11.19 illustrates the sequential process for growing Al_2O_3 by ALD. First the substrate is dosed with trimethyl aluminum, $Al(CH_3)_3$, until a self-limiting monolayer forms on the surface. An inert gas, such as Ar or N_2, then purges the reaction chamber, after which H_2O, the oxygen source, is admitted and also forms a monolayer. The two layers react at temperature to form Al_2O_3 and the gas reaction products are purged with inert gas. A film of Al_2O_3 grows by repeating this process or cycle to achieve a specified film thickness.

ALD growth of thin films on plastic is relatively unexplored and challenging. One obvious challenge is that plastics restrict the upper process temperature. Another is that polymer surfaces, with a variety of chemistries and morphology, may retard or inhibit nucleation of ALD films. Since ALD film growth of oxide dielectrics is established on a Si surface, we will first discuss the performance of ZnO TFTs on ALD dielectrics deposited on rigid Si substrates and then later on plastic.

Table 11.2 compares performance of ZnO transistors, sputtered at room temperature, on HfO_2, $HfSiO_x$ and Al_2O_3 gate dielectrics [33], grown by ALD on heavily doped n-type Si, which serves as the gate electrode. Comparison is made with similar devices on thermally grown SiO_2. The last column in this table compares the gate leakage *current density*, measured at the same gate electric field, 1 MV cm^{-1} in ZnO transistors. On ALD Al_2O_3 grown at higher temperature (200 °C and 450 °C), the gate leakage current *density*

Table 11.2 Summary of ZnO TFT properties on ALD gate dielectrics. T_S is the preparation temperature of the dielectric, μ is the field effect mobility of the ZnO TFT, V_{th} is the TFT threshold voltage, d is the dielectric thickness, and J_{GL} is the dielectric leakage current density at 1 MV cm^{-1}. Reprinted with permission from [33]. Copyright (2006) American Institute of Physics

Dielectric	T_S(°C)	μ(cm^2 V^{-1} s^{-1})	V_{th}(V)	d(nm)	J_{GL}(A cm^{-2})
HfO_2	300	12.2	2.5	25	2.5×10^{-4}
$HfSiO_x$	400	4.5	1.7	25	2.5×10^{-3}
SiO_2	~1000	0.3	28.4	200	3.0×10^{-7}
Al_2O_3	450	0.5	8.7	25	5.0×10^{-7}
Al_2O_3	200	17.6	6.0	100	3.0×10^{-7}
Al_2O_3	125	3.1	1.1	25	4×10^{-3}

($3–5 \times 10^{-7}$ A cm^{-2}) was comparable with thermally grown SiO_2. The leakage current was higher (4×10^{-3} A cm^{-2}) on thin ALD Al_2O_3 grown at 120 °C and also high (2.5×10^{-4} A cm^{-2}) on HfO_2. For all of these ALD dielectrics the ZnO TFT threshold voltage was much smaller than on thermal SiO_2, so ALD dielectrics enable lower voltage operation through a combination of higher dielectric constant, a thin dielectric layer and low trap density.

The transfer curves for ZnO TFTs on Al_2O_3 and HfO_2 are shown in Figures 11.20 and 11.21. ZnO TFTs on ALD HfO_2 grown at 300 °C have a mobility of 12 cm^2 V^{-1} s^{-1}, and on Al_2O_3 grown at 200 °C the mobility is 17.6 cm^2 V^{-1} s^{-1}. Hysteresis for ZnO devices on both dielectrics is negligibly small. For HfO_2 the threshold voltage was only 2.6 V, the subthreshold slope was 0.5 V decade^{-1}, and the on/off ratio was ~10^7. The high output current (~mA) at less than 10 V for this device is particularly attractive for devices needing high drive current at low voltage, such as OLEDs.

The very small hysteresis for devices on ALD HfO_2 and Al_2O_3 is consistent with a low trap density. We can estimate the magnitude of unoccupied surface trap density for

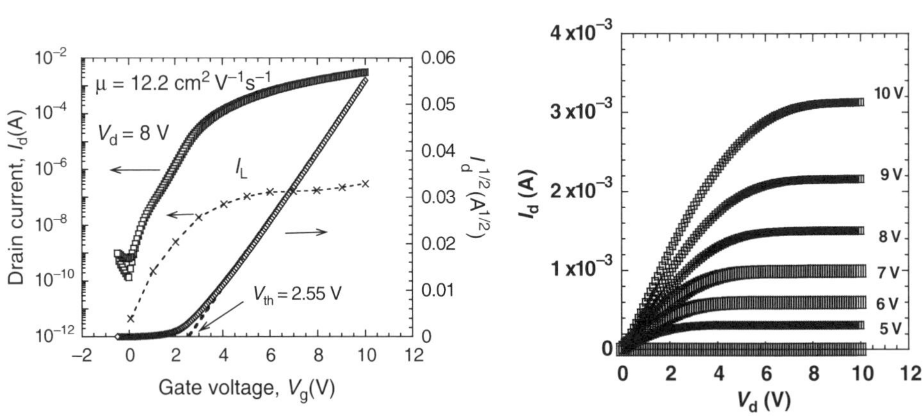

Figure 11.20 (a) Transfer characteristic with gate leakage current for ZnO thin film transistor fabricated on HfO_2 atomic layer deposition dielectric, 25 nm thick. Gate sweep with drain voltage at 8 V is from −1 to 10 V back to −1 V. Field effect mobility (12.2 cm^2 V^{-1} s^{-1}) and threshold voltage (V_{th} = 2.6 V) were determined from the slope of $I_d^{1/2}$ versus V_g. (b) Corresponding transistor output curve. Reprinted with permission from [33]. Copyright (2006) American Institute of Physics

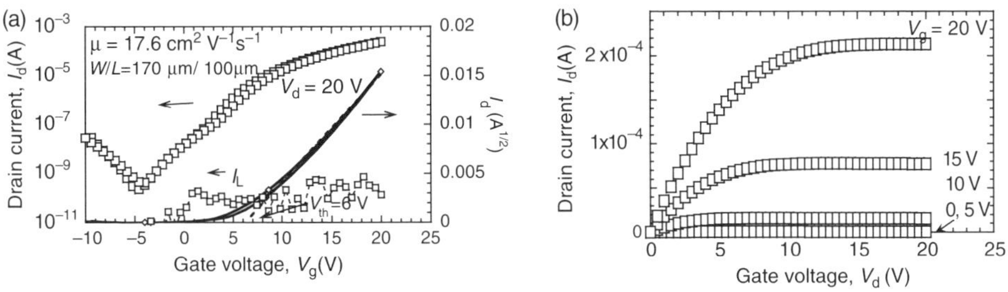

Figure 11.21 *(a) Transfer characteristic for ZnO TFT made at* $pO_2 = 2 \times 10^{-5}$ *Torr on* Al_2O_3 *deposited by atomic layer deposition, 100 nm thick. Drain voltage was 20 V. Field effect mobility (17.6* $cm^2\ V^{-1}\ s^{-1}$*) and threshold voltage (*$V_{th} = 6$ *V) were determined from slope of* $I_d^{1/2}$ *versus* V_g*. Substrate is Si. (b) Output current characteristic for ZnO TFT in (a) for drain and gate voltages 0–20 V and gate step increment, Δ(*$V_g = 5$ *V. Substrate is Si. Reprinted with permission from [33]. Copyright (2006) American Institute of Physics*

ZnO TFTs on HfO_2 from the hysteresis shift in gate voltage ($\Delta V_g = 50$ mV) at $I_d = 10^{-8}$ A from $N_t \sim C_g\ (\Delta V_g)/e$, where C_g is 0.89 μF cm^{-2} for 25 nm thick HfO_2. N_t is then ~2.7 10^{11} cm^{-2}, or about 10 times smaller than we determined for ZnO TFTs on thermally grown SiO_2. While the gate leakage current was relatively high for HfO_2, increasing its thickness could substantially reduce this leakage. Clearly the choice of dielectric and the deposition process, in this case ALD, can substantially improve the device properties of ZnO TFTs. Although these results were on rigid and not flexible substrates, 200 °C used for the ALD Al_2O_3 is already compatible with heat-stabilized PEN polyester film and Kapton® polyimide. As a further endorsement of ALD for oxide transistor, Levi *et al.* [61] report stable, high mobility (~13 $cm^2\ V^{-1}\ s^{-1}$) ZnO TFTs, when both the ZnO semiconductor and the Al_2O_3 gate dielectric are deposited by an atmospheric ALD process at 200 °C on glass.

11.6 Transistors on Plastic Substrates

11.6.1 Plastic Substrates

While the substrate demands of organic circuitry for RFID tags and active-matrix backplanes can be met with inexpensive polymer substrates, backplanes based on inorganic circuitry and displays have more demanding substrate requirements. The list of properties a plastic substrate needs for display applications to compete with or replace glass is formidable. They include optical clarity for light transmission, low thermal shrinkage and low coefficient of thermal expansion for good dimensional reproducibility, moisture and solvent resistance, low surface roughness, mechanical stiffness for ease of handling, especially in batch processing, and high upper use temperature for critical processes. MacDonald and coworkers at DuPont-Teijin Films have, in fact, developed heat-stabilized polyester films, which are engineered specifically for high-end display applications and meet or approach most of those requirements [1, 2]. One of these films is optical grade PEN (PEN Teonix® Q65), which can be processed up to 200 °C with good dimensional reproducibility. Both heat-stabilized PEN and heat-stabilized PET (HS-PET) have an upper-use temperature well above their glass transition temperature ($T_g = 78$ °C for PET

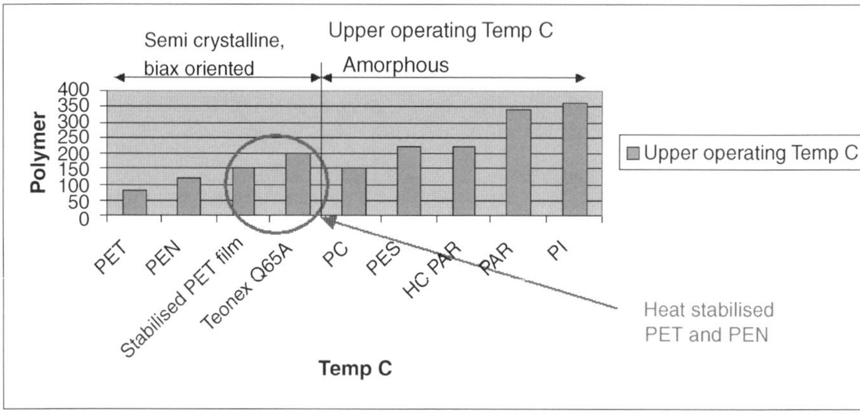

Figure 11.22 *Upper processing temperature of film substrates of interest for flexible electronics applications: poly(ethylene terphthalate) (PET), poly(ethylene naphthalate) (PEN), polycarbonate (PC), polyether sulphone (PES), polyarylate (PAR) and polyimide (PI)*

and 120 °C for PEN). Figure 11.22 compares the upper operating temperature for a number of polymer films under consideration as substrates for flexible electronics. When optical transparency is not essential in the base substrate, such as in top emissive displays, polyimide films such as DuPont's Kapton®, which allow processing above 300 °C, can be used. For those applications where optical clarity and high temperature processing are needed together, DuPont is developing clear polyimide films with low thermal expansion ($<10\,ppm\,°C^{-1}$) and a use temperature above 300 °C. A comparison of optical transmission for Kapton® E film (yellow) and experimental clear polyimide is shown in Figure 11.23.

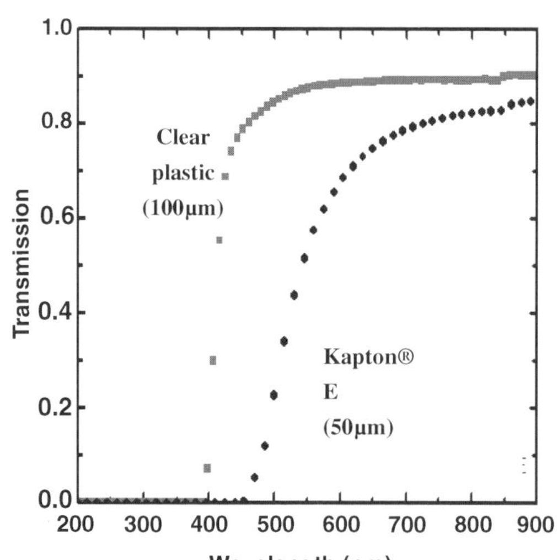

Figure 11.23 *Comparison of the optical transmission spectra for Kapton® E (DuPont) polyimide film and clear polyimide film under development at DuPont*

11.6.2 ZnO Transistors with a Fluoropolymer Gate Dielectric on Kapton® Polyimide Substrate

An example of a flexible ZnO transistor fabricated with a polymer dielectric is shown in Figure 11.24. The gate dielectric is a copolymer of vinylidene fluoride (85.5 mol%) and perfluoromethylvinylether (14.5 mol%), which was bar coated to a 2 μm film thickness. This experimental fluoropolymer dielectric has a low frequency, dielectric constant ~10. The transistor is fabricated on flexible copper-coated Kapton® polyimide (DuPont Pyralux®), 50 μm thick. Cu source-drain electrodes are patterned using a dry photopolymer resist, DuPont Riston®, and then the ZnO channel is sputtered through a shadow mask. A narrow strip of the fluoropolymer gate dielectric film is laminated between source-drain electrodes at ~120 °C. The device is then completed by evaporating a top Al gate pad through a shadow mask. Although the field effect mobility is modest (0.4 $cm^2 V^{-1} s^{-1}$), this is, to our knowledge, the first demonstration of a flexible oxide transistor on a polymer dielectric [62].

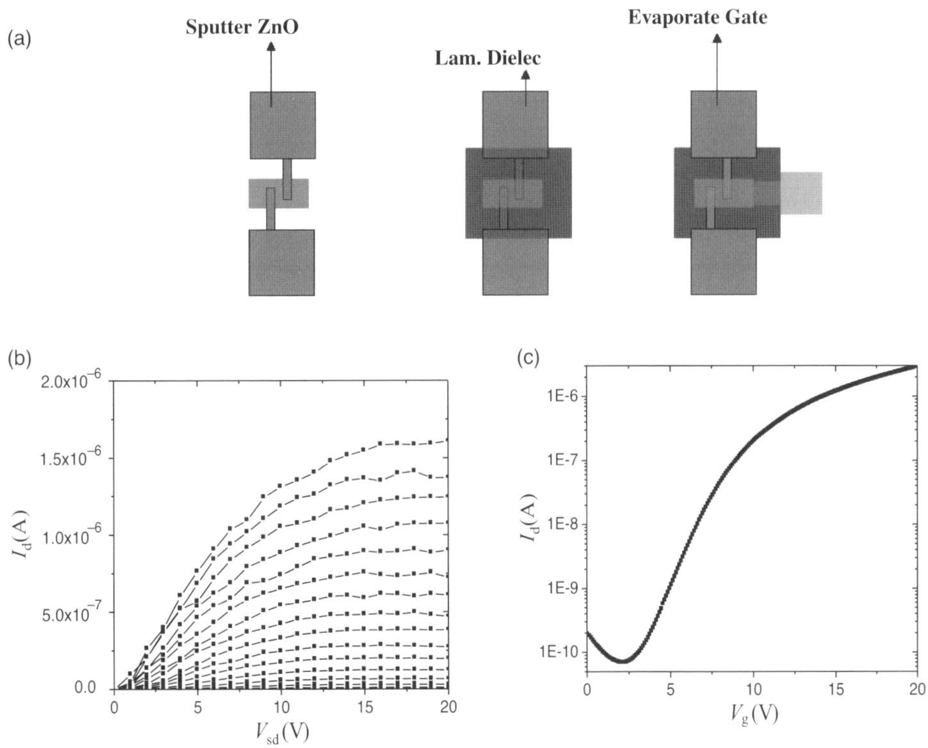

Figure 11.24 (a) Process steps in fabricating ZnO thin film transistors with laminated fluoropolymer gate dielectric on a flexible substrate (DuPont Pyralux®). (b) Output and (c) transfer curves

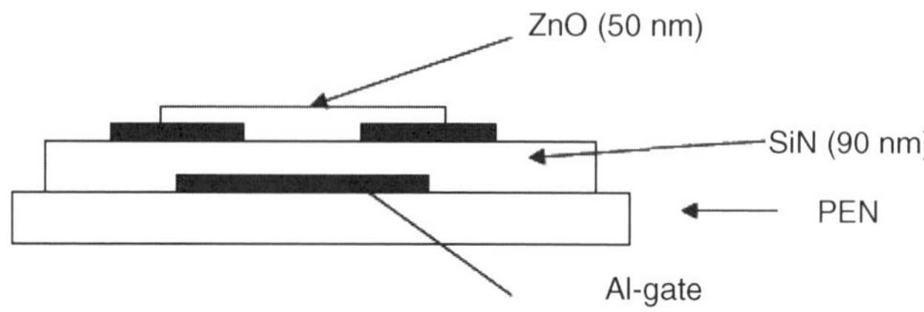

Figure 11.25 Structure of bottom-gate ZnO transistors, fabricated on flexible PEN polyester substrate, 125 μm thick. Source, drain and gate electrodes were thermally evaporated Al, ~100 nm thick. Gate dielectric was 90 nm thick sputtered SiN_x. Reprinted with permission from [34]. Copyright (2007) American Institute of Physics

11.6.3 ZnO Transistors with a Sputtered SiN_x Gate Dielectric on PEN Polyester Substrate

Flexible ZnO transistors can also be fabricated on a polyester substrate, such as 125 μm thick heat stabilized PEN (Q65A, DuPont-Tejein Films) [34]. A cross-section of the device is given in Figure 11.25. The gate dielectric is 90 nm thick SiN_x, reactively sputtered from a Si target in an Ar/60% N_2 atmosphere on evaporated Al bottom-gate pads. All layers, including Al source-drain electrodes and the ZnO semiconductor, are defined with shadow masks and deposited at room temperature. A photograph of 40 TFTs on a 1 in. flexible PEN substrate and the transfer curve for a representative device are shown in Figure 11.26. ZnO TFTs on sputtered SiN_x operate at low voltage and the threshold voltage is close to zero, although the gate leakage current with a sputtered SiN_x gate dielectric is moderate (~1®nA) at 5 V.

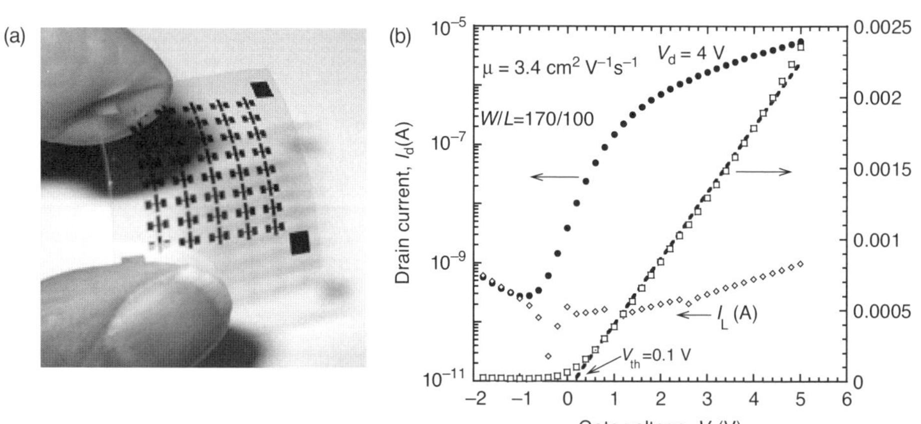

Figure 11.26 (a) Photograph of 40 ZnO thin film transistors fabricated on 1 in. square, flexible PEN polyester substrate. (b) Transfer curve of ZnO transistors. Saturation mobility determined from plot of $I_d^{1/2}$ versus gate voltage for $V_d = 4$ V. Reprinted with permission from [34]. Copyright (2007) American Institute of Physics

11.6.4 ZnO Transistors with an Evaporated Al_2O_3 Gate Dielectric on Paper-Like Tyvek® Substrate

DuPont Tyvek® is a tear resistant paper-like product with physical characteristics that combine the properties of paper, film and cloth. Tear-resistant envelopes and house-wrap are common applications. Tyvek® is formed by a continuous process from very fine (0.5–20 μm) fibers of polyethylene. These nondirectional fibers (plexifilaments) are first spun and then bonded together by heat and pressure, without binders. Figure 11.27(a) is a high magnification scanning electron micrograph of a Tyvek® substrate, showing individual polyethylene fibers (∼10–20 μm).

A scanning electron microscopy (SEM) cross-section of a ZnO TFT, fabricated with all room temperature processes using metal shadow masks on Tyvek®, is shown in Figure 11.28. The device has a bottom gate (Al) with a thick (425 nm) Al_2O_3 gate dielectric, e-beam evaporated, Al source-drain electrodes, and sputtered (∼50 nm) ZnO semiconductor. The output curve for a ZnO transistor with a 170 μm width and 100 μm channel length, shown in Figure 11.27(b), clearly exhibits modulation of the drain current with gate voltage, validating the feasibility of fabricating oxide electronics on a paper-like substrate.

Figure 11.27 *(a) SEM micrograph (1000x) of fiber structure of paper-like DuPont Tyvek® substrate. (b) Output curve of ZnO thin film transistor on flexible Tyvek®*

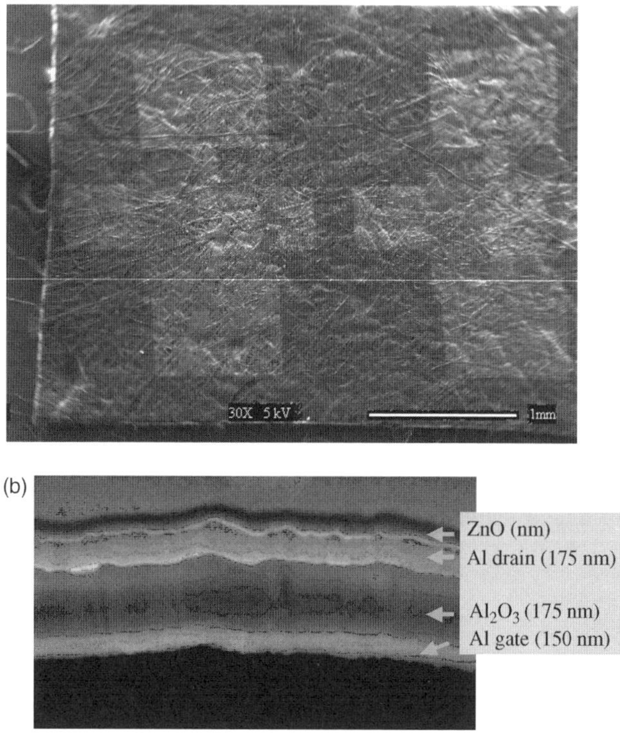

Figure 11.28 *(a) Low magnification (300x) SEM image of ZnO thin film transistors on Tyvek®. (b) High resolution, TEM cross-section of ZnO transistor layers on Tyvek®*

11.6.5 ZnO Transistors with an Evaporated Al_2O_3 Gate Dielectric on Kapton® Polyimide Substrate

Kapton® polyimide is a flexible substrate which can tolerate high processing temperature, although its yellow color precludes display applications requiring high optical clarity. ZnO TFTs were fabricated using DuPont Kapton® 200EZ film, 50 μm thick using metal shadow masks at each level of the device construction [19]. First we thermally evaporated Al gate electrodes, followed by e-beam evaporation of 430 nm thick Al_2O_3 dielectric, then evaporated Al source-drain electrodes, and finally 50 nm of sputtered ZnO semiconductor. Representative transfer and output characteristics for these ZnO TFTs with Al_2O_3 dielectric on flexible polyimide are shown in Figure 11.29. These devices have high mobility ($\sim$50 $cm^2\,V^{-1}\,s^{-1}$) with a low threshold voltage of 3.2 V. However, hysteresis for ZnO TFTs on thick, e-beam evaporated Al_2O_3 is large [19], while gate leakage is moderate at $\sim$5 nA, suggesting that a high interfacial trapping and low insulation resistance are deficiencies of this dielectric.

11.6.6 ZnO Transistors with an Al_2O_3 Gate Dielectric Grown by Low Temperature ALD on PEN Polyester Substrate

While the previous examples demonstrate feasibility to fabricate oxide devices on plastic substrates, the gate leakage current contributes to relatively high device off-state current,

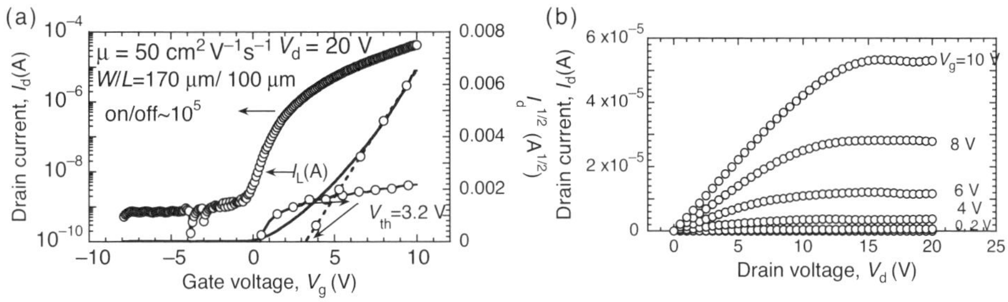

Figure 11.29 *(a) Transfer characteristic for ZnO thin film transistor on flexible, 50 μm thick Kapton® polyimide film with 400 nm thick Al_2O_3 gate dielectric. Field effect mobility ($\sim$50 cm^2 V^{-1} s^{-1}) and threshold voltage ($V_{th} = 3.2$ V). (b) Output current characteristic for ZnO TFT on Kapton®. Reprinted with permission from [19]. Copyright (2005) SID*

which is unacceptable for battery-powered portable devices and many display applications. For ZnO TFTs fabricated on a 100 nm thick Al_2O_3 gate dielectric, grown by ALD on Si at 200 °C, the leakage current is less than 10 pA, even at 20 V. This is a much smaller than for thick Al_2O_3 grown by e-beam evaporation or for sputtered SiN_x on flexible substrates. However, 200 °C is at the upper-use temperature for PEN, so we investigated the dependence of leakage current for ALD Al_2O_3 dielectrics on growth temperature. Leakage current density versus ALD growth temperature is plotted in Figure 11.30 for Al_2O_3 on heavily doped Si and with 1 mm^2 Al top electrodes. The salient result is that leakage current is lower for Al_2O_3 grown at 150 °C and 200 °C than at 50 °C, 80 °C and 120 °C. The electric field where the current density becomes 10^{-6} A cm^{-2} also increased with higher ALD synthesis temperature of Al_2O_3, approaching 5 MV cm^{-1} for both 150 °C and 200 °C.

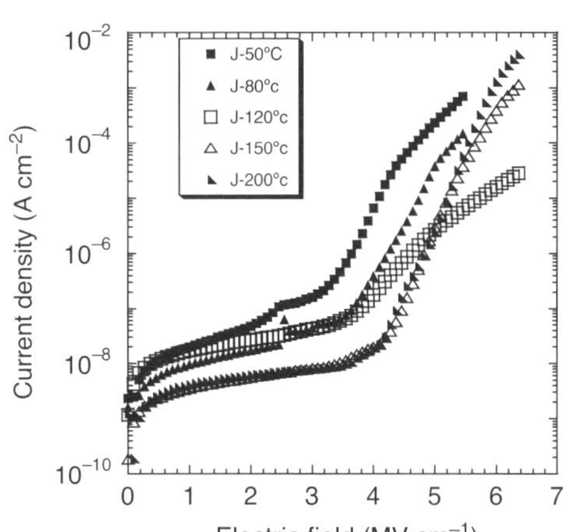

Figure 11.30 *Current density versus electric field for Al_2O_3 dielectric, nominally 50 nm thick, grown by atomic layer deposition at the temperatures indicated in the inset*

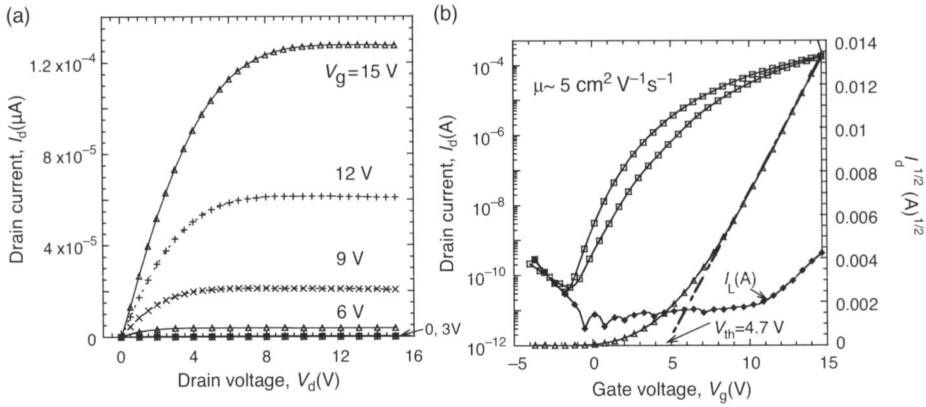

Figure 11.31 *(a) Output and (b) transfer characteristics of ZnO thin film transistors with 73 nm thick Al_2O_3 gate dielectric, grown by atomic layer deposition at 150 °C on PEN substrate with array of Al bottom-gate electrodes*

Representative transfer and output curves for a ZnO TFT on 73 nm thick ALD Al_2O_3 grown at 150 °C on PEN are shown in Figure 11.31. The saturation mobility is ~8 cm^2 V^{-1} s^{-1}, with an on/off ratio >10^6, a threshold voltage of 4.7 V, an inverse subthreshold slope *S* of 0.875 V decade^{-1}, while the gate leakage current was less than 10 pA up to 10 V. There was some small hysteresis ($\Delta V \sim 1$ V) in the gate sweep, indicative of charge trapping. This device, with modest dimensions (170 μm × 80 μm), produced an output current >100 μA for drain and gate voltage less 10 V. These devices on PEN have properties comparable with those reported by Lim *et al.* [63] for N-doped ZnO TFTs ($\mu = 6.7$ cm^2 V^{-1} s^{-1}, $V_{th} = 4.1$ V, $S = 0.67$ V decade^{-1}, on/off ratio ~10^6) on a rigid substrate with an ALD Al_2O_3 gate dielectric grown at 150 °C. In the work by Lim *et al.* those devices have thicker (97 nm) Al_2O_3 and operate at higher voltage for the same output current. ZnO:N is deposited by ALD at 125 °C, whereas ZnO in our devices is sputtered at room temperature.

11.7 Patterning

Finally, numerous schemes have been proposed for fabricating electronic circuits on flexible substrates using organic semiconductors, but not oxides. Many of these methods involve some form of printing. In this section we discuss a photolithographic technique that uses a dry photoresist to fabricate circuits on flexible substrates with an oxide semiconductor. This method is based on a solid rather than liquid photopolymer resist, such as DuPont's Riston® dry film photoresist, a technology that is currently practiced for fabricating printed circuit boards. Advantages of dry resist circuit fabrication include low processing temperature, use of innocuous nontoxic chemistry and low capital investment. Figure 11.32 illustrates the patterning steps with DuPont Riston® dry photoresist.

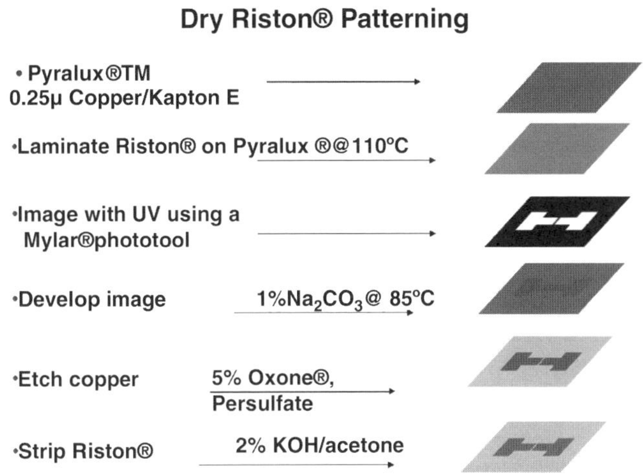

Figure 11.32 *Process steps for patterning source-drains using DuPont Riston® dry photoresist on Pyralux® (Cu-coated Kapton® polyimide film)*

The process begins with laminating Riston® dry photoresist film (~30 μm thick) to Kapton® polyimide film with a 0.25 μm thick Cu coating (DuPont Pyralux®). (Al can also be used with a corresponding change in etch chemistry.) The stencil mask or phototool can be Mylar® polyester, which is held by vacuum in close contact with Pyralux®. UV exposure forms a positive image in photoresist film. The image is developed in a weak sodium carbonate solution at 85 °C and the Cu is etched in a potassium peroxymonosulfate solution. Finally the Riston® photoresist is removed leaving patterned Cu (or Al). Figure 11.33 illustrates Cu gate and source-drain electrodes patterned with a 15 μm thick Riston® dry photoresist.

Using the process flow shown schematically in Figure 11.34, we have fabricated ZnO backplanes, 6 in. × 6 in. In those structures, all deposition processes were at room temperature, including the SiN_x:H gate dielectric, deposited by PECVD, while the top gates were screen-printed Ag. However, the SiN_x:H dielectric, deposited at room temperature, has too low breakdown voltage, and frequent shorting occurs between metal lines separated by this dielectric. A solution to this problem may be to use a dielectric grown by ALD. ALD Al_2O_3 grown as low as 50 °C, which should be compatible with a Riston® shadow mask, exhibits relatively low leakage current up to 2 MV cm^{-1}. At this stage the application of a dry photoresist approach for fabricating oxide electronics on plastic substrates is still nascent, however, it does have the advantage that the materials, tools, and chemistries are mature, commercially available, compatible with flexible substrates, and the initial capital investment is less than for traditional lithography.

11.8 Conclusions

Oxides represent a relatively new class of semiconductor materials applied to active devices, such as TFTs. The combination of high field effect mobility and low processing temperature

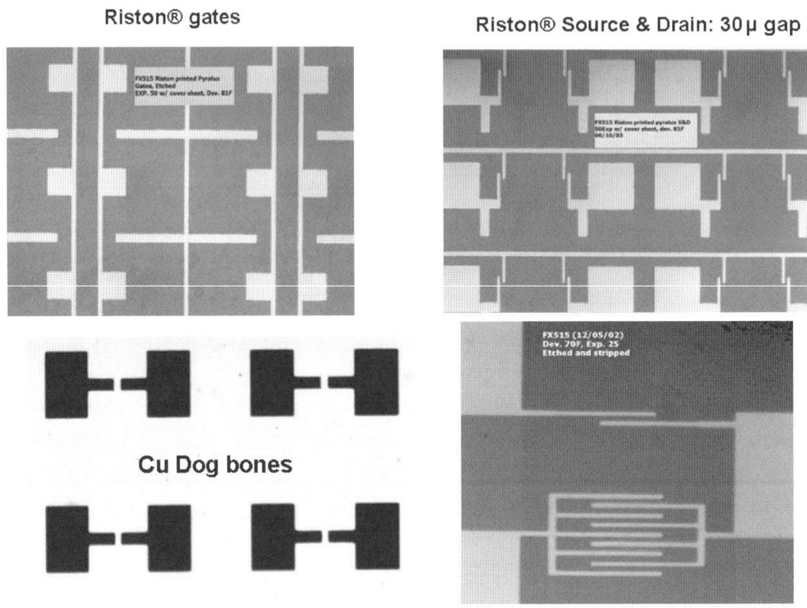

Figure 11.33 *Examples of source-drain and gate structures patterned with Riston® dry photoresist on Pyralux® (Cu-coated Kapton® polyimide film)*

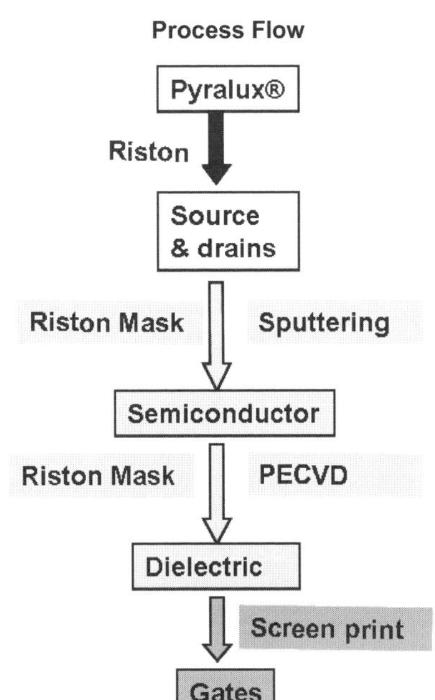

Figure 11.34 *Process steps used to fabricate thin film transistor backplane with sputtered ZnO semiconductor, PECVD SiN_x:H gate dielectric, Cu source-drains and screen-printed, thick film Ag top-gate electrode*

for oxide semiconductors makes them attractive for high performance electronics on flexible plastic substrates. In this chapter we summarized fabrication of oxide transistors on a variety of plastic substrates. We have also identified ALD as a preferred process for the gate dielectric on flexible substrates. This work described in the chapter represents the initial steps toward developing a robust flexible electronics technology based on oxide semiconductors.

Acknowledgements

The author wishes to thank his colleagues at DuPont who contributed to the results discussed in this chapter. Scott McLean and Mike Reilly did most of the device fabrication, and Dr Ken Sharp and Seema Agrawal did the patterning work with dry photoresist.

References

[1] W. A. MacDonald, Engineered films for display technologies, *J. Mater. Chem.* **14**, 4–10 (2004).
[2] W. A. MacDonald, M. K. Looney, D. MacKerron, R. Eveson, R. Adam, K. Hashimoto, and K. Rakos, Latest advances in substrates for flexible electronics, *J. SID* **15**, 1075–1083 (2007).
[3] I. McCulloch, Rolling out organic electronics, *Nature Mater.* **4**, 583–584 (2005).
[4] V. Subramanian, J. Frechet, P. Chang, D. Huang, J. Lee, S. Molesa, A. Murphy, D. Redinger, and S. Volkman, Progress toward development of all-printed RFID tags: materials, processes, and devices, *Proc. IEEE* **93**, 1330–1338 (2005).
[5] R. H. Reuss, D. G. Hopper, and J.-G. Park, Macroelectronics *MRS Bull.* **31**, 447–450 (2006).
[6] S. Wagner, H. Gleskova, I.-C. Cheng, and M. Wu, Silicon for thin-film transistors, *Thin Solid Films* **430**, 15–19 (2003).
[7] J. Kanicki and S. Martin, Hydrogenated amorphous silicon thin-film transistors, in *Thin Film Transistors*, C. R. Kagan and P. Andry (Eds), Marcel Dekker, New York, 2003, pp. 71–137.
[8] F. R. Libsch, Thin-film transistors in active-matrix liquid crystal displays (AMLCDs), in Thin Film Transistors, C. R. Kagan and P. Andry (Eds), Marcel Dekker, New York, 2003, pp. 209–299.
[9] A. T. Voutsas and M. K. Hatalis, Technology of polysilicon thin-film transistors, in *Thin Film Transistors*, C. R. Kagan and P. Andry (Eds), Marcel Dekker, New York, 2003, pp. 139–207.
[10] T. Tsujimura, Amorphous/microcrystalline silicon thin film transistor characteristics for large size OLED television driving, *Jpn. J. Appl. Phys.* **43**, 5122–5128 (2004).
[11] A. R. Brown, C. P. Jarrett, D. M. de Leeuw, and M. Matters, Field-effect transistors made from solution-processed organic semiconductors, *Synth. Met.* **88**, 37–55 (1997).
[12] C. D. Dimitrakopoulos and D. J. Mascaro, Organic thin-film transistors: a review of recent advances, *IBM J. Res. Dev.* **45**, 11–27 (2001).
[13] Z. Bao, A. J. Lovinger, and J. Brown, New air-stable n-channel organic thin film transistors, *J. Am. Chem. Soc.* **120**, 207–208 (1998).
[14] H. E. Katz, J. Johnson, A. J. Lovinger, and W. Li, Naphthalenetetracarboxylic diimide-based n-channel transistor semiconductors: structural variation and thiol-enhanced gold contacts, *J. Am. Chem. Soc.* **122**, 7787–7792 (2000).
[15] G. H. Gelinck, H. E. A. Huitema, E. van Veenendaal, E. Cantatore, L. Schrijnemakers, J. van der Putten, T. Geuns, M. Beenhakkers, J. Giesbers, B. Huisman, E. Meijer, E. Benito, F. Touwslager, A. Marsman, B. van Rens, and D. de Leeuw, Flexible active-matrix displays and shift registers based on solution-processed organic transistors, *Nature Mater.* **3**, 106–110 (2004).
[16] S. Masuda, K. Kitamura, Y. Okumura, and S. Miyatake, Transparent thin film transistors using ZnO as an active channel layer and their electrical properties, *J. Appl. Phys.* **93**, 1624–1630 (2003).

[17] R. L. Hoffman, B. J. Norris, and J. F. Wager, ZnO-based transparent thin-film transistors, *Appl. Phys. Lett.* **82**, 733–735 (2003).
[18] P. F. Carcia, R. S. McLean, M. H. Reilly, and G. NunesJr, Transparent ZnO thin-film transistor fabricated by RF magnetron sputtering, *Appl. Phys. Lett.* **82**, 1117–1119 (2003).
[19] P. F. Carcia, R. S. McLean, and M. H. Reilly, Oxide engineering of ZnO thin-film transistors for flexible electronics, *J. SID* **13**, 547–554 (2005).
[20] B. Yaglioglu, H. Y. Yeom, R. Beresford, and D. C. Paine, High-mobility amorphous In_2O_3-10 wt % ZnO thin film transistors, *Appl. Phys. Lett.* **89**, 062103-1-3 (2006).
[21] R. Martins, P. Barquinha, I. Ferreira, G. Goncalves, and E. Fortunato, Role of order and disorder on electronic performance of oxide thin film transistors, *J. Appl. Phys.* **101**, 044505-1-7 (2007).
[22] J. F. Wager, Transparent electronics, *Science* **300**, 1245–1246 (2003).
[23] D. Pribat and F. Plais, Matrix addressing for organic electroluminescent displays, *Thin Solid Films* **383**, 25–30 (2001).
[24] P. Servati, S. Prakash, A. Nathan, and C. Py, Amorphous silicon driver circuits for organic light-emitting diode displays, *J. Vac. Sci Technol.* **A20**, 1374–1378 (2002).
[25] See articles on transparent conducting oxides in *Mater Res. Bull.* **25**, (2000).
[26] R. A. Swalin, *Thermodynamics of Solids*, John Wiley & Sons, Ltd, New York, 1972.
[27] P. F. Carcia, R. S. McLean, M. H. Reilly, Z. G. Li, L. J. Pillione, and R. F. Messier, Influence of energetic bombardment on stress, resistivity, and microstructure of indium tin oxide films grown by radio frequency magnetron sputtering on flexible substrates, *J. Vac. Sci. Technol.* **A21**, 745–751 (2003).
[28] P. F. Carcia, R. S. McLean, M. H. Reilly, Z. G. Li, L. J. Pillione, and R. F. Messier, Resistivity and microstructure issues in indium-oxide based films grown by RF magnetron sputtering on flexible polyester substrates, Soc. *Vacuum Coaters, 46th Annual Technical Conf. Proc* 21–25. (2003).
[29] A. Zunger, Practical doping principles, *Appl. Phys. Lett.* **83**, 57–59 (2003).
[30] T. Minami, New n-type transparent conducting oxides, *MRS Bull.* **25**, 38–44 (2000).
[31] M. Orita, H. Ohta, M. Hirano, S. Narushima, and H. Hosono, Amorphous transparent conductive oxide $InGaO_3(ZnO)_m$ ($m \leq 4$): a Zn 4s conductor, *Philos. Mag. B* **81**, 501–515 (2001).
[32] H. Hosono, Ionic amorphous oxide semiconductors: material design, carrier transport, and device application, *J. Non-Crystall. Solids* **352**, 851–858 (2006).
[33] P. F. Carcia, R. S. McLean, and M. H. Reilly, High performance ZnO thin-film transistors on gate dielectrics grown by atomic layer deposition, *Appl. Phys. Lett.* **88**, 123509–3 (2006).
[34] P. F. Carcia, R. S. McLean, M. H. Reilly, M. K. Crawford, and E. N. Blanchard, A comparison of zinc oxide thin-film transistors on silicon oxide and silicon nitride gate dielectrics, *J. Appl. Phys.* **102**, 074512–7 (2007).
[35] H. Q. Chiang, J. F. Wager, R. L. Hoffman, J. Jeong, and D. A. Keszler, High mobility transparent thin-film transistors with amorphous zinc tin oxide channel layer, *Appl. Phys. Lett.* **86**, 013503–3 (2005).
[36] N. L. Dehuff, E. S. Kettenring, D. Hong, H. Q. Chiang, J. F. Wager, R. L. Hoffman, C.-H. Park, and D. A. Keszler, Transparent thin-film transistors with zinc indium oxide channel layer, *J. Appl. Phys.* **97**, 064505–5 (2005).
[37] H. Yabuta, M. Sano, K. Abe, T. Aiba, K. Nomura, T. Kamiya, and H. Hosono, High-mobility thin-film transistor with amorphous $InGaZnO_4$ channel fabricated by room temperature rf-magnetron sputtering, *Appl. Phys. Lett.* **89**, 112123 (2006).
[38] J. Jang, Preparation and properties of hydrogenated amorphous silicon thin-fim transistors, in *Thin Film Transistors*, C. R. Kagan and P. Andry (Eds), Marcel Dekker, New York, 2003, pp. 35–69.
[39] Landolt-Bornstein, *Semiconductors 17*, Springer-Verlag, Berlin, (1982).
[40] K. Tominaga, S. Iwamura, Y. Shintani, and O. Tada, Energy analysis of high-energy neutral atoms in sputtering of ZnO and $BaTiO_3$, *Jpn. J. Appl. Phys.* **21**, 688–695 (1982).
[41] H.-H. Hsieh and C.-C. Wu, Amorphous ZnO transparent thin-film transistors fabricated by full lithographic and etching processes, *Appl. Phys. Lett.* **91**, 013502 (2007).
[42] B. D. Cullity, *Elements of X-ray Diffraction*, Addison-Wesley, Reading, MA, 1967, p. 436.
[43] M. E. Thomas, M. P. Hartnett, and J. E. McKay, The use of surface profilometers for measurement of wafer curvature, *J. Vac. Sci. Technol.* **A6**, 2570–2571 (1988).

[44] D. W. Greve, *Field-Effect Devices and Applications*, Prentice Hall, Upper Saddle River, NJ, 1998, Chapter 7, p. 273.
[45] J. Levinson, F. R. Shepherd, P. J. Scanlon, W. D. Westwood, G. Este, and M. Rider, Conductivity behavior in polycrystalline semiconductor thin film transistors, *J. Appl. Phys.* **53**, 1193–1202 (1982).
[46] K.-S. Weissenreider and J. Muller, Conductivity model for sputtered ZnO-thin film gas sensors, *Thin Solid Films* **300**, 30–41 (1997).
[47] A. Ambrosini, A. Duarte, K. Poppelmeier, M. Lane, and C. R. Kannewurf, and T.O Mason, Electrical, optical, and structural properties of tin-doped In_2O_3-M_2O_3 solid solutions (M = Y, Sc), *J. Solid State Chem.* **153**, 41–47 (2000).
[48] J. R. Bellingham, W. A. Phillips, C. J. Adkins, Electrical and optical properties of amorphous indium oxide, *J. Phys.: Condens. Matter* **2**, 6207–6221 (1990).
[49] M. R. Graham, J. R. Bellingham, C. J. Adkins, Hopping conduction near the metal-insulator transition in amorphous indium oxide, *Philos. Mag. B* **65**, 669–673 (1992).
[50] P. F. Carcia, R. S. McLean I. Malajovich, and M. H. Reilly, Oxide semiconductors for flexible electronics, in *Thin Film Transistor Technologies (TFTT VII)*, Proceedings Vol. 2004-15, Y. Kuo (Ed.), The Electrochemical Society, Pennington, NJ, 2005, pp. 178–183.
[51] D. Zhang, C. Li, S. Han, X. Liu, T. Tang, W. Jin, and C. Zhou, Electronic transport studies of single-crystalline In_2O_3 nanowires, *Appl Phys. Lett.* **82**, 112–114 (2003).
[52] M. Meitine, A. Sazonv, Low temperature PECVD silicon oxide for devices and circuits on flexible substrates, *Mater. Res. Soc. Proc.* **769**, H6.6.1–H6.6.6 (2003).
[53] G. Bersuker, P. Zeitzoff, G. Brown, and H. R. Huff, Dielectrics for future transistors, *Materials Today* 26–33 (2004).
[54] K. Vanheusden, W. L. Warren, C. H. Seager, D. R. Tallant, J. A. Voigt, B. E. Gnade, Mechanisms behind phootoluminescence in ZnO phosphor powders, *J. Appl. Phys.* **79**, 7983–7990 (1996).
[55] Y.-J. Lin, C.-L. Tsai, Y.-M. Lu, C.-J. Liu, Optical and electrical properties of undoped ZnO films, *J. Appl. Phys.* **99**, 093501–4 (2006).
[56] H. Gleskova, S. Wagner, V. Gasparik, P. Kovac, Low-temperature silicon nitride for thin-film electronics on polyimide foil substrates, *Appl. Surf. Sci.* **175–176**, 12–16 (2001).
[57] J. W. Elam, M. D. Groner, S. M. George, Viscous flow reactor with quartz crystal microbalance for thin film growth by atomic layer deposition, *Rev. Sci Instrum.* **73**, 2981–2987 2002.
[58] R. G. Gordon, D. Hausmann, E. Kim, and J. Shepard, A kinetic model for step coverage by atomic layer deposition in narrow holes or trenches, *Chem. Vap. Deposition* **9**, 73–78 (2003).
[59] L. Niinisto, J. Paivasaari, J. Niinisto, M. Putkonen, M. Nieminen, Advanced electronic and optoelectronic materials by atomic layer deposition: an overview with special emphasis on recent progress in processing high-k dielectrics and other oxide materials, *Phys. Status Solidi* A **201**, 1443–1452 (2004).
[60] J. Kim, S. Kim, H. Jeon, M.-H. Cho, K.-B. Chung, C. Bae, Characteristics of HfO_2 thin films grown by plasma atomic layer deposition, *Appl. Phys. Lett.* **87**, 053108–3 (2005).
[61] D. H. Levi, D. Freeman, S. F. Nelson, P. J. Cowdery-Corvan, and L. M. Irving, Stable ZnO thin film transistors by fast open air atomic layer deposition, *Appl. Phys. Lett* **92**, 192101–3 (2008).
[62] P. F. Carcia, R. S. McLean, M. H. Reilly, I. Malajovich, K. G. Sharp, S. Agrawal, and G. NunesJr, ZnO thin film transistors for flexible electronics, *Mater. Res. Soc. Proc.* **769**, H7.2.1–H.7.2.6 (2003).
[63] S. J. Lim, S. Kwon, H. Kim, and J.-S. Park, High performance thin film transistor with low temperature atomic layer deposition nitrogen-doped ZnO, *Appl. Phys. Lett.* **91**, 183517–3 (2007).

12

Transparent OLED Displays

Thomas Riedl

Advanced Semiconductors Group, Institute of Electronic Devices, Bergische Universität Wuppertal, Germany

12.1 Introduction

Today, displays based on a variety of technologies play an important role as the human–machine interface for a plethora of applications. Compared with well-established liquid crystal displays (LCDs) and plasma screens, displays based on organic light emitting diodes (OLEDs) [1] are envisaged to offer more brilliant images with high levels of contrast. Concomitantly, a reduction in both energy consumption and also production cost are expected. OLED displays are currently in the market predominantly for mobile consumer products (cellular phones, MP3 players, etc.).

There are applications where transparent displays or signage are very attractive and bring about additional benefits. If switched off, a transparent display may appear as an ordinary window which allows a clear view of everything behind it. On demand, this see-through device may be used to directly display arbitrary information without using elaborate projection techniques. Among others, designers predominantly in the automotive industry are currently creating 'concept cars' having transparent display elements. However, applications range considerably beyond plain design: The automotive industry is looking at see-through displays directly integrated into the windshield of a car as an important means of increasing security in daily road traffic. In critical situations, the driver could be supplied with important – in some cases probably life-saving – information or warnings directly in his field of view. For security or defense, head-mounted displays integrated into goggles or visors of helmets are discussed. In medicine, there are studies that use so-called augmented reality systems based on transparent displays that supply the surgeons, e.g. during brachytherapy [2], with important data without them having to change their line of vision.

Transparent Electronics: From Synthesis to Applications Edited by Antonio Facchetti and Tobin J. Marks

For the realization of transparent displays, LCD technology offers only limited possibilities. With the requirement of an external light source and additional optical components such as polarizing filters, the overall transparency is severely compromised.

Transparent electroluminescent (EL) displays based on inorganic phosphors were proposed several years ago [3]. However, they have fundamental shortcomings such as high operating voltages (~200 V) and low levels of achievable brightness (~100 cd m^{-2}), limiting their day-time visibility. Alternatively, OLEDs with their low operation voltages (~2–4 V), high brightness levels of up to 100 000 cd m^{-2}, and efficiencies of more than 70 lm W^{-1} appear to be promising building blocks of future transparent displays [4]. Owing to the large achievable Stokes shifts between absorption and emission in organic light emitting materials, OLEDs can be designed to transmit more than 80% of visible light.

Transparent OLEDs were introduced in 1996 [5], Entirely transparent passive matrix OLED (PMOLED) displays have also been demonstrated [6]. However, in PMOLED displays with more than 100 rows and desired brightness levels of 500 cd m^{-2} and beyond, the individual OLED pixels would have to be driven at more than 50 000 cd m^{-2}, limiting both the efficiency and the lifetime of the display. Therefore, for larger area high resolution displays the active matrix (AM) driving scheme is mandatory. Since OLEDs are current driven devices, each AM pixel circuit consists of at least two thin film transistors (TFTs) and a storage capacitor (2T1C). If a-Si:H TFTs are used, instabilities in the threshold voltage (V_{th}) and the field effect mobility (μ_{FE}) [7] need to be compensated by more sophisticated approaches, e.g. involving four TFTs [8]. Depending on the desired display brightness, the OLED efficiency and the basic TFT parameters, the driving circuit will occupy a considerable amount of the pixel area. For a 3.4″ QVGA OLED display, the area occupied by the four a-Si:H TFTs will account for ca. 50% of the pixel area [9]. For this reason, opaque silicon-based TFTs are of limited use for transparent AMOLED displays. Previous reports using a-Si TFTs and transparent OLEDs led to 'transparent' displays with a transmissivity of only 20% [10, 11]. These figures are not sufficient for several applications (e.g. automotive windshields). Recently, pixel driver electronics based on transparent TFTs has been envisaged [12–14]. This novel branch of electronic thin film devices relies on metal oxide semiconductors which allow for TFTs with a transmissivity of more than 80% throughout the visible part of the spectrum.

Before we deal with transparent AMOLED displays we will first address the area of individual transparent OLEDs.

12.2 Transparent OLEDs

12.2.1 The Transparent Top Electrode

Typically, OLEDs emit their light through the glass substrate which comprises an electrode of a transparent conducting oxide (TCO), e.g. indium tin oxide (ITO). The top electrode (typically the cathode) is usually an opaque low-work-function metal layer. For a transparent OLED, the top electrode needs to be see-through as well (Figure 12.1).

First attempts to realize transparent OLEDs employed very thin, semi-transparent, top electrodes of Mg:Ag [15], LiF/Al or Ca [16]. Considerable optical losses due to the metal films limit the transparency of the devices in the visible spectral region (450–750 nm) to

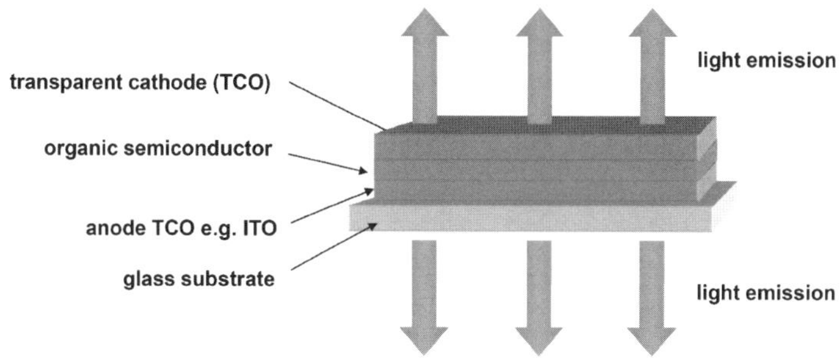

Figure 12.1 *Layer structure of a transparent OLED*

60–70%. Alternatively, one may construct transparent OLEDs with an inverted layer structure, namely with the anode side up [17]. Here, a semitransparent Au anode has been used. Efficient electron injection via the TCO bottom electrode has been accomplished by the well established concept of electrochemically doped organic electron transport layers [18]. However, the overall device transparency was only 40% due to losses of the Au electrode. In view of these results, a metal top electrode (even a very thin one) is not compatible with highly transparent devices. Therefore the use of a TCO top electrode appears mandatory. A 100 nm thick layer of ITO shows a transparency above 80% throughout the visible part of the spectrum and provides a sufficiently low sheet resistivity ($20\,\Omega\,sq^{-1}$) for OLED applications. Typically, these high quality ITO films are prepared with plasma-assisted deposition techniques, e.g. sputtering, which is the premier technique for large area TCO coatings. Unfortunately, the high energy particles emerging during the sputtering process have been identified to significantly damage the underlying organic layers [19, 20]. As a result, an increased leakage current, reduced efficiency and reduced lifetime of the OLEDs that have undergone such a process have been reported [5, 19]. Therefore, a buffer layer is needed to protect the organic materials against the particle bombardment during the deposition of the TCO top electrode.

There are several requirements for such a buffer layer in order to qualify for this purpose: In addition to an efficient protection of the organic layers against impinging particles, a high transparency in the visible part of the spectrum and reasonable electrical conductivity are desired. Moreover, the buffer layer should facilitate efficient charge injection of carriers from the ITO into the adjacent organic charge transport material. Furthermore, a nonreactive, low-damage deposition technique is needed for the buffer material. The process should not affect the underlying organic layers themselves and should allow for the formation of dense thin films. These specifications are hardly met in one material all at the same time.

Thin metal films, for example, offer reasonable protecting properties and some metals also allow for ohmic charge injection into organic charge transport materials, but the maximum transparency is strongly limited. Even if the layer thickness of e.g. a Mg:Ag alloy is reduced to only 8 nm a total transmittance of 50% cannot be exceeded [16]. However, transparent displays, e.g. in automobile windshields, require a minimum total transmittance of 75% for safety reasons. On the contrary, organic buffer layers can provide an adequate transmittance, for instance copper phthalocyanine (CuPc) or pentacene. Parthasarathy *et al.* attributed the effective sputter protection by CuPc to the extended π system in these molecules, which distributes the energy of impinging particles over several bonds [21].

Planar molecules like CuPc or pentacene, however, tend to grow in polycrystalline structures with grain sizes of more than 5 μm [22]. Therefore, the surface morphology of the polycrystalline films deteriorates the quality of the ITO film deposited on top of these layers. Thus, the conductivity of the transparent top-electrode drops significantly [23]. Furthermore, a spin-coated layer of poly(3,4-ethylenedioxythiophene) poly(styrenesulfonate) (PEDOT:PSS) on top of the pentacene has been used to level the roughness and to form a very efficient hybrid buffer with ohmic hole injection properties [24, 25]. Unfortunately, the hybrid buffer approach is not favorable for serious manufacturing because of the necessary vacuum break and the water introduced by the PEDOT:PSS wet process. Aside from that, the strong absorption of CuPc in the orange-red region limits the applicability of this concept for see-through displays in the entire visible spectrum. The incorporation of a transparent organic buffer of 2,9-dimethyl-4,7-diphenyl-1,10-phenanthroline (BCP) with an ultra-thin film of Li led to devices with improved transparency [26]. At the same time significant damages to the BCP layer due to the sputter process are observed.

Recently, a class of transition metal oxides (TMOs), such as MoO_3, WO_3, V_2O_5, ReO_3 and NiO, have been rediscovered for OLED applications owing to their efficient hole injection properties [27–29], stability improvements [30] and their use in charge generation layers [31–34]. It turns out that these materials also qualify for highly efficient sputter protection layers. This can be verified in an OLED structure as shown in Figure 12.2. Here an inverted set-up (bottom electrode = cathode) is used. A Li doped bathophenanthroline (BPhen) layer has been chosen for the injection of electrons. The emission zone consists of a 1,3,5-tris(phenyl-2-benzimidazolyl) benzene (TPBi) layer doped with 7 vol% of *fac*-tris(2-phenylpyridine) iridium [$Ir(ppy)_3$]. As hole transport material 4,4′,4″-tris(*N*-carbazolyl) triphenylamine (TCTA) is used. The tungsten oxide (WO_3) layer has been thermally evaporated on top and serves as hole injection and buffer layer at the same time. The ITO top electrode (60 nm) has been prepared by radio frequency (RF) magnetron sputtering.

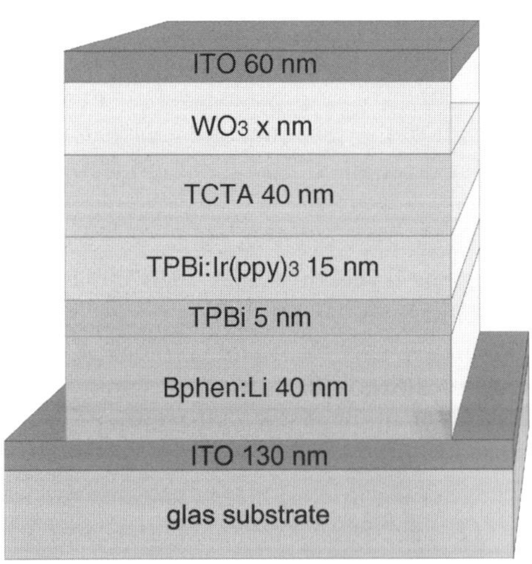

Figure 12.2 *Layer sequence of a transparent inverted OLED structure with ITO bottom cathode and ITO top anode. The thickness of the* WO_3 *buffer has been varied*

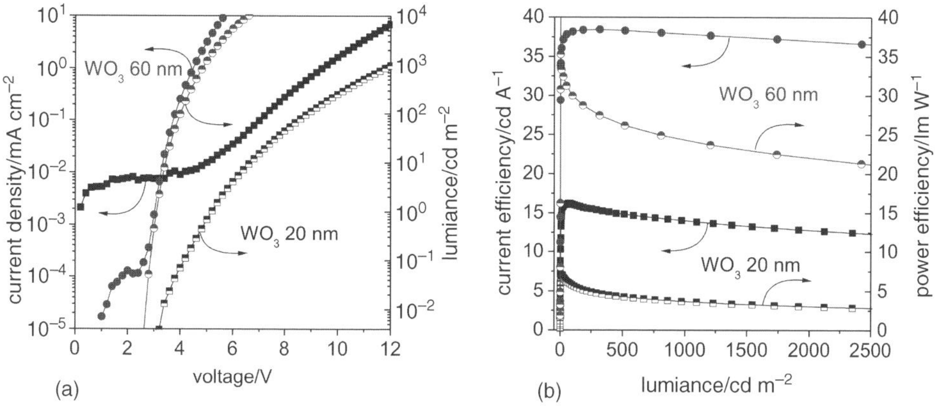

Figure 12.3 *(a) Current density–voltage and (b) current efficiency–luminance characteristics of transparent OLEDs with 20 nm and 60 nm WO_3 buffer layer thickness*

Without the WO_3 sputter buffer most of the devices are shorted or show high leakage current densities after ITO deposition (not shown here). This is a consequence of the sputter damage to the organic layer structure. Figure 12.3 displays the characteristics of OLEDs with a 20 nm and a 60 nm WO_3 buffer, respectively. Even for a 20 nm thick buffer there is still a significant leakage current density which is obvious in Figure 12.3(a) at low voltages (below the onset of light emission). For a 60 nm thick buffer the leakage current is significantly smaller and corresponds to that of a device that has not undergone a sputter process. As a result, the OLED with the thicker buffer layer shows a significantly higher current and power efficiency of 30 lm W^{-1} and 38 cd A^{-1} at 100 cd m^{-2}, respectively [Figure 12.3(b)]. At the same time, the transmissivity of the entire device is higher than 78% throughout the visible spectrum (Figure 12.4).

Atomic force microscopy (AFM) and transmission electron microscopy (TEM) have revealed that WO_3 grows in a nanocrystalline structure with grain sizes of around 3–5 nm.

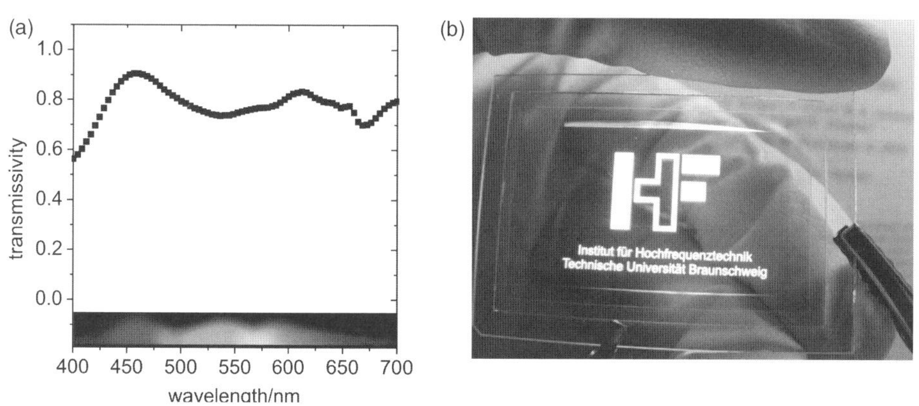

Figure 12.4 *(a) Overall transmissivity of a device with a 60 nm thick WO_3 buffer layer and (b) photograph of a transparent white light OLED. See color plate section*

The average roughness (root mean square deviation) was found to be as low as 1 nm. Thus, the WO_3 buffer is a reasonably smooth underlayer for the subsequent deposition of the ITO top electrode.

12.2.2 In-Free Transparent OLEDs

As we have seen above, TCOs play a paramount role in the realization of transparent OLEDs. Indium-tin-oxide (ITO) is certainly the most widely used TCO for display applications. However, indium is a relatively scarce material and it might become difficult to satisfy the future demand [35]. Therefore, abundant and low cost alternatives are vigorously pursued. Among them, Al doped ZnO (AZO) seems to be a highly promising candidate as electrode material for OLEDs [36, 37]. Transparent OLEDs using AZO for both bottom and top electrode have not been studied so far. In principle, a device structure similar to the one in Figure 12.2 could be prepared in which the ITO layers are entirely substituted by AZO. AZO may be prepared by various deposition techniques. Figure 12.5 shows the characteristics of a series of AZO films deposited by pulsed laser deposition (PLD) at various substrate temperatures between room temperature and 400 °C [25]. The Al content in the PLD target was about 4 wt%. As can be seen in Figure 12.5(a) an optimum of the conductivity is found around 200 °C with $\sigma = 4000\ \mathrm{S\ cm^{-1}}$. At room temperature, which is relevant for deposition on top of an OLED, conductivities on the order of $2800\ \mathrm{S\ cm^{-1}}$ are achieved. For comparison, commercially available ITO coated glass substrates (Merck) feature a conductivity of $4700\ \mathrm{S\ cm^{-1}}$. The spectral transmissivity of a 130 nm thick AZO film deposited at 200 °C on a glass substrate is displayed in Figure 12.5(b). A transmissivity well above 90% throughout the visible spectral region is found [with a modulation due to Fabry Perot (FP) interference]. Measurements of the absorption of AZO have been conducted in waveguide direction (to circumvent FP interference) and an absorption coefficient at 550 nm of $750\ \mathrm{cm^{-1}}$ has been determined for layers deposited by PLD at room temperature [38]. Thus, even a 1 μm thick layer of AZO will transmit more than 92%

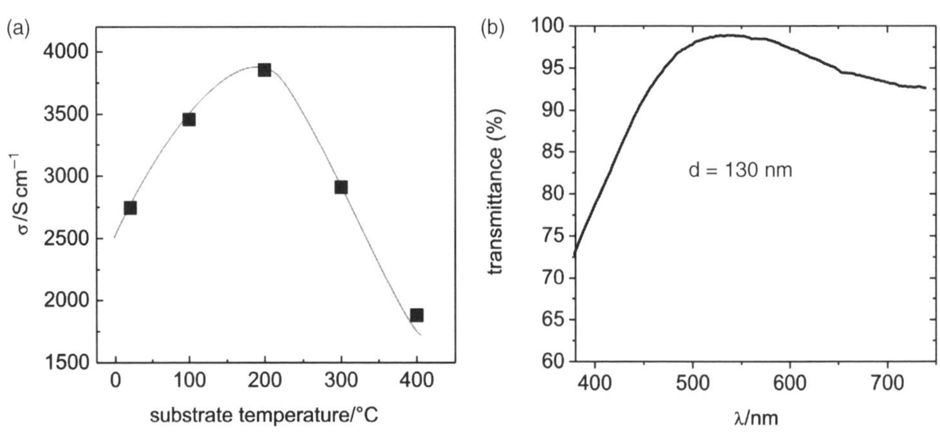

Figure 12.5 (a) Conductivity of PLD deposited AZO vs. substrate temperature and (b) spectral transmissivity of a 130 nm thick AZO layer, deposited at 200 °C. Reprinted with permission from [25]. Copyright (2007) SPIE

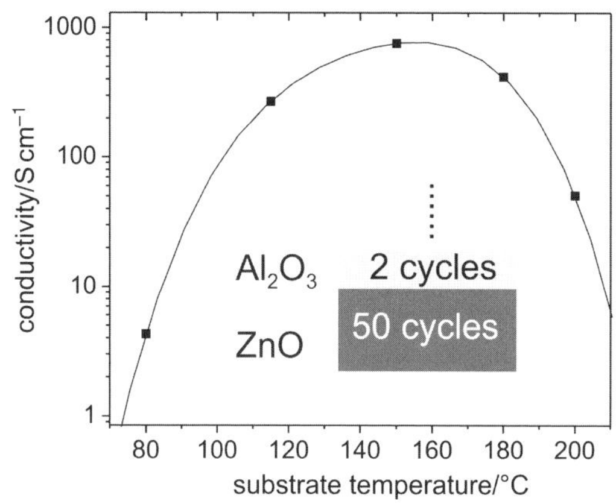

Figure 12.6 *Conductivity of ALD deposited AZO vs. substrate temperature. The cycle sequence is shown in the inset*

of light in this spectral region. At the same time the sheet resistance would be around $4\,\Omega\,sq^{-1}$.

Aside from PLD, atomic layer deposition (ALD) is a useful technique to prepare very homogeneous and smooth TCO films on large area substrates [39]. ALD relies on the sequential exposition of a surface that needs to be coated to a metal organic precursor and a reactant (e.g. H_2O, O_3, NH_3, etc.) [40]. For ZnO based systems, most commonly diethylzinc (DEZn) and water are used as precursors. It is also possible to achieve aluminum doped ZnO layers by the repetitive inclusion of a few monolayers of Al_2O_3 using trimethylaluminum (TMA) and water as precursors [41–43].

Figure 12.6 shows the conductivity of a series of AZO layers prepared by ALD at varied substrate temperatures. Here, after the deposition of 50 cycles of ZnO (9 nm) two monolayers of Al_2O_3 have been deposited. This has been done repetitively until the integral film thickness corresponded to 180 nm. As can be seen, an optimum in the conductivity close to $800\,S\,cm^{-1}$ has been reached at around 150 °C. For lower temperatures presumably incomplete reaction of the precursors caused a dramatically reduced conductivity of the films. The optimum conductivity for the ALD films is somewhat lower than that for PLD deposited AZO. The highest conductivity for boron doped ZnO films deposited by ALD is around $1300\,S\,cm^{-1}$ [44]. Nevertheless, owing to the excellent transparency of the material and the massive parallelism of the technique, ALD might be attractive to realize large area TCO coatings for various organic applications. For the preparation of a top electrode for transparent OLEDs the decrease in the conductivity towards lower deposition temperatures may become an issue. However, temperatures of around 150 °C are hardly tolerable for many organic compounds used in typical OLEDs.

For the preparation of an entirely In-free transparent OLED a bottom AZO electrode (cathode) prepared by ALD has been used. For the work function of AZO a value of 4.3 eV was determined by Kelvin Probe analysis. Thus, an electrochemically *n*-doped BPhen:Cs_2CO_3 layer was used to facilitate electron injection from the AZO cathode. For the top electrode the AZO was deposited by PLD at room temperature. To minimize particle impact, the energy density of the pulses from the KrF excimer laser used for the PLD

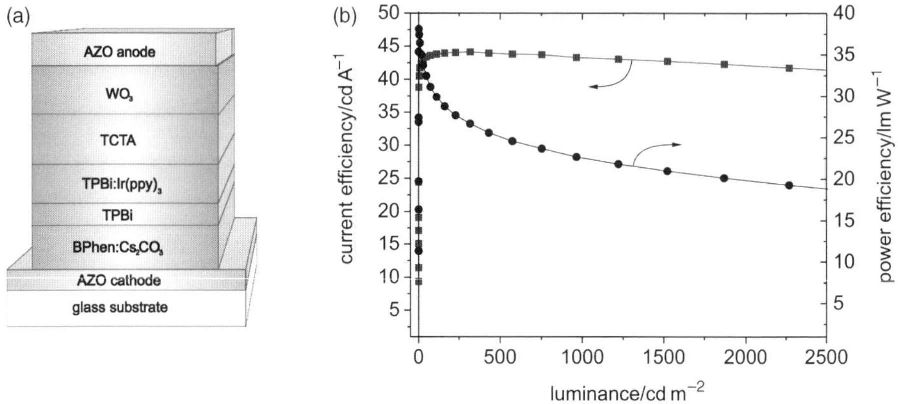

Figure 12.7 *(a) Layer sequence of the In-free transparent OLED and (b) efficiency vs. luminance characteristics for the devices*

process was tuned close to the ablation threshold for the AZO. Moreover, the concept of the WO_3 buffer layer has been employed (see above). The device set-up is displayed in Figure 12.7(a). Highly efficient In-free, transparent OLEDs (43 cd A^{-1}, 30 lm W^{-1} at 100 cd m^{-2}) have been prepared via this approach [Figure 12.7(b)].

The transmissivity of the In-free OLEDs is on the order of 80% throughout the visible spectral region, predominantly governed by the reflectance of the devices rather than any absorption due to the electrodes.

12.2.3 Stacked Transparent OLEDs

Recently, a very attractive means to increase the OLED current efficiency by the use of multiply stacked OLED structures has been demonstrated. The main concept relies on a series connection of *n* vertically stacked OLEDs with the result of an *n*-fold increased operating voltage but also an *n*-fold increased current efficiency. Ideally, in these structures one injected electron-hole pair creates *n* photons. For the practical realization of stacked OLEDs, two or more emission units (OLEDs) are connected by introducing thin metal interlayers [45] or by employing so-called charge generation layers (CGLs) [46–50]. Several different CGL architectures have been proposed, but in general they consisted either of doped organic/inorganic layer sequences [46–48] or doped organic/organic *p-n* junctions [49, 50]. These organic *p-n* heterostructures have been shown to behave in analogy to tunnel diodes known from inorganic semiconductor device physics [51]. The large electric field drop over the few nanometer thin CGL heterostructure leads to the tunnel transfer of an electron from the HOMO of the *p*-doped material to the LUMO of the *n*-doped layer. Thus an electron-hole pair is generated in the reverse biased CGL structure.

Particularly in the case where daytime readability or overall high brightness levels are required, stacked architectures are very attractive for transparent OLEDs. Figure 12.8(a) shows the device architecture of a stacked transparent OLED with three emissive units (EMUs). Here, the CGLs consist of the electrochemically *p*-/*n*-doped layers tetrafluorotetracyanoquinodimethane (F4-TCNQ) doped 4,4′,4″-tris(*N*-1-naphtyl-*N*-phenylamino)-triphenylamine (1-TNATA) and Li doped 1,3,5-tri(phenyl-2-benzimidazole)benzene

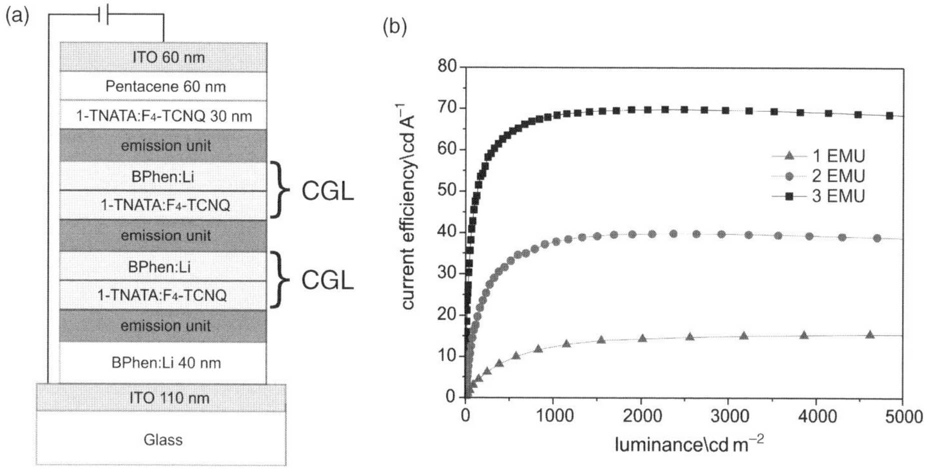

Figure 12.8 *(a) Layer sequence of a stacked transparent OLED with three light emitting units and (b) current efficiency vs. luminance characteristics for OLEDs with one, two and three EMUs. Reprinted with permission from [23]. Copyright (2007) SPIE*

(TPBi). In contrast to previous reports where the use of more than one EMU leads to a multiplication of the current efficiency by the number of EMUs, here, the current efficiency increases superlinearly with each additional EMU. For instance, at a luminance of 1000 cd m^{-2} a single EMU reaches an efficiency of 12 cd A^{-1}. For two EMUs values of 37.5 cd A^{-1} are found and with three EMUs the current efficiency reaches values of 70 cd A^{-1}. The power efficiencies reach values of 11 lm W^{-1} at a luminance of 1000 cd m^{-2}. The reason for the superlinear increase is due to some sputter damage to the topmost device due to a less than optimized sputter protection layer that consists of 60 nm pentacene in this case [52]. This circumstance, however, affects the device with only one EMU more severely than those with two and three EMUs.

The concept of stacking is also very attractive to attain mixed color OLEDs with varied spectral properties of the individual EMUs. Transparent, white light emitting devices may become an attractive light source for the future integration in architectural windows.

12.2.4 Light Extraction

Directly linked to the design of transparent OLEDs is the desire to control the ratio of light emitted through the top and bottom electrode, respectively. For applications of transparent OLEDs as lighting elements or in heads-up displays it would be required to direct as much light as possible to the user and to avoid light emission in the opposite direction. For other applications, e.g. signage, etc., a balanced emission through both electrodes may be required. Regardless of the desired light extraction ratio, the OLEDs should at the same time be as transparent as possible. Hence, the sequence and the dielectric properties of the layer set-up constituting a transparent OLED deserves particular attention in terms of light extraction. For an experimental simulation of the light extraction through the top and bottom electrode upon modification of the OLED layer structure a layer sequence as shown in Figure 12.9(a) has been studied. Specifically, the hole transport layer (HTL) of a transparent

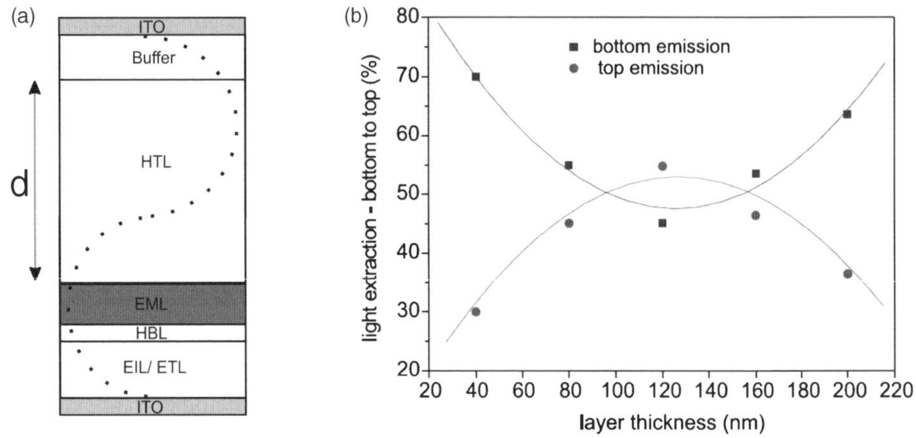

Figure 12.9 *(a) Transparent inverted OLED structure with varied thickness of the hole transport layer (HTL) and (b) extracted light through the top and bottom electrode vs. HTL thickness. Reprinted with permission from [23]. Copyright (2007) SPIE*

inverted OLED structure has been varied in thickness. For a HTL thickness between 40 nm and 200 nm about 65–70% of the light emission is extracted via the bottom electrode. For a HTL thickness between 80 nm and 160 nm the emission through the top and bottom electrode is almost equal [Figure 12.9 (b)]. This example demonstrates the design considerations relevant for effective transparent OLEDs fitted to the needs of the respective application. Aside from a variation of the layer thickness of charge transport layers in the OLED, one may also use a dielectric capping layer on top of the upper electrode to tune the light extraction ratio. This capping layer may at the same time function as an encapsulation to protect the OLED against water and oxygen.

12.3 Transparent Thin Film Transistors

Transparent OLED displays driven in the passive matrix mode are considered as an approach for small-sized, low information content displays with moderate pixel counts. To meet the demand for larger area, high resolution OLED displays with an active matrix addressing scheme is mandatory [53]. Conventional α-Si:H or poly-Si TFT backplanes are not suitable as drivers for transparent displays because they are opaque in the visible part of the spectrum. Locating pixel and driving electronics next to each other would severely compromise the displays filling factor and its overall transparency.

In the past few years, the research on transparent thin film transistors (TTFTs) with channels made from oxide semiconductors has seen some significant progress [54–60]. Devices based on single crystalline materials [59] show high channel mobilities of $80\,cm^2\,V^{-1}\,s^{-1}$ but require either high deposition temperatures (700 °C) or some high temperature post growth treatment (1400 °C). This not only imposes severe restrictions on the choice of substrates but also compromises the homogeneity on a larger area and hampers the overall suitability for prospective cost-effective fabrication. On the contrary, devices based on nanocrystalline ZnO or amorphous In-Ga-Zn-O may be prepared at room temperature, even on flexible plastic substrates [56, 58]. Chiang *et al.* have used the amorphous semiconductor zinc tin oxide (ZTO) deposited by RF magnetron sputtering as channel material and reported field effect mobilities between $5\,cm^2\,V^{-1}\,s^{-1}$ and $50\,cm^2\,V^{-1}\,s^{-1}$, depending on the post deposition annealing temperature between 300 °C and 600 °C [57]. Despite the material being amorphous relatively high electron mobilities have been attributed to the electronic structure $(n-1)d^{10}\,ns^0$ $(n \geq 4)$ of the incorporated heavy metal cations [56, 61]. As evidenced by the work of Minami *et al.*, oxygen deficient deposition conditions cause a degradation of the films and result in reduced carrier mobilities [62]. Post deposition annealing in oxygen/air atmosphere at elevated temperatures is thus commonly applied to improve film quality in terms of reduced defect density and reduce background carrier concentration [57].

12.3.1 Channel Material for Transparent TFTs

A typical device structure of a bottom-gate TTFT on glass substrate is shown in Figure 12.10(a). The gate electrode can be made from TCOs like ITO or AZO. The gate dielectric in this example is a 220 nm thick superlattice of Al_2O_3 and TiO_2 (ATO) with a capacity per unit area of $60\,nF\,cm^{-2}$. The dielectric has been deposited by ALD. As this so-called nanolaminate is extremely difficult to pattern, neat Al_2O_3 dielectric layers prepared by ALD might be favored for the realization of transparent circuits.

The channel semiconductor is composed of ZTO which can be deposited by sputtering [57] or by PLD in the present case. In general, the targets for the PLD process have been prepared by cold pressing and subsequent sintering a mixture of ZnO and SnO_2 powders with the desired molar ratio of Zn and Sn in an oxygen atmosphere at 1100 °C for 12 h. A KrF excimer laser ($\lambda = 248$ nm) has been used as ablation light source with an energy density in the pulse of ~ 1–$2\,J\,cm^{-2}$. As the stoichiometry in the pellet and the deposited film may differ, the composition of the ZTO layer is routinely checked by energy

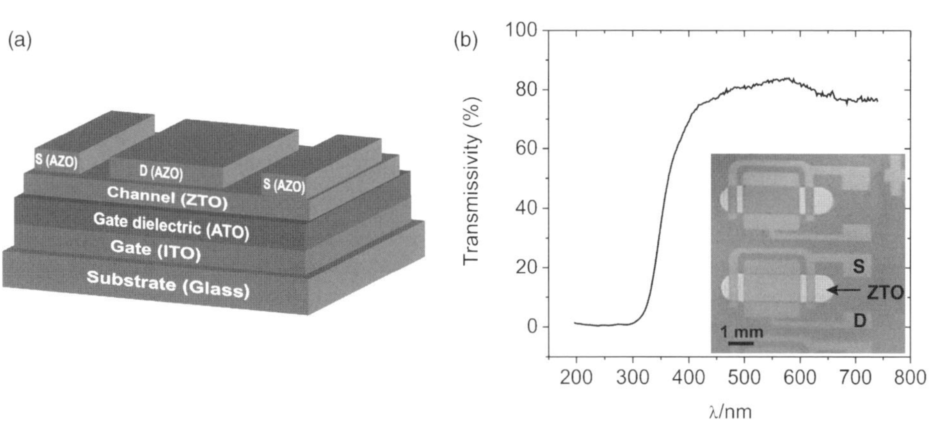

Figure 12.10 (a) Device structure of a ZTO based TTFT and (b) spectral transmissivity of the TTFT. Inset: Photograph of the devices under tuned illumination conditions to make the transistor structure visible. Reprinted with permission from [14]. Copyright (2007) Wiley-VCH

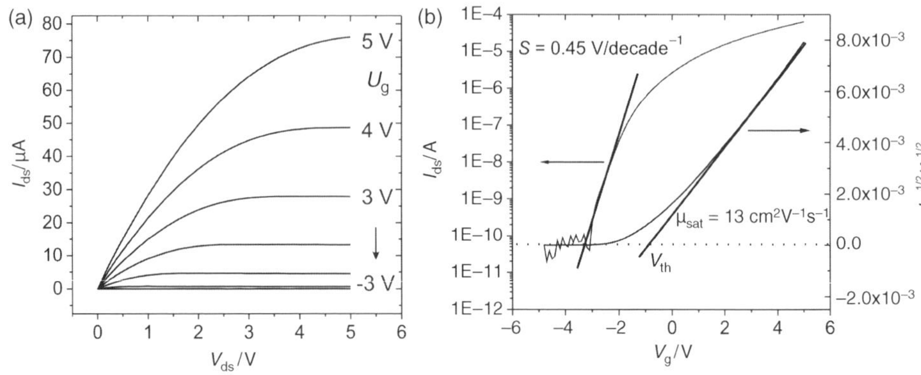

Figure 12.11 *(a) Output characteristics (drain-source current I_{ds} vs. drain-source voltage V_{ds}) of a transparent TFT and (b) TTFT transfer characteristics. In this case the [Zn]/[Sn] ratio is approximately 1:1 [25]*

dispersive X-ray spectroscopy (EDX). Again, AZO or ITO may be used as transparent drain and source electrodes. The resulting TTFTs show a transmissivity of more than 80% in the visible spectral region [Figure 12.10(b)]. Specifically, we use an oxygen-plasma-assisted PLD (PA-PLD) process for the preparation of the active ZTO layer. This deposition technique relies on the additional supply of radical oxygen species via an RF oxygen plasma source. This improves the overall material quality and prevents oxygen deficiency (and thus unintentional doping) in the films. Unlike the reports of sputter-deposited films [57], no post-deposition annealing in oxygen or air is necessary to improve the electronic properties.

For exemplary devices with a [Zn]:[Sn] ratio of 1:1 (prepared at 350 °C) the output characteristics and the transfer characteristics are displayed in Figure 12.11. The drain-source current I_{ds} shows a clear pinch-off and saturation. I_{ds} reaches values up to 80 μA at moderate values of V_g and V_{ds} [Figure 12.11(a)]. The I_{on}/I_{off} ratio is on the order of 10^6 and the saturated field effect mobility (μ_{sat}) is around 13 $cm^2\ V^{-1}\ s^{-1}$, approximately an order of magnitude higher than that of typical a-Si TFTs. In general, the threshold voltage (V_{th}) is between −1 V and 1 V for this kind of TTFT. The subthreshold slope is around $S = 0.45$ V decade^{-1}. No hysteresis in the transfer characteristics can be observed in these devices, indicating a low trap density at the semiconductor/insulator interface. The channel width to length ratio was $(W/L) = 1\ mm/200\ \mu m = 5$ in this particular example.

Transmission electron microscopy can be used to verify the amorphous nature of the ZTO channel layer [Figure 12.12(a) and (b)]. Amorphous channel TFTs have been shown to allow an improved homogeneity on larger area substrates as opposed to their polycrystalline or crystalline counterparts [63]. The [Zn]:[Sn] ratio is found to strongly affect the TTFT mobility in the ZTO system, whereas the influence of the deposition temperature is not so critical in this case. (It will affect other properties, however, e.g. stability and light sensitivity.) Figure 12.12(c) shows the dependence of μ_{sat} on the composition of the films. While there is a broad region of compositions around [Zn]:([Sn] + [Zn]) = 50% where μ_{sat} is ∼10 $cm^2\ V^{-1}\ s^{-1}$ the primary oxides yield TTFTs with dramatically reduced mobilities.

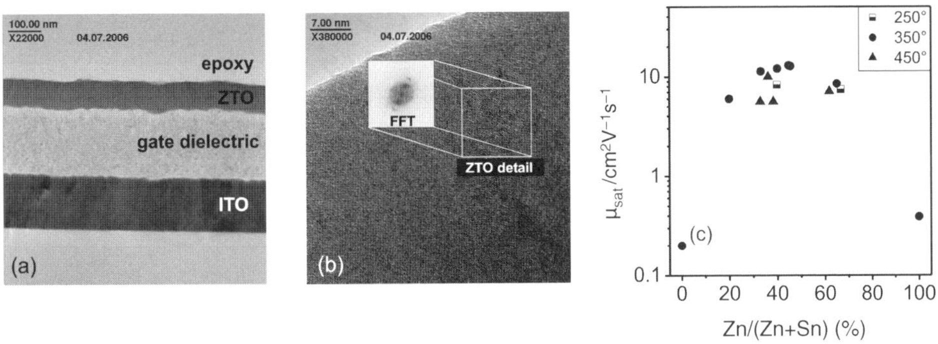

Figure 12.12 (a) Transmission electron microscopy image of the TTFT layer stack, (b) magnified view of the ZTO layer, and fast Fourier transformation (FFT) of a selected region and (c) saturation mobility of the TFTs vs. composition at various deposition temperatures. Reprinted with permission from [67]. Copyright (2007) American Institute of Physics

12.3.2 Stability versus Bias Stress

Unlike in LCDs, the TFTs in AMOLED displays are not only used as switches but also as analog current drivers for the diodes. In this case, long term stability plays a critical role for analog devices because a shift of threshold voltage (V_{th}) concomitantly leads to a change of the individual pixel brightness. There have been many investigations of bias-induced instabilities in α-Si:H [64–66], where a shift of V_{th} of the order of several volts after a few hours of bias stress is quite common. Only recently the question of operational stability under stress has been tackled for some oxide based TFT structures like ZTO [67, 68] and α-IGZO [69, 70].

Figure 12.13 shows the effect of bias stress on two ZTO TTFTs with a Zn content of 45.5% [Figure 12.13(a)] and 62% [Figure 12.13(b)]. The gate bias has been 10 V throughout 1000 min with regular measurements of the transfer characteristics. Interestingly, one

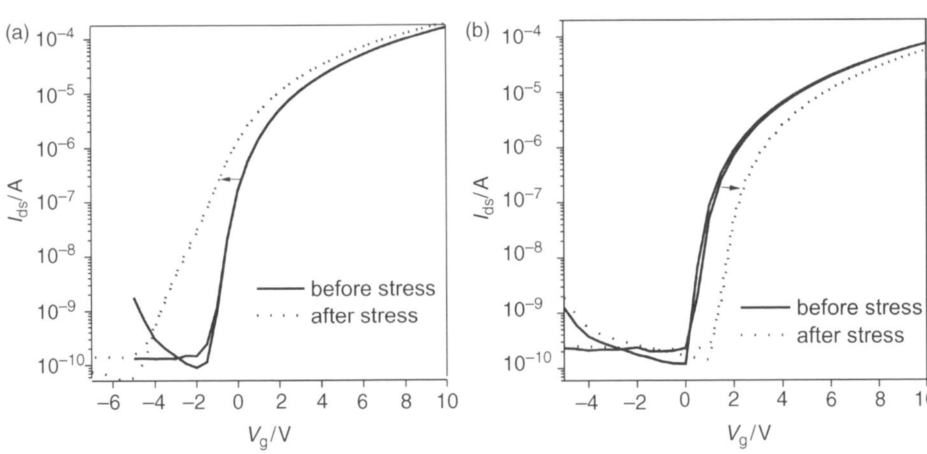

Figure 12.13 Results of 1000 min of bias stress. Transfer characteristics for ZTO-TFTs with (a) negative (Zn content 45.5%) and (b) positive threshold voltage shift (Zn content 62%). Reprinted with permission from [67]. Copyright (2007) American Institute of Physics

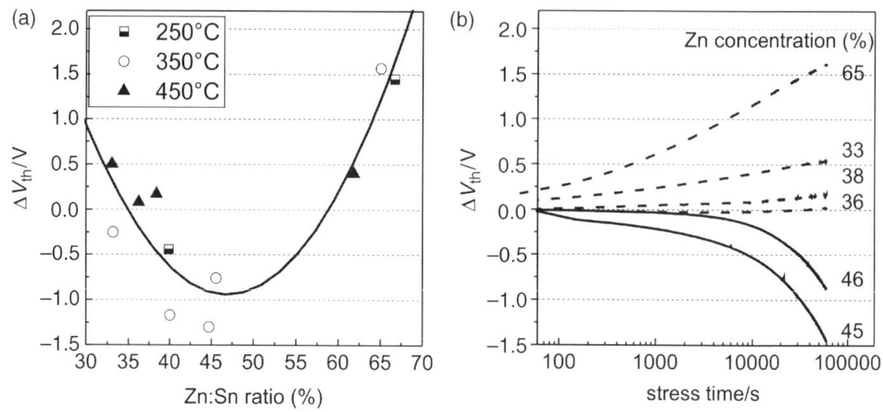

Figure 12.14 *(a) Composition and deposition temperature dependent threshold voltage shifts (ΔV_{th}) after 60 000 s of gate bias stress ($V_g = 10$ V) (the line is a guide to the eye) and (b) the corresponding time dependencies of V_{th} for selected samples. Reprinted with permission from [67]. Copyright (2007) American Institute of Physics*

observes positive as well as negative shifts of V_{th} upon stress. Substantially in agreement with silicon based TFTs, in the case of positive V_{th} no change in subthreshold slope can be seen [Figure 12.13(b)]. Defects in the dielectric serving as charge trapping centers have been suspected to be responsible for this kind of response. In contrast, samples with negative V_{th} (a behavior of V_{th} unknown for silicon TFTs) show a gradual decrease from the beginning and a significant change of subthreshold slope [Figure 12.13(a)]. Here, the origin for this behavior has been attributed to the dielectric/semiconductor interface and the semiconductor bulk [67].

Obviously the composition of the channel plays an important role in the question of which kind of V_{th} shift is obtained upon stress. A more complete study of ZTO TFTs with varied Zn content and varied deposition temperatures reveals the results displayed in Figure 12.14. ΔV_{th} is found between −1.4 V and 1.5 V for all samples in this study [67]. Devices with Zn contents below 36% and above 65% tend to exhibit positive shifts of V_{th} similar to a-Si:H whereas transistors with nearly equal fractions of Zn and Sn show negative shifts of V_{th}. Extremely stable devices with a minimum shift of 30 mV could be realized for an atom ratio of 36:64 ([Zn]:[Sn]). A clear impact of the substrate temperature during deposition is not derived from the data [Figure 12.14(a)]. Figure 12.14(b) shows the temporal dependence of the shift in V_{th}. The different dynamics for positive and negative shifts allows the origin of the shifts to be attributed to charge trapping in the dielectric and the bulk or dielectric/semiconductor interface, respectively [67].

Devices with a [Zn]:[Sn] ratio of 36:64 have been subject to electrical stress via a gate-source bias of 10 V and a drain-source bias of 10 V leading to a drain-source current of 188 μA which is about three orders of magnitude higher than that required to achieve a typical display brightness of 250 cd m^{-2} in a 100 × 300 μm^2 sized OLED pixel with a current efficiency of 40 cd A^{-1} [a quite reasonable value even for transparent OLEDs (see above)]. As can be seen in Figure 12.15(a) even under these very demanding stress conditions the threshold voltage of the TTFT shifts by only 320 mV within 1000 h. The regular oscillations that appear in the graph have been attributed to changes in the ambient

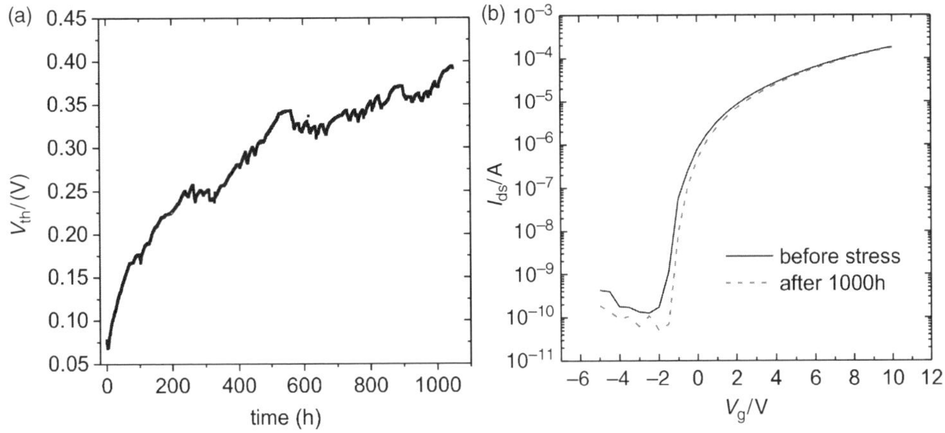

Figure 12.15 *(a) V_{th} vs. stressing time for a ZTO TTFT with a [Zn]:[Sn] ratio of 36:64 and (b) transfer characteristics of a fresh device and after 1000 h of stressing. Reprinted with permission from [68]. Copyright (2007) Wiley-VCH*

temperature in the laboratory (day/night, weekends) are reflected in the tiny ripples with the corresponding periodicity. A more detailed study of the temperature dependence of V_{th} yielded a shift of –300 mV when the sample temperature has been increased from 20 to 80 °C. In Figure 12.15(b) the transfer characteristics before and after 1000 h of stress appear almost identical. The average saturation mobility in these devices is around 12.5 $cm^2\,V^{-1}\,s^{-1}$. Remarkably, the mobility remains almost unaltered within a range of ±1% even after 1000 h of continuous stressing. Very recently, for α-IGZO a relative change in both V_{th} and μ_{sat} of less than 1.5% during 500 h of stress with V_{ds} has been reported [69].

12.3.3 Sensitivity to (Visible) Light

Although we are dealing with transparent electronics and the semiconductors like ZnO, ZTO, etc., are all wide gap materials a significant contribution of defect states to the absorption of the semiconductor channel is expected. This holds particularly for amorphous materials with a rich variety of sub band gap states [71, 72]. For display applications, especially, the influence of irradiation with visible light is important. When highly transparent TFTs are required, it will not be acceptable to use shielding layers in order to reduce the influence of light. While (ultra)violet radiation with $\lambda = 430$ nm or below may be blocked using UV absorbing coatings which are transparent in the visible, the TFTs themselves need to be highly insensitive towards visible light.

In a study on Zn-In-O TFTs, Barquinha and coworkers have seen a tremendous shift in the threshold voltage of the devices under illumination. At an irradiance of 1 mW cm^{-2} of blue light with $\lambda = 450$ nm V_{th} shifted by approximately –20 V compared with V_{th} in darkness [73]. To elucidate the principal impact of light irradiation on the key parameters of ZTO TTFTs, the devices have been exposed to the light of an LED with $\lambda = 425$ nm and $I = 250$ μW cm^{-2} in air. A decrease of the threshold voltage V_{th} and the saturation field effect mobility μ_{sat} has been the general result. Simultaneously, the off-current is increased. Figure 12.16 illustrates these effects for devices deposited at 450 °C. V_{th} and μ_{sat} begin to

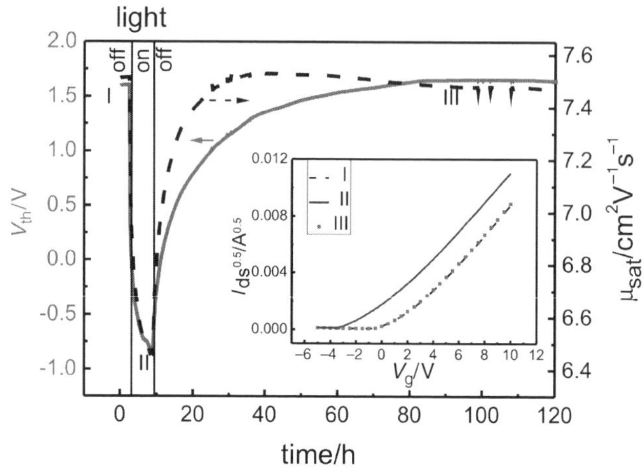

Figure 12.16 Effect of violet light irradiation ($\lambda = 425$ nm and $I = 250\ \mu$W cm^{-2}) on V_{th} and μ_{sat} of ZTO TFTs. Inset: transfer characteristics with and without irradiation. Reprinted with permission from [74]. Copyright (2007) American Institute of Physics

decrease as the light is turned on [74]. After several hours a steady state is reached. The absolute shifts of V_{th} and μ_{sat} between the steady state in darkness and after reaching a stationary value upon illumination are called the total change of these parameters. As soon as the light is turned off V_{th} and μ_{sat} start to increase and recover to their initial values. All light-induced changes of the transistor characteristics were found to be totally reversible. The time constant for the recovery of V_{th} was on the order of 20 h.

In the dark, a negligible hysteresis was found in the $I_{ds}-V_g$ characteristics of less than 50 mV for all investigated samples. Even under strong illumination with violet light ($\lambda = 425$ nm and $I = 1$ mW cm^{-2}) the hysteresis at $V_g = V_{th}$ has been less than 200 mV. The total change of μ_{sat} was found to strongly depend on the substrate temperature during the preparation of the ZTO channel. While TFTs deposited at 250 °C showed a decrease of μ_{sat} by up to 80%, in TFTs processed at temperatures of 350 °C or above the change in μ_{sat} upon illumination has been always less than 20% in these studies.

An overview of the total change in V_{th} versus the wavelength of the illuminating light is given in Figure 12.17(a). Samples with varied processing temperatures and [Zn]:[Sn] ratios have been studied. As light sources, LEDs have been used with spectra peaked at 628, 525 and 470 nm at an intensity of 1 mW cm^{-2}, which was equivalent to brightness levels of approximately 23 000, 50 000 and 6000 cd m^{-2}, respectively. The emission spectra represent the red, green, and blue color coordinates according to the National Television Systems Committee standard. Blue light was found to have the strongest impact on the TFT performance. The processing temperature was identified as the dominating parameter for the light sensitivity of the resulting devices. Whereas TFTs with a channel material deposited at 250 °C appeared to be very sensitive to visible light, increased deposition temperatures led to significantly smaller changes in V_{th}. On the contrary, the zinc content of the semiconductor only had a minor effect on the light sensitivity of the TFTs. The dependence of the total threshold voltage shift on the intensity of the blue light ($\lambda = 470$ nm) is shown in Figure 12.17(b). The TFTs processed with

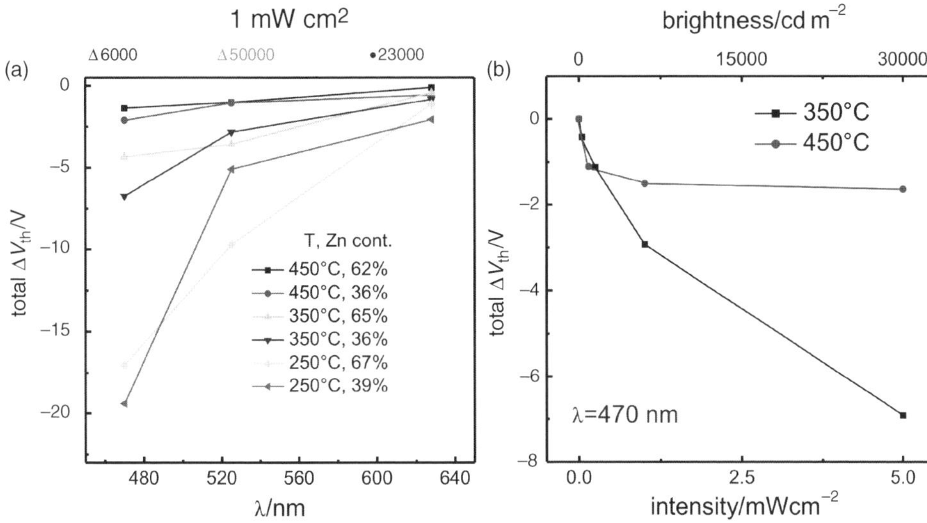

Figure 12.17 *(a) Influence of the processing temperature and the Zn to Sn composition ratio on the total ΔV_{th} at light irradiation of an intensity of 1 mW cm^{-2} and a wavelength of 628, 525, and 470 nm and (b) total change of threshold voltage vs. light intensity ($\lambda = 470$ nm) for TFTs of different processing temperatures. Reprinted with permission from [74]. Copyright (2007) American Institute of Physics*

higher substrate temperatures, which have been found to be less sensitive to light, show a very pronounced saturation behavior in V_{th} at intensities around 1 mW cm^{-2}. This has been interpreted in terms of a lower density of defects in the material owing to the elevated deposition temperature [74].

Trapping of photogenerated charges in the dielectric or the semiconductor/dielectric interface as the origin for the shift in the threshold voltage and the resulting time scale for the recovery have been ruled out by experiments on simple resistor structures [74]. It is well known that persistent photoconductivity in ZnO and related compounds is associated with oxygen chemisorbed on the free surface of the semiconductor. The adsorbed oxygen molecule can capture an electron and therefore a negative surface charge may lead to a depletion of the semiconductor channel. If free carriers are generated (e.g. by light), the negatively charged oxygen may be neutralized and leave the surface. As a result, the channel conductivity increases. The time constant for the recovery after the light is switched off would then be governed by the readsorption of oxygen. In preliminary studies, where a TTFT has been illuminated by violet light (at 425 nm) in vacuum, the return of V_{th} to its initial value before illumination has been studied (Figure 12.18). Here, the light has been switched off at time, $t = 0$ and the TFT characteristics have been monitored in ambient oxygen at various partial pressures. It turned out, that the O_2 partial pressure strongly governs the dynamics of the restoring of V_{th}. In vacuum (10^{-8} mbar, limit of the chamber) the return of V_{th} is very slow – even after 60 h V_{th} has only recovered by 0.5 V. With increased oxygen background pressure the dynamics of the recovery is considerably accelerated. Thus the adsorption of oxygen and the subsequent partial depletion of the channel appears to be a key process for oxide TFTs. With the surface being of paramount importance for oxide

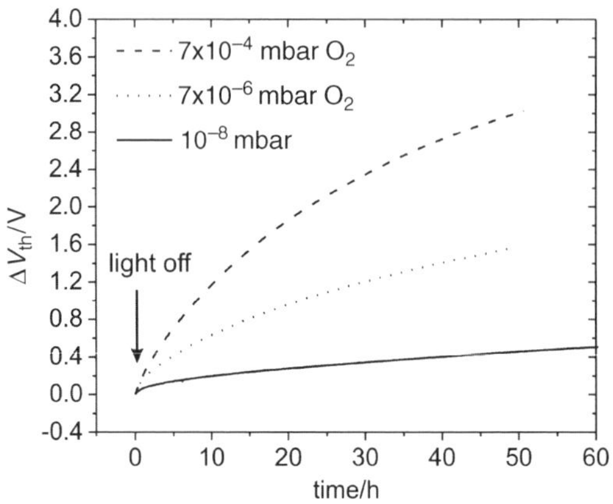

Figure 12.18 *Return of the threshold voltage to its initial value in ambient oxygen at various partial pressures after exposure to violet light (425 nm) in vacuum*

TFTs, questions like surface passivation and encapsulation will have to be addressed when aiming at integration of these devices into OLED displays.

12.4 Transparent Active Matrix OLED Displays

As stated in the introduction, transparent OLED displays using oxide electronics as drivers have recently moved into focus [12–14]. The first report about the use of transparent oxide TFTs as drivers of transparent OLEDs was published in 2006 [14].

12.4.1 Active OLED Pixels

Figure 12.19(a) shows the integration of a ZTO TFT and a transparent inverted OLED. The circuit is displayed in Figure 12.19(b) along with the output characteristics (luminance vs. V_{op}) for various gate bias levels between –5 V and 5 V. The luminance vs. V_g characteristics [Figure 12.19(c)] clearly demonstrate the control of the brightness via the gate of the driving TFT. The average transmissivity of the entire set-up is more than 70% in the visible part of the spectrum. A demonstration of the pixel in operation ($V_g = 4$ V, 500 cd m^{-2}) is shown in Figure 12.19(d). In the off-state the pixels are largely invisible and the underlying seal of the Technical University of Braunschweig can be clearly seen through the sample without interference, as necessary for a transparent display.

The results presented in Figure 12.19 provide the initial building block on the way to transparent driving electronics for OLEDs and ultimately a transparent AMOLED display. Until now several reports have addressed the AMOLED displays driven by transparent amorphous oxide semiconductor (TAOS) TFTs. The prominent TAOS α-IGZO has very recently been used in active OLED pixels. However, the TFTs in this report were not exposed to the illumination of the OLEDs [13]. Even though α-IGZO is a highly promising material for transparent TFTs, knowledge about its sensitivity to light is very limited. Moreover, no transparent IGZO circuitry for *transparent* AMOLED displays has been demonstrated yet.

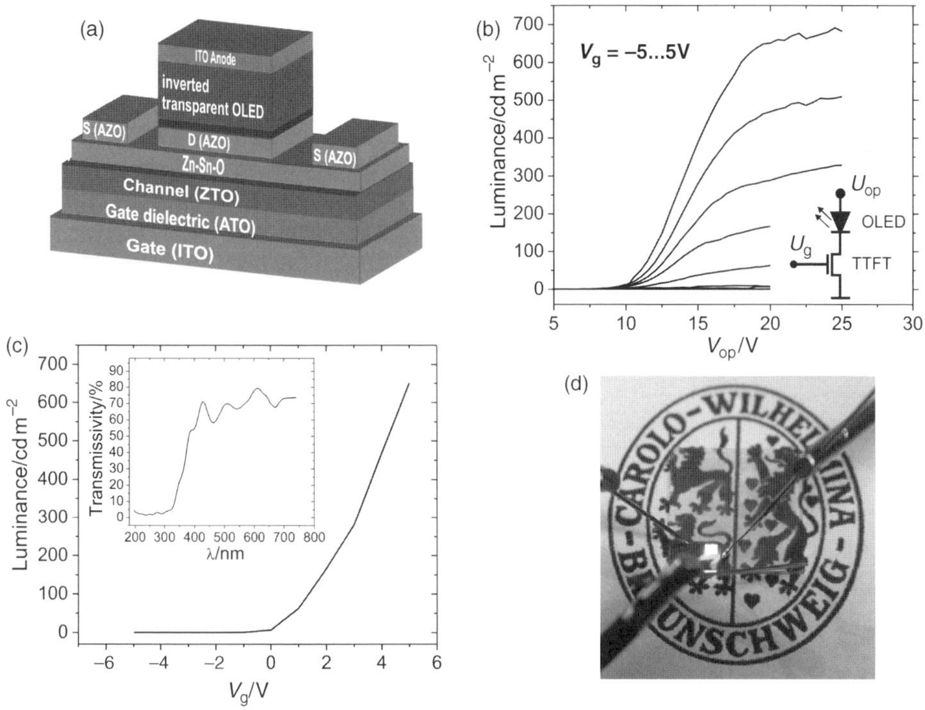

Figure 12.19 *(a) Active pixel structure: transparent inverted OLED on top of a TTFT structure. (b) Luminance when viewed from one side vs. V_{op} (V_g was varied between –5 V and 5 V in 1 V steps). The inset shows a circuit scheme of the active pixel. (c) Transfer characteristics (luminance vs. V_g) at V_{op} = 20 V. The inset shows the spectral transmissivity of the entire active pixel stack. (d) Photograph of a substrate with 10 active pixels with one pixel in the 'on' state (V_g = 4 V). Reprinted with permission from [14]. Copyright (2007) Wiley-VCH*

12.4.2 Simple Transparent AMOLED Driver Circuits

For a real AMOLED application, at least a simple 2T1C (2 transistors and 1 capacitor) set-up is required [75]. An example of such a circuit diagram for an AMOLED pixel driver and a microscopy image of a corresponding transparent circuit based on ZTO TFTs is shown in Figure 12.20. In this example, the ZTO TFTs T_1 (switching or digital TFT) and T_2 (driving or analog TFT) have equal dimensions ($W/L = 10$); the capacitance, $C = 12.3$ pF. Separate TFTs were placed on the same substrates next to the pixel drivers. The threshold voltage and saturation field effect mobility were determined to be –2 V and 5.0 cm^2 V^{-1} s^{-1}, respectively. The size of the entire pixel is 180 × 240 μm^2, which is sufficiently small for the realization of high resolution displays. However, this kind of high resolution display is very demanding in terms of driver performance.

The first challenge in this sense is the switching speed. For a refresh rate of 100 Hz, all rows have to be addressed within 10 ms, and thus the switching time for the pixels is around 10 μs, if a number of 1000 rows is assumed. (Full HD equals a number of 1080 rows.) With the *select* pulse a row is selected (T_1 is switched from the depleted to enhanced state). That is

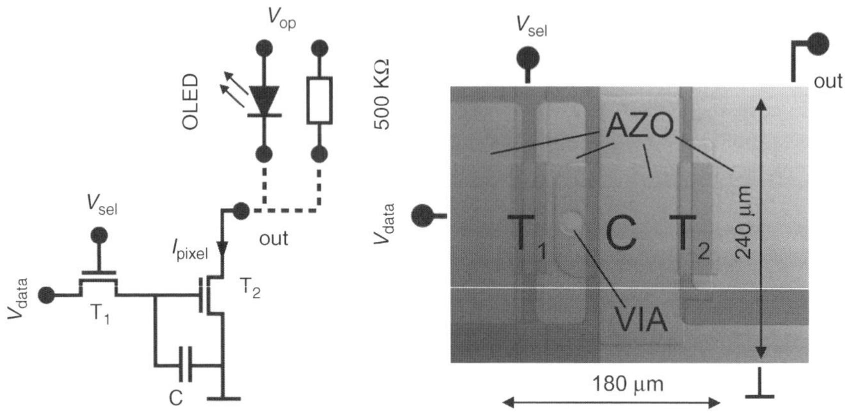

Figure 12.20 *Circuit diagram of the active matrix OLED pixel driver and microscopic image of the realized transparent device*

why charging of C according to the *data; on* level ($V_{data;on}$) coincides with the *select* pulse. Figure 12.21(a) illustrates this behavior. The *data; on* level defines the voltage at C and hence the gate voltage of T_2. Thus, a change in the charge of C results in a change of the pixel current. The charging saturates for a longer duration of the *select* pulse. Figure 12.21(a) shows a pixel current of 1.83 μA after a charging time of 40 μs with $V_{data;on} = 4$ V. A decrease of the charge time to 5 μs leads to a reduction of the pixel current by less than 10% to 1.69 μA. It can, therefore, be stated that the charging of C to the desired voltage is nearly completed after 5 μs. Thus, for the pixel addressing the *select* and *data* pulses can be as short as 5 μs. Thus, the pixel addressing can be by far *fast* enough for the realization of high resolutions and fast refresh rates.

Another important point is cross-talk between neighboring rows. Thus, the influence of *select* pulses that do not coincide with the *data; on signal* has to be investigated. In Figure 12.21(b), the pixel current is shown for input pulses with a width of 10 μs. The impact of the temporal shift between the pulses is given. As expected, without relative delay the maximum pixel current is observed. Shifting the *select* pulse by –5 μs reduces the charging time by 50% and so slightly decreases the current. In case of a positive shift of 5 μs no output signal is seen. Here, a charging of C for 5 μs is followed by a discharging to the *data; off* level. For input pulses without temporal overlap no pixel current can be measured. Translated to the AMOLED display application this means that the desired time period of 10 μs for row addressing can be realized with input pulse widths of 10 μs and without influence of cross talk. As these widths can be easily reduced to 5 μs even higher resolutions and refresh rates are within reach.

After C is charged according to the *data; on* levels and consequently the pixel is set to a certain brightness level, this information should be stored until the next addressing. At a refresh rate of 100 Hz that corresponds to a hold time of 10 ms. During this time, C is discharged through leakage currents (in C, as off-current of T_1 and gate leakage of T_1 and T_2). At *data; on* levels of 5 V (2 μA, 960 cd m^{-2}) and below, the maximum decrease of the output current is about 1.4%, which corresponds to an overall leakage current of 170 pA. At higher levels the drop is around 3.8%. Owing to the small loss of charge, the exponential

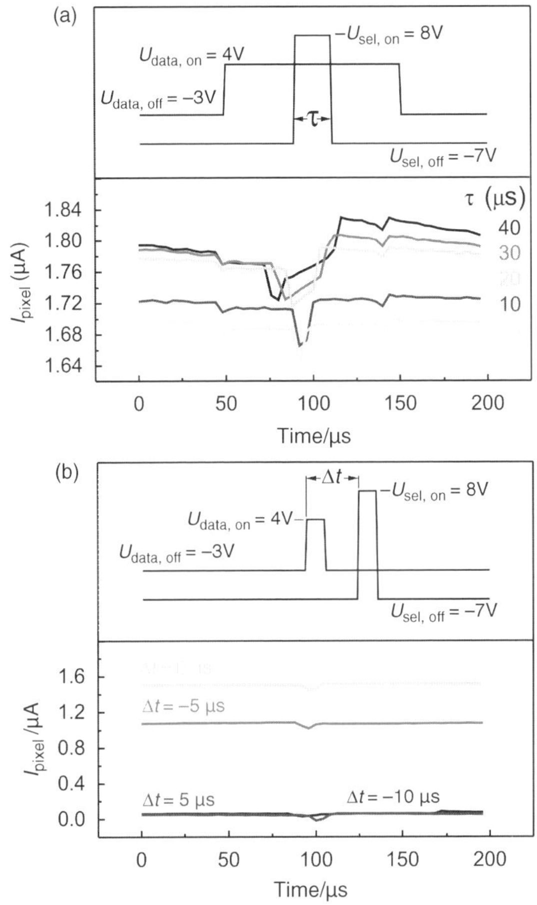

Figure 12.21 *(a) Pixel current for different select pulse widths. After 5 μs the charging of C is nearly completed and (b) influence of temporal shift between the data and select pulses*

discharge of C can be approximated by a linear function. In addition, the current efficiency of the OLED may be considered constant in this range. Hence, the maximum decrease of the time dependent brightness of 3.8% is translated to a decrease of the average brightness of 1.9%. As the human eye is unable to distinguish the small brightness fluctuations within the hold time of 10 ms average brightness levels are perceived. Consequently, not the decrease in the start values of the pixel brightness, but rather the accuracy of the average values will limit the number of possible gray levels in the resulting display.

12.5 Conclusions

The marriage of two rapidly evolving areas of research, OLEDs and transparent electronics, enables the realization of novel transparent OLED displays. This appealing class of see-through devices will have great impact on the human–machine interaction in the near future. Benefits not only for consumer applications but also for medicine and defense purposes will become a reality. The enabling technology for high-end transparent AMOLED displays is directly linked to transparent driver electronics based on transparent TFTs. Transparent TFTs have matured significantly both in terms of the fundamental understanding of material properties of the oxide semiconductors and device technology. Ultra-high stability against bias stress and the ability to realize devices with a drastically reduced sensitivity to visible light are clear advantages of oxide based TFTs. With many applications for see-through displays at hand, the transfer of this technology to real products is now vigorously pursued by industry.

Acknowledgements

The dedicated work of my collaborators and colleagues at the Institute of High Frequency Technology (TU Braunschweig) deserves particular acknowledgement. Moreover, the fruitful cooperation with our partners from academia and industry is highly appreciated. Special thanks go to the German Research Foundation (DFG) and the German Federal Ministry of Education and Research (BMBF) for the generous funding of our research.

References

[1] C. W. Tang and S. A. vanSlyke, Organic electroluminescent diodes, *Appl. Phys. Lett.*, **51**, 913 (1987).

[2] S. Wesarg, E. A. Firle, B. Schwald, H. Seibert, P. Zogal, and S. Roeddige, Accuracy of needle implantation in brachytherapy using a medical AR system - a phantom study, *Proc. SPIE Med. Imag.*, **5367**, 341 (2004).

[3] M. Antikainen, J. Haaranen, J. Honkala, M. Lahonen, V.-M. Liias, A. Pakkala, T. Pitkänen, E. Soininen, and R. Törnqvist, Transparent emissive thin-film electroluminescent display, *SID Int. Symp. Dig. Tech. Pap.*, **31**, 885 (2000).

[4] G. He, M. Pfeiffer, K. Leo, M. Hofmann, J. Birnstock, R. Pudzich, and J. Salbeck, High-efficiency and low-voltage p-i-n electrophosphorescent organic light-emitting diodes with double-emission layers, *Appl. Phys. Lett.*, **85**, 3911 (2004).

[5] G. Gu, V. Buloviç, and S. R. Forrest, Transparent organic light emitting devices, *Appl. Phys. Lett.*, **68**, 2606 (1996).

[6] J. K. Mahon, T. X. Zhou, P. E. Bourrows, S. R. Forrest, M. E. Thompson, Small-molecule organic light emitting devices in flat panel display applications, *Proc. SPIE*, **3279**, 87 (1998).

[7] M. J. Powell, S. C. Deane, W. I. Milne, Bias-stress-induced creation and removal of dangling-bond states in amorphous silicon thin-film transistors, *Appl. Phys. Lett.*, **60**, 207 (1992).

[8] Y. He, R. Hattori, J. Kanicki, Improved a-Si:H TFT circuits for active-matrix organic light emitting displays, *IEEE Trans. Electron. Devices*, **48**, 1322 (2001).

[9] A. Kumar, A. Nathan, G. E. Jabbour, Does TFT mobility impact pixel size in AMOLED back-planes?, *IEEE Trans. Electron. Devices*, **52**, 2386 (2005).

[10] K. H. Lee, S.-Y. Ryu, J. H. Kwon, S. W. Kim, and H. K. Chung, 2.2 inch QCIF full color transparent AMOLED display, *SID Int. Symp. Dig. Tech. Pap.*, **34**, 104 (2003).

[11] Y.-J. Tung, R. Hewitt, A. Chwang, M. Hack, J. Brown, K.-M. Kim, D. S. Kim, J. H. Hur, and J. Jang, A 200-dpi transparent a-Si TFT active-matrix phosphorescent OLED display, *SID Int. Symp. Dig. Tech. Pap.*, **36**, 1546 (2003).

[12] S. K. Park, C. Hwang, J. Lee, S. M. Chung, Y. S. Yang, L. Do, and H. Y. Chu, Transparent ZnO thin film transistor array for the application of transparent AM-OLED display, *SID Int. Symp. Dig. Tech. Pap.*, **37**, 25 (2006).
[13] R. Hayashi, M. Ofuji, N. Kaji, K. Takahashi, K. Abe, H. Yabuta, M. Sano, H. Kumomi, K. Nomura, T. Kamiya, M. Hirano, and H. Hosono, Circuits using uniform TFTs based on amorphous In-Ga-Zn-O, *J. SID*, **15**, 915 (2007).
[14] P. Görrn, M. Sander, J. Meyer, M. Kröger, E. Becker, H.-H. Johannes, W. Kowalsky, and T. Riedl, Towards see-through displays: fully transparent thin-film transistors driving transparent organic light-emitting diodes, *Adv. Mater.*, **18**, 738 (2006).
[15] G. Gu, P. E. Burrows, S. Venkatesh, S. R. Forrest, and M. E. Thompson, Vacuum-deposited, nonpolymeric flexible organic light-emitting devices, *Opt. Lett.*, **22**, 172 (1997).
[16] P. E. Burrows, G. Gu, S. R. Forrest, E. Vicenzi, and T. Zhou, Semitransparent cathodes for organic light emitting diodes, *J. Appl. Phys.*, **87**, 3080 (2000).
[17] X. Zhou, M. Pfeiffer, J. S. Huang, J. Blochwitz-Nimoth, D. S. Qin, A. Werner, J. Drechsel, B. Maennig, and K. Leo, Low-voltage inverted transparent vacuum deposited organic light-emitting diodes using electrical doping, *Appl. Phys. Lett.*, **81**, 922 (2002).
[18] J. Kido and T. Matsumoto, Bright organic electroluminescent devices having a metal-doped electron-injecting layer, *Appl. Phys. Lett.*, **73**, 2688 (1998).
[19] L.-S. Hung and J. Madathil, Radiation damage and transmission enhancement in surface-emitting organic light-emitting diodes, *Thin Solid Films*, **410**, 101 (2002).
[20] L. S. Hung, L. S. Liao, C. S. Lee, and S. T. Lee, Sputter deposition of cathodes in organic light emitting diodes, *J. Appl. Phys.*, **86**, 4607 (1999).
[21] G. Parthasarathy, P. E. Burrow, V. Khalfin, V. G. Gorkhale, S. R. Forrest, and M. E. Thomson, A metal-free cathode for organic semiconductor devices, *Appl. Phys. Lett.*, **72**, 2138 (1998).
[22] T. Minari, T. Nemoto, and S. Isoda, Fabrication and characterization of single-grain organic field-effect transistor of pentacene, *J. Appl. Phys.*, **96**, 769 (2004).
[23] J. Meyer, T. Winkler, S. Hamwi, S. Schmale, M. Kröger, P. Görrn, H.-H. Johannes, T. Riedl, E. Lang, D. Becker, T. Dobbertin, and W. Kowalsky, Highly efficient fully transparent inverted OLED, *Proc. SPIE*, **6655**, 66550L (2007).
[24] T. Dobbertin, O. Werner, J. Meyer, A. Kammoun, D. Schneider, T. Riedl, E. Becker, H.-H. Johannes, and W. Kowalsky, Inverted hybrid organic light-emitting device with polyethylene dioxythiophene-polystyrene sulfonate as an anode buffer layer, *Appl. Phys. Lett.*, **83**, 5071 (2003).
[25] W. Kowalsky, P. Görrn, J. Meyer, M. Kröger, H.-H. Johannes, and T. Riedl, See-through OLED Displays, *Proc. SPIE*, **6486**, 64860L (2007).
[26] G. Parthasarathy, C. Adachi, P. E. Burrows, and S. R. Forrest, High-efficiency transparent organic light-emitting devices, *Appl. Phys. Lett.*, **76**, 2128 (2000).
[27] T. Matsushima, Y. Kinoshita, and H. Murata, Formation of Ohmic hole injection by inserting an ultrathin layer of molybdenum trioxide between indium tin oxide and organic hole-transporting layers, *Appl. Phys. Lett.*, **91**, 253504 (2007).
[28] J. Meyer, S. Hamwi, T. Bülow, H.-H. Johannes, T. Riedl, and W. Kowalsky, Highly efficient simplified organic light emitting diodes, *Appl. Phys. Lett.*, **91**, 113506 (2007).
[29] I.-M. Chan, T.-Y. Hsu, and F. C. Hong, Enhanced hole injections in organic light-emitting devices by depositing nickel oxide on indium tin oxide anode, *Appl. Phys. Lett.*, **81**, 1899 (2002).
[30] X.-Y. Jiang, Z.-L. Zhang, J. Cao, M. A. Khan, K.-U. Haq, and W.-Q. Zhu, White OLED with high stability and low driving voltage based on a novel buffer layer MoOx, *J. Phys. D: Appl. Phys.*, **40**, 5553 (2007).
[31] C.-C. Chang, J. F. Chen, S.-W. Hwang, and C. H. Chen, Highly efficient white organic electroluminescent devices based on tandem architecture, *Appl. Phys. Lett.*, **87**, 253501 (2005).
[32] F. Guo and D. Ma, White organic light-emitting diodes based on tandem structures, *Appl. Phys. Lett.*, **87**, 173510 (2005).
[33] C.-W. Chen, Y.-J. Wu, E. H.-E. Wu, C.-W. Chu, and Y. Yang, Effective connecting architecture for tandem organic light-emitting devices, *Appl. Phys. Lett.*, **87**, 241121 (2005).

[34] T. Tsutsui and M. Terai, Electric field-assisted bipolar charge spouting in organic thin-film diodes, *Appl. Phys. Lett.*, **84**, 440 (2004).
[35] T. J. Coutts, D. L. Young, X. Li, W. P. Mulligan, and X. Wu, Search for improved transparent conducting oxides: a fundamental investigation of CdO, Cd_2SnO_4, and Zn_2SnO_4, *J. Vac. Sci. Technol. A*, **18**, 2646 (2000).
[36] H. Kim, C. M. Gilmore, J. S. Horwitz, A. Piquée, H. Murata, G. P. Kushto, R. Schlaf, Z. H. Kafafi, and D. B. Chrisey, Transparent conducting aluminum-doped zinc oxide thin films for organic light-emitting devices, *Appl. Phys. Lett.*, **76**, 259 (2000).
[37] Y. Tomita, C. May, M. Toerker, J. Amelung, M. Eritt, F. Loeffler, C. Luber, K. Leo, K. Walzer, K. Fehse, and Q. Huang, Highly efficient *p-i-n*-type organic light emitting diodes on ZnO:Al substrates, *Appl. Phys. Lett.*, **91**, 063510 (2007).
[38] P. Görrn, T. Rabe, T. Riedl, and W. Kowalsky, Loss reduction in fully contacted organic laser waveguides using TE_2 modes, *Appl. Phys. Lett.*, **91**, 041113 (2007).
[39] B. S. Sang and M. Konagai, Growth of transparent conductive oxide ZnO films by atomic layer deposition, *Jpn. J. Appl. Phys., Part 2*, **35**, L602 (1996).
[40] T. Suntola, Surface chemistry of materials deposition at atomic layer level, *Appl. Surf. Sci.*, **100–101**, 391 (1996).
[41] J. W. Elam and S. M. George, Growth of ZnO/Al_2O_3 alloy films using atomic layer deposition techniques, *Chem. Mater.*, **15**, 1020 (2003).
[42] S.-H. Ko Park, J.-I. Lee, C.-S. Hwang, and H.-Y. Chu, Characteristics of organic light emitting diodes with Al-doped ZnO anode deposited by atomic layer deposition, *Jpn. J. Appl. Phys., Part 2*, **44**, L242 (2005).
[43] S. J. Kwon, H. J. Lee, Y. W. Seo, and H. S. Jeong, Variation in the properties of ZnO:Al films grown by using atomic layer deposition with the precursor-pulse ratio, *J. Kor. Phys. Soc.*, **43**, 709 (2003).
[44] A. Yamada, B. Sang, and M. Konagai, Atomic layer deposition of ZnO transparent conducting oxide, *Appl. Surf. Sci.*, **112**, 216 (1997).
[45] G. Gu, G. Parthasarathy, P. E. Burrows, P. Tian, I. G. Hill, A. Kahn, and S. R. Forrest, Transparent stacked organic light emitting devices. I. Design principles and transparent compound electrodes, *J. Appl. Phys.*, **86**, 4067 (1999).
[46] T. Tsutsui and M. Terai, Electric field-assisted bipolar charge spouting in organic thin-film diodes, *Appl. Phys. Lett.*, **84**, 440 (2004).
[47] M. Terai, K. Fujita, and T. Tsutsui, Capacitance measurement in organic thin-film device with internal charge separation zone, *Jpn. J. Appl. Phys., Part 2*, **44**, L1059 (2005).
[48] T.-Y. Cho, C.-L. Lin, and C.-C. Wu, Microcavity two-unit tandem organic light-emitting devices having a high efficiency *Appl. Phys. Lett.*, **88**, 111106 (2006).
[49] L. S. Liao, K. P. Klubek, and C. W. Tang, High-efficiency tandem organic light-emitting diodes, *Appl. Phys. Lett.*, **84**, 167 (2004).
[50] C. W. Law, K. M. Lau, M. K. Fung, M. Y. Chan, F. L. Wong, C. S. Lee, and S. T. Lee, Effective organic-based connection unit for stacked organic light-emitting devices, *Appl. Phys. Lett.*, **89**, 133511 (2006).
[51] M. Kröger, S. Hamwi, J. Meyer, T. Dobbertin, T. Riedl, W. Kowalsky, and H.-H. Johannes, Temperature-independent field-induced charge separation at doped organic/organic interfaces: Experimental modeling of electrical properties, *Phys. Rev. B*, **75**, 235321 (2007).
[52] T. Dobbertin, O. Werner, J. Meyer, A. Kammoun, D. Schneider, T. Riedl, E. Becker, H.-H. Johannes, and W. Kowalsky, Inverted hybrid organic light-emitting device with polyethylene dioxythiophene-polystyrene sulfonate as an anode buffer layer, *Appl. Phys. Lett.*, **83**, 5071 (2003).
[53] R. M. A. Dawson, Z. Shen, D. A. Furst, S. Connor, J. Hsu, M. G. Kane, R. G. Stewart, A. Ipri, C. N. King, P. J. Green, R. T. Flegal, S. Pearson, W. A. Barrow, E. Dickey, K. Ping, C. W. Tang, S. Van Slyke, F. Chen, J. Shi, J. C. Sturm, and M. H. Lu, Design of an improved pixel for a polysilicon active-matrix organic led display, *SID Int. Symp. Dig. Tech. Pap.*, **29**, 11 (1998).
[54] L. Wang, M.-H. Yoon, G. Lu, Y. Yang, A. Facchetti, and T. J. Marks, High-performance transparent inorganic–organic hybrid thin-film n-type transistors, *Nat. Mater.*, **5**, 893 (2006).

[55] J. F. Wager, Transparent electronics, *Science*, **300**, 1245 (2003).
[56] K. Nomura, H. Ohta, A. Takagi, T. Kamiya, M. Hirano, and H. Hosono, Room-temperature fabrication of transparent flexible thin-film transistors using amorphous oxide semiconductors, *Nature (London)*, **432**, 488 (2004).
[57] H. Q. Chiang, R. L. Hoffman, J. Jeong, J. F. Wager, and D. A. Keszler, High mobility transparent thin-film transistors with amorphous zinc tin oxide channel layer, *Appl. Phys. Lett.*, **86**, 013503 (2005).
[58] E. M. C. Fortunato, P. M. C. Barquinha, A. C. M. B. G. Pimentel, A. M. F. Gonçalves, A. J. S. Marques, L. M. N. Pereira, and R. F. P. Martins, Fully transparent ZnO thin-film transistor produced at room temperature, *Adv. Mater.*, **17**, 590 (2005).
[59] K. Nomura, H. Ohta, K. Ueda, T. Kamiya, M. Hirano, and H. Hosono, Thin film transistor fabricated in single-crystalline transparent oxide semiconductor, Science, **300**, 1269 (2003).
[60] H. Hosono, Transparent amorphous oxide semiconductors for high performance TFT, *SID Int. Symp. Dig. Tech. Pap.*, **38**, 1830 (2007).
[61] S. Narushima, M. Orita, M. Hirano, and H. Hosono, Electronic structure and transport properties in the transparent amorphous oxide semiconductor 2 $CdO{\cdot}GeO_2$, *Phys. Rev. B*, **66**, 035203 (2002).
[62] T. Minami, S. Takata, H. Sato, and H. Sonohara, Properties of transparent zinc-stannate conducting films prepared by radio frequency magnetron sputtering, *J. Vac. Sci. Technol. A*, **13**, 1095 (1995).
[63] R. Hayashi, M. Ofuji, N. Kaji, K. Takahashi, K. Abe, H. Yabuta, M. Sano, H. Kumomi, K. Nomura, T. Kamiya, M. Hirano, and H. Hosono, Circuits using uniform TFTs based on amorphous In-Ga-Zn-O, *J. SID*, **15**, 915 (2007).
[64] H. Gleskova and S. Wagner, DC-gate-bias stressing of a-Si:H TFTs fabricated at 150°C on polyimide foil, *IEEE Trans. Electron Devices*, **48**, 1667 (2001).
[65] M. Powell, Charge trapping instabilities in amorphous silicon-silicon nitride thin-film transistors, *Appl. Phys. Lett.*, **43**, 597 (1983).
[66] K. S. Karim, A. Nathan, M. Hack, and W. I. Milne, Drain-bias dependence of threshold voltage stability of amorphous silicon TFTs, *IEEE Electron Device Lett.*, **25**, 188 (2004).
[67] P. Görrn, P. Hölzer, T. Riedl, W. Kowalsky, J. Wang, T. Weimann, P. Hinze, and S. Kipp, Stability of transparent zinc tin oxide transistors under bias stress, *Appl. Phys. Lett.*, **90**, 063502 (2007).
[68] T. Riedl, P. Görrn, P. Hölzer, and W. Kowalsky, Ultra-high long-term stability of oxide-TTFTs under current stress, *Phys. Status Solidi (RRL)*, **1**, 175 (2007).
[69] Wantae Lim, S.-H. Kim, Yu-Lin Wang, J. W. Lee, D. P. Norton, S. J. Pearton, F. Ren, and I. I. Kravchenko, Stable room temperature deposited amorphous $InGaZnO_4$ thin film transistors, *J. Vac. Sci. Technol. B*, **26**, (2008).
[70] A. Suresh and J. F. Muth, Bias stress stability of indium gallium zinc oxide channel based transparent thin film transistors, *Appl. Phys. Lett.*, **92**, 033502 (2008).
[71] T. Minami, S. Takata, H. Sato, and H. Sonohara, Properties of transparent zinc-stannate conducting films prepared by radio frequency magnetron sputtering, *J. Vac. Sci. Technol. A*, **13**, 1095 (1995).
[72] M. K. Jayaraj, K. J. Saji, K. Nomura, T. Kamiya, and H. Hosono, Optical and electrical properties of amorphous zinc tin oxide thin films examined for thin film transistor application, *J. Vac. Sci. Technol. B*, **26**, 495 (2008).
[73] P. Barquinha, A. Pimentel, A. Marques, L. Pereira, R. Martins, and E. Fortunato, Effect of UV and visible light radiation on the electrical performances of transparent TFTs based on amorphous indium zinc oxide, *J. Non-Cryst. Solids*, **352**, 1756 (2006).
[74] P. Görrn, M. Lehnhardt, T. Riedl, and W. Kowalsky, the influence of visible light on transparent zinc-tin-oxide thin film transistors, *Appl. Phys. Lett.*, **91**, 193504 (2007).
[75] W. F. Aerts, S. Verlaak, and P. Heremans, Design of an organic pixel addressing circuit for an active-matrix OLED display, *IEEE Trans. Electron Devices*, **49**, 2124 (2002).

13

Oxide-Based Electrochromics

Claes G. Granqvist

Department of Engineering Sciences, The Angstrom Laboratory, Uppsala University, Sweden

13.1 Introduction

Electrochromic (EC) device technology can be used for modulating the transmittance of visible light and solar radiation in see-through applications, including in architectural windows. This chapter gives a broad overview of oxide-based EC technology with particular emphasis on the large energy savings that can be achieved by its implementation in the built environment; the savings can be made simultaneously with achieving improved indoor comfort for the users of the building. Manufacturing aspects are considered with focus on methods compatible with low-cost roll-to-roll methods. Recent work on foil-type devices embodying sputter deposited WO_3 and NiO-based films joined by a polymer electrolyte is then discussed in particular detail, as are some recent research results of particular relevance for the indicated applications.

The current research and development on oxide-based electrochromics is to a large extent tied to the possibilities of this technology being able to provide energy efficiency in buildings and hence contribute to combating the harmful effects of global warming. A brief discussion of these issues serves as an appropriate background to this chapter. Global warming is presently receiving worldwide attention, and means to alleviate its harmful influences are urgently needed [1]. The effects are real and with us today; thus, for example, it has been stated that the warming and precipitation trends due to anthropogenic climate change during the past three decades has already claimed over 150 000 human lives per year [2, 3]. Furthermore, these changes may be accompanied by more frequent and/or extreme events such as heatwaves, heavy rainfall, storms and coastal flooding. It is also predicted that nonlinear climate responses can lead to breakdown of ocean 'conveyor belt'

Transparent Electronics: From Synthesis to Applications Edited by Antonio Facchetti and Tobin J. Marks

circulation, collapse of major ice sheets, and/or release of methane at high latitudes thereby aggravating global warming [4]. Major changes in energy technology are necessary, which will influence the world's economy [5]. It is important to realize too that the activities to mitigate global warming must account for an increasing population whose growing concentration in mega cities leads to 'heat islands', which tend to enhance the warming [6].

In order counteract global warming it is necessary to cut down radically on the use of fossil fuel, which will influence energy use in all sectors of society: in industry, transport, and in buildings. A focus on the built environment is natural considering the fact that this sector accounts for as much as 40% of the primary energy in industrialized countries such as in the EU(15) [7] and the USA; the corresponding number for the entire world lies between 30% and 40%, as found from a recent detailed study by the United Nations Environment Programme [8]. This energy is used predominantly for heating, cooling, ventilation and lighting. The energy spent on air conditioning has grown especially rapidly–by about 17% per year–in the EU(15) [9], and today electrically driven air conditioning dominates the peak power during the summer in parts of Europe and the USA, and in an extreme climate such as in Kuwait more than 75% of the electrical peak load is currently used for air conditioning [10], largely as a consequence of contemporary building practices using large glass façades [11].

The increase in energy for air conditioning occurs as a consequence of growing demands for indoor comfort. Part of this increase ensues from an unwillingness of the users of buildings to accept thermal discomfort due to too high or too low perceived temperatures, and another reason is found in the desire to experience good indoors–outdoors contact via large windows and glass façades. Large glazed areas tend to have cooling requirements, at least in commercial buildings in most parts of the world, but small windows lead to bad indoor comfort and hence poor job satisfaction and accompanying poor job performance. One way to improve the situation is to have buildings with variable throughput of visible light and solar energy, i.e. 'smart windows', as discussed further below. The same technology can be combined with light-guiding, which then opens possibilities to obtain energy efficient day-lighting via new concepts such as 'light balancing' [12]. In principle, one can use several different 'chromogenic' technologies to accomplish the variable optical properties [13]. The most widely known of these make use of photochromic materials (which darken under ultraviolet irradiation), thermochromic coatings (which change their optical properties reversibly at a temperature-dependent structural phase transition), and EC devices whose optical properties can be changed electrically. The EC technology allows user operation and is by far the one most suited to building applications.

The energy savings inherent in the EC 'smart windows' (also referred to as 'switchable windows') technology has been much discussed over the past several years. A simple 'back-of-an-envelope' analysis illustrates this savings as an analogy: we consider a surface with arbitrary orientation but facing the Sun; letting this surface be a 'smart window' leads to a certain amount of saved energy, and letting it be covered with today's best solar cells for terrestrial applications leads to energy production of a magnitude that is the same as the savings in the case of a 'smart window' [14]. More elaborate evaluations have been conducted as well; the most detailed investigation so far was reported in recent work for the California Energy Commission [15, 16]. The summary of this report states verbatim that [15]:

Switchable variable-tint electrochromic windows preserve the view out while modulating transmitted light, glare, and solar gains and can reduce energy use and peak demand./. . ./. Compared to an efficient low-e window with the same day-lighting control system, the electrochromic window showed annual peak cooling load reductions from control of solar heat gains of 19–26% and lighting energy use savings of 48–67% when controlled for visual comfort. Subjects strongly preferred the electrochromic window over the reference window, with preferences related to perceived reductions in glare, reflections on the computer monitor, and window luminance.

The detailed numbers on the energy savings, given above, are likely to be underestimated since they account neither for day-lighting strategies based on 'light balancing' [12] nor for novel transparency control strategies founded on physical presence [14]. The most important conclusion of the quote is not the precise numbers but their overall order of magnitude which proves that one is dealing here with a new energy technology capable of radical energy savings.

The present chapter summarizes parts of two recent treatises [17, 18]. Its format resembles that of a recent conference paper [19], and it includes some recent research results that are presented in more detail elsewhere [20].

13.2 Electrochromic Devices

13.2.1 Overall Design and Materials

There are several principles that can be exploited to accomplish variable transmittance of visible light and solar energy by use of electrochromics [18]. Figure 13.1 illustrates the most widely investigated of these. It comprises five superimposed layers on a transparent substrate, typically of glass or flexible polyester [poly(ethylene terephthalate), PET] foil, or positioned between two such substrates in a laminate arrangement [17, 21]. The

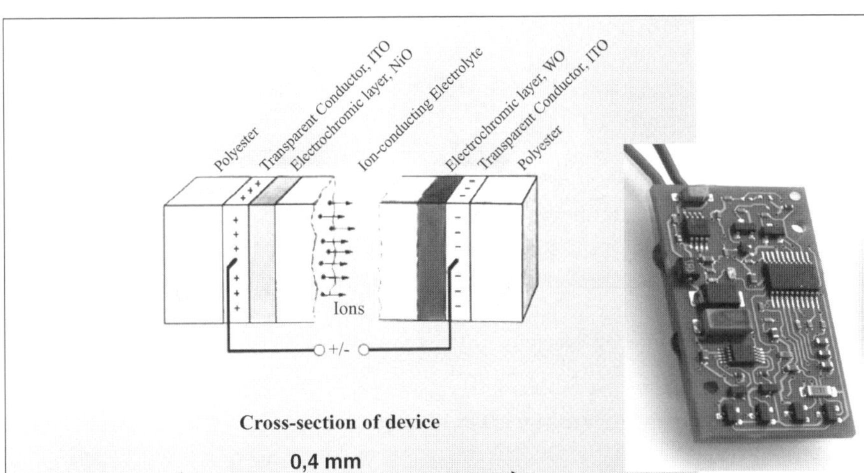

Figure 13.1 Sketch of an EC foil-type device and the unit for supplying charge

resemblance to a thin film battery is obvious. The outermost layers are transparent electrical conductors, typically of In_2O_3:Sn (i.e. indium tin oxide, ITO) [18, 22]. One of these layers is coated with an EC film, and the other is coated with an ion storage film, with or without EC properties. The two films must consist of nanomaterials with well developed nanoporosities (analogously with the requirements for battery electrodes). A transparent ion conductor (electrolyte) is at the middle of the device and joins the EC and ion storage films. A voltage applied between the transparent electrodes leads to charge being transported between the EC and ion storage films, and the overall transparency is then changed. A voltage pulse with opposite polarity—or, with suitable materials, short circuiting—makes the device regain its original properties. The optical modulation requires a DC voltage of 1 to 2 V. The charge insertion into the EC film(s) is balanced by electron inflow from the transparent electrode(s); these electrons can produce intervalency transitions, which is the basic reason for the optical absorption [17]. Importantly, the devices do not display visible haze irrespective of their absorption [23, 24].

The five-layer battery-type EC device in Figure 13.1 is not the only possibility [18], and recent work has considered an alternative wherein the EC material is a metal hydride [18, 25, 26]. Other possibilities are solution/gel redox systems (such as those in antidazzling rear view mirrors for automobiles), suspended particle devices, liquid-crystal-based devices, and arrangements based on reversible electroplating. Limited durability, visible haze, and/or the necessity to supply a continuous current/voltage tend to limit the usefulness of these alternatives, though.

Regarding materials in the five-layer battery-type EC devices, the ITO can be replaced by ZnO:Al, SnO_2:F, or similar oxides [18, 27]–or, possibly, by carbon nanotubes [28] or graphene [29]–if the cost and availability of indium turn out to be problematic. In fact, the global indium resource appears to be ample and about as large as that of silver [30], but it is also true that the cost of indium has risen greatly in recent years as a consequence of the booming display market and due to production constraints.

The EC film is WO_3-based in almost all devices for window applications, while there are numerous options for the counter electrode [17, 18, 31, 32]. Among the latter, films based on IrO_2 and NiO have enjoyed much interest recently. IrO_2-based alternatives are inherently expensive, but good EC properties are maintained after dilution with cheaper Ta_2O_5 [33]. NiO-based films combine moderate cost with good optical properties; their slight inherent absorption can be minimized if the NiO is mixed with another oxide characterized by a wide band gap such as MgO or Al_2O_3 [17, 34]. There are many possibilities as to the electrolyte of EC devices, with hydrous oxides exhibiting proton conduction and polymers with ion conduction due to added salts being the most common.

EC technology has been discussed for many years, ever since display-type devices were presented shortly after the discovery of electrochromism in WO_3 films [35–37]. Generally speaking, the progress of this technology has been slow, which is likely to be associated with the fact that a range of nonstandard procedures must be applied successfully, as follows [20, 38]:

(i) The ITO, or an analogous transparent and conducting material, must combine electrical resistivity in the $10^{-4}\,\Omega\,cm$ range with very low luminous absorptance, which requires special thin film deposition technology in the case of temperature sensitive substrates such as polymers [18].

(ii) The EC and counter electrode films must have uniform nanoporosities over large areas, which again calls for nonstandard coating technologies.
(iii) Viewing the EC device as a thin film battery points at the fact that charge insertion/extraction and charge balancing are necessary and must be accomplished by properly controllable and industrially viable techniques, with gas treatment during or after film deposition being suitable options [39].
(iv) The electrolyte must combine good ion conductivity with adhesiveness and high transparency for ultraviolet irradiation.
(v) Long-term durability demands adequate strategies for voltage and current control during coloration/bleaching, just as it does for charging/discharging of batteries.

All of these challenges can be successfully met, however, and EC technology finally may emerge as appropriate for large-area, large-scale applications [40].

Film porosity on the nanoscale is demanded for successful operation of EC devices, as stressed in (ii) above. Virtually any thin film technology may be capable of achieving the required properties, though with more or less difficulty. In the case of sputtering [38], the deposition parameters should be confined to those giving 'zone 1' films in the well known 'Thornton diagram'. Figure 13.2 shows schematically what a sputter deposited film looks like under an electron microscope [41]. Thin films for most other applications are prepared under conditions such as the ones of the 'transition zone' denoted T. Those films are compact and it is then possible to minimize grain-boundary scattering of the conduction electrons in a metallic film; the transparent and electrically conducting ITO films in typical EC devices, for example, should be of this type [42]. Nanocrystallinity and nanoporosity are found at comparatively high pressures in the sputter plasma, such as in 'zone 1', and it is then possible to have ion conduction in intercolumnar spaces, which is highly advantageous in EC films and in films for solid state ionics in general.

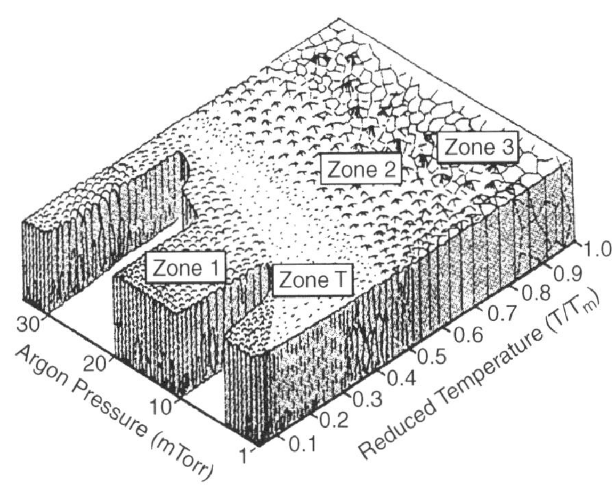

Figure 13.2 'Thornton diagram' showing nanostructures of thin sputter deposited films prepared at different argon pressures and substrate temperatures. The melting point of the material is denoted T_m. Reprinted with permission from [41]. Copyright (2007) Annual Reviews

13.2.2 Discussion of Flexible Devices

This section treats the preparation and properties of a specific EC device. It is prepared by reactive DC magnetron sputtering onto flexible PET foil from targets based on 99.95% pure tungsten and nickel [17, 43]. A target of $V_{0.07}Ni_{0.93}$ was used in many experiments, which made the target nonmagnetic and hence convenient for magnetron sputtering. Deposition took place in $Ar + O_2 + H_2$ with optimized mixing ratio under conditions that avoided excessive substrate heating. The target–substrate distance typically was 20–25 cm, and the total gas pressure was in the 30–40 mTorr range. These conditions yielded nanoporous oxide films characteristic of 'zone 1' of the 'Thornton diagram' in Figure 13.2.

Device manufacturing was implemented in a manner consistent with the schematic in Figure 13.3 [44]. The PET foil was cut to $30 \times 30\,\text{cm}^2$ pieces which were transferred to a sputter coater where a conducting pattern ('bus bars'), ITO films and EC films were deposited. The WO_3-based film was deposited in the presence of hydrogen so that it became charged and had a blue appearance [45], and the NiO-based film was post-treated in ozone [39] and was then discharged and exhibited a brownish color. Coated foils of the two types were laminated together, cut to shape, equipped with electrical contacts, sealed, and tested. The particular devices produced in accordance with Figure 13.3 are employed in variable-tint visors for motorcycle helmets [46]. Their dark-state appearance can be neutral gray. The technology is compatible with roll-to-roll manufacturing. The most sensitive part of the manufacturing may be the bending of the ITO coatings, but detailed work has shown convincingly that such films are surprisingly strain resistant [47, 48].

Powering the devices using drive units such as the one shown on the right-hand side of Figure 13.1 led to ions being shuttled between the two EC films and to concomitant insertion

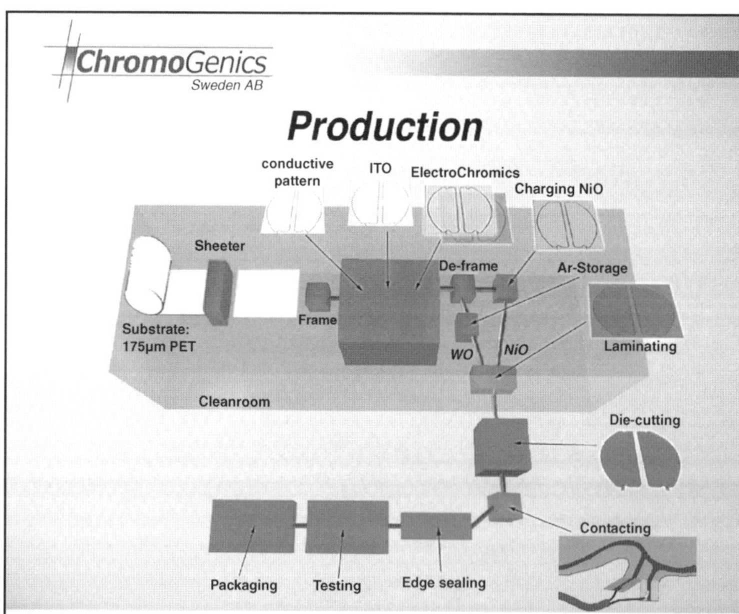

Figure 13.3 Production diagram for EC foil devices

and extraction of electrons. The corresponding electrochemical 'color reactions' can be represented as [17, 49, 50]:

$$[Ni(OH)_2]_{bleached} \leftrightarrow [NiOOH + H^+ + e^-]_{colored} \qquad (13.1)$$

$$[NiO + Ni(OH)_2]_{bleached} \leftrightarrow [Ni_2O_3 + 2H^+ + 2e^-]_{colored} \qquad (13.2)$$

and

$$[WO_3 + H^+ + e^-]_{bleached} \leftrightarrow [HWO_3]_{colored} \qquad (13.3)$$

The reactions in (13.1) and (13.2) represent an extension of the well known Bode reaction scheme [51, 52].

Figure 13.4 shows the optical modulation of a 240 cm^2 visor-type EC foil device subjected to extended coloring and bleaching [19]. Figure 13.4(a) shows the transmittance

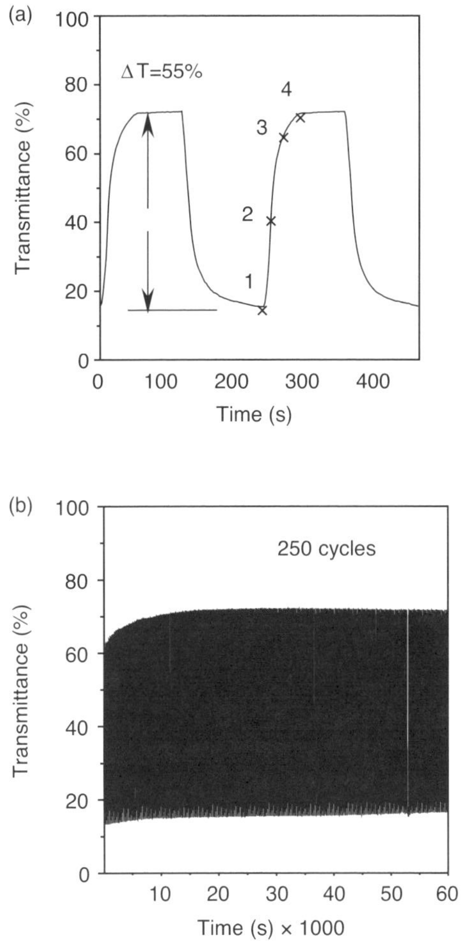

Figure 13.4 *Mid-luminous transmittance for a 240 cm^2 EC foil-type device. Reprinted with permission from [19]. Copyright (2007) SPIE*

T for two consecutive color/bleach cycles adjusted so that the transmittance difference ΔT is 55%. The separation between points 1 and 2 corresponds to 10 s and half of this optical modulation, the separation between points 1 and 3 corresponds to 20 s and 0.9 of the modulation, and the separation between points 3 and 4 corresponds to 30 s. Figure 13.4(b) shows the evolution of the transmittance during the first 250 color/bleach cycles. This is by no means the performance limit, though, but the optical modulation persisted practically unchanged for tens of thousands of cycles.

13.3 Some Recent Research Results

There is vigorous and worldwide research on electrochromism and EC-based devices. This section summarizes a few recent results specifically aimed at improving foil-based devices and testing their properties. The presentations are based on earlier articles [16, 53, 54].

13.3.1 Enhanced Transmittance

Antireflection (AR) coatings can be used to boost the transmittance through optical devices, and a number of low refractive index, nanoporous, and/or nanostructured coatings can be applied to glass and plastic substrates [18]. For example, coating both sides of a glass pane with self-assembled silica nanoparticle films made it possible to obtain a transmittance as large as 99.5% in the middle of the luminous spectrum [55, 56].

A 175 μm thick PET foil was AR coated on both sides by dip-coating using sol-gel technology, essentially according to a previously reported procedure [55]. The foil was immersed in a sol, based on an earlier reported recipe [57], comprised of ~50 nm diameter SiO_2 nanoparticles in ethanol; the foil was then withdrawn and was subsequently heat treated in air at ~200 °C. This heat treatment improved the mechanical properties of the films and made them strongly adherent to the PET foil. The coated foils exhibited high transmittance over the entire luminous spectrum, and the mid-luminous transmittance increased from 88.8% for the untreated material to 98.6% [16].

Figure 13.5 shows optical data for an EC device of the type shown in Figure 13.1 and manufactured according to Figure 13.3, incorporating two laminated 175 μm thick PET foils, in its bleached state. The mid-luminous transmittance was enhanced from 73.1 to 79.1%, i.e. by as much as 6%, as a consequence of the AR treatment. This EC device had a vanadium-containing counter electrode, which tended to limit the transmittance to a significant degree; other oxide admixtures would have boosted the transmittance [17, 34].

13.3.2 Enhanced Contrast Ratio

It is often of interest to have a large contrast ratio ζ, which is conveniently defined as

$$\zeta = T_{\text{colored}}/T_{\text{bleached}} \qquad (13.4)$$

A value between three and six, typical for the luminous or solar properties of EC devices, has been reported several times [21, 31, 32], although $\zeta > 10$ has been claimed in some earlier work [58–63]. Requiring T_{bleached} to be high–which is often a strong demand for architectural uses [64]–tends to make T_{colored} so large that the window is unable to become dark

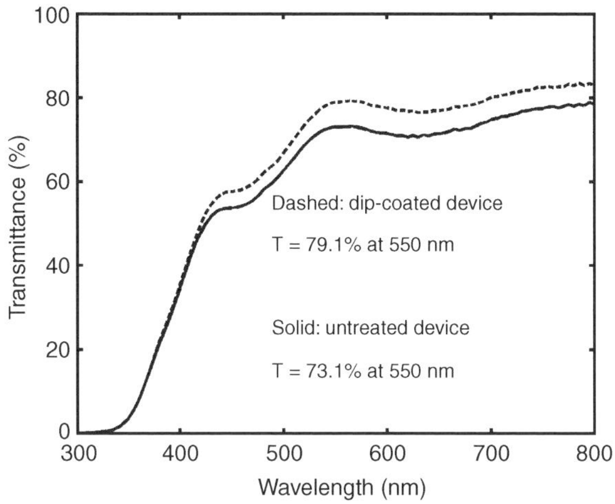

Figure 13.5 *Spectral transmittance of an EC foil-type device with and without antireflection treatment*

enough. For other applications, such as automotive roof windows, $T_{bleached} \approx 15\%$ may be desirable [65] in which case $T_{colored}$ becomes low, possibly even low enough for achieving privacy, but a low $T_{bleached}$ leads to severe limitations for most other applications. Putting EC devices in tandem is possible, and an arrangement of this type gives as a first approximation that

$$\zeta_{total} = \zeta_1 \times \zeta_2 \times \ldots \times \zeta_n \tag{13.5}$$

for n superimposed EC devices. Practical demands based on weight and/or cost restrict this option, though, particularly for glass-based devices. However, these limitations are much less stringent for foil-based EC devices.

Figure 13.6 shows mid-luminous transmittance for color/bleach cycling of two superimposed EC foils of the type described earlier, incorporating films of tungsten oxide and nickel-vanadium oxide. The transmittance could be varied between 45% and 0.8%, implying $\zeta \approx 56$, which surpasses earlier results on ζ by a wide margin. The minimum value of $T_{colored}$ may be low enough to provide privacy. Irrespective of whether privacy can be accomplished or not, the results prove that very large modulation ranges, and very low values of $T_{colored}$, can be achieved with tandem EC foils.

13.3.3 Enhanced Electrochromism Under Ultraviolet Irradiation

EC materials are well known to display photochromism [21], and it is expected that light irradiation with sufficiently high energy will influence the properties of EC devices. Considering WO_3, there has been detailed work recently to investigate photostimulated hydrogen atom transfer quantum reactions [66] consistent with resonant tunneling [67]; this work elucidates the theoretical underpinning for understanding the interplay of photochromism and electrochromism in WO_3 and similar oxides.

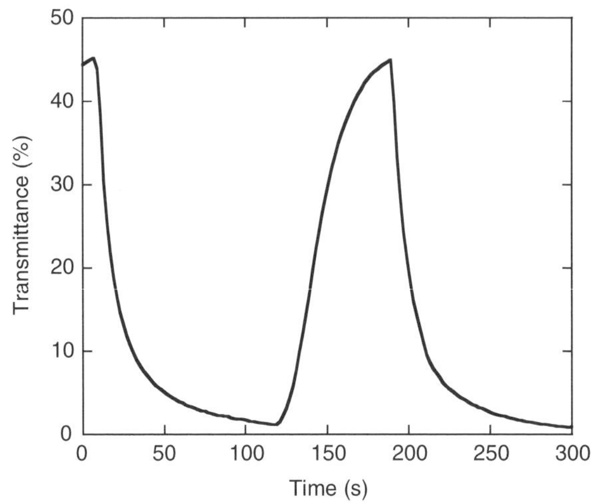

Figure 13.6 Optical modulation of an EC device with two superimposed foils

Figure 13.7 shows data on insertion/extraction of charge in a 240 cm^2 EC foil-type device for visor applications in the dark and subjected to weak UV irradiation due to exposure to an overcast sky at an ambient temperature of 15 °C. Charging at –1.6 V was performed over 1 min; the device was then rested for 1 min, and discharge took place at 1.4 V for 1 min. It is evident that measurements under UV irradiation produced a charge exchange of 1.45 C, whereas the corresponding number was 1.3 C without UV exposure. Charge exchange and coloration are intimately connected, and the optical modulation was between 25% and 65% in the latter case and noticeably larger in the former.

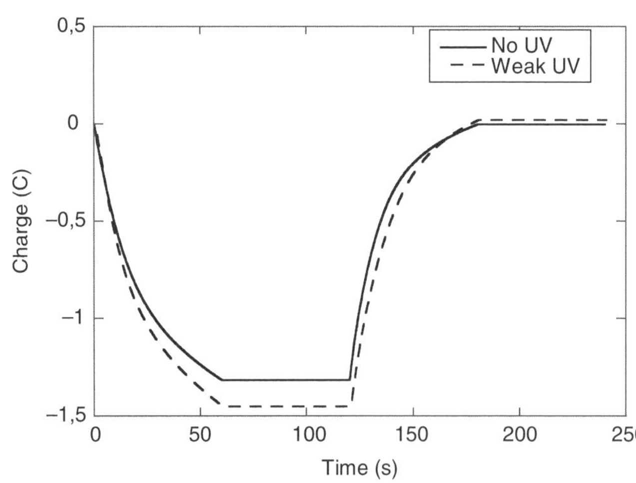

Figure 13.7 Insertion/extraction of charge as a function of time in a 240 cm^2 EC device in the presence and absence of ultraviolet (UV) irradiation

13.3.4 Durability Assessment Based on Noise Spectroscopy

Long term durability is of the greatest importance for most applications of EC devices, and, just as in the case of battery technology, durability issues are challenging to handle. A number of studies have concerned longevity [68, 69] but the need to have expedient methods for quality assurance has remained. Electromagnetic noise is well known for being able to give information related to degradation in EC devices [53, 54], as discussed below.

Electrical noise is a general feature for conductors, semiconductors, electrolytes, and electrode–electrolyte interfaces [70]; it can originate from thermal equilibrium as well as nonequilibrium processes. These mechanisms can lead to various types of noise, referred to as thermal noise, shot noise, burst noise, generation-recombination noise, $1/f$ noise, and $1/f^2$ noise [71]. With regard to EC devices, information can be gained from $1/f$ noise, where f is frequency, during discharge as well as with changes of this noise related to device degradation. It is known from prior work that $1/f$ noise can give information on corrosion processes [72–74].

Specifically, current noise was measured as a foil-type EC device was discharged via a $\sim$1 kΩ resistor, and current I was recorded. Data were taken at room temperature during the final stage of the discharging with $I < 100\,\mu A$. Fluctuations were sampled at a frequency of 15 kHz within an interval of less than 5 s in order to register 2^{16} samples each time. The DC current changed by less than 2% over this interval. The recorded quantity of samples led to reliable data with a sufficiently small random error of the power spectrum [53, 54].

The experimental conditions led to easily measurable $1/f$ noise. Figure 13.8(a) shows the current power spectral density S_i at three DC currents for an as-prepared foil-type EC device. Low-frequency noise clearly dominates at $f > 1$ kHz; its frequency dependence can be approximated by

$$S_i \sim 1/f^x \tag{13.6}$$

with x lying between 1.0 and 1.2. Figure 13.8(b) shows the corresponding S_i data for a severely cycled and degraded device. It is evident that the low-frequency noise has increased significantly, and $1/f$ noise now dominates in the full investigated frequency range even for the smallest current.

$S_i(f)$ vs. I was studied at $f = 700$ Hz. Figure 13.9 shows that

$$S_i(f) \sim I^2 \tag{13.7}$$

is obeyed and that the cycling treatment has increased S_i by about a factor of four irrespective of I. This relationship suggests that the noise originates from resistance/conductance fluctuations and that their magnitude is a measure of the device degradation.

The data in Figures 13.8 and 13.9 do not provide any information on the origin of the noise. In order to shed light on this issue, 'symmetrical' devices were produced [54]; they incorporated two laminated WO_3-based films or two laminated NiO-based films. $S_i(f)$ vs. I was investigated at $f = 175$ Hz, and it was found that relation (13.7) was still obeyed. Interestingly, the NiO-based device had a noise that was about four orders of magnitude larger than the noise in the WO_3-based device, which points to very significant differences between the two designs. Investigations of this type are expected to be useful for optimizing materials to avoid degradation at oxide/electrolyte interfaces and elsewhere and hence provide guidelines for strategies to improve the long term durability of EC devices.

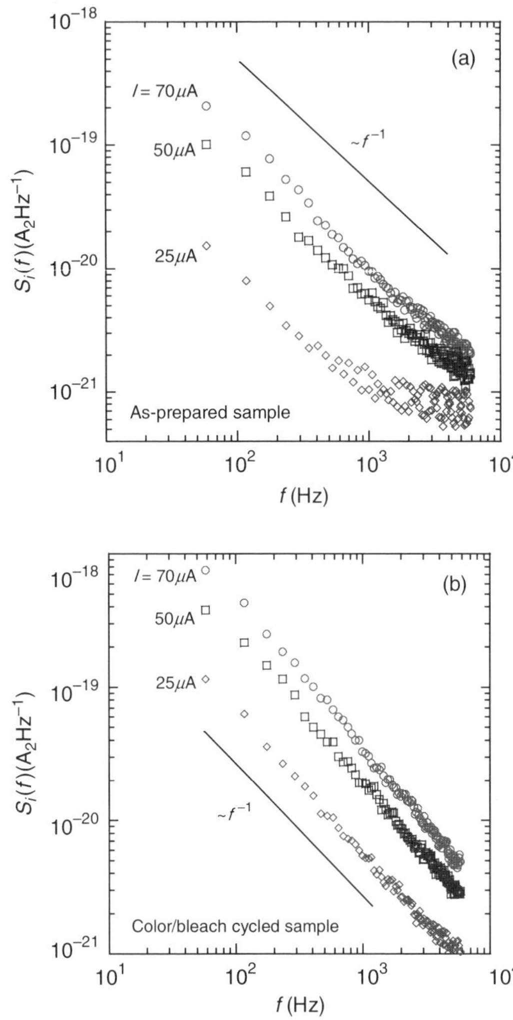

Figure 13.8 *Noise spectra, denoted $S_i(f)$ where f is frequency, of a $2.5 \times 2.5\,cm^2$ EC device before (a) and after (b) extensive color/bleach cycling. Data were taken at the current I. The line indicates f^{-1} dependence*

13.4 Summary and Concluding Remarks

EC device technology for the built environment may emerge as one of the keys to combating the effects of global warming, and this novel technology may also serve as an example of the business opportunities arising from the challenges caused by climate changes [5]. This chapter gave an introduction to EC technology and introduced novel EC foils, which appear to offer possibilities to accomplish low-cost manufacturing of a product enabling energy savings jointly with comfort improvements in new and existing buildings. Manufacturing aspects and data were presented regarding foil-type devices. The chapter also contained

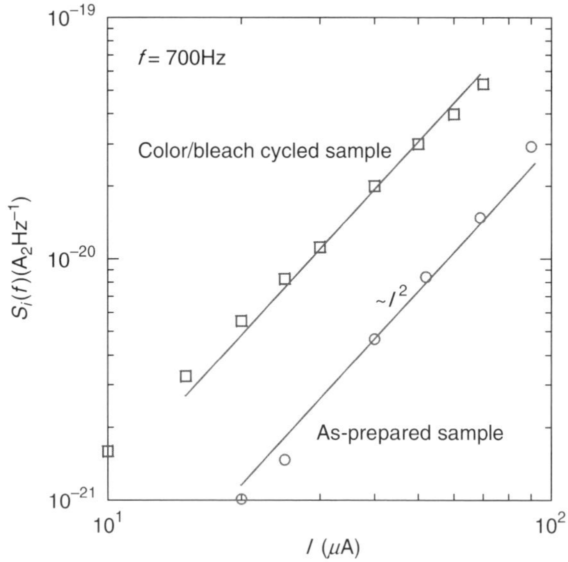

Figure 13.9 *Current noise power spectra, denoted $S_i(f)$ where $f = 700$ Hz, vs. current I of a $2.5 \times 2.5\,cm^2$ EC device before and after extensive color/bleach cycling. The lines denote I^2 dependence*

some new results showing that: double-sided antireflection coatings based on dip coating can enhance the transmittance significantly; that tandem foils can yield a ratio between bleached-state and colored-state transmittance exceeding fifty; that solar irradiance onto an EC device can enhance its charge insertion dynamics and thereby its optical modulation; and that electromagnetic noise spectroscopy may be used for quality assessment of EC devices.

It is speculated that membrane architecture [75–77] may in the future be merged with EC foil technology in order to allow light-weight buildings with little embodied energy. Ethylene tetrafluoroethylene (ETFE) is a well-proven polymer for such applications [78, 79]. These notions are consistent with contemporary concepts such as 'intelligent buildings' [80] and 'smart skins' [81]. One can envisage huge membranes allowing the flow of visible light and solar energy to be controlled and optimized, thereby leading to resource-lean buildings. The possibilities offered by such membranes–although then based on glass technology–were pointed out more than 50 years ago by the great visionary Buckminster Fuller [82].

References

[1] IPCC, *Climate Change 2007: Mitigation. Contribution of Working Group III to the Fourth Assessment Report of the Intergovernmental Panel on Climate Change*, edited by B. Metz, O. R. Davidson, P. Bosch, R. Dave, and L. A. Meyer, Cambridge University Press, Cambridge, UK and New York, NY, USA, 2007.

[2] A. J. McMichael, D. Campbell-Lendrum, P. Wilkinson, R. Nicholls, S. Hales, F. Tanser, D. Le Sueur, M. Schlesinger, and N. Andronova, Global climate change, in: *Comparative*

Quantification of Health Risks: Global and Regional Burden of Disease due to Selected Major Risk Factors, edited by M. Ezzati, A.D. Lopez, A. Rodgers, and C.J.L. Murray, World Health Organization, Geneva, Switzerland, Ch. 20, pp. 1543–1649, 2004.

[3] J. A. Palz, D. Campbell-Lendrum, T. Holloway, and J. A. Foley, Impact of regional climate change on human health, *Nature* **438**, 310–317 (2005).

[4] J. F. B. Mitchell, J. Lowe, R. A. Wood, and M. Vellinga, Extreme events due to human-induced climate change, *Philos. Trans R. Soc. A* **364**, 2117–2133 (2006).

[5] N. Stern, *The Economics of Climate Change: The Stern Report*, Cambridge University Press, Cambridge, UK, 2007.

[6] M. Santamouris, Solar and natural resources for a better efficiency in the built environment, in: *Solar Energy: The State of the Art*, edited by J. Gordon, James & James, London, UK, Ch. 1, pp. 1–28, 2001.

[7] C.F. Reinhart, Energy efficient solar buildings, in: *The Future for Renewable Energy 2: Prospects and Directions*, James & James, London, UK, Ch. 4, pp. 79–114, 2002.

[8] UNEP *Buildings and Climate Change: Status, Challenges and Opportunities*, United Nations Environment Programme, Paris, France, 2007.

[9] A. Jäger-Waldau, editor, *REF-SYST Status Report 2004*, EUR 21297 EN; JRC, Ispra, Italy, 2004.

[10] M. A. Darwich, Energy efficient air conditioning: case study for Kuwait, *Kuwait J. Sci. Eng.* **32**, 209–222 (2005).

[11] M. M. Aboulnaga, Towards green buildings: glass as a building element—the use and misuse in the Gulf region, *Renew. Energy* **31**, 631–653 (2006).

[12] C. G. Granqvist, Materials for good day-lighting and clean air: new vistas in electrochromism and photocatalysis, in: *Technical Conference Proceedings of the Society of Vacuum Coaters' 50th Conference*, Society of Vacuum Coaters, Albuquerque, NM, USA, pp. 21–27, 2007.

[13] C. M. Lampert and C. G. Granqvist,editors, *Large-Area Chromogenics: Materials and Devices for Transmittance Control*, SPIE Press, Bellingham, WA, USA, 1990.

[14] A. Azens and C. G. Granqvist, Electrochromic smart windows: energy efficiency and device aspects, *J. Solid State Electrochem.* **7**, 64–68 (2003).

[15] E. S. Lee, S. E. Selkowitz, R. D. Clear, D. L. DiBartolomeo, J. H. Klems, L. L. Fernandes, G. J. Ward, V. Inkarojrit, and M. Yazdanian, *Advancement of Electrochromic Windows, California Energy Commission*, PIER, CEC-500-2006-052, 87 pages, 2006.

[16] R. D. Clear, V. Inkarojrit, and E. S. Lee, Subject responses to electrochromic windows, *Energy Buildings* **38**, 758–779 (2006).

[17] G. A. Niklasson and C. G. Granqvist, Electrochromics for smart windows: thin films of tungsten oxide and nickel oxide, and devices based on these, *J. Mater Chem.* **17**, 127–156 (2007).

[18] C. G. Granqvist, Transparent conductors as solar energy materials: a panoramic review, *Solar Energy Mater. Solar Cells* **91**, 1529–1598 (2007).

[19] C. G. Granqvist and E. Avendaño, Electrochromic device technology based on nanoporous nanocrystalline thin films on PET foil, *Proc. SPIE* **6800**, 680002 1-12 (2008).

[20] C. G. Granqvist, S. Green, E. K. Jonson, R. Marsal, G. A. Niklasson, A. Roos, Z. Topalian, A. Azens, P. Georén, G. Gustavsson, R. Karmhag, J. Smulko, and L. B. Kish, Electrochromic foil-based devices: optical transmittance and modulation range, effect of ultraviolet irradiation, and quality assessment by $1/f$ current noise, *Thin Solid Films* **516**, 5921–5926 (2008).

[21] C. G. Granqvist, *Handbook of Inorganic Electrochromic Materials*, Elsevier, Amsterdam, The Netherlands, 1995.

[22] C. G. Granqvist and A. Hultåker, Transparent and conducting ITO films: new developments and applications, *Thin Solid Films* **411**, 1–5 (2002).

[23] D. Rönnow, L. Kullman, and C. G. Granqvist, Spectroscopic light scattering from electrochromic tungsten-oxide-based films, *J. Appl. Phys.* **80**, 423–430 (1996).

[24] T. Lindström, L. Kullman, D. Rönnow, C.-G. Ribbing, and C. G. Granqvist, Electrochromic control of thin film light scattering, *J. Appl. Phys.* **81**, 1464–1469 (1997).

[25] K. Tajima, Y. Yamada, S. Bao, M. Okada, and K. Yoshimura, Flexible all-solid-state switchable mirror on plastic sheet, *Appl. Phys. Lett.* **92**, 041912 1-3 (2008).
[26] K. Tajima, Y. Yamada, S. Bao, M. Okada, and K. Yoshimura, Near colorless all-solid-state switchable mirror based on magnesium-titanium thin film, *J. Appl. Phys.* **103**, 013512 1-5 (2008).
[27] K. G. T. Hollands, J. L. Wright, and C.G. Granqvist, Glazings and coatings, in: *Solar Energy: The State of the Art*, edited by J. Gordon, James & James, London, UK, Ch. 2, pp. 29–107, 2001.
[28] Z.-C. Wu, Z.-H. Chen, X. Du, J. M. Logan, J. Sippel, M. Nikolou, K. Kamaras, J. R. Reynolds, D. B. Tanner, A. F. Hebard, and A. G. Rinzler, Transparent, conductive carbon nanotube films, *Science* **305**, 1273–1276 (2004).
[29] X. Wang, L. Zhi, and K. Müllen, Transparent, conductive graphene electrodes for dye-sensitized solar cells, *Nano Lett.* **8**, 323 (2008).
[30] U. Schwartz-Schampera and P. M. Herzig, *Indium: Geology, Mineralogy, and Economics*, Springer, Berlin, Germany, 2002.
[31] C. G. Granqvist, Electrochromic tungsten oxide films: review of progress 1993–1998, *Solar Energy Mater. Solar Cells* **60**, 201–262 (2000).
[32] C. G. Granqvist, E. Avendaño, and A. Azens, Advances in electrochromic materials and devices: survey of some recent advances, *Thin Solid Films* **442**, 201–211 (2003).
[33] J. Backholm, A. Azens, and G. A. Niklasson, Electrochemical and optical properties of sputter deposited Ir-Ta and Ir oxide thin films, *Solar Energy Mater. Solar Cells* **90**, 414–421 (2006).
[34] E. Avendaño, A. Azens, G. A. Niklasson, and C. G. Granqvist, Electrochromism in nickel oxide films containing Mg, Al, Si, V, Zr, Nb, Ag, or Ta, *Solar Energy Mater. Solar Cells* **84**, 337–350 (2004).
[35] S. K. Deb, Optical and photoelectric properties and colour centres in thin films of tungsten oxide, *Philos. Mag.* **27**, 801–822 (1973).
[36] S. K. Deb, Reminiscences on the discovery of electrochromic phenomena in transition metal oxides, *Solar Energy Mater. Solar Cells* **39**, 191–201 (1995).
[37] S. K. Deb, Opportunities and challenges in science and technology of WO_3 for electrochromic and related applications, *Solar Energy Mater. Solar Cells* **92**, 245–258 (2008).
[38] C. G. Granqvist, Electrochromic materials and devices prepared by magnetron sputtering, in: *Reactive Sputter Deposition*, edited by D. Depla and S. Maheu, Springer, Berlin, Germany, pp. 487–498, 2008.
[39] A. Azens, L. Kullman, and C. G. Granqvist, Ozone coloration of Ni oxide and Cr oxide films, *Solar Energy Mater. Solar Cells* **76**, 147–153 (2003).
[40] C. G. Granqvist, Electrochromic materials: out of a niche, *Nature Mater.* **5**, 89–90 (2006).
[41] J. A. Thornton, High rate thick film growth, *Ann. Rev. Mater. Sci.* **7**, 239–260 (1977).
[42] I. Hamberg and C. G. Granqvist, Evaporated Sn-doped In_2O_3 films: basic optical properties and applications to energy efficient windows, *J. Appl. Phys.* **60**, R123–R159 (1986).
[43] A. Azens, G. Gustavsson, R. Karmhag, and C. G. Granqvist, Electrochromic devices on polyester foil, *Solid State Ionics* **165**, 1–5 (2003).
[44] www.chromogenics.com.
[45] A. P. Giri and R. Messier, Physical structure and the electrochromic effect in tungsten oxide films, *Mater. Res. Soc. Symp. Proc.* **24**, 221–227 (1984).
[46] M. Buyan, P. A. Brühwiler, A. Azens, G. Gustavsson, R. Karmhag, and C. G. Granqvist, Facial warming with tinted helmet visors, *Int. J. Ind. Ergon.* **36**, 11–16 (2006).
[47] D. R. Cairns, R. P. Witte II, D. K. Sparacin, S. M. Sachsman, D. C. Paine, G. P. Crawford, and R. R. Newton, Strain-dependent electrical resistance of tin-doped indium oxide on polymer substrates, *Appl. Phys. Lett.* **76**, 1425 (2000).
[48] Y. Leterrier, L. Médico, F. Demarco, J.-A. E. Månson, U. Betz, M. F. Escolà, M. Kharrazi Olsson, and F. Atamny, Mechanical integrity of transparent conductive oxide films for flexible polymer-based displays, *Thin Solid Films* **460**, 156 (2004).
[49] A. Avendaño, A. Azens, G. A. Niklasson, and C. G. Granqvist, Proton diffusion and electrochromism in hydrated NiO_y and $Ni_{1-x}V_xO_y$ thin films, *J. Electrochem. Soc.* **152**, F203–F212 (2005).

[50] E. Avendaño, H. Rensmo, A. Azens, A. Sandell, G. de Azevedo, H. Siegbahn, G. A. Niklasson, and C. G. Granqvist, Coloration mechanism in proton intercalated electrochromic hydrated NiO_y and $Ni_{1-x}V_xO_y$ thin films, *J. Electrochem. Soc.* **156**, P132–P138 (2009).
[51] H. Bode, K. Dehmelt, and J. Witte, Zur Kenntnis der Nickelhydroxidelektrode—I. Über das Nickel(II)-Hydroxidhydrat, *Electrochim. Acta* **11**, 1079–1087 (1966).
[52] H. Bode, K. Dehmelt, and J. Witte, Zur Kenntnis der Nickelhydroxidelektrode. II. Über die Oxydationsprodukten von Nickel(II)-Hydroxiden, *Z. Anorg. Allg. Chem.* **366**, 1–21 (1969).
[53] J. Smulko, A. Azens, L. B. Kish, and C. G. Granqvist, Quality assessments of electrochromic devices: the possible use of 1/*f* current noise, *Ionics* **13**, 179–182 (2007).
[54] J. Smulko, A. Azens, R. Marsal, L. B. Kish, S. Green, and C. G. Granqvist, Application of 1/*f* current noise for quality and age monitoring of electrochromic devices, *Solar Energy Mater. Solar Cells* **92**, 914–918 (2008).
[55] P. Nostell, A. Roos, and B. Karlsson, Optical and mechanical properties of sol-gel antireflective films for solar energy applications, *Thin Solid Films* **351**, 170–175 (1999).
[56] S. E. Yancey, W. Zhong, J. R. Heflin, and A. L. Ritter, The influence of void space on antireflection coatings of silica nanoparticle self-assembled films, *J. Appl. Phys.* **99**, 034313 1-10 (2006).
[57] K. J. Cathro, D. C. Constable, and T. Solaga, Silica low-reflection coatings for collector covers, by a dip-coating process, *Solar Energy* **32**, 573–579 (1984).
[58] P. V. Ashrit, K. Benaissa, G. Bader, F. E. Girouard, and V.-V. Truong, Lithiation studies on some transition metal oxides for an all-solid thin film electrochromic system, *Solid State Ionics* **59**, 47–57 (1993).
[59] P. Schlotter, G. Baur, R. Schmidt, and U. Weinberg, Laminated electrochromic device for smart windows, *Proc. Soc. Photo-Opt. Instrum. Eng.* **2255**, 351–362 (1994).
[60] F. Michalak, K. von Rottkay, T. Richardson, J. Slack, and M. Rubin, Electrochromic lithium nickel oxide thin films by RF-sputtering from a $LiNiO_2$ target, *Electrochim. Acta* **44**, 3085–3092 (1999).
[61] Y. Nishikitani, T. Asano, S. Ushida, and T. Kubo, Thermal and optical behavior of electrochromic windows fabricated with carbon-based counterelectrode, *Electrochim. Acta* **44**, 3211–3217 (1999).
[62] L.-C. Chen and K.-C. Ho, Design equations for complementary electrochromic devices: application to the tungsten oxide–Prussian blue system, *Electrochim. Acta* **46**, 2151–2158 (2001).
[63] A. Hauch, A. Georg, S. Baumgärtner, U. Opara Krašovec, and B. Orel, New photoelectrochromic device, *Electrochim. Acta* **46**, 2131–2136 (2001).
[64] M. Wigginton, *Glass in Architecture*, Phaidon, London, UK, 1996.
[65] J. Schütt, J.-C. Giron, F. Béteille, X. Fanton, Electrochromic automotive sunroofs, in: *Proceedings of the 4th International Conference on Coatings on Glass (4th ICCG)*, edited by C.-P. Klages, H. J. Gläser, and M. A. Aegerter, Fraunhofer-Institut für Schicht- und Oberflächentechnik, Braunschweig, Germany, pp. 661–663, 2002.
[66] A. Gavrilyuk, U. Tritthart, and W. Gey, Photo-stimulated proton-coupled electron transfer in quasi-amorphous WO_3 and MoO_3 thin films, *Philos. Mag.* **87**, 4519–4553 (2007).
[67] V. I. Gol'danskii, L. I. Trakhtenberg, and V. N. Fleurov, *Tunneling Phenomena in Chemical Physics*, Gordon & Breach, New York, NY, USA, 1989.
[68] J. M. Bell and I. L. Skryabin, Failure modes of sol-gel deposited electrochromic devices, *Solar Energy Mater. Solar Cells* **56**, 437–448 (1999).
[69] C. M. Lampert, A. Agrawal, C. Baertlien, and J. Nagai, Durability evaluation of electrochromic devices—an industry perspective, *Solar Energy Mater. Solar Cells* **56**, 449–463 (1999).
[70] N. G. van Kampen, *Stochastic Processes in Physics and Chemistry*, Elsevier, Amsterdam, The Netherlands, 1992.
[71] L. K. J. Vandamme, Noise as diagnostic tool for quality, *IEEE Trans. Electron. Devices* **41**, 2176–2187 (1994).
[72] R. A. Cottis, Sources of electrochemical noise in corroding systems, *Eletrokhim.* **42**, 557–566 (2006) [English translation: *Russ. J. Electrochem.* **42**, 497–505 (2006)].

[73] J. M. Smulko, K. Darowicki, and A. Zieliński, Evaluation of reinforcement corrosion rate in concrete structures by electrochemical noise measurements, *Eletrokhim.* **42**, 611–616 (2006) [English translation: *Russ. J. Electrochem.* **42**, 546–550 (2006)].
[74] J. Smulko, Methods of electrochemical noise analysis for investigation of corrosion processes, *Fluctuation Noise Lett.* **6**, R1–R9 (2006).
[75] K. Ishii, editor, *Membrane Designs and Structures in the World*, Shinkenchikusa-sha, Tokyo, Japan, 1999.
[76] K.-M. Koch,editor, *Membrane Structures*, Prestel, Munich, Germany, 2004.
[77] W. Nerdinger, editor, *Frei Otto: Complete Works*, Birkhäuser, Basel, Switzerland, 2005.
[78] S. Robinson-Gayle, M. Kolokotroni, A. Cripps, and S. Tanno, ETFE foil cushions in roofs and atria, *Construction Building Mater.* **15**, 323–327 (2001).
[79] J. E. Fernández, Materials for aesthetic, energy-efficient, and self-diagnostic buildings, *Science* **315**, 1807–1810 (2007).
[80] H. W. Kua and S. E. Lee, Demonstration intelligent building–a methodology for the promotion of total sustainability in the built environment, *Building Environ.* **37**, 231–240 (2002).
[81] D. De Rossi, F. Carpi, and E. P. Scilingo, Polymer-based interfaces as bioinspired 'smart skins', *Adv. Colloid Interface Sci.* **116**, 165–178 (2005).
[82] M. J. Gorman, *Buckminster Fuller: Designing for Mobility*, Skira, Milan, Italy, 2002.

14

Transparent Solar Cells Based on Organic Polymers

Jinsong Huang, Gang Li, Juo-Hao Li, Li-Min Chen and Yang Yang
Department of Materials Science and Engineering, UCLA, USA

14.1 Introduction

Organic electronics carry several prominent advantages over inorganic semiconductors, including light weight, flexibility, low cost, being environmentally friendly and transparent. After two decades of development, organic light emitting diode (OLED) [1] devices have started to be an industry [2]. Polymer solar cell (PSC) research has achieved several breakthroughs [3–6] in the first few years of the 21st century. These fast improvements in performance have distinguished this technology as a promising cost-effective alternative or complimentary technology to inorganic solar cells. The regio-regular poly(3-hexylthiophene) (RR-P3HT) [7, 8] and [6,6]-phenyl C_{60} butyric acid methyl ester (PCBM) [9] bulk heterojunction (BHJ) structure [10, 11] is a representative system for high efficiency PSCs [4–6]. While an external quantum efficiency (EQE) of over 70% [4–6, 12, 13] has been achieved in these PSCs, approaching that of their inorganic semiconductor counterparts, limited absorption in the solar spectrum remains a major limitation to achieving high efficiency. For example, only $\sim$40% of the solar energy is in the wavelength range shorter than 650 nm – the cut-off wavelength of RR-P3HT. While pursuing high efficiency is important, PSCs can provide other useful applications with their intrinsic properties like flexibility and transparency. Transparent or more precisely translucent solar cells could transfer the disadvantages of PSCs into advantages and will enable energy generation in various applications such as window and portable electronics. In addition, transparent solar cells provide an approach to further enhance PSC efficiency.

Transparent Electronics: From Synthesis to Applications Edited by Antonio Facchetti and Tobin J. Marks

It is generally accepted that multiple solar cells utilizing a tandem structure, each absorbing different spectral regions, will provide more efficient solar energy harvesting and thus higher efficiency [14–16]. Efficient light harvesting in the near-infrared region has recently been realized in low band gap polymers such as poly[2,6-(4,4-bis-(2-ethylhexyl)-4*H*-cyclopenta-[2,1-b;3,4-b′]-dithiophene)-*alt*-4,7- (2,1,3-benzothiadiazole)] (PCPDTBT) [17]. In addition to tandem approaches with multiple PSCs on one substrate, an alternative approach is to fabricate efficient transparent PSCs, utilizing (semi-) transparent electrodes, with different spectral responses and stack them together [18]. Several methods have been reported to obtain semitransparent cathodes for polymer light emitting diodes (PLEDs), primarily focusing on two common techniques: sputtering transparent conducting oxides (TCOs) as the top cathode [19–24], and using a multilayer stacked metal cathode [25, 26]. In them, sputtering deposition of TCOs is more likely to bring severe damage to the underlying organic layers.

In this chapter, we summarize our recent work on different approaches to achieve transparent solar cells. In the first section, we discuss the fabrication of multilayer semitransparent cathodes utilizing multiple metal layers, and its application in transparent PLEDs and PSCs. PSC performance enhancement through the multiple-device stacking approach is also described. The second section focuses on device interface engineering. We will first showcase transition metal oxides (V_2O_5, MoO_3) as an effective hole buffer layer replacement for the commonly used **poly(3,4-ethylenedioxythiophene) poly(styrenesulfonate)** (PEDOT:PSS). We will then demonstrate efficient inverted PSCs realized by varying the positions of the hole injection (transition metal oxide V_2O_5) and electron injection (Cs_2CO_3) layers, irrespective of the top and bottom electrodes. These highly transparent transition metal oxide buffer layers potentially will provide protection to the organic layer and enables the application of TCOs to achieve transparent solar cells. We also modified PEDOT:PSS to function as electronic glue, which exhibits excellent mechanical and electrical properties. The fabrication of flexible organic devices using the electronic glue via a lamination process was also demonstrated. Finally, by combining with interface modification, a semitransparent, vacuum-free, self-encapsulated, one-step process was shown for PSCs with 3% efficiency, which represents a critical step towards the ultimate goal of low-cost PSCs.

14.2 Multiple Metal Layer Structure as Transparent Cathode

14.2.1 Single Layer of Semi-Transparent Metal Thin Film

When used to fabricate a transparent cathode, sputtering may either damage the organic layers beneath the cathode, or it has to be at a very slow rate to avoid damage. However, the evaporated metal layers were either reactive metals that may oxidize in ambient environment or a capping layer is required to enhance the transparency of the cathode. The structure, composition and deposition technique of the cathode dictate its final physical properties and device performance. Here, we demonstrate techniques to control the optical properties of the cathode through controlling the morphology of stacked metal layers. Using simple evaporation methods and nonreactive common metals, the optical transparency of a stacked metal cathode was manipulated.

Table 14.1 Optical constants of some bulk metals. Besides Ag, all values of refraction index, extinction coefficient and skin depth are taken at a wavelength of around 563 nm

Metal	eV	Wavelength, λ (nm)	Refraction index, n	Extinction coefficient, k	Skin depth, δ (nm) $= \lambda/4\pi k$
Al	2.2	563.1818182	1.018	6.846	**6.549710303**
Au	2.2	563.1818182	0.24	2.54	**17.6532743**
Ag	2.26	548.2300885	0.06	3.586	**12.172028**
Cu	2.2	563.1818182	0.83	2.6	**17.24589105**
Ca	2.2	563.1818182	0.6	2.1	**21.35205559**
Ni	2.2	563.1818182	1.8	3.33	**13.46526028**
Rh	2.2	563.1818182	2	5.11	**8.774817365**
Zr(polycrystal)	2.2	563.1818182	1.87	0.97	**46.22609973**

Several factors (skin depth, thickness dependence and the evaporation process) are known to have obvious influences on the optical properties of metal thin film. Conventionally, when radiation is shone onto a metal surface, a significant part of the light will travel a short distance, δ, into the metal surface before it is reflected. This distance, known as the skin depth, is the distance where the transmitted light amplitude is reduced to about 1/e, which is about 36.8 % of the incident light amplitude. Table 14.1 shows some suggested values for δ of several metals. Around a wavelength of 563 nm, δ for Al, Au, Cu and Ca is about 6.55, 17.65, 17.25 and 21.35 nm, respectively. Ag has a δ of 12.17 nm around a wavelength of 548 nm. These values were obtained from measurements of bulk metals and may also change depending on the deposition methods of the metal thin films.

Since the skin depth is usually short for metal, the thickness dependence on optical properties is important.

Figure 14.1 shows the transmission vs. wavelength curves for Ag and Al. The transmission graphs are compared with air, which is designated as 100% transmission.

When the metal thicknesses are less than δ, at least 63.2% of the light can penetrate through them. It is interesting to find that the transparency of the film does not necessarily decrease with increasing thickness. This is especially true in the case of Ag and similar traits were observed with Au. This is believed to be caused by the island formation during deposition. At very thin thicknesses, the metal has not yet formed a continuous film and is therefore composed of islands. These islands act as nanoparticles that will scatter and absorb light directed at the metal layer surface. This effect is strongest with Ag because Ag has a lower surface tension γ compared with Au or Cu. Therefore, Au and Cu tend to form a better film on glass compared with Ag due to higher surface tension.

In addition to the skin depth, the morphology of thin film deposited on the substrate is also found vital to the transparency. The morphology of metal thin film can vary depending on the fabrication process, especially the evaporation rate.

Figure 14.2 shows the transmission graphs of 8 nm of silver with different evaporation rates. It is interesting to see that although all films share the same thickness, their transmission curves differ widely.

Figure 14.2 also shows the electrical resistance of the films vs. different evaporation rates. The electrical resistance decreases dramatically as the evaporation rate of the film increases.

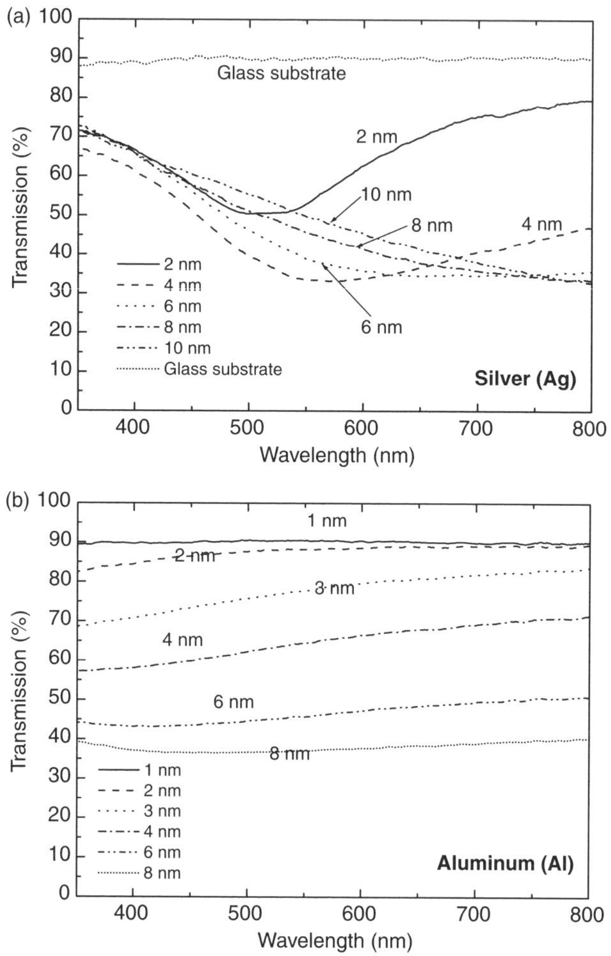

Figure 14.1 *Optical transmission spectra of (a) various thicknesses of Ag with evaporation rate of 0.5 Ås^{-1} and (b) various thicknesses of Al with evaporation rate of 1 Ås^{-1}*

Atomic force microscopy (AFM) images are also used to observe the surface morphology. It was found that the transparency results are consistent with the appearance of the films – as the evaporation rate increases from 0.5 to 4 Ås^{-1}, transparency and the size of the clusters increase gradually. When the size of these clusters falls within a certain optimal range for wavelength scattering and absorption, the transparency of the film will decrease. These voids inside the clusters may not only act as light scattering sites but also account for higher electrical resistance for films with lower evaporation rates. There is more energy during evaporation with higher evaporation rates, and therefore the material has more energy to move around during deposition. The increased movement will reduce voids in the films and cause the films to be more continuous. However, it is observed that the transparency drops at even higher evaporation rate. This is possibly because the time period for the metal vapor is too short to form clusters.

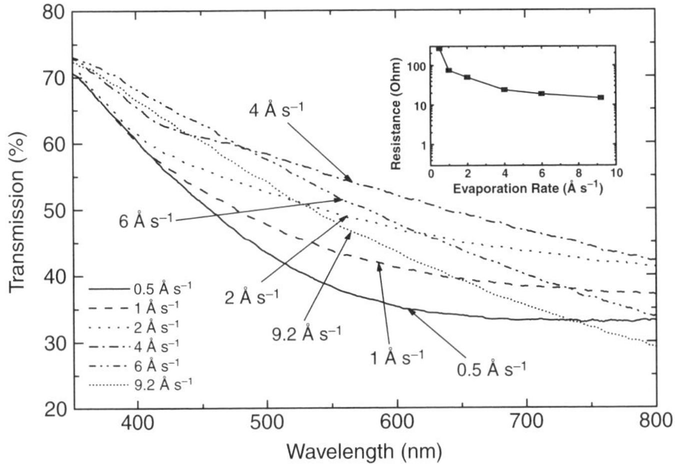

Figure 14.2 Optical transmission spectra of 8 nm of Ag with evaporation rates of 0.5, 1, 2, 4, 6 and 9.2 $\text{Å}s^{-1}$. Inset: electrical resistance of the films vs. different evaporation rates

14.2.2 Stacked Metallic Thin Film for Polymer Light Emitting Devices

Typical PLEDs consist of one or two thin (typically ~100 nm) polymer layers sandwiched between a conductive surface electrode and a transparent substrate. The surface electrode, normally a reflective metallic layer, is the cathode which can be specially designed to facilitate the electron injection, whilst the substrate is typically a standard ITO coated glass, through which the electroluminescence from the polymer layer is emitted [27]. Fabricating a transparent cathode can be utilized for: (1) top-emitting PLEDs for display and communication applications [28]; and (2) transparent displays where light can be observed from both the anode and the cathode [29].

A successful transparent metal cathode needs to have both high optical transparency and electrical conductivity and we solved the problem by utilizing surface properties, controlling evaporation rates (and related morphology), and adjusting individual layer thickness of different metals. The PLEDs integrated with the stacked transparent cathode even without a capping layer provide 47% of the light through the cathode surface and 53% through the ITO surface. Only two types of nonreactive metal are introduced; this method requires no modification of the existing evaporation system.

The transparent cathode PLEDs displayed high transparency when the devices were turned off. Figure 14.3 shows the transmission spectra of the fluorine-containing copolymer (green PF) device with and without the cathodes. All the transmission spectra were normalized with respect to that of ITO on glass. The green PF device without the cathode showed nearly 90% transmittance in the visible range. To illustrate the strong dependence of the transmittance on the constituent and thickness of each individual layer of the cathode, three devices with different cathode structures were fabricated: (1) a single layer of Au; (2) a bilayer structure of Al/Au; and (3) a trilayer structure of Au/Al/Au. The device with only Au as cathode (labeled Au) showed a significant reduction of transmittance (20%) in the visible spectrum. Both the bilayer and trilayer cathode devices (labeled Al/Au and Au/Al/Au)

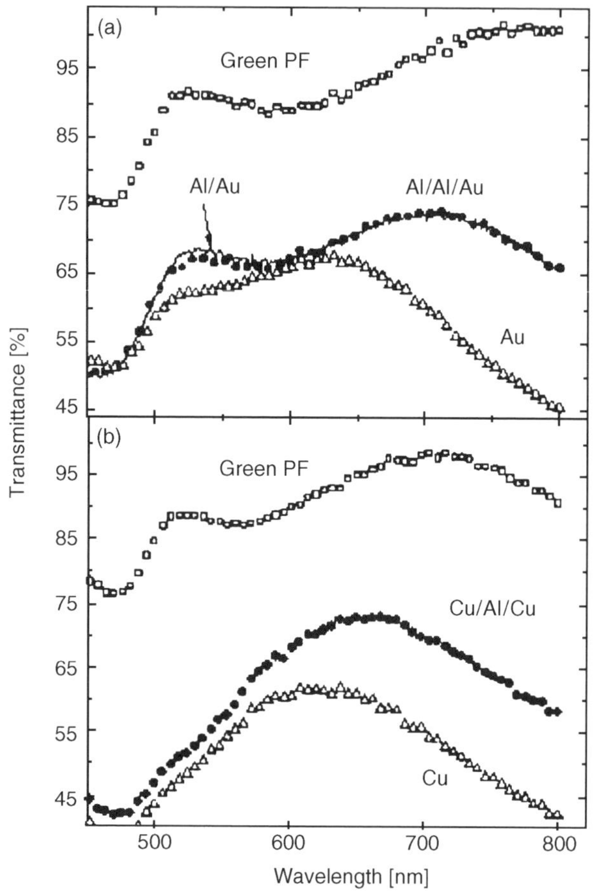

Figure 14.3 *Optical transmission spectra of PLED with various cathode structures in (a) Au and Al combinations and (b) Cu and Al combinations. All the transmission spectra were normalized with respect to that of ITO glass*

demonstrated an enhancement of transmittance compared with the Au only device even though the overall thickness of the cathode increases. A similar effect was also observed in the Ca/Ag bilayer cathode [26]. The enhancement is more pronounced in the spectrum ranging from 600 to 800 nm. The underlying physics of using a trilayered structure for the cathode should not be restricted to the use of Al and Au. We have tried other metal combinations and the effect was similar. Figure 14.3(b) shows the device transmittance when using Cu/Al/Cu and 8 nm Cu as the cathode. Again, the PLEDs showed enhanced transmittance when using a trilayer cathode. These data revealed that the skin depth and morphology of different layers significantly influenced the transmittance behavior of the transparent cathode. The optimum thicknesses of individual layers in the cathode stack for the highest transmittance were found to be 0.2–0.5, 2.0–4.0 and 9.0–11.0 nm for bottom Au, Al and top Au protective layers, respectively. For the case of the Cu/Al/Cu structure, the thickness ratio is 0.2–0.5/2.0–4.0/8.0–11.0 nm, respectively. The overall

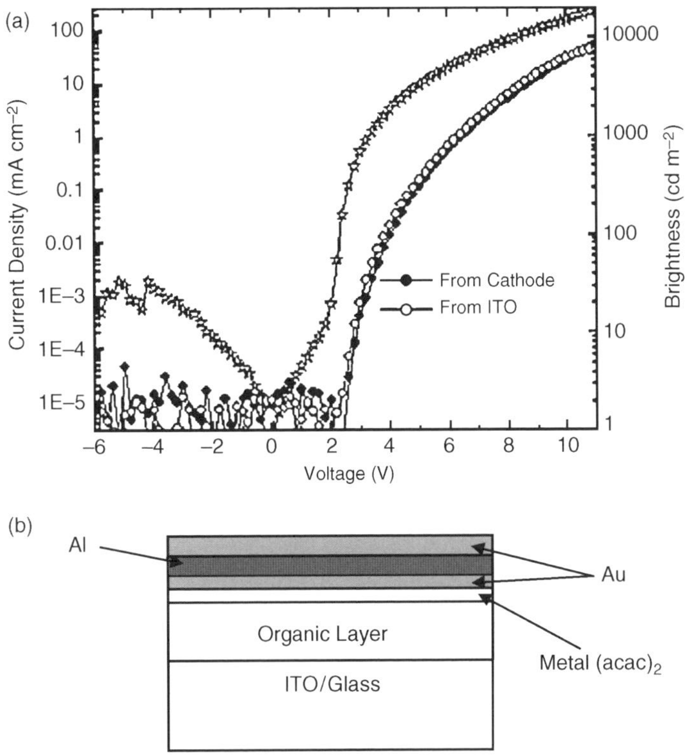

Figure 14.4 *(a) Brightness–current density–voltage characteristics of the PLED with semi-transparent cathode. (b) Cathode structure of PLED device*

transmittance of the cathode determining by optical interference effects, absorption losses and electrical properties can in principle be controlled by tailoring each metal thickness in the cathode.

Figure 14.4 shows the current density vs. voltage characteristics and the electroluminescence of the PLED with a triple Au/Al/Au structured cathode. There was about 10% difference in brightness between the top and bottom emission when the device was turned on. A luminance of 832 and 789 cd m^{-2} were achieved, respectively, from the ITO and cathode surfaces at a current density of 25 mA cm^{-2}. The semi-transparent PLED provided 47% of the electroluminescence through the surface cathode and 53% through the bottom ITO. The electrical sheet resistance of the multilayer cathode was below 10 ohm $square^{-1}$. The total efficiency of the device in Figure 14.4 is about 6.7 cd A^{-1}, which is lower than that (8 cd A^{-1}) of a control device with Al as the cathode. This reduction in efficiency may be due to the weak optical absorbance of the cathode, since the cathode is not completely transparent.

We studied the morphology change of the cathode with AFM as $Ba(acac)_2$, Au and Al layers were deposited. Each modified layer was observed to be on the order of 1 nm. Figure 14.5 shows the AFM images of the surface of the Au film (9 nm), the Al (2 nm)/Au (9 nm) film, and the Au (0.25 nm)/Al (2 nm)/Au (9 nm) film, all evaporated on a $Ba(acac)_2$

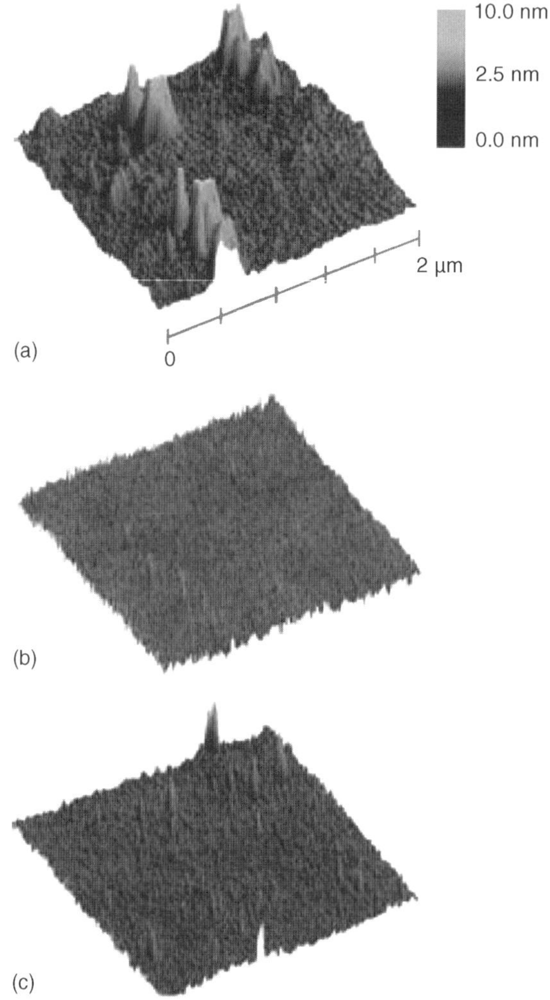

Figure 14.5 *AFM images on the Au surface with (a) [Ba(acac)$_2$] underneath, (b) Al and [Ba(acac)$_2$] underneath and (c) complete cathode structure Au/Al/Au. All the image sizes are 2 μm × 2 μm*

surface. The AFM results showed that the surface of plain Au evaporated on was very rough (surface roughness 2.500 nm) compared with the same Au surface with Al underneath (surface roughness 1.159 nm). The cathodes with Au (9 nm) and with Al (2 nm)/Au (9 nm) have electrical resistivities of 18.1 and 13.15 ohm square^{-1}, respectively. It is worth noting that although the cathode with structure Al (2 nm)/Au (9 nm) is thicker than the cathode with structure Au (9 nm), the transparency of the thicker cathode was higher (Figure 14.3). The role of the 2 nm Al layer was: (1) to interact with the layer so it can lower the work function and improve the electron injection property of the cathode [30]; and (2) to change the surface morphology between the Al and the Au layer.

Comparing the wettability of Au to Al, since Al has shorter skin depth and lower wettability than Au, in principle, it should not have helped the transparency of the cathode. It is believed that the bonding of the metal–metal interface is stronger than the bonding of the metal–organic interface, therefore the wettability of metal on metal is better than the wettability of metal on organic. From the wettability point of view, although Au is better than Al, the formation of Au film on Al is better than Au film on organic surface. Concerning the effect of skin depth, it is speculated that the Al layer may have readily oxidized into transparent aluminum oxide during deposition and as the device was transferred in air to the measuring glove box.

The Al layer enhances the wettability of the deposited Au layer which leads to forming a better Au film with reduction in surface roughness [cf. Figure 14.5(a) and (b)]. This reduction in surface roughness is likely to be the cause of the enhancement in the optical transparency of the cathode. The surface roughness may be an indication of the voids and scattering sites within the film. As the surface roughness increases, it is likely that the voids and scattering sites increase as well. These voids and scattering sites will cause light to scatter and render the film less transparent. It also means that the film is less continuous which makes the cathode less electrically conductive. The device stability is relatively low using either Au or two stacked metal structures as cathode. In fact, both Au and Al/Au stacked metal cathodes decay readily as the device lit up in the glove box. However, a tremendous improvement on the performance and stability can be achieved by evaporating an ultrathin layer of Au before the deposition of Al and the successive Au layer. This ultrathin Au layer is thought to be small Au particles. By virtue of the higher wettability of Au compared with Al, these Au particles acts as a wetting layer for the Al layer and make the later deposited Au layer an even better film. The thickness of this ultrathin Au buffer layer is critical. If the layer is too thick, it may setback the interaction between Al and Ba $(acac)_2$ and if it is too thin then it may not serve as an efficient wetting layer. It is observed that within a certain range of Au thicknesses, the overlaid cathode surface roughness (0.941 nm) is reduced, as shown in Figure 14.5(c), and it is expected that the device performance is improved accordingly [with better interaction between Al and $Ba(acac)_2$]. However, the overall transparency of the cathode is reduced when increasing the buffer layer thickness. The critical thickness of this Au layer was found to be 0.25 nm. The overall cathode thickness with structure Au/Al/Au is 11.25 nm; it has a significant transparency in the visible region, having in addition a low sheet resistance of 10 ohm $square^{-1}$. Thus, the results indicated that the skin depth and morphology evolution were the two dominating factors responsible for the high transmittance in the triple metal layer structure.

The thickness of each individual layer in the trilayer structure is very dependent on the different electron injection buffer layers deposited underneath. In our case, there may be different surface roughnesses due to the preparation method [30]. The above data revealed the fact that the skin depth and morphology of different layers significantly influenced the transmittance behavior of the transparent cathode. From the results presented here, the optimum thicknesses of individual layers in the cathode stack for the highest transmittance were found to be 0.2–0.5, 2.0–4.0 and 9.0–11.0 nm for bottom Au, Al, and top Au protective layers, respectively. The overall transmittance of the cathode determined by optical interference effects, absorption losses and electrical properties can be in principle be controlled by tailoring each metal thickness in the cathode.

14.3 Transparent Metal Oxide for Anode of High Performance Transparent Solar Cell

Gold metal has been demonstrated to be a good transparent electrode in the transparent solar cell due to its high transparency. However, its high evaporation temperature inevitably damages the polymer. The gold is easy to diffuse inside the polymer, which might result in device short or failure. In this section, we introduce several metal oxides, vanadium oxide (V_2O_5) and molybdenum oxide (MoO_3), which can be evaporated onto polymer at low temperature. These metal oxides serve as anode buffer layers and protect the underneath polymer layer from evaporation damage. We also demonstrated that these metal oxide can be candidates to replace PEDOT:PSS for higher device stability.

14.3.1 Transition Metal Oxides as Hole Buffer Layers in Organic Photovoltaics

It has been demonstrated that using an ITO surface coated with a buffer layer as the anode results in enhanced device performance [31]. The external quantum efficiency of polymer photovoltaic (PV) cells is greatly improved when PEDOT:PSS is used as a buffer layer [32]. However, the interface between ITO and PEDOT:PSS is not stable and the chemical reaction between ITO and PEDOT can result in degraded device performance [33]. Several metal oxides have been demonstrated as efficient hole injection materials for organic electroluminescent devices [34]. Here transition metal oxides, V_2O_5 and MoO_3, are demonstrated to effectively substitute PEDOT:PSS as the buffer layer in polymer PV cells [35], which makes high performance transparent solar cells feasible.

V_2O_5 and MoO_3 layers were thermally evaporated onto ITO substrates under a vacuum of $\sim 10^{-6}$ Torr. A PEDOT:PSS layer of about 25 nm thickness, formed by standard procedure on ITO, was used for comparison with metal oxide buffer layers. The energy level diagrams of the different materials used in the device fabrication are shown in Figure 14.6.

The current density–voltage characteristics under illumination for PV devices with various buffer layers are shown in Figure 14.7. For the device with UV-ozone-treated ITO as the anode, the power conversion efficiency (PCE) is less than 2.0%. Although the short circuit current (J_{SC}) is high ($\sim$7.8 mA cm^{-2}), the open circuit voltage (V_{OC}) and fill factor (FF) are comparatively low at 0.49 V and 51%, respectively. Inserting a metal oxide buffer layer between ITO and polymer results in increased V_{OC} and FF for both V_2O_5 and MoO_3. The V_{OC} and FF values are 0.59 V and 59%, respectively, for V_2O_5 ($\sim$3 nm), and 0.60 V and 62% for MoO_3 ($\sim$5 nm). The J_{SC} also increases to 8.83 and 8.94 mA cm^{-2} for V_2O_5 and MoO_3, respectively. As a result, the PCE of devices with metal oxide buffer layers increases to more than 3.0%. The highest PCE of 3.3% is obtained for the devices with MoO_3 buffer layer thickness of $\sim$5 nm. Also shown in Figure 14.7 is the current density–voltage curve for a device with an ITO/PEDOT:PSS anode for comparison. The PCE of this device is 3.2%, with $J_{SC} = 8.95$ mA cm^{-2}, $V_{OC} = 0.59$ and FF = 60%. These results show that both metal oxides, V_2O_5 and MoO_3, can effectively substitute PEDOT:PSS as the buffer layer with comparable or even better device performance. Interestingly, V_{OC} is the same for all devices with different buffer layers, irrespective of the work function of the buffer layer material [4.7 eV (V_2O_5), 5.3 eV (MoO_3) and 5.2 eV (PEDOT:PSS)]. (The work functions of V_2O_5 and MoO_3 were measured using ultraviolet photoelectron spectroscopy.) This suggests that the V_{OC} of the device is independent of the work function of the anode material. It has been

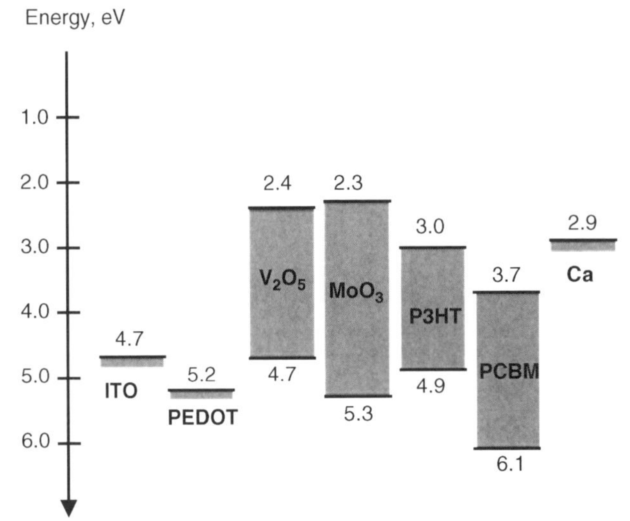

Figure 14.6 *A schematic diagram showing energy levels of the bottom electrode ITO and various other layer materials (PEDOT:PSS, V_2O_5, MoO_3, donor polymer P3HT, acceptor PCBM and top electrode Ca)*

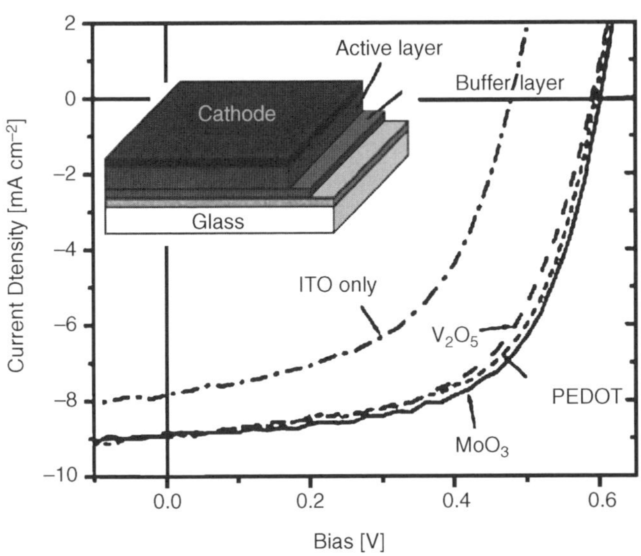

Figure 14.7 *Current density–voltage characteristics of PV devices under illumination for a device with different types of anode, namely, ITO only, ITO with PEDOT:PSS (25 nm), ITO with V_2O_5 (3 nm) and ITO with MoO_3 (5 nm)*

Table 14.2 *Device operation parameters for devices with different types of anodes fabricated in this study. The numbers in bold represent devices whose characteristics are plotted in Figure 14.7*

Anode	I_{SC} (mA cm^{-2})	V_{OC} (V)	FF (%)	PCE (%)
ITO only	**7.82**	**0.49**	**51.1**	**1.96**
ITO/PEDOT:PSS (25 mn)	**8.95**	**0.59**	**59.6**	**3.18**
ITO/V_2O_5 (1 nm)	8.86	0.59	47.5	2.48
ITO/V_2O_5(3 nm)	**8.83**	**0.59**	**59.1**	**3.10**
ITO/V_2O_5 (5 nm)	8.54	0.59	57.2	2.88
ITO/V_2O_5 (10 nm)	8.16	0.59	57.9	2.79
ITO/MoO_3 (1 nm)	8.75	0.53	42.3	1.98
ITO/MoO_3 (3 nm)	8.86	0.59	58.3	3.06
ITO/MoO_3(5 nm)	**8.94**	**0.60**	**61.9**	**3.33**
ITO/MoO_3 (10 nm)	8.73	0.60	59.8	3.13
ITO/MoO_3 (20 nm)	8.19	0.58	59.9	2.86

reported earlier [36] that the work function of the cathode metal slightly affects the V_{OC} of the PV cell. One possible explanation for the observed independence of V_{OC} on the buffer layer in our case is the Fermi-level pinning [37] of the metal oxide work function at the HOMO level of *p*-type polymer P3HT due to the presence of donor surface states. It is widely believed that the V_{OC} of polymer BHJ PV devices depends mainly on the relative energy levels of the donor and acceptor [38]; thus the V_{OC} being independent of the buffer layer work function is understandable. A summary of the various device parameters for all types of devices made in this study is given in Table 14.2. It is important to take into consideration the energy levels of the materials, including the buffer layer, when designing the PV device. The holes, which transport through the P3HT network, will be collected at the anode. As shown in Figure 14.6, the HOMO level of P3HT is 4.9 eV. V_2O_5 (HOMO ~4.7 eV) and MoO_3 (HOMO~5.3 eV) both form efficient hole injection/collection contact with the active layer. For comparison, the HOMO level of PEDOT:PSS is 5.2 eV.

Introducing V_2O_5 or MoO_3 buffer layers will contribute to the series resistance (R_S) of the device and larger thickness may result in lower current. An increase in R_S will also lower the FF of the devices [39]. However, a sufficient oxide layer thickness is required to form a uniform contact with the polymer to reduce leakage current. Therefore, it is necessary to optimize the metal oxide layer thickness in order to obtain optimum device performance. The current density–voltage characteristics of devices with different oxide layer thickness are shown in Figure 14.8. For V_2O_5 the optimum thickness is 3 nm [Figure 14.8(a)], and for MoO_3 it is 5 nm [Figure 14.8(b)]. For both oxides, when the thickness is about 1 nm or less, the ITO substrate is only partially covered with the oxide, forming small islands instead of a uniform film. As a result, the anode/polymer contact has two different interfaces: ITO/polymer and ITO/oxide/polymer. The current density–voltage curves for devices with 1 nm oxide layer can be thought of as two separate curves overlapping each other, with a different characteristic corresponding to the ITO/polymer and ITO/oxide/polymer contacts. Since the V_{OC} values for these two curves are different, the overlap of the two competing curves will give the visible concave feature in the resultant curve of the device and as a result, a poor FF.

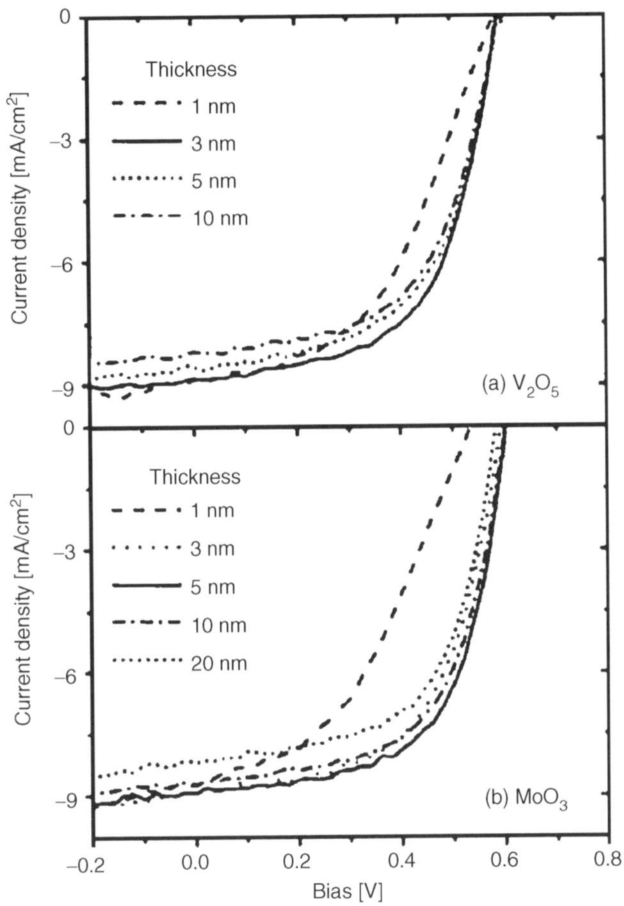

Figure 14.8 *Current density–voltage characteristics under illumination for devices with different thicknesses of (a)* V_2O_5 *and (b)* MoO_3 *used as the buffer layer between the ITO and the active layer*

To summarize, transition metal oxides, V_2O_5 and MoO_3, have been demonstrated to be an efficient buffer layer for polymer PV devices. The oxide layer plays an important role in preventing an unwanted chemical reaction between the ITO and the active layer. The V_{OC} of the devices was observed to be independent of the work function of the metal oxide used. Also, devices with an oxide buffer layer had a similar, or better, performance than the devices with a PEDOT:PSS buffer layer. The highest power conversion efficiency of 3.3% was achieved for the PV devices with MoO_3 as the buffer layer.

14.3.2 Inverted and Transparent Polymer Solar Cells Using Metal Oxide Anodes

It is interesting from both scientific and application points of view to invert PSCs. In the invert configuration, a dielectric capping layer can be used to enhance light trapping. As described above, the metal oxides (V_2O_5 and MoO_3) can provide sufficient protection to the active organic layer while maintaining good device performance. Here an enhancement in PSC efficiency by using cesium carbonate (Cs_2CO_3) interfacial buffer layer at the cathode is demonstrated. Then it is shown that the polarity of solar cells can be reversed by changing the position of V_2O_5 (hole injection/extraction) and Cs_2CO_3 (electron injection/extraction) interfacial layers, independent of the top and bottom electrodes. An efficient inverted PSC is fabricated with the device structure ITO/Cs_2CO_3/P3HT:PCBM/V_2O_5/metal. Based on this inverted structure, a transparent PSC was enabled by simply using a Au thin layer as transparent metal electrode [40]. The combination of transition metal oxide and Au as anode is superior to previous LiF/Au and Ca/Au transparent electrodes due to matched energy levels and higher stability.

LiF is an effective cathode interfacial layer for both PLEDs [41] and PSCs [39]. Cs_2CO_3 is a relatively new interfacial material, first reported for OLED applications by the Canon group [42]. Unlike LiF, in an OLED the function of Cs_2CO_3 is insensitive to the metal electrode above it. Our group has demonstrated a white PLED with 16 lm W^{-1} efficiency using this method [43]. In Figure 14.9, the current density–voltage curves for four different PSCs are shown, with different interfacial layers at the ITO and Al interfaces. In the device with no buffer layer (ITO/blend/Al), reasonable PV effect was observed with J_{SC} of 4.75 mA cm^{-2}. However, V_{OC} and FF are poor at 0.22 V and 28.5%, resulting in a PCE of only 0.23%. Modifying the ITO anode by PEDOT:PSS provides significant improvement in the device performance where J_{SC} increases to 7.44 mA cm^{-2}, V_{OC} to 0.42 V, FF to 51.8%, and the overall efficiency to 1.25%. Furthermore, insertion of 1 nm thermally evaporated Cs_2CO_3 layer at the polymer/Al interface leads to a reduction in J_{SC} to 5.95 mA cm^{-2}. However, the V_{OC} increases to 0.52 V, and an excellent FF of 65.6% is achieved. This results in a PCE of

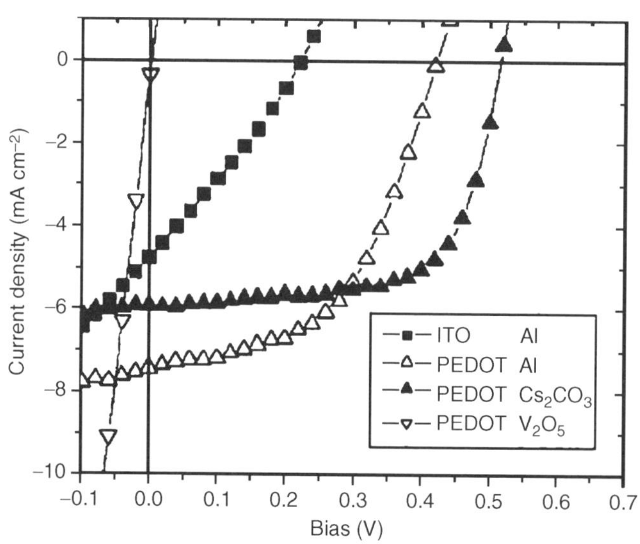

Figure 14.9 *The effect of interfacial layers on the performance of conventional polymer solar cells. The interfacial layers at ITO/blend and blend/Al interfaces are: none, none (solid square); PEDOT:PSS, none (open triangle); PEDOT:PSS, Cs_2CO_3 (solid triangle); and PEDOT:PSS, V_2O_5 (open inverted triangle). The light intensity is 130 mW cm^{-2}*

1.55%, a 25% improvement. These results clearly show that Cs_2CO_3 can act as a functional interfacial layer to enhance PSC efficiency. The work function of PEDOT:PSS is 5.0 eV, which is 0.3 eV higher than that of ITO (4.7 eV). This work function increase can explain the increase in V_{OC} by 0.2 eV according to the metal-insulator-metal (MIM) model [44]. This apparently contradicts the common belief that the energy level difference between the donor HOMO and the acceptor LUMO levels dominates [38] the V_{OC} in the BHJ PSC. However, the contact changes from nonohmic in the case of ITO to ohmic for PEDOT:PSS, and both electrodes being ohmic [45] is a necessary condition for the above belief to be valid. An earlier study on Cs_2CO_3 indicated that during thermal evaporation, Cs_2CO_3 decomposes into cesium oxide. Depending on the film thickness, the resulting cesium oxide has a field emission work function of ~1.1 eV [46]. Ultraviolet photoemission spectroscopy (UPS) measurements conducted in our laboratory on thermally evaporated Cs_2CO_3 films show a work function of 2.2 eV [47]. The polymer/Cs_2CO_3 contact is therefore ohmic. An increase in V_{OC} by only 0.1 V, despite the work function difference between Cs_2CO_3 and Al of 2 eV, agrees well with the earlier observation by Brabec *et al.* [38] and indicates Fermi-level pinning. The V_{OC} for the ITO/V_2O_5/blend/Al device (current density–voltage curve not shown here) is 0.38 eV, also significantly higher compared with bare ITO. The HOMO level of a thermally evaporated V_2O_5 film was determined by UPS to be 4.7 eV, which is identical to that of ITO. The most plausible reason for V_{OC} enhancement is the formation of surface dipoles between V_2O_5 and the active layer, which causes an upward shift in work function of at least 0.2 eV. These results indicate that V_2O_5 can be an effective hole injection layer, like PEDOT:PSS, with a similar effective work function. This is further evidenced by the current density–voltage curves for an ITO/PEDOT:PSS/polymer blend/V_2O_5/Al device, where no PV effect was observed (Figure 14.9). The anode contact therefore has an important effect on PSC performance.

Figure 14.10 shows current density–voltage curves for various inverted PSC structures. The ITO/polymer blend/V_2O_5 (10 nm)/Al inverted solar cell has $J_{SC} = 6.97$ mA cm^{-2}, $V_{OC} = 0.30$ V, FF = 41.2% and PCE of 0.66%. This provides further evidence for the presence of surface dipoles that enhance the V_2O_5 work function by ~0.3 eV. An efficient inverted PSC can be achieved with the structure ITO/Cs_2CO_3/polymer blend/V_2O_5/Al, where Cs_2CO_3 was either thermally evaporated or spin-coated. The current density–voltage curves in Figure 14.10 for solar cells with thermally evaporated (1 nm, open circle) and solution processed (solid triangle) Cs_2CO_3 clearly demonstrate efficient inverted solar cells. The J_{SC}, V_{OC} and FF are very similar for the evaporated (8.42 mA cm^{-2}, 0.56 V and 62.1%) and the solution processed (8.78 mA cm^{-2}, 0.55 V and 56.3%) devices, with overall efficiencies 2.25% and 2.10%, respectively. Therefore, inserting V_2O_5 and Cs_2CO_3 interfacial layers can result in efficient conventional as well as inverted PSCs. LiF (1 nm, electron injection layer) and V_2O_5 (10 nm, hole injection layer) were also used to fabricate inverted solar cells. This device has a current density comparable with the device with Cs_2CO_3 and V_2O_5, but an antidiode behavior results in low V_{OC} (0.39 V), FF (40.7%) and PCE (0.99%). A thin LiF layer has been reported to work well with a wide range of metals (Ca, Al and Au) in conventional device configuration. The presence of an antidiode in the inverted configuration indicates that the LiF growth pattern (on ITO versus on polymer) might have a significant impact on device performance.

Recently we have increased the efficiency of an inverted BHJ PSC based on RR-P3HT: PCBM with a low temperature annealed interfacial buffer layer, cesium carbonate

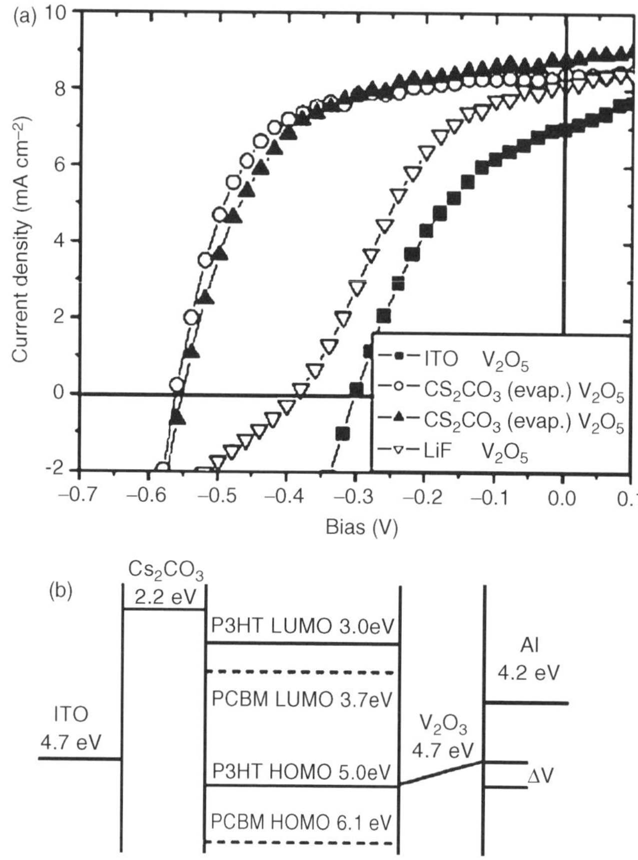

Figure 14.10 *(a) Current density–voltage characteristics of inverted polymer solar cells. The interfacial layers at ITO/blend and blend/Al interfaces are: none, V_2O_5 (solid square); Cs_2CO_3 (evaporated), V_2O_5 (open circle); Cs_2CO_3 (solution processed), V_2O_5 (solid triangle); and LiF, V_2O_5 (open inverted triangle). (b) Energy level diagrams for various materials in the inverted solar cells*

(Cs_2CO_3) [48]. This approach improves the PCE of the inverted cell from 2.3 to 4.2%, with J_{SC} of 11.17 mA cm^{-2}, V_{OC} of 0.59 V, and FF of 63% under AM 1.5G 100 mW cm^{-2} irradiation, which is comparable with the regular structure device on the same system. Contact angle measurement indicates the surface property of Cs_2CO_3 varies from hydrohphilic to hydrophobic upon annealing, which is beneficial for polymer film growth. UPS shows that the work function of the annealed Cs_2CO_3 layer decreases from 3.45 to 3.06 eV. Further X-ray photoelectron spectroscopy (XPS) reveals that Cs_2CO_3 can decompose into Cs_2O (*n*-type) doped with Cs_2O_2 upon annealing, which accounts for the improved device efficiency due to a lower interface resistance.

In the conventional device structure, introducing 1 nm thermally evaporated Cs_2CO_3 reduces the photocurrent but improves V_{OC} and FF significantly, indicating possible physical damage. However, in the inverted structure, where Cs_2CO_3 is deposited on ITO substrates, all three parameters improved. Based on the information collected, V_2O_5 can be treated as a hole injection layer with an 'effective' work function of ~5.0 eV and Cs_2CO_3 as an electron injection layer with very low work function, both of which provide ohmic contacts. The polarity of the device is determined by the relative positions of these two interfacial layers and is insensitive to the conducting electrodes. The energy level diagrams for various inverted configurations are illustrated in Figure 14.10(b).

Due to the presence of a sufficiently thick (10 nm) V_2O_5 layer which protects the underlying polymer, the inverted configuration is especially suitable for making transparent solar cells. A thick Al top electrode was replaced with 12 nm of Au in the inverted structure. When illuminated from the ITO side, the device shows an overall efficiency of 0.85%. However, when illuminated from the semitransparent Au electrode, the PCE is 0.52%. The difference between the two current density–voltage curves is due to the partial loss by the reflection and absorption at the semi-transparent Au electrode. In a transparent solar cell, the light absorption is less than that in a device with a reflecting metal electrode. When designing a tandem PSC, the optical losses due to the first transparent solar cell should be reduced. To provide sufficient electrical conductance, Au layer thickness has to be sufficient and the optical loss at the Au electrode becomes significant. However, the inverted solar cell structure has a metal oxide layer that is not only transparent but also provides effective protection for the polymer layer. A transparent conductive oxide electrode, such as ITO, can therefore be deposited without compromising device performance. This structure thus provides a very efficient method for realizing tandem PSCs for improving device efficiency.

To summarize, efficient conventional and inverted PSCs using different functional interfacial layers have been fabricated. Efficiency up to 4.2% has been achieved for an inverted PSC with FF as high as 62.1%. Preliminary efforts have also demonstrated transparent PSCs with 0.85% efficiency.

14.4 Transparent Solar Cell Fabricated by Lamination

It is desirable that solar cell devices can be fabricated using a low-cost roll-to-roll fabrication process. One critical issue in this fabrication process is how to form the active layer/cathode mechanical and electronic contacts. The lamination process is one very promising technique to fulfill this requirement owing to its simplicity and low cost. Some methods have been demonstrated for laminating two organic films [49, 50] but they are either only suitable for special materials or are impractical. In this section, an electronic glue-based lamination process combined with interface modification is presented as a one-step process for semi-transparent PSC fabrication. We first introduce a simple approach to develop solvent-free electric glue using a conducting polymer. This electric glue exhibits a conductivity of 10^2 S cm^{-1} and could effectively laminate various materials electrically and mechanically. Then we report the fabrication of a transparent PSC using this conducting electronic glue by a direct lamination process.

14.4.1 Conducting Polymer as Electronic Glue

Many applications have been discovered for conducting polymers but electric glue from conducting polymers has not been developed so far. The potential for developing electric glue using conducting polymers is high, since most types of glue are made of polymers. In addition to application in a roll-to-roll fabrication process, electric glue using conducting polymers may replace conventional toxic lead solders in flexible electric circuits. It also has better compatibility with organic devices and the plastic substrates. PEDOT:PSS has emerged as the most important conducting polymer because of its availability in aqueous solution, high transparency in the visible range and excellent thermal stability [51]. It has been extensively used in organic electronic devices, particularly in optoelectronic devices [52]. Recently, it has been discovered that the conductivity of a PEDOT:PSS film could be enhanced by about two orders of magnitude by adding an organic compound with multiple polar groups into the PEDOT:PSS aqueous solution or by treating the PEDOT:PSS film with an organic compound having multiple polar groups [53, 54]. When the organic compound with multiple polar groups is solid at room temperature, it needs to be heated at a temperature higher than its melting point for conductivity enhancement.

A new application for high-conductivity PEDOT:PSS films has been discovered when an appropriate organic compound is added into its aqueous solution or when the film is treated with an appropriate compound, such as D-sorbitol. The high conductivity PEDOT:PSS film can adhere various films together, so that the PEDOT:PSS film can serve as an electric glue with high transparency. This adhesive function was not observed with an untreated PEDOT: PSS film. When epoxy, which is a widely used glue, was blended into PEDOT:PSS, the PEDOT:PSS film either lost its conductivity or acted as a glue. The PEDOT:PSS films modified by the three approaches all have the same function, but the method of thermally depositing D-sorbitol on the PEDOT:PSS film is the most controllable; this is represented as PEDOT:PSS(D-sorbitol).

Three approaches were developed to modify PEDOT:PSS as a transparent electric glue. The first approach was to add D-sorbitol, or other organic compounds with multiple hydroxyl groups, into the PEDOT:PSS aqueous solution. The polymer film with a thickness of 30–50 nm was formed by spin-coating this blended solution on various substrates. After baking at 90 °C for 60 min, the blended film was used for lamination. In the second approach, PEDOT:PSS films were first formed by spin-coating the aqueous PEDOT:PSS solution onto the substrates. Then, a thin layer of D-sorbitol with a thickness of 10 nm was formed on the PEDOT:PSS film. The third approach is similar to the second approach except that the D-sorbitol layer on PEDOT:PSS was formed by thermal evaporation.

The PEDOT:PSS(D-sorbitol) film is able to laminate two substrates well, and the lamination process is free of solvent. One substrate can be a flexible substrate, such as plastic or plastic coated with indium tin oxide (plastic/ITO), and the other substrate can be flexible or rigid, such as plastic, plastic/ITO, glass or glass/ITO. The PEDOT:PSS(D-sorbitol) can be deposited on either of the two laminated substrates. The plastic substrates used in our experiments were poly(ethylene terephthalate) (PET). Moreover, lamination can also take place between a plastic or glass substrate (with or without ITO coating) coated with a PEDOT:PSS(D-sorbitol) film and a film of a polymeric semiconductor, such as poly-(2-methoxy-5-(2′- ethylhexyloxy)-*p*-phenylene vinylene) (MEH-PPV). Figure 14.11(a)–(c) shows the laminated structures. The laminated structure of Figure 14.11(c) has been fabricated by laminating PET/ITO/PEDOT:PSS(D-sorbitol) with PET/Al/MEH-PPV and D-sorbitol in contact with MEH-PPV. This structure is described by PET/ITO/PEDOT:PSS (D-sorbitol)//MEH-PPV/Al/PET. [The double slash (//) is used to indicate the lamination of two parts.] The laminated structures presented in Figure 14.11(a) and (b) have been fabricated

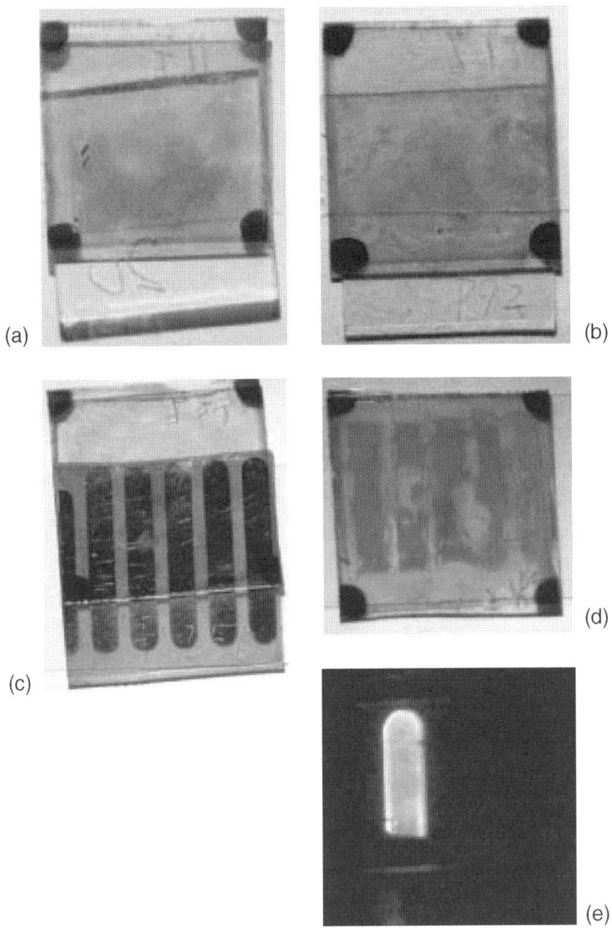

Figure 14.11 Laminated structures (the double slash is used to indicate the lamination of two parts): (a) PET/ITO/PEDOT:PSS(D-sorbitol)//ITO/PET; (b) glass/PEDOT:PSS(D-sorbitol)//ITO/PET; (c) plastic/ITO/PEDOT: SS(D-sorbitol)//MEH-PPV/Al/plastic; (d) PET/ITO/PEDOT:PSS/D-sorbitol/MEH-PPV separated from the laminated structure of PET/ITO/PEDOT:PSS(D-sorbitol)//MEH-PPV/glass; (e) laminated polymer light emitting diode, plastic/ITO/PEDOT:PSS (D-sorbitol)//MEH-PPV/Al/PET, under an electric field

through a similar process but a PET/ITO substrate and glass substrate were used in place of the PET/Al/MEH-PPV in Figure 14.11(a) and (b), respectively.

The glue-like properties of the PEDOT:PSS(D-sorbitol) film are due to the presence of D-sorbitol. After treatment at a temperature higher than its melting point (98–100 °C), PEDOT:PSS(D-sorbitol) is able to laminate two films very well mechanically, which has been further evidenced by two experiments. In the first, a glass/ITO substrate was laminated with a plastic/ITO substrate coated with PEDOT:PSS(D-sorbitol). The lamination was so effective that the PEDOT:PSS layer was transferred from the ITO layer of the plastic/ITO substrate to the ITO coating of the glass/ITO substrate when the two laminated substrates were separated by mechanical force. The second experiment was the lamination of a plastic/ITO substrate coated with PEDOT:PSS(D-sorbitol) with a glass substrate coated with a MEH-PPV film.

The lamination results in good electrical contact between the two parts as well. When two plastic/ITO substrates coated with a 50 nm PEDOT:PSS film were laminated with a 10 nm layer of D-sorbitol between the two PEDOT:PSS films, the electric contact between the two ITO layers was so good that the resistance from one side of the bottom ITO layer to the opposite side of the top ITO layer was almost the same as that from opposite sides of a single ITO layer. The PEDOT:PSS(D-sorbitol) film exhibited a conductivity of $10^2\,S\,cm^{-1}$, as determined by the four-point contact technique. The PEDOT:PSS(D-sorbitol) film exhibited good conductivity along the vertical direction as well. D-sorbitol melts and penetrates into the polymer at temperatures above 98–100 °C, and the presence of a thin layer of D-sorbitol does not significantly affect the charge transport. A PLED has been fabricated by using PEDOT:PSS(D-sorbitol) as the buffer layer. This PEDOT:PSS(D-sorbitol) layer was heated at 130 °C for 10 min. The device exhibits similar performance as the device using PEDOT:PSS as the buffer layer (Figure 14.12).

The excellent lamination of an organic film with a conductive film renders possible the fabrication of various electronic devices through a lamination process. The PEDOT:PSS film has been widely used as the buffer layer or electrode in optoelectronic devices, such as in PLEDs or OLEDs and PV cells [12, 13]. We have demonstrated that these optoelectronic devices can be fabricated through a lamination process using PEDOT:PSS-based electric glue.

Figure 14.11(c) shows a PLED fabricated through a lamination process as presented in Figure 14.13. The device has an area of about 2 mm × 6 mm. Figure 14.11(e) shows the device under an electric field in the dark, which emits light homogeneously. The current density–voltage and light–voltage curves of the device are presented in Figure 14.14, and indicate good diode behavior. The turn-on voltage is 3.6 V and the efficiency is $0.017\,cd\,A^{-1}$. This efficiency, although two orders of magnitude lower than that of a device made by the regular fabrication process, is almost the same as that of the glass/ITO/PEDOT:PSS/MEH-PPV/Al device fabricated by the conventional bottom-up process [55] and indicates a good contact through the lamination process. The higher turn-on voltage and the lower efficiency of the laminated devices compared with the devices fabricated through the bottom-up process with calcium as the cathode [56] are not due to the electric glue of PEDOT:PSS or the lamination process but to the high work function of the aluminum cathode.

14.4.2 Lamination of Transparent Polymer Solar Cell

The lamination process was applied in PSC fabrication, and a semi-transparent solar cell was formed by using two transparent metal oxide electrodes. The fabrication process is illustrated in Figure 14.15(a), and can be described by the following steps. In Step I, two transparent substrates coated with a transparent conductor such as ITO, fluorine-doped tin oxide (FTO) or a high conductivity polymer are selected. In Step II, one substrate is coated with a very thin buffer layer (Cs_2CO_3 [43, 57]) to act as the low work function cathode, followed by coating of the active polymer layer. Step III involves the coating of conductive polymer glue to the other transparent substrate. Step IV is the lamination process: after drying both the substrates, they were laminated together by exerting force so that the two substrates are tightly glued together.

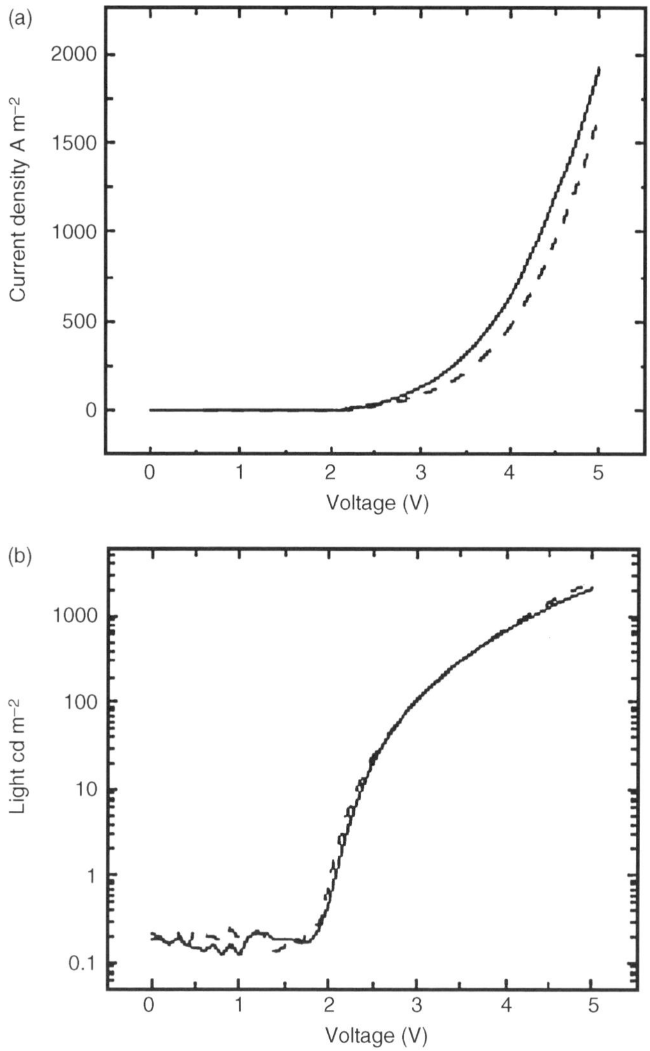

Figure 14.12 *(a) Current density–voltage and (b) light–voltage curves of glass/ITO/PEDOT:PSS/MEH-PPV/Ca/Al (solid line) and glass/ITO/PEDOT:PSS/D-sorbitol (10 nm)/MEH-PPV/Ca/Al (dashed line) fabricated by the traditional bottom-up process*

During this lamination, a plastic rod with proper hardness rolls the plastic substrate to remove air bubbles. Both substrates were heated to a temperature of 105–120 °C during the lamination process, and the finished devices were then kept on the hotplate for 5–10 min for the final heat treatment [5]. A 200 nm thick RR-P3HT:PCBM (1:1 w/w ratio) polymer blend film was deposited by the slow-growth method (or solvent annealing) to enhance device efficiency. Figure 14.15(b) shows an all-plastic solar cell, and the device area is ca. 40 mm^2. With both cathode and anode being transparent, a semi-transparent PSC is formed. The transparency (T%) of the device is shown in Figure 14.15(c), together with the solar

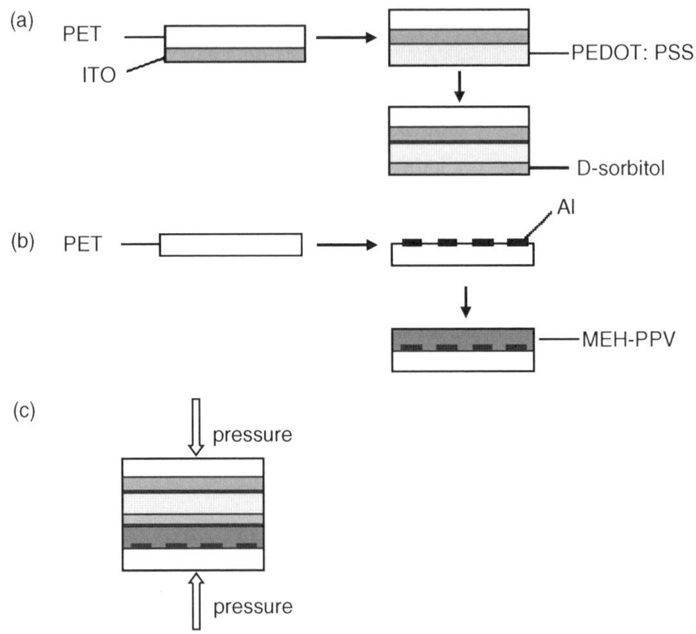

Figure 14.13 *Schematic demonstration of the fabrication process of a polymer light emitting diode by lamination, PET/ITO/PEDOT:PSS(D-sorbitol)//MEH-PPV/Al/PET: (a) preparation of the anode; (b) preparation of the cathode; (c) the lamination process*

illumination spectrum. A transparency of around 70% was obtained in the wavelength range where polymer/PCBM has no absorption, which makes this device suitable for application in stacking devices to make full use of the solar spectrum.

This device fabrication method has many advantages over the regular procedure. First of all, no thermal evaporation process is involved in the process, and each layer is coated by a low-cost and easy solution process. Secondly, in contrast to the reactive metal cathode in regular devices, the cathode in this device is very stable in air. Thirdly, these devices are self-encapsulated if proper substrates are used. Fourthly, this method is potentially easy to apply to large-area device fabrication through a roll-to roll process. Therefore, all-plastic devices can be realized by using plastic materials for both the substrates.

The devices show high-quality mechanical and electrical contact between the laminated components. For a device area of ca. 40 mm^2, the rectification ratio is two to three orders of magnitude at ± 2 V, demonstrating good diode behavior. The series resistance of $12 \pm 3\ \Omega\ cm^2$ is derived from the slope of dark current at a high driving voltage of 2 V, which is several times higher than that of a regular device. One possible origin of this higher series resistance is the high resistance of the ITO on plastic substrate used in this work (150 ohm $square^{-1}$ compared with ca. 15 ohm $square^{-1}$ for glass substrate). Despite the high sheet resistance, our laminated device has surprisingly high performance. Photocurrents were measured with light shining from either side of the device, as well as with a piece of white paper placed behind the device. Figure 14.16(a) shows the photocurrent density of the device under AM 1.5 simulated illumination of 100 mW cm^{-2} (spectral mismatch corrected). The active area is defined by a

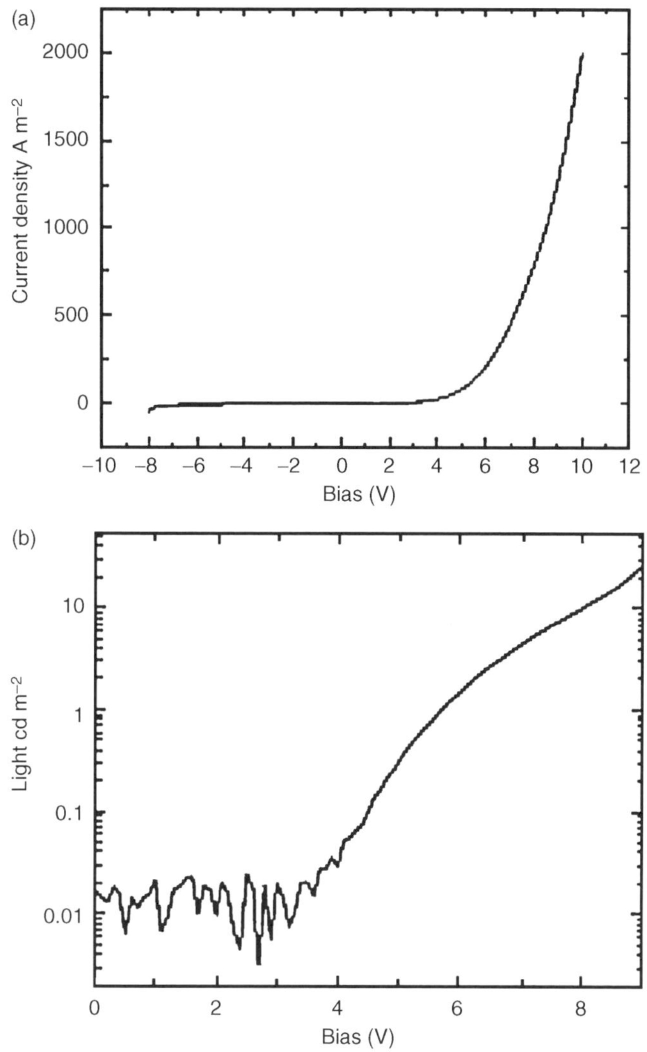

Figure 14.14 *(a) Current density–voltage curve and (b) light–voltage curve of a laminated PET/ITO/PEDOT:PSS(D-sorbitol)//MEH-PPV/Al/PET polymer light emitting diode*

photo mask of 4 mm^2. J_{SC} of 11 mA cm^{-2} from the ITO/PEDOT:PSS side and 10 mA cm^{-2} from the ITO/Cs_2CO_3 side were obtained. The J_{SC} values were actually a little higher than those obtained from regular devices [5]. Figure 14.16(b) shows the EQE of the device photoexcited from the PEDOT:PSS side with white paper on the back. The purpose of placing a piece of white paper behind the device is to reflect the unused light back to the device for secondary absorption. The maximum EQE obtained is 67% at 514 nm for the transparent device, which further increases to almost 70% with a piece of white paper behind. One possible reason for the higher EQE than that of a regular device is the high conductivity PEDOT:PSS anode penetrates into the polymer layer and forms a high surface area structure

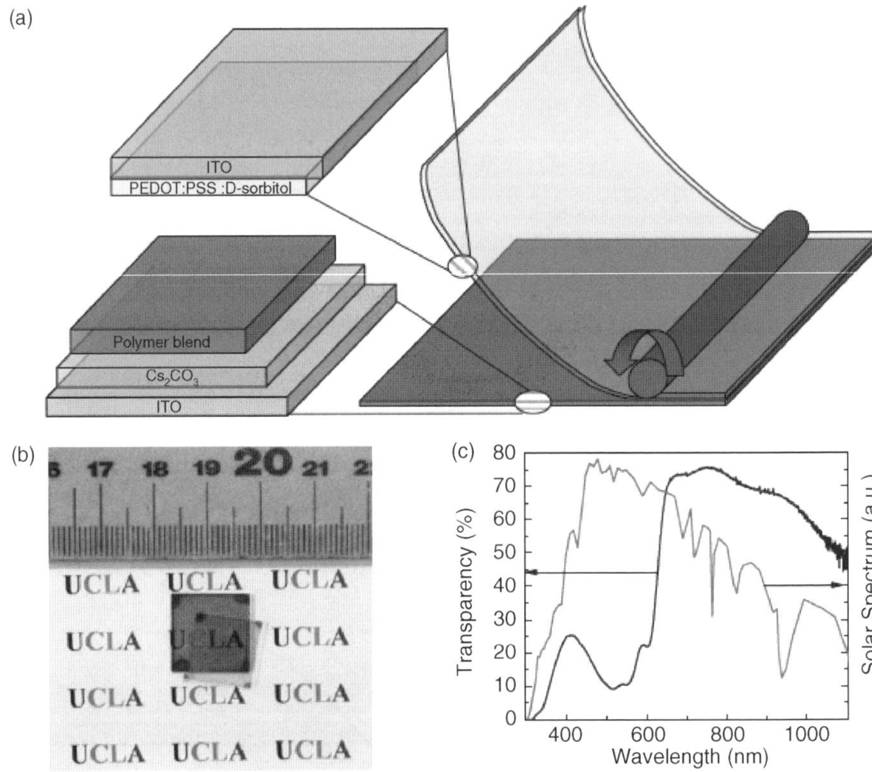

Figure 14.15 *(a) Scheme for the device fabrication, (b) an all-plastic solar cell device and (c) transparency spectrum of the device and the solar spectrum*

during the lamination process. In this scenario, the hole collection will be further enhanced by the increased interface area in this interpenetrating structure. A relatively good fill factor of 55% indicates that the contact between the polymer blend and PEDOT:PSS layer is ohmic. The V_{OC} of the device is 0.48 V, which is lower than the regular device (0.6 V), possibly from the reduced work function of PEDOT:PSS by D-sorbital doping, the high resistance of plastic ITO, and/or extra dark current in the unilluminated area. The obtained preliminary power efficiency reaches 3% under AM 1.5 with spectral mismatch correction. Higher solar cell performance is feasible upon further optimization of parameters, such as the ITO conductivity, the PEDOT:PSS conductivity and the work functions.

The Cs_2CO_3 nanolayer plays a crucial role in replacing the low work function reactive metal cathode, which makes the all-solution processing possible. Besides Cs_2CO_3, a number of other salts were also evaluated, including alkali carbonates from Li_2CO_3 to Cs_2CO_3, and Cs-containing salts such as CsF and cesium acetylacetonate [Cs(acac)]. The salts were dissolved in either water or polar organic solvent such as 2-ethoxyethanol to form dilute (0.2 wt%) solutions for spin-coating. The inverted PSCs have the structure of ITO/ buffer layer/P3HT:PCBM/V_2O_5 (10 nm)/Al, where V_2O_5 acts as the anode side and the buffer layer as the cathode. The characteristic parameters of these solar cells with different buffer layers are summarized in Table 14.3. It is easy to deduce from the table that a thin

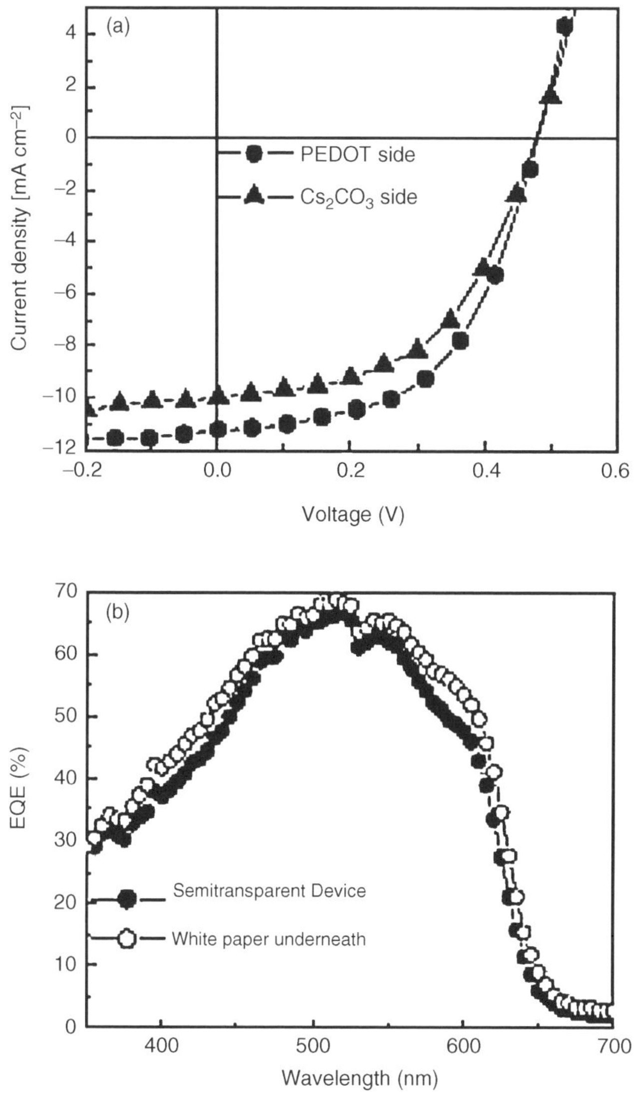

Figure 14.16 *(a) Photocurrents under AM 1.5 simulated illumination from front and back and (b) EQE of device with and without a piece of white paper underneath*

Table 14.3 *V_{OC}, J_{SC} and R_s of inverted solar cells using different buffer layers*

	Cs (asac)	CsF	Cs_2CO_3	K_2CO_3	Na_2CO_3	Li_2CO_3	ITO
V_{OC} (V)	0.55	0.54	0.56	0.56	0.50	0.26	0.20
J_{SC} (mA cm^{-2})	9.64	8.34	9.70	9.62	8.91	8.57	7.01
R_s (Ω cm^{-2})	3.56	4.19	3.50	3.78	8.60	20.3	18.2

layer of each of these materials can reduce the work function of ITO but to different extents. Two distinct behaviors were clearly observed: first, for devices with buffer layer from Cs_2CO_3 to Li_2CO_3, the V_{OC} drops, the J_{SC} increases and the series resistance reduces monotonously; secondly, for the devices with Cs-containing salt buffer layer with different anions [Cs_2CO_3, CsF and Cs(acac)], these characteristics are almost the same within experimental error.

The principle to obtain a low work function surface should be completely different from the case of Cs_2CO_3 in PLED devices, where the reaction of Cs_2CO_3 with subsequent deposited Al forms the low work function product Cs-O-Al [47]. In order to find out the detailed mechanism of the buffer layers, XPS/UPS analysis was conducted on the ITO/buffer layer interfaces. The sample preparation duplicates real device fabrication steps, i.e. the salts were spin-coated from the dilute aqueous solutions. The thickness of the layers was measured by the decay of X-ray signal intensity of indium according to:

$$I = I_0 e^{-L/L_0} \tag{14.1}$$

where I_0 and I are the indium signals before and after spin-coating the buffer layer, respectively, L_0 is the free electron path length at a specific energy and L is the thickness of the buffer layer. The thickness of the spin-coated salts is estimated to be approximately 0.6 to 3 nm depending on solution concentration and spin speed. It is clear that there are only a few monolayers of salt molecules on the ITO surface. Figure 14.17(a) shows the secondary electron edge of ITO covered by these buffer layers, which provide information about the work function modification of ITO. The work function variation follows one clear trend: the work function of ITO substrates reduces with buffer layers from Li-containing to Cs-containing salts, and buffer layers containing the same metal ion give almost the same work function, which agrees well with the inverted cell performance, inferring that the work function is only related to the metal. As there is a large change in the work function with the addition of only a few monolayers of molecules, it is believed that a strong dipole layer forms at the interface. The formation of such a dipole layer can be explained by the scenario illustrated in Figure 14.17(b). A thin layer of O-M is formed at the ITO or FTO surface with metal ions in the vacuum side (where O and M represent oxygen and metal species, respectively). The dipole points from ITO to vacuum and reduces the ITO surface work function, and the dipole moment is directly related to the electron-donating ability of the metal species. Since the change in ITO work function is proportional to the dipole moment, the ITO surface with O-Cs dipole layer has the lowest effective work function owing to the highest dipole moment. The characteristics of the inverted solar cell can be well correlated to this dipole layer: J_{SC} increases and series resistance decreases because of enhanced interfacial charge transfer rate at the cathode, and V_{OC} [58] increases because of the lower cathode work function before Fermi level pinning [38]. There are two possibilities for the formation of an O-M structure: (1) positive metal ions are directly attracted by the negative oxygen ions from ITO itself, particularly after the UV ozone process; and (2) O-M replaces the hydroxyl group on the ITO surface by surface chemical reaction: $-OH + M+ \rightarrow -O\text{–}Cs + H+$, similar to chemisorption of Cl-terminated molecules onto ITO [47]. In order to confirm the possibility of chemisorption, the atom ratio of Cs and F was examined by XPS. CsF was chosen because it does not contain either O or C species which might overlap signals from ITO substrate or ambient contamination. Another reason is that

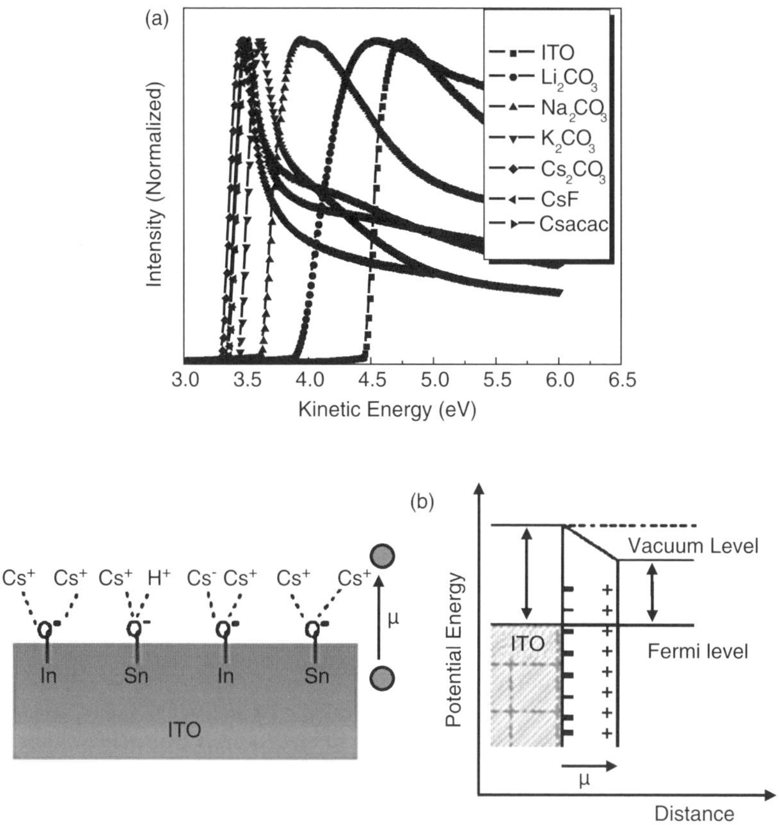

Figure 14.17 *(a) Evolution of secondary electron edge with different buffer layers on ITO and (b) scheme for the formation of a dipole layer on ITO and its effect on reducing the work function of ITO*

both Cs and F have high sensitivity in the X-ray signal. The Cs/F ratio of 4.6–4.9 was obtained for CsF on ITO surface, which agrees with the second possibility.

To summarize, an important method is reported to take full advantage of the solution fabrication process for plastic electronic devices. A lamination process was invented by using the conductive polymer glue as the medium to form the device. This method has the advantage of being low cost, self-encapsulating and has high transparency for various applications. The preliminary results on solar cells show promising efficiency comparable with devices produced by the regular processing method.

14.5 Conclusion and Remarks

In this chapter, we have reviewed our recent important progress in transparent PSC research and development. It should be apparent from the discussion that although much progress has been made in developing new materials and devices for high performance transparent solar cells, there is still plenty of opportunity to study and improve device performance and fabrication techniques compared with the nontransparent solar cell devices. In particular, the stability of transparency solar cells has not been studied yet.

Solution-processable transparent PSCs have become a promising emerging technology for tandem solar cell application to increase energy conversion efficiency. The transparency of solar cells at a specific light band will also lead to new applications such as solar windows. The field of energy harvesting is gaining momentum by the increases in gasoline price and environment pollution caused by traditional techniques. Continued breakthroughs in materials and device performance, accelerate and establish industrial applications. It is likely that new scientific discoveries and technological advances will continue to cross-fertilize each other for the foreseeable future.

References

[1] C. W. Tang and S. Van Slyke, Organic electroluminescent diodes, *Appl. Phys. Lett.* **51**, 913–915 (1986).

[2] S. R. Forrest, The path to ubiquitous and low cost organic electronic appliances on plastic, *Nature* **428**, 911–918 (2004).

[3] S. E. Shaheen, C. J. Brabec, F. Padinger, T. Fromherz, J. C. Hummelen and N. S. Sariciftci, 2.5% efficient organic plastic solar cells, *Appl. Phys. Lett.* **78**, 841–843 (2001).

[4] F. Padinger, R.S. Rittberger and N.S. Sariciftci, Effects of postproduction treatment on plastic solar cells, *Adv. Funct. Mater.* **13**, 85–88 (2003).

[5] G. Li, V. Shrotriya, J. Huang, Y. Yao, T. Moriarty, K. Emery and Y. Yang, High-efficiency solution processable polymer photovoltaic cells by self-organization of polymer blends, *Nat. Mater.* **4**, 864–868 (2005).

[6] W. Ma, C. Yang, X. Gong, K. Lee and A. J. Heeger, Thermally stable, efficient polymer solar cells with nanoscale control of the interpenetrating network morphology, *Adv. Funct. Mater.* **15**, 1617–1622 (2005).

[7] T. Chen, X. Wu and R. D. Rieke, Regiocontrolled synthesis of poly(3-alkylthiophenes) mediated by Rieke zinc: their characterization and solid-state properties, *J. Am. Chem. Soc.* **117**, 233–244 (1995).

[8] R. D. McCullough, S. Tristram-Nagle, S. P. Williams, R. D. Lowe and M. Jayaraman, Self-orienting head-to-tail poly(3-alkylthiophenes): new insights on structure-property relationships in conducting polymers, *J. Am. Chem. Soc.* **115**, 4910–4911 (1993).

[9] J. C. Hummelen, B. W. Knight, F. Lepeo, F. Wudl, J. Yao and C. L. Wilkins, Preparation and characterization of fulleroid and methanofullerene derivatives, *J. Org. Chem.* **60**, 532–538 (1995).

[10] N. S. Sariciftci, L. Smilowitz, A. J. Heeger and F. Wudl, Photoinduced electron transfer from a conducting polymer to Buckminsterfullerene, *Science* **258**, 1474–1476 (1992).

[11] G. Yu, J. Gao, J. C. Hummelen, F. Wudl and A. J. Heeger, Polymer photovoltaic cells: enhanced efficiencies via a network of internal donor-acceptor heterojunctions, *Science* **270**, 1789–1791 (1995).

[12] M. S. White, D. C. Olson, S. E. Shaheen, N. Kopidakis and D. S. Ginley, Inverted bulk-heterojunction organic photovoltaic device using a solution-derived ZnO underlayer, *Appl. Phys. Lett.* **89**, 143 517–143 519 (2008).

[13] H. Liao, L. Chen, Z. Xu, G. Li and Y. Yang, Highly efficient inverted polymer solar cell by low temperature annealing of Cs_2CO_3 interlayer, *Appl. Phys. Lett.* **92**, 173303–173305 (2008).

[14] K. Kawano, N. Ito, T. Nishimori and J. Sakai, Open circuit voltage of stacked bulk heterojunction organic solar cells, *Appl. Phys. Lett.* **88**, 073 514–073 516 (2006).

[15] J. Gilot, M.M. Wienk and R.A. Janssen, Double and triple junction polymer solar cells processed from solution, *Appl. Phys. Lett.* **90**, 143512–143514 (2007).

[16] J.Y. Kim, K. Lee, N.E. Coastes, D. Moses, T.Q. Nguyen, M. Dante and A.J. Heeger, Efficient tandem polymer solar cells fabricated by all-solution processing, *Science* **317**, 222–225 (2007).
[17] D. Muehlbacher, M. Scharberm, M. Morana, Z. G. Zhu, D. Waller, R. Gaudianam and C. Brabec, High photovoltaic performance of a low-bandgap polymer, *Adv. Mater.* **18**, 2884–2889 (2006).
[18] V. Shrotriya, E. H. Wu, G. Li, Y. Yao and Y. Yang, Efficient light harvesting in multiple-device stacked structure for polymer solar cells, *Appl. Phys. Lett.* **88**, 064 104–064 106 (2006).
[19] J. A. Dobrowolski, B. T. Sullivan and R. C. Bajcar,US Patent 5 049 780, 1989.
[20] L.S. Hung and M.G. Mason, 10th International Workshop on Inorganic and Organic Electroluminescence, Hamamatsu, Japan, 2000.
[21] V. Bulovic, P.E. Burrows, S.R. Forrest and M.E. Thompson, Transparent organic light emitting devices, *Appl. Phys. Lett.* **68**, 2606–2608 (1996).
[22] L.S. Hung and C.W. Tang, Interface engineering in preparation of organic surface emitting diodes, *Appl. Phys. Lett.* **74**, 3209 (1999).
[23] G. Pathasarathy, P.E. Burrows, V. Khalfin, V.G. Kozolov and S.R. Forrest, A metal-free cathode for organic semiconductor devices, *Appl. Phys. Lett.* **72**, 2138 (1998).
[24] M. E. Thompson, S. R. Forrest and P. Burrows,US Patent 5 986 401, 1997..
[25] L. S. Hung, C. W. Tang, M. G. Mason, P. Raychaudhuri and J. Madathil, Application of an ultrathin LiF/Al bilayer in organic surface-emitting diodes, *Appl. Phys. Lett.* **78**, 544–546 (2001).
[26] R. B. Pode, C. J. Lee, D. G. Moon and J. I. Han, Transparent conducting metal electrode for top emission organic light-emitting devices: Ca–Ag double layer, *Appl. Phys. Lett.* **84**, 4614–4616 (2004).
[27] J. H. Burroughes, D. D. C. Bradley, A. R. Brown, R. N. Marks, K. Mackay and R. H. Friend, Light-emitting diodes based on conjugated polymers, *Nature* **347**, 539–541 (1990).
[28] C. N. King, R. H. Mauch and H.-E. Gumlich, *Inorganic and Organic Electroluminescence*, edited by R. H. Mauch and H.-E. Gumlich, Berlin Wissenschaft und Technik, Berlin, p. 375 (1996).
[29] A. Yamamori, S. Hayashi, T. Koyama and Y. Taniguchi, Transparent organic light-emitting diodes using metal acetylacetonate complexes as an electron injective buffer layer, *Appl. Phys. Lett.* **78**, 3343–3345 (2001).
[30] Q. Xu, J. Ouyang and Y. Yang, Ultrahigh efficiency green polymer light-emitting diodes by nanoscale interface modification, *Appl. Phys. Lett.* **83**, 4695–4697 (2003).
[31] L. S. Roman, M. Berggren and O. Inganäs, Polymer diodes with high rectification, *Appl. Phys. Lett.* **75**, 3557–3559 (1999).
[32] L. S. Roman, W. Mammo, L. A. A. Petersson, M. R. Andersson and O. Inganäs, High quantum efficiency polythiophene, *Adv. Mater.* **10**, 774 (1998).
[33] M. P. de Jong, L. J. van Ijzendoorn and M. J. A. de Voigt, Stability of the interface between indium-tin-oxide and poly(3,4-ethylenedioxythiophene)/poly(styrenesulfonate) in polymer light-emitting diodes, *Appl. Phys. Lett.* **77**, 2255–2257 (2000);K. W. Wong, H. L. Yip, Y. Luo, K. Y. Wong, W. M. Lau, K. H. Low, H. F. Chow, Z. Q. Gao, W. L. Yeung and C. C. Chang, Blocking reactions between indium-tin oxide and poly (3,4-ethylene dioxythiophene):poly (styrene sulphonate) with a self-assembly monolayer, *Appl. Phys. Lett.* **80**, 2788–2790 (2002).
[34] G. L. Frey, K. J. Reynolds and R. H. Friend, Novel electrodes from solution-processed layer-structure materials, *Adv. Mater.* **14**, 265 (2002); K. J. Reynolds, J. A. Barker, N. C. Greenham, R. H. Friend and G. L. Frey, Inorganic solution-processed hole-injecting and electron-blocking layers in polymer light-emitting diodes, *J. Appl. Phys.* **92**, 7556–7560 (2002).
[35] V. Shrotriya, G. Li, Y. Yao, C. W. Chu and Y. Yang, Transition metal oxides as the buffer layer for polymer photovoltaic cells, *Appl. Phys. Lett.* **88**, 073508 (2006).
[36] V. D. Mihailetchi, L. J. A. Koster and P. W. M. Blom, Effect of metal electrodes on the performance of polymer:fullerene bulk heterojunction solar cells, *Appl. Phys. Lett.* **85**, 970–972 (2004).
[37] J. M. Bharathan and Y. Yang, Polymer/metal interfaces and the performance of polymer light-emitting diodes, *J. Appl. Phys.* **84**, 3207–3212 (1998).

[38] C. J. Brabec, A. Cravino, D. Meissner, N. S. Sariciftci, T. Fromherz, M. T. Rispens, L. Sanchez and J. C. Hummelen, Origin of the open circuit voltage of plastic solar cells, *Adv. Funct. Mater.* **11**, 374–380 (2001).
[39] C. J. Brabec, S. E. Shaheen, C. Winder and N. S. Sariciftci, Effect of LiF/metal electrodes on the performance of plastic solar cells, *Appl. Phys. Lett.* **80**, 1288–1290 (2002).
[40] G. Li, C. W. Chu, V. Shrotriya, J. Huang and Y. Yang, Efficient inverted polymer solar cells, *Appl. Phys. Lett.* **88**, 253503–253505 (2006).
[41] T. M. Brown, R. H. Friend, I. S. Millard, D. J. Lacey, J. H. Burroughes and F. Cacialli, Efficient electron injection in blue-emitting polymer light-emitting diodes with LiF/Ca/Al cathodes, *Appl. Phys. Lett.* **79**, 174–176 (2001).
[42] T. Hasegawa, S. Miura, T. Moriyama, T. Kimura, I. Takaya, Y. Osato and H. Mizutani, *Dig. Tech. Papers–Soc. Inf. Disp. Int. Symp.* **35**, 154–157 (2004).
[43] J. Huang, G. Li, E. Wu, Q. Xu and Y. Yang, Achieving high-efficiency polymer white-light-emitting devices, *Adv. Mater.* **18**, 144–147 (2006).
[44] I. D. Parker, Carrier tunneling and device characteristics in polymer light-emitting diodes, *J. Appl. Phys.* **75**, 1656 (1994).
[45] V. D. Mihailetchi, P. W. M. Blom, J. C. Hummelen and M. T. Rispens, Cathode dependence of the open-circuit voltage of polymer: fullerene bulk heterojunction solar cells, *J. Appl. Phys.* **94**, 6849–6851 (2003).
[46] T. R. Briere and A. H. Sommer, Low-work-function surfaces produced by cesium carbonate decomposition, *J. Appl. Phys.* **48**, 3547–3548 (1977).
[47] J. Huang, Z. Xu and Y. Yang, low-work-function surface formed by solution processed and thermally deposited nanoscale layers of cesium carbonate, *Adv. Funct. Mater.* **17**, 1966–1973 (2007).
[48] H. H. Liao, L.-M. Chen, Z. Xu, G. Li and Y. Yang, Highly efficient inverted polymer solar cell by low temperature annealing of Cs_2CO_3 interlayer, *Appl. Phys. Lett.* **92**, 173303–173305 (2008).
[49] M. Granstrom, K. Petritsch, A. C. Arias, A. Lux, M. R. Andersson and R. H. Friend, Laminated fabrication of polymeric photovoltaic diodes, *Nature* **395**, 257–260 (1998).
[50] T. F. Guo, S. Pyo, S. C. Chang and Y. Yang, high performance polymer light-emitting diodes fabricated by a low temperature lamination process, *Adv. Funct. Mater.* **11**, 339–343 (2001).
[51] L. Groenendaal, F. Jonas, D. Freitag, H. Pielartzik and J. R. Reynolds, Poly(3,4-ethylenedioxythiophene) and its derivatives: past, present, and future, *Adv. Mater.* **12**, 481–494 (2000).
[52] Y. Cao, G. Yu, C. Zhang, R. Menon and A. J. Heeger, Polymer light-emitting diodes – sulfonate as the transparent anode, *Synth. Met.* **87**, 171–174 (1997).
[53] J. Ouyang, Q. Xu, C. W. Chu, Y. Yang, G. Li and J. Shinar, On the mechanism of conductivity enhancement in poly(3,4-ethylenedioxythiophene):poly(styrene sulfonate) film through solvent treatment, *Polymer* **45**, 8443–8450 (2004).
[54] W. H. Kim, A. J. Mäkinen, V. Nikolov, R. Shashidhar, H. Kim and Z. H. Kafafi, Molecular organic light-emitting diodes using highly conducting polymers as anodes, *Appl. Phys. Lett.* **80**, 3844–3846 (2002).
[55] T.-F. Guo, F.-S. Yang, Z.-J. Tsai, T.-C. Wen, S.-N. Hsieh and Y.-S. Fu, High performance polymer light-emitting diodes utilizing modified Al cathode, *Appl. Phys. Lett.* **87**, 13504–13506 (2005).
[56] J. Ouyang, T. F. Guo, Y. Yang, H. Higuchi, M. Yoshioka and T. Nagatsuka, High-performance, flexible polymer light-emitting diodes fabricated by a continuous polymer coating process, *Adv. Mater.* **14**, 915–918 (2002).
[57] J. Huang, T. Watanabe, K. Ueno and Y. Yang, Highly efficient red emission polymer phosphorescent lighting emitting diodes based on two novel $Ir(piq)_3$ derivatives, *Adv. Mater.* **19**, 739–743 (2007).
[58] S. Khodabakhsh, B. M. Sanderson, J. Nelson and T. S. Jones, Using self-assembling dipole molecules to improve charge collection in molecular solar cells, *Adv. Funct. Mater.* **16**, 95–100 (2006).

15

Organic Electro-Optic Modulators with Substantially Enhanced Performance Based on Transparent Electrodes

Fei Yi[1], *Seng-Tiong Ho*[1] *and Tobin J. Marks*[2]

[1]*Department of Electrical Engineering and Computer Science, Northwestern University, USA*

[2]*Department of Chemistry and the Materials Research Center, Northwestern University, USA*

This chapter describes the application of transparent electrodes to enhance the performance of electro-optic (EO) modulators. Transparent electrodes are typically made from transparent conducting oxides (TCOs). They are electrically conductive while having significantly lower optical absorption loss compared with metals. Their optical refractive indices are typically around $n_{TC} = 1.6–2.0$, which makes them compatible for use as optical waveguiding materials in organic EO modulators. In this chapter, we show that with use of transparent electrodes as 'bridge electrodes' together with a proper design of a metallic transmission line, high-speed organic EO modulators with substantially lower switching voltages can be realized. In this chapter, we describe in detail the design, fabrication, and characterization of an organic EO modulator based on transparent electrodes. We show that transparent-conductor (TC)-based EO modulator structures can be used to reduce modulator switching voltages by over 5–15x compared with organic modulators based on conventional organic EO modulator structures, while still achieving broad modulation bandwidths of 10–100 GHz, depending on the properties of the transparent conducting materials. The TC-based EO modulator structures are particularly advantageous when

Transparent Electronics: From Synthesis to Applications Edited by Antonio Facchetti and Tobin J. Marks

applied to organic EO materials having high EO coefficients. This approach opens the possibility of realizing organic EO modulators with ultra-compact sizes or ultra low switching voltages, and broad modulation bandwidths. In Section 15.1, we introduce the idea of using the transparent electrodes in EO modulators and review the typical design considerations for a high-speed conventional modulator. We then compare the voltage performance of TC-based EO modulators with their conventional counterparts. In Section 15.2, we describe TC-based low-voltage, high-speed organic EO modulator structures in detail. We also discuss the basic design algorithm of the TC-based organic EO modulators. In Section 15.3, we discuss a detailed design of a TC-based organic EO modulator and calculate the frequency response of the device. In Section 15.4, we will discuss the fabrication process and characterization of a TC-based organic EO modulator.

15.1 Introduction

15.1.1 Interest in Low-Voltage, High-Speed Optical Intensity Modulators

An EO optical intensity modulator is often used when one needs to convert electrical signals to optical signals. This can be achieved by using the modulator to turn on and off the optical power from a continuous-wave (CW) laser beam passing through the modulator. Optical intensity modulators are thus used widely in optical fiber communications to modulate light from semiconductor lasers to produce the required optical pulses carrying digital data for transmission through optical fibers [1]. At data rates below 2.5 Gigabits per second (Gbits s^{-1}), it is much more convenient to simply directly modulate the injection current of the semiconductor laser to produce the required optical pulses [2]. However, direct current modulation can produce frequency chirping (change in the lasing frequency at the leading and trailing edges of the pulse) caused by the carrier-induced modulation of the laser material's refractive index and hence, the resonant frequency of the laser cavity. At high modulation rates (above 10 Gbits s^{-1}), a frequency chirped pulse will be rapidly broadened in width after propagating through a long length of optical fiber due to the fiber's group velocity dispersion, causing serious degradation in the signal integrity [3]. Thus, external optical intensity modulators are needed for long-distance optical communications at bit rates at or above 10 Gbits s^{-1}.

Current commercially available high-speed EO optical intensity modulators are based on lithium niobate crystals ($LiNbO_3$) [4]. These modulators typically have on/off switching voltages (also called half-wave voltage or π-phase-shift voltage, V_π) of ~5 V and electrical terminal impedance Z of ~50 Ω [5]. This means that the electrical power required to drive the modulator given by $P = V_\pi^2/Z$ will be about 0.5 W, which is high and quite inefficient in terms of electrical-to-optical signal-power conversion, considering that typical semiconductor laser power emitted into optical fibers is less than 10 mW. Since electrical power is proportional to the voltage squared, there is much interest in reducing the driving voltage of the modulator. For example, a 0.5 V modulation voltage will reduce the drive power to 5 mW, making it closer to achieving a one-to-one electrical-to-optical signal-power conversion when used with a typical semiconductor laser emitting 1–10 mW power into an optical fiber.

Based on the capability of modulating the light intensity in optical fibers with analog high-frequency signals by using EO modulators, there has been much recent interest in using optical fibers to transmit analog radio frequency (RF) electrical signals by first intensity modulating a laser beam with the RF signals and then recovering the RF signals with a photodetector that converts optical power back to electrical voltage. This area of pursuit is referred to as RF photonics [6]. The interest in RF photonics arises because the traditional way to transmit RF signals is by using a metallic RF transmission line (or coax cable), which can become quite lossy at frequencies above 10 GHz due to skin-depth effects (electrical current confinement to a thin layer at the metal surface) that becomes more serious at higher frequencies [7]. High-frequency metallic transmission lines are also costly and heavy. For reasons similar to those discussed above, in order for an RF signal of power P_{RF} to be transmitted through an optical fiber and be converted back to an RF signal of power $\sim P_{RF}$ at the photodetector, without adding significant excess noise (i.e in order to realize a near-lossless and low-noise RF signal transmission with near-unity noise figure), EO intensity modulators with V_{π} less than 0.5 V are required (when used with a typical 1–10 mW semiconductor laser). Otherwise, semiconductor lasers and detectors designed for significantly higher optical powers (e.g. 100 mW–1 W) would be required, resulting in significantly higher total electrical power consumption and higher cost. Such high-speed, low-voltage modulators are difficult to realize using $LiNbO_3$ due to its relatively low EO coefficient.

Recently, it has been shown that organic EO materials can be engineered to have over 5x higher EO coefficients than $LiNbO_3$, leading to the realization of sub 1 V organic EO modulators (for 2-cm-long devices) [8–11]. In this section, we show that with the use of a novel organic EO modulator structure based on transparent electrodes, it is possible to substantially reduce the modulation voltage of organic modulator by another factor of 5–15x. We discuss how this technology will lead to a whole new generation of ultra-low-voltage, high-speed EO modulators with $V_{\pi} < 0.5$ V and modulation speeds >40 GHz. We believe that such TC-based organic modulators will play important roles in enabling RF photonics and next-generation higher-bit-rate optical fiber communication systems.

15.1.2 Conventional Organic EO Modulator Structures and the Concept of TC-Based Electrode Structures

An EO material has the property that the refractive index can be altered under an applied electric field, resulting in an optical phase shift for a laser beam propagating through the material [1]. The intrinsic material response is generally electronic in nature and can be fast (100 GHz or faster), which enables the refractive index to be modulated at high frequencies by an RF signal. The optical phase shift can be converted to an intensity change using a Mach Zehnder interferometer (MZI), which has a typical waveguide geometry shown in Figure 15.1 in which an input laser beam is split into two waveguide arms by a Y-junction waveguide, forming an input 50/50 beam splitter. One arm of the MZI contains the EO material while the other arm has no EO material. The beams from the two arms are then recombined using another Y-junction waveguide that merges the beams into an output waveguide. The optical path lengths going through the two arms are made equal so that the two beams will interfere constructively at the output Y-junction and transfer all the power into the output waveguide. An applied voltage on the EO arm causing a π phase shift changes the constructive interference to a destructive interference, resulting in no light going into the output waveguide (it will be dispersed to the side areas outside the waveguide).

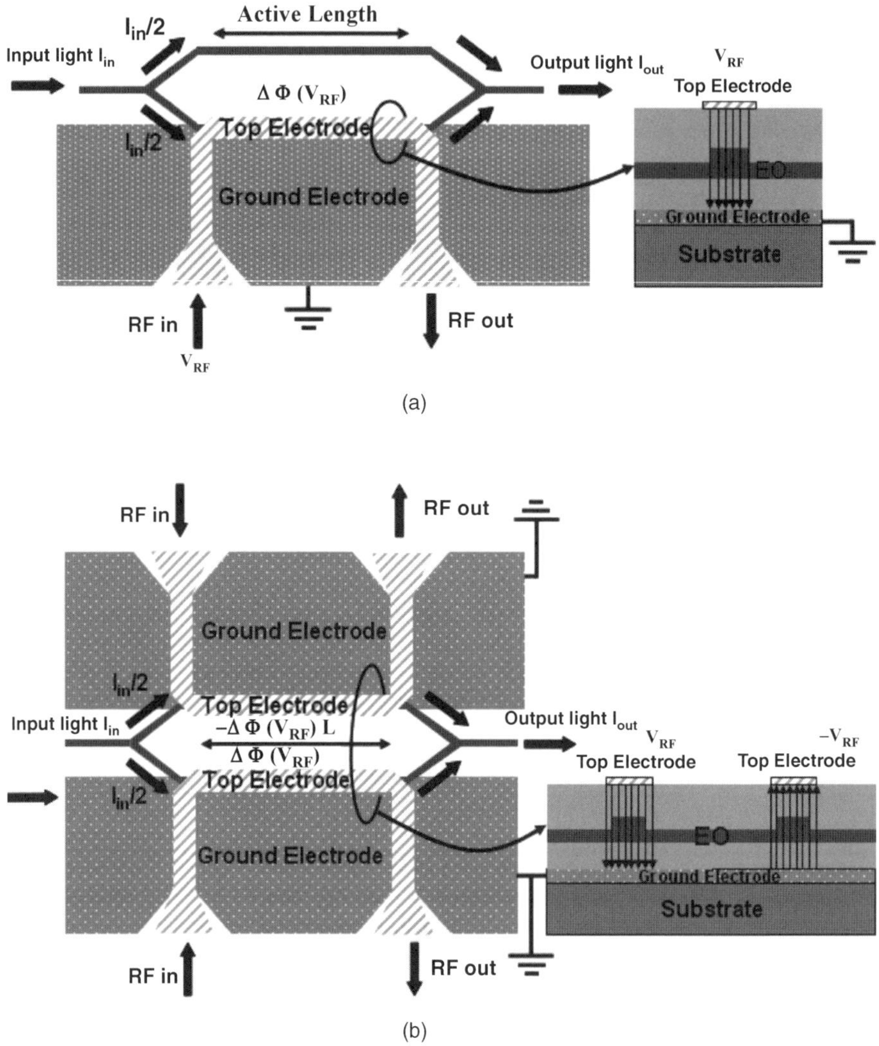

Figure 15.1 *Mach Zehnder Interferometer (MZI) geometry for an EO intensity modulator: input light energy is split into two halves by a Y-junction, and each half attains a different optical phase shift through EO modulation. These beams are recombined through another Y-junction at the output waveguide. Two designs are possible: (a) a single-arm MZI design for which the optical beam in only one of the two arms is phase modulated; and (b) a push-pull MZI design in which optical beams in the two arms experience opposite phase modulations*

This π-phase-shift voltage is referred to as V_π. In an alternative 'push-pull' geometry, both arms have EO materials with applied voltages, however the voltages are applied in such a way that the phase shifts in the two arms are equal and opposite in direction. In this case, to turn the light power in the output waveguide to off requires an applied voltage of $V_\pi/2$, which is half the V_π value. This is termed a push-pull MZI modulator geometry.

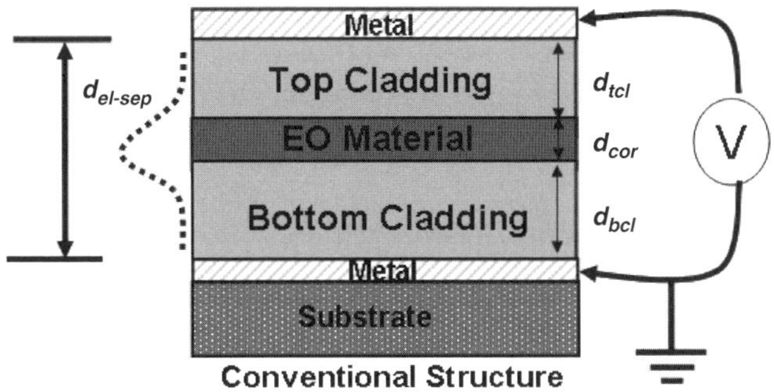

Figure 15.2 *Conventional organic EO modulator structure using metal electrodes*

The V_π achieved depends on the EO coefficient of the EO material as well as the actual device geometry with respect to the electrodes and the vertical and horizontal optical mode confinement structures. The typical cross-section of a conventional organic modulator structure is shown in Figure 15.2 in which the optical mode propagates in the center waveguide core with refractive index n_{cor} and thickness d_{cor}. The waveguide core is surrounded above and below by a top cladding with refractive index n_{tcl} and a bottom cladding with refractive index n_{bcl}, for which n_{tcl} and n_{bcl} are lower than n_{cor}. The thicknesses of the top and bottom claddings are d_{tcl} and d_{bcl}, respectively. The waveguide core is filled with an organic EO material. An electric field is applied to the EO material through a top metal electrode (M_t) and a bottom metal electrode (M_b) above and below the top and bottom claddings, respectively. The net separation between the two electrodes is given by $d_{el\text{-}sep} = d_{cor} + d_{tcl} + d_{bcl}$. For an EO material with EO coefficient r, refractive index n, the optical phase shift $\Delta\Phi$ induced by an applied electric field E is given by [5]:

$$\Delta\Phi = k_0 \Delta n L = \frac{\pi}{\lambda} n^3 r \Gamma_{EO} E L \tag{15.1}$$

where L is the length that the optical beam propagates through the EO medium under the applied field (length of interaction), r is the EO coefficient relevant to the direction of the applied electric field (for organics, it is typically the coefficient r_{33}), and λ is the optical wavelength. In the organic EO modulator described above, the electric field is related to the applied voltage via $E = V/d_{el\text{-}sep}$. In an actual waveguide structure, the optical beam may not have 100% overlap with the active EO material or the applied field may not be uniform across the EO material. This is taken into account of by the EO mode-overlapping factor Γ_{EO} in Equation (15.1). Solving for the π-phase-shift voltage, we obtain the following formula for the modulator V_π ($= E \times d_{el\text{-}sep}$ at $\Delta\Phi = \pi$):

$$V_\pi = \frac{\lambda d_{el-sep}}{n^3 r \Gamma_{EO} L} \tag{15.2}$$

We see from Equation (15.2) that V_π is also proportional to the electrode separation distance $d_{el\text{-}sep}$. For conventional polymer waveguide EO modulator structures, $d_{el\text{-}sep}$ is much larger than the EO layer thickness due to the existence of two thick cladding layers which are needed to separate the metal electrodes from the EO film to avoid metal induced

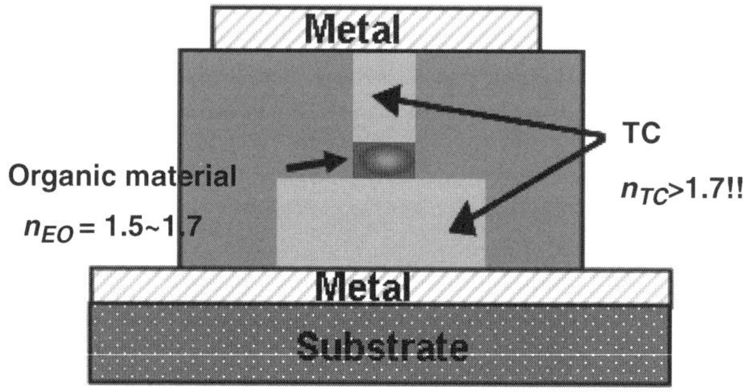

Figure 15.3 *Top/down conduction structure of a TC-based organic EO modulator. The two conductive cladding layers directly introduce the switching voltage from the metal electrode to the EO layer*

optical loss, which means the switching voltage of the device is not fully driving the EO layer but is wasted across the two thick cladding layers (i.e. $d_{el\text{-}sep} = d_{cor} + d_{tcl} + d_{bcl}$) as shown in Figure 15.2. The conventional organic modulator structure for 1550 nm wavelength operation has $d_{el\text{-}sep}$ values ranging from 7.5 to 15 μm, with a typical value of about 12 μm, and the waveguide core thickness is typically about 1.5 μm [9–11]. However, if the two cladding layers are both optically transparent and electrically conductive, they can introduce the switching voltage from the metal electrode directly to the EO layer and thus greatly reduce the modulation voltage. This is the basic concept behind the TC-based organic EO modulator. One simple way to realize such a TC-based EO modulator is by replacing the top and bottom cladding with TCs as illustrated in Figure 15.3 (it turns out that this is not the preferred way as will be explained below), as 'bridges' to conduct the voltage from the metal electrodes to the top and bottom parts of the EO material that forms the waveguide core, thereby reducing $d_{el\text{-}sep}$ from $d_{el\text{-}sep} = d_{cor} + d_{tcl} + d_{bcl}$ to $d_{el\text{-}sep} = d_{cor}$. The TC electrodes are also called the bridge electrodes. A common transparent conductor material is indium-tin oxide (Sn-doped In_2O_3, ITO) used widely in the electronic display industry. There are various metal oxides that are TCs, which will be referred to as transparent conducting oxides (TCOs). ITO is a good TCO for visible light wavelengths, but not for the infrared (IR) wavelength range of 1550 nm due to its high optical absorption. Since modulators of interest for fiber-optic applications operate at 1550 nm, TCOs that have low optical absorption at 1550 nm are preferred. For such applications, metal oxides such as indium oxide (In_2O_3), zinc oxide (ZnO) and cadmium oxide (CdO) are the preferred TCOs over ITO for the IR wavelength range (see Section 15.2.2).

In a TC-based organic EO modulator, $d_{el\text{-}sep}$ is equal to d_{cor}, which usually ranges from less than 0.9 μm to around 2 μm instead of the typical 12 μm $d_{el\text{-}sep}$ in a conventional EO modulator, resulting in a huge reduction in V_π by 5–15x. However, as will be discussed in detail below, one often needs to trade off lower V_π values for smaller electrical modulation bandwidths. Table 15.1 shows how different EO materials are required to achieve the desired $V_\pi L$ (voltage–length product) values based on (a) the conventional modulator structure and (b) the TC-based modulator structure. Note that the usual inorganic modulator such as $LiNbO_3$ modulator has $r = 30\,\text{pm V}^{-1}$ and the usual, more commonly available organic

Table 15.1 Comparison of the required EO coefficient for TCO electrodes and metal electrodes to achieve given $V_\pi L$ product under specific device conditions ($\lambda = 1.55$ μm, $n = 1.7$, $d = 1.5$ μm for transparent electrodes, $d_{el\text{-}sep} = 12$ μm for metal electrodes, $\Gamma_{EO} = 90\%$, push-pull structure)

$V_\pi L$ (V cm)	r for TC-based structure (assuming $d_{el\text{-}sep} = 1.5$ μm) (pm V^{-1})	r for conventional structure ($d_{el\text{-}sep} = 12$ μm) (pm V^{-1})
5	5	40
1	25	200
0.5	50	400
0.25	100	800
0.1	250	2000

modulator material has $r \sim 10$–30 pm V^{-1}. As depicted in Table 15.1, we see that the conventional modulators (employing a push-pull MZI configuration) typically result in a switching voltage $V_\pi/2$ of about 5 V for a 2-cm- long device (assuming $d_{el\text{-}sep} = 12$ μm). The use of TC offers a reduction in the voltage to less than 1 V for a 1-cm-long device, even using an $r = 25$ pm V^{-1} material. The more recently synthesized organic chromophoric materials with high EO coefficients of $r = 150$–300 pm V^{-1} promise to either reduce the switching voltage $V_\pi/2$ to less than 0.1 V or to reduce a 1 V modulator length to 1 mm, making it far cheaper to manufacture and/or more easily to integrate with other devices.

15.1.3 High Frequency Operation: Effect of RF Propagation Loss

It is important to ensure that low modulator voltages can be achieved while still maintaining broad modulation bandwidths. There are a few common factors that can affect the modulation bandwidth of an EO modulator, which we review below. From Equation (15.2), we see that V_π is inversely proportional to the length of interaction, L. Given the device structure and EO material, in order to achieve a sufficiently low switching voltage of a few volts, L usually must be long. For $LiNbO_3$ devices, it is typically a few centimeters. Such a long L has consequences for the modulator frequency response. When the modulation frequency enters the RF and microwave ranges (above 10 GHz), at which point the RF wavelength λ_{RF} becomes smaller than L (e.g. at 40 GHz, λ_{RF} is only 5 mm), the modulation voltage along the electrodes will no longer be a uniform voltage but will behave like a traveling wave with a sinusoidally varying voltage pattern, and the electrodes must then be designed as an RF transmission line in order to efficiently propagate the modulation voltage applied at one end of the electrodes (at the end close to the optical input so that the RF wave can co-propagate with the optical beam energy). The transmission line then allows the RF modulation voltage to travel with the optical beam in the waveguide. In addition, due to higher metal conductor loss at high frequency, the amplitude of the RF traveling wave will decay along the transmission line. As a result, the switching voltage along the RF transmission-line electrodes will be given by:

$$V_{RF}(z,t) = V_{appl}e^{-\alpha_{RF}z}\cos(\omega_{RF}t - \beta z) \tag{15.3}$$

where V_{appl} is the RF source voltage applied at the starting point of the transmission line, α_{RF} is the RF wave loss coefficient, and β is the RF wave propagation constant. The first factor $V_{appl}e^{-\alpha_{RF}z}$ means that the amplitude of the modulation voltage will decay along the

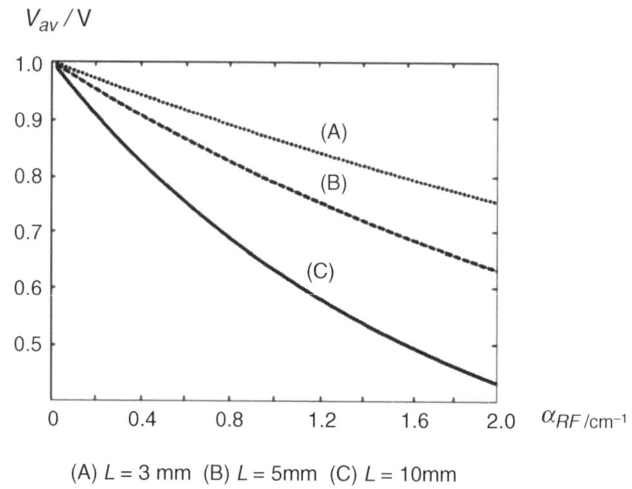

Figure 15.4 *Averaged amplitude V_{av} of the switching voltage versus RF loss coefficient α_{RF} for different device lengths L, from 3 to 10 mm. We assume $V_{appl} = 1$ V*

transmission line exponentially, and its averaged amplitude (V_{av}) will be given by an integration over the modulator length L:

$$V_{av} = \frac{\int_0^L V_{appl} e^{-\alpha_{RF} z}}{L} = \frac{(1 - e^{-\alpha_{RF} L})}{\alpha_{RF} L} V_{appl} \triangleq r_{RFloss} V_{appl} \tag{15.4}$$

Figure 15.4 shows this averaged voltage V_{av} versus α_{RF}. Note that we have denoted the rate of V_{av} over V_{appl} as r_{RFloss}, so that $r_{RFloss} = 1$, or $V_{av} = V_{appl}$ indicates no voltage drop. When the decay α_{RF} is greater than 2 cm^{-1}, the averaged amplitude of the switching voltage V_π will be below 0.5 V for a 1-cm-long device assuming $V_{appl} = 1$ V (i.e. less than half the applied voltage). In conventional metallic transmission lines, $\alpha_{RF}(f)$ is a function of frequency due to the skin depth effect of high frequency current in the metal as discussed above. Later we will see that in a TC-based structure, due to the finite conductivity and small thickness of the TC bridge electrodes, when the operation frequency f increases, $\alpha_{RF}(f)$ will increase to a significant value, and the averaged voltage $V_{av} = r_{RFloss} V_{appl}$ will then drop from the applied voltage V_{appl} at the modulator RF input terminal, resulting in an important frequency cut-off factor.

15.1.4 High Frequency Operation: Effect of Velocity Matching

Besides above voltage decay effect due to RF loss, another important effect is due to the difference between the propagation speed of the light pulse and the propagation speed of the RF traveling wave. The velocity at which an RF pulse travels along the transmission line is given by $v_{RF} = c/n_{RF}$, where n_{RF} is the effective propagating refractive index of the transmission line and c is the speed of light in vacuum. Similarly, the group velocity for the propagation of the optical energy in the waveguide is given by $v_{opt} = c/n_{gopt}$, where n_{gopt} is the effective waveguide group-velocity refractive index for the propagating optical beam. Therefore, even if we assume $\alpha_{RF} = 0$, the amount of phase modulation experienced by the optical field still decays along z if n_{RF} and n_{gopt} are not equal. This is equivalent to a decrease in the effective modulation voltage. As a result, the effective voltage seen by the optical field traveling at velocity v_{opt} at traveling distance z along the electrode, given by the voltage applied to the modulator RF input terminal (V_{appl}), is obtained by replacing time t by $t = z/v_{opt}$ in Equation (15.3):

$$\begin{aligned} V_{RF}(z) &= V_{appl} e^{-\alpha_{RF} z} \cos\left(\omega_{RF} \frac{z}{v_{opt}} - \beta z\right) \\ &= V_{appl} e^{-\alpha_{RF} z} \cos\left[\frac{\omega_{RF}}{c} z(n_{gopt} - n_{RF})\right] \end{aligned} \tag{15.5}$$

which is dependent on the propagating refractive index difference ($n_{RF} - n_{gopt}$) between the RF wave and optical wave. Assuming zero RF loss ($\alpha_{RF} = 0$) for simplicity, the average effective voltage V_{av} seen by the optical field in a device with interaction length L is then given by integrating Equation (15.5) over L to be:

$$\begin{aligned} V_{av} &= \frac{\int_0^L V_{RF}(z)}{L} = V_{appl} \frac{\sin x}{x} \triangleq r_{v-mismatch} V_{appl} \\ x &= \frac{\omega_{RF}}{c} (n_{RF} - n_{gopt}) L \end{aligned} \tag{15.6}$$

This effect is called velocity mismatch or velocity walk-off because physically the difference between n_{RF} and n_{gopt} causes the RF wave and optical energy to travel at different speeds, resulting in a higher switching voltage V_π compared with the applied voltage V_{appl}. From Figure 15.5 we see that since the velocity mismatch has a significant impact on the modulator frequency bandwidth, the ideal design for a low voltage EO modulator is to have zero velocity mismatch, or to have perfect match between the n_{RF} and n_{gopt}. For example, for a 1-cm-long device operating at 40 GHz, even a slight ($n_{RF} - n_{gopt}$) = 0.3 will cause V_{av} to drop below half of V_{appl}.

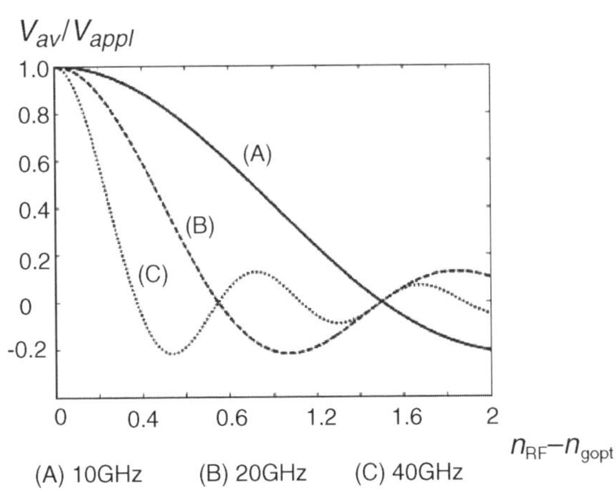

Figure 15.5 *The ratio of average effective voltage V_{av} over the terminal voltage V_{appl} denoted by $r_{v\text{-}mismatch}$ versus the velocity mismatch assuming no RF loss for a 1-cm-long device*

15.2 TC-Based Low-Voltage, High-Speed Organic EO Modulators

15.2.1 TC-Based Organic EO Modulator Structures

The use of TC materials as modulator's bridge electrodes discussed above is not as straightforward as it might first seem because the TCs are never as conductive as metallic electrodes and hence, careful device structural design based on realistic TC optical and electrical properties must be made in order to realize modulators with low switching voltages, high modulating frequencies and low optical losses. There are two basic device geometries for TC-based traveling-wave EO modulators: (1) a top-down conduction geometry (TDCG); and (2) a side conduction geometry (SCG). The schematic for the TDCG design is shown in Figure 15.3. In TDCG, the TC is used both as the waveguide cladding layers to separate the metal electrodes (optically lossy) from the waveguide core and as the conducting layer to introduce the driving voltage directly to the waveguide core. This geometry requires that the TC have an optical refractive index (n_{TC}) lower than that of the EO material used as the waveguide core layer in order to confine the optical mode inside the waveguide core region. However, typical organic EO materials have refractive indices of 1.5~1.7, lower than the refractive indices of most TC materials (typically larger than 1.7). As a result, TDCG is in general not suitable for use with organic EO materials. This motivates the concept of a SCG device, which is illustrated in Figure 15.6. In this geometry, vertical, the optical energy is still confined by a conventional waveguide cladding layer which has a lower refractive index than the refractive index of the EO core layer, and the TC is designed to be a thin layer with thickness $d_{TC} \sim 50$–100 nm (thin compared with the optical mode size) between the cladding layer and the waveguide core layer. This geometry does not require the TC material to have a lower refractive index than that of the waveguide core layer. Therefore, the SCG design is more suitable than the TDCG design for organic EO materials.

Laterally, there are three ways to confine the mode. The first approach (top ridge structure) is to form a rib (or ridge) waveguide on top of the EO layer using materials with lower optical refractive indices than that of the EO material as shown in Figure 15.6(a). The second approach (bottom ridge structure) is to form a trench on the bottom cladding material with low refractive index and fill in the EO material in the trench as shown in Figure 15.6(b), so that it forms an 'inverted rib waveguide' structure to provide the lateral mode confinement. The third approach (buried structure) is to etch through the EO material and fill the side of the EO material with materials having low refractive indices so as to form side waveguide claddings.

As will be depicted below, another advantage of the SCG design is that it has a lower fraction of optical energy confined in the TC material than in the case of the TDCG device, and hence allows the TC material to have higher optical loss, which makes it easier to engineer the TC material. The thin thickness for the TC material in the SCG case also makes the TC material easier to deposit and encounter fewer problems in terms of thermal expansion matching with its surrounding materials.

15.2.2 Materials for the TCs and their Requirements

The TC materials should have as low optical absorption as possible since the guided optical mode energy overlaps with them spatially. In order to reach high modulation speeds, the TCs

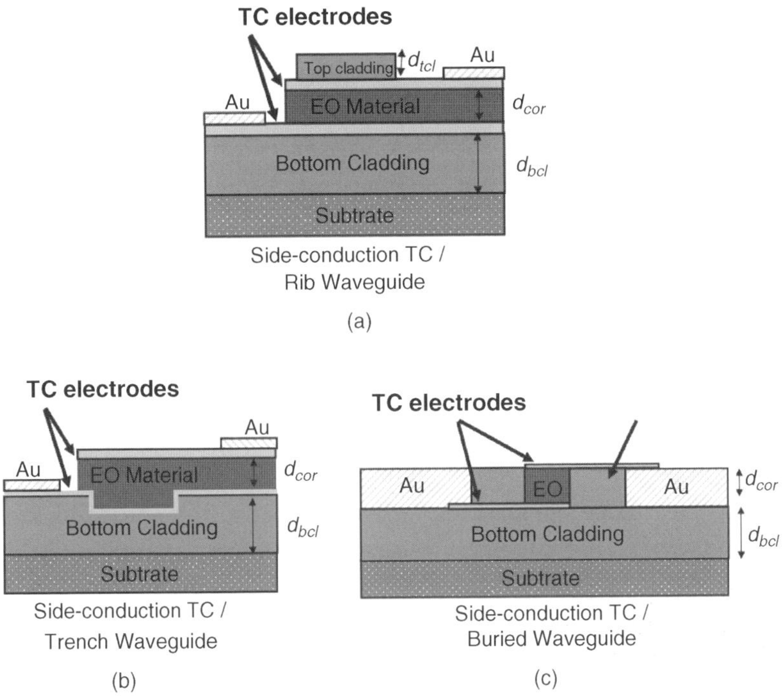

Figure 15.6 *Side conduction geometry: the TC is a thin layer (~50–100 nm) between the conventional cladding layer and the EO core layer. The optical mode is still confined by the conventional claddings. Horizontally, the optical mode can be confined using three different geometries, shown as: (a) top rib waveguide geometry; (b) bottom trench waveguide geometry; (c) buried waveguide geometry*

should be as conductive as possible without trading off their optical losses. While the TC conductivity can be enhanced with the increase of free carrier concentration, a fundamental limit is the absorption of optical energy due to the free carriers, called free-carrier absorption. For TC materials, optical absorption typically increases with the free carrier concentration, so the optical absorption loss per unit length (or 'optical absorption coefficient', denoted by α_{TCopt} with units of cm^{-1}) is roughly proportional to its electrical conductivity (denoted by σ_{TC} with units of $\mathrm{S\,cm}^{-1}$), within typical material types if the optical absorption is dominated by free-carrier absorption: $\Delta\alpha = (e^3\lambda^2/4\pi^2c^3\varepsilon_0 n)\times [\Delta N_e/m_{ce}^{*2}\mu_e + \Delta N_h/m_{ch}^{*2}\mu_h]$, where e is the electronic charge, ε_0 is the permittivity of free space, n is the refractive index of the material, m_{ce}^* is the conductivity effective mass of electrons, m_{ch}^* is the conductivity effective mass of holes, μ_e is the electron mobility and μ_h is the hole mobility[12].

For the modulator applications, it is desirable to use TCs with high conductivity to optical absorption loss ratios, which we will refer to as the TC figure of merit ($F_{TC} = \sigma_{TC}/\alpha_{TCopt}$, having units of S). There are various growth methods for TC materials, including metal-organic chemical vapor deposition (MOCVD), ion-assisted deposition (IAD), pulsed laser deposition (PLD) and sputtering [13–22]. Table 15.2 shows a few of the TCs of interest, such

Table 15.2 Some typical TC figure of merit values achieved experimentally by us

TC	σ_{TC} (S cm^{-1})	α_{TCopt} (cm^{-1})	F_{TC} (S)
IAD-In_2O_3-1	300	2000	0.15
IAD-In_2O_3-2	108	800	0.13
IAD-ZnO	15	100	0.15
IAD-ZnO/In_2O_3/ZnO	98	150	0.65
MOCVD-ZnO	54	20	2.7

as ZnO, In_2O_3, and TCP multi-layers, which have better transparency than the Sn-doped In_2O_3 (ITO) at the fiber-optic communication wavelength of 1550 nm, their deposition methods, and typical F_{TC} values.

15.2.3 Basic Modulator Design Considerations

In this section, we describe the basic design considerations for a low-voltage, high-speed EO modulator. Let us illustrate the design algorithm for the case of a TC-based modulator with SCG. The basic geometry is shown in Figure 15.7. In the SCG design with a buried optical waveguide structure, two metal electrodes, left electrode (M_L) and right electrode (M_R), form a coplanar slot transmission line. Two TC electrodes, top electrode (TC_T) and bottom electrode (TC_B), form a pair of parallel plates to apply the switching voltage directly across the EO waveguide core layer. The geometry for the metallic transmission line can be in various forms such as in the form of a coplanar transmission line, parallel-plate transmission line, or other RF transmission line geometry.

15.2.3.1 Minimum Vertical Optical Mode Size and Half-Wave Voltage

The half-wave voltage V_π of the TC-based organic EO modulator structure is given by Equation (15.2) with $d_{el\text{-}sep} = d_{cor}$ and V_π is related to the ratio $d_{el\text{-}sep}/\Gamma_{EO}$ in Equation (15.2). We can define an 'effective thickness' $d_{eff} = d_{el\text{-}sep}/\Gamma_{EO}$ for Equation (15.2). The EO mode overlapping factor Γ_{EO} can be divided into the vertical and horizontal components:

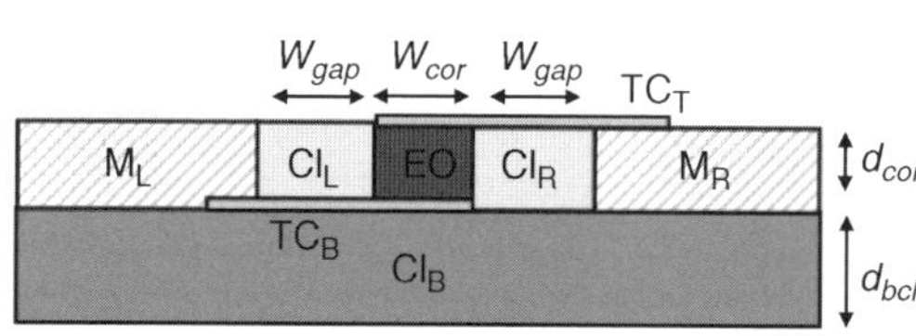

Figure 15.7 A side conduction modulator geometry with a buried waveguide structure. Two thin layers of TC material, TC_T and TC_B, are used to form a pair of 'bridge electrodes' to side conduct the voltage from the metallic transmission line and to apply the switching voltage directly across the waveguide core with organic EO material. In this case, the metallic electrodes, M_L and M_R, form a coplanar slot line structure. The two side claddings, CL_L and CL_R, help to confine the optical mode in the horizontal direction. The structure is located on top of a bottom cladding CL_B

$\Gamma_{EO} = \Gamma_{VEO} \times \Gamma_{HEO}$. The vertical component Γ_{VEO} is given by the amount of the optical mode energy overlapping with the EO slab in the vertical direction (the EO slab includes the EO core and the two side claddings that confine the mode horizontally), which accounts for the fact that optical mode energies in the upper and lower claddings are not being modulated as there is no EO material or modulating electric field there. The horizontal component Γ_{HEO} accounts for the remaining horizontal mode confinement in the EO core and will reach unity ($\Gamma_{HEO} = 1$) if the EO core (and its associated applied electric field) is sufficiently wide to horizontally cover the entire optical mode in the slab region. Γ_{VEO} will drop from unity when the EO core layer thickness d_{cor} is thin and the optical mode begins to spread into the top and bottom cladding layers. The minimum V_π is determined by the minimum of d_{eff}, which is the only geometry dependent factor in Equation (15.2).

If $\Gamma_{HEO} = 1$, d_{eff} will depend only on Γ_{VEO}. As discussed above, Γ_{VEO} varies with d_{cor}. Assuming $\Gamma_{HEO} = 1$, in Figure 15.8 we plot d_{eff} as a function of d_{cor} with $\lambda = 1.55\ \mu m$, $n_{cor} = 1.7$, $n_{scl} = 1.45$, $n_{bcl} = 1.35$ (solid line) and $n_{bcl} = 1.55$ (dotted line). From Figure 15.8 we see that for $n_{bcl} = 1.35$, the TC-based organic modulator structure can have a minimum effective thickness $d_{eff} = 0.88\ \mu m$ when the real core thickness $d_{cor} = 0.6\ \mu m$. At $d_{cor} = 0.6\ \mu m$ (for lowest voltage operation), the overlap $\Gamma_{VEO} = 69\%$, and if $d_{cor} = 1.5\ \mu m$ (for voltage-frequency response balance operation), Γ_{VEO} can be as high as 97%. However, for $n_{bcl} = 1.35$, $d_{cor} = 1.5\ \mu m$, which is no longer a single mode. To achieve single mode operation with $d_{cor} = 1.5\ \mu m$, we would need $n_{bcl} = 1.55$, at which $\Gamma_{VEO} = 92\%$. As an example, suppose $r = r_{33} = 150\ pm\ V^{-1}$, then the corresponding $V_\pi L = 0.17$ V cm (assuming push-pull operation). Thus, the switching voltage can be as low as 1 Veven with a short (1.7-mm-long) device. However, there is also a trade-off between the low half-wave voltage and large electrical bandwidth because of the increased capacitance C when d_{cor} is thin, resulting in resistor-capacitor (RC) cut-off at lower frequency. In addition, there is increased RF loss and optical loss in the TC region when d_{cor} decreases, while the thickness–conductivity product of the TC

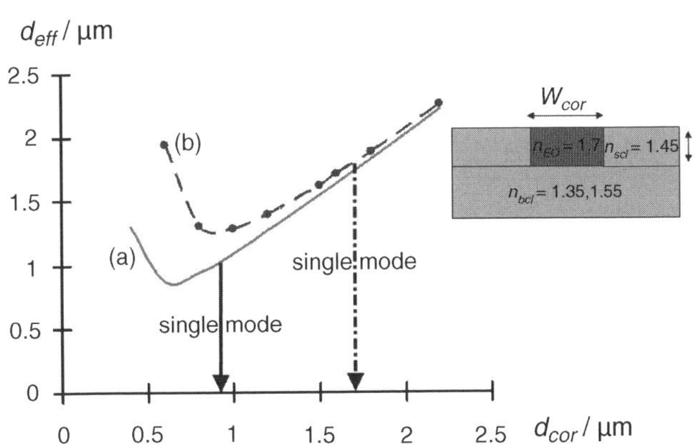

Figure 15.8 *Calculated d_{eff} as a function of d_{cor} ($\lambda = 1.55\ \mu m$, $n_{cor} = 1.7$, $n_{scl} = 1.45$) for two cases: (a) $n_{bcl} = 1.35$ (solid line) and (b) $n_{bcl} = 1.55$ (dotted line). We assume $W_{cor} = 3\ \mu m$ so that $\Gamma_{HEO} = 1$ to determine the effect of d_{eff} on Γ_{VEO}. Arrows indicate vertical single mode cut-off points*

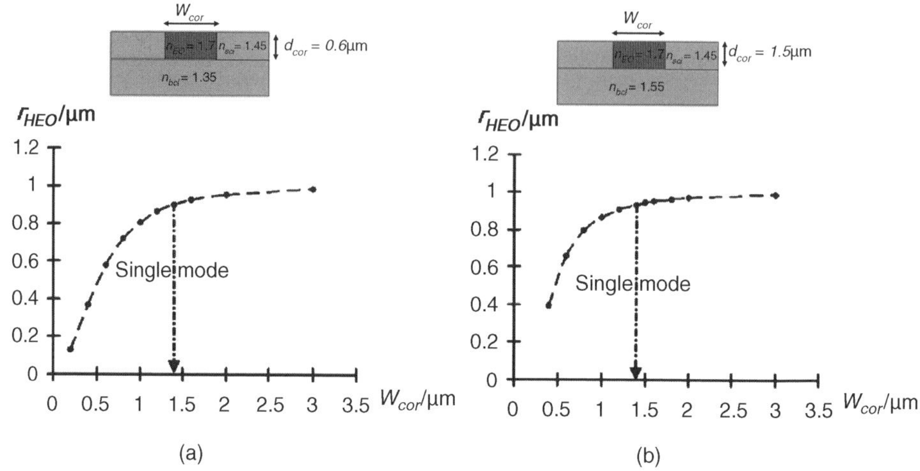

Figure 15.9 *Dependence of Γ_{HEO} on W_{cor}: (a) $d_{cor} = 0.6\ \mu m$, $n_{bcl} = 1.35$; (b) $d_{cor} = 1.5\ \mu m$, $n_{bcl} = 1.55$. Here we assume $\lambda = 1.55\ \mu m$, $n_{EO} = 1.7$, and $n_{scl} = 1.45$*

layer remains unchanged. We will discuss the voltage–bandwidth trade-off in detail below (see Section 15.2.3.5).

15.2.3.2 Minimum Horizontal Optical Mode Size and Frequency Response

For a certain EO thickness d_{cor}, Γ_{VEO} is fixed, and Γ_{HEO} is determined by the width of the core, W_{cor} (assuming uniform electric field). A wider waveguide core will have a larger Γ_{HEO} and therefore a larger Γ_{EO} which gives lower V_{π}. However, this also requires wider overlapping TC parallel plates, which will result in larger capacitance C and higher RF loss, and both of them will limit the frequency response of the device. To find the optimal W_{cor}, we next analyze Γ_{HEO} trends with increasing W_{cor}.

From Figure 15.9, we can see that Γ_{HEO} increases rapidly to 0.8–0.9 when W_{cor} increases from 0.5 to 1 μm, and then increases gradually to 1 when W_{cor} increases from 1 to 2 μm. A single optical guided mode (in horizontal direction) would require W_{cor} to be smaller than 1.4 μm. Therefore, to have sufficiently high Γ_{HEO} without limiting the bandwidth too greatly and to maintain a single mode, we can choose $W_{cor} = 1$ μm.

15.2.3.3 Minimum Gap Width and Frequency Response

Another important factor in the modulator horizontal direction that will determine the bandwidth of the device is the gap between the EO core and the metal electrode (W_{gap}). In order to decrease the electrical resistance R from the metal electrode to the TC parallel plate region and increase the RC cut-off frequency, W_{gap} must be minimized, and the minimal value of the W_{gap} is determined by the metal induced optical loss.

Figure 15.10 shows an example of the numerical simulation of the relationship between the metal induced loss coefficient α_{Met} versus the distance between the metal electrode and the buried optical waveguide. As an optimal design, we will choose a W_{gap} so that the metal

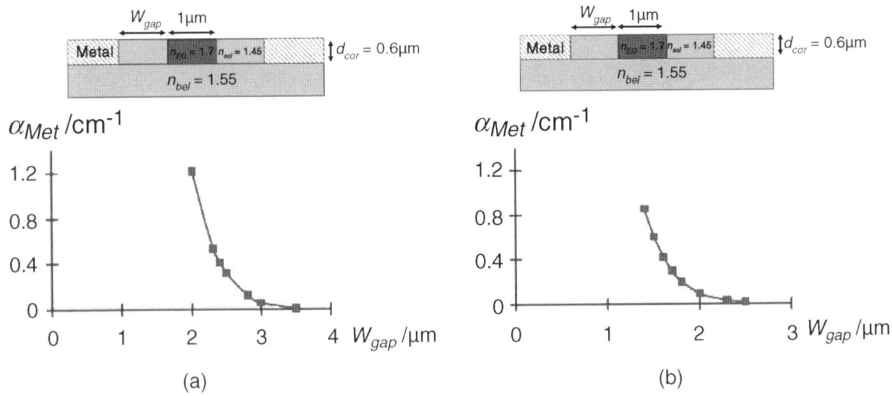

Figure 15.10 *Example of the metal induced loss coefficient (both left and right side included) vs. gap size: (a) low n_{bcl}, where $d_{cor} = 0.6$ um, $n_{bcl} = 1.35$; (b) high n_{bcl}, where $d_{cor} = 1.5$ um, $n_{bcl} = 1.55$*

induced optical loss is approximately the same value as the optical loss induced by the TC layers (see Sections 15.2.3.4 and 15.2.4).

15.2.3.4 Modulator Length, TC Thickness, and Optical Loss

For a TC material with a given figure of merit F_{TC}, and a given optical loss coefficient α_{TCopt}, an important device consideration is that the modulator must have low optical loss for an optical beam propagating through the modulator. The total loss of an optical waveguide device with TC layers is given by: $I_{out} = I_{in} \times T_{TC} \times T_{Met} \times T_{oth}$, where T_{TC} is the optical power transmission coefficient accounting for the optical loss caused by the TC layer alone, which can be further described by: $T_{TC} = \exp(-\alpha_{TCopt}\Gamma_{TC}L)$, where L is the length of TC layer in the device (it is also the modulator interaction length) and Γ_{TC} is the percentage of optical mode energy overlapping with the TC layer (the TC optical-mode overlapping factor). T_{Met} is the transmission coefficient accounting for the optical loss due to the optical power touching the metal transmission line on both sides and is given by $T_{Met} = \exp(-\alpha_{Met}L)$. As mentioned above, for an optimal design, we can let $T_{Met} = T_{TC}$. T_{oth} is the transmission efficiency accounting for the optical-fiber coupling efficiency and other waveguide propagation losses assuming that the TC layer has no optical loss. For the same TC material, increasing the thickness of the TC layer, d_{TC}, will increase the TC mode overlapping Γ_{TC} and therefore cause higher optical loss.

A typical commercial $LiNbO_3$ EO modulator has a device optical insertion loss of lower than 6 dB (<75% loss in optical power). The typical fiber coupling loss at the input and output ports can typically be lower than 30% per port, yielding a total coupling loss of less than 50%. Assuming that other propagation losses, including EO material absorption loss, total up to be less than 20%, T_{oth} will be no less than $(1-0.3) \times (1-0.3) \times (1-0.2) \approx 0.4$. In order to achieve a similar total device insertion loss of 6 dB for our modulator design, it is desirable to keep the optical propagation loss due to TC and metal to be less than 40% [i.e. keep $(T_{TC} \times T_{Met}) > 0.6$ or $T_{TC} > 0.77$ if $T_{TC} = T_{Met}$], so that $(T_{oth} \times T_{TC} \times T_{Met})$ will be greater than 0.25 (<75% loss or <6 dB total device insertion loss). For here and all the examples we give below, we will assume an RF-optical interaction length $L = 0.5$ cm, so that the optical transmission T_{TC} will be over 0.8 if

$$\alpha_{TCopt}\Gamma_{TC} < \ln(0.8)/L = 0.22/L \; (\text{for } L \text{ in cm}) \tag{15.7}$$

which requires $\Gamma_{TC} < 0.44/\alpha_{TCopt}$ when $L = 0.5$ cm. For a given L, the allowed α_{TCopt} is not fixed, but the factor $\alpha_{TCopt}\Gamma_{TC}$ is fixed. Since Γ_{TC} is approximately proportional to d_{TC}, the TC figure of merit $F_{TC} = \sigma_{TC}/\alpha_{TCopt}$ will then give a certain 'fixed' conductivity–thickness product ($\sigma_{TC}d_{TC}$) for the TC layer. As discussed below, the conductance per unit length given by $\sigma_{TC}d_{TC}/W_{TCeff}$ (W_{TCeff} is the effective conductance width of the TC) has an impact on the modulator RC frequency cut-off. Since the allowed $\alpha_{TCopt}\Gamma_{TC}$ is inversely proportional to the modulator length L, then for a given F_{TC} there is a specific modulator length (or switching voltage) versus frequency bandwidth trade-off. To achieve this generally low overlapping factor, d_{TC} must be small, approximately 50–100 nm, compared with the vertical optical mode size D_{VMfwhm} [defined as the full-width half-maximum (fwhm) of the mode energy]. The TC overlapping factor is approximately given by $\Gamma_{TC} \sim d_{TC}/D_{VMfwhm}$. Since Γ_{TC} may be solved exactly computationally using Mode Solver, we can use this to define an effective vertical mode size D_{VMeff} to be exactly given by $D_{VMeff} = d_{TC}/\Gamma_{TC}$. Taking $D_{VMeff} \sim D_{VMfwhm}$ is only an approximation.

15.2.3.5 TC RC Cut-Off Frequency

From the requirement that the TC optical loss factor $\alpha_{TCopt}\Gamma_{TC} < 0.22/L$, we have $\alpha_{TCopt}d_{TC}/D_{VMeff} = 0.22/L$. With use of the TC figure of merit $F_{TC} = \sigma_{TC}/\alpha_{TCopt}$, one can find the highest allowed TC conductance per unit length:

$$\sigma_{TC}d_{TC}/W_{TCeff} = (0.22/L)F_{TC}D_{VMeff}/W_{TCeff} \tag{15.8}$$

Since σ_{TC} and α_{TCopt} can be tuned simultaneously by controlling the dopant density of the TC material, for a chosen d_{TC} (and the given F_{TC}, L, and D_{VMeff}), we can engineer the dopant density so that Equation (15.8) is satisfied. Typically, σ_{TC} is not as high as that of typical metals such as gold or copper. Hence, TC cannot be used to form a low-loss RF transmission line, but can only be used as a bridge electrode to conduct voltage from the metallic transmission line to the EO waveguide core.

As an effective circuit model, the TC bridge-electrode material can be modeled as a resistor connecting the transmission line to the EO region, and the EO region with top and bottom TC electrodes can be modeled as an effective capacitor as shown in Figure 15.11.

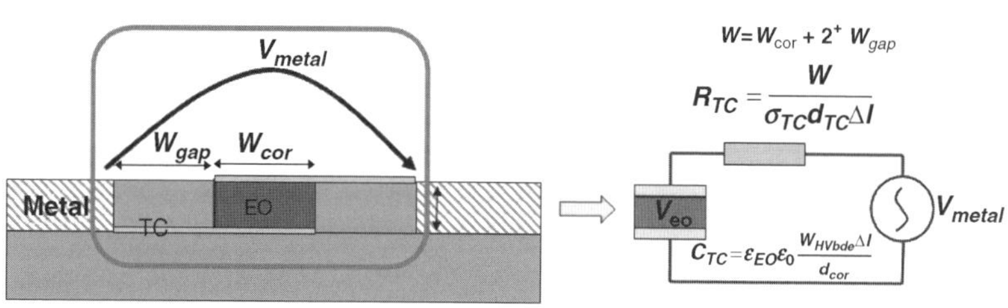

Figure 15.11 *Effective circuit model for TC-based EO modulator structure. Here $W_{TCeff} = W_{cor}/2 + W_{gap}$ is the effective conductance width of the TC resistor*

This effective RC element gives a TC RC cut-off frequency to a TC-based modulator. There are also other frequency limiting factors. The TC RC cut-off is only one of the factors and is due to the material parameters and the geometry of the TC-based bridge electrodes.

More specifically, based on the effective circuit model, the voltage applied across the EO layer V_{EO} versus the voltage at the metallic transmission line V_{Met} is given by:

$$\left|\frac{V_{eo}}{V_{metal}}\right| = \frac{1}{\sqrt{(2\pi f R_{TC} C_{TC})^2 + 1}} \tag{15.9}$$

where R_{TC} is the combined resistance–length product of the top and bottom TC bridge electrodes and C_{TC} is the capacitance per unit length, as defined in Figure 15.11.

If we define the frequency bandwidth f_{BW} to be the RF frequency at which $V_{eo} = V_{metal}/2$ (when the optical power drops by half), then

$$f_{BW-TC-RC} = \frac{\sqrt{3}}{2\pi R_{TC} C_{TC}} \tag{15.10}$$

This will be referred to as the 3 dB optical power modulation bandwidth limitation due to the TC bridge electrode ($f_{BW\text{-}TC\text{-}RC}$). From our deposition experience, TC films sputtered at room temperature can typically be engineered to achieve F_{TC} between 0.1 S and 1 S. With high temperature growth methods (e.g. MOCVD), $F_{TC} \sim 1$–10 S can also be achieved. For the discussion below, we assume a mean F_{TC} value of 1 S. R_{TC} can be obtained from $R_{TC} = 2W_{TCeff}/\sigma_{TC} d_{TC}$ (the factor of 2 arises from adding top and bottom TCs) and C_{TC} is simply given by $C_{TC} = \varepsilon_{EO} W_{cor}/d_{cor}$.

15.2.4 Basic Design Examples and Regions of Operation

In terms of design, there are two main regions of operation: (1) lowest voltage case (LV case), where the design will push towards the lowest switching voltage. From Figure 15.8, we see that in this case it is advantageous to work just above the minimal d_{eff} with d_{cor} of about 0.6–0.8 μm to achieve close to the lowest voltage but still having a reasonably high frequency response; (2) voltage–frequency balance case (VFB case), where the design will balance between the achievements of low enough voltage and high enough frequency response. From Figure 15.8, we see that in this case, it is advantageous to work at $d_{cor} \sim 1.2$–1.7 μm. The refractive index of the bottom cladding n_{bcl} must be chosen to be close enough to that of the EO core n_{cor} to make the waveguide a single-mode optical waveguide. It turns out that for the waveguide structure of Figure 15.8, with $n_{cor} = 1.7$ and $n_{bcl} = 1.35$, the waveguide vertical mode is single mode (at $\lambda = 1550$ nm) up to d_{cor} of ~0.9 μm and with $n_{cor} = 1.7$ and $n_{bcl} = 1.55$, the waveguide vertical mode is single mode (at $\lambda = 1550$ nm) up to d_{cor} of ~1.7 μm. This single mode d_{cor} value can be made thicker (up to ~2.5 μm) by reducing the refractive index difference between n_{cor} and n_{bcl} by using appropriate material for the bottom cladding. Thus, in general, d_{cor} must not be much thicker than 2.5 μm because of single mode considerations.

Below, we will give specific examples for the LV case (case 1) assuming $d_{cor} = 0.6$ μm with buried waveguide structure, and the VFB case (case 2) assuming $d_{cor} - 1.5$ μm with a buried waveguide structure.

- Step 1: Basic structural parameters
  We take the bottom cladding to be the polymer Cytop or benzocyclobutene (BCB) with refractive index $n_{bcl} = 1.35$ or 1.55, and RF dielectric constant $\varepsilon_{RFbcl} = 2.1$ or 2.5, respectively, and the EO waveguide core layer to have refractive index $n_{EO} = 1.7$ and $\varepsilon_{RFEO} = 3$. The optical wavelength is $\lambda = 1550\,\text{nm}$, and the EO waveguide core layer has an EO coefficient of $r = r_{33} = 150\,\text{pm V}^{-1}$. We assume that the TC is ZnO with refractive index $n_{TC} = 1.7$, $F_{TC} = 1\,\text{S}$, optical loss coefficient $\alpha_{TCopt} = 10\,\text{cm}^{-1}$, and conductivity $\sigma_{TC} = 10\,\text{S cm}^{-1}$. We assume $L = 0.5\,\text{cm}$.
- Step 2: Optical loss considerations
  From Equation (15.7), we require: $\alpha_{TCopt}\Gamma_{TC} < 0.22/L$. For $L = 0.5\,\text{cm}$ and $\alpha_{TCopt} = 10\,\text{cm}^{-1}$, we obtain $\Gamma_{TC} = 0.044$ (4.4%). The required TC thickness (top and bottom) d_{TC} to give a total $\Gamma_{TC} = 0.044$ for $d_{cor} = 0.6\,\mu\text{m}$ is 50 nm, and for $d_{cor} = 1.5\,\mu\text{m}$ it is 130 nm.
- Step 3: Choosing W_{cor} and the corresponding V_π
  From Figure 15.9, we choose $W_{cor} = 1\,\mu\text{m}$ so $\Gamma_{HEO} = 0.8$ when $d_{cor} = 0.6\,\mu\text{m}$ and $\Gamma_{HEO} = 0.85$ when $d_{cor} = 1.5\,\mu\text{m}$. From Figure 15.8, we find $V_\pi/2 = 0.26$ or 0.43 V (push-pull) when $d_{cor} = 0.6$ or $1.5\,\mu\text{m}$, respectively.
- Step 4: Choosing W_{gap}
  To optimize the design, we shall also choose W_{gap} so that metal optical loss is approximately equal to the TC optical loss, which means $\alpha_{Met} = 0.44\,\text{cm}^{-1}$. From Figure 15.10, we obtain $W_{gap} = 2.4$ or $1.6\,\mu\text{m}$ when $d_{cor} = 0.6$ or $1.5\,\mu\text{m}$, respectively.
- Step 5: Computing TC RC frequency cut-off
  From the design, we tabulate in Table 15.3 the various parameters for computing the TC RC frequency cut-off ($f_{BW\text{-}TC\text{-}RC}$).

From Table 15.3 we can see that thinner EO layers will give lower V_π but will limit the RC cut-off bandwidth because they will increase the capacitance of the parallel plates and also decrease the allowed TC conductivity due to the increase of optical mode energy in the TC layer.

15.2.5 High Frequency Design Considerations: Transmission Line RF Loss, Impedance Matching and Velocity Matching

Besides the TC RC voltage drop, the full frequency response is also determined by three other factors. First, in the direction of propagation, the RF loss at high frequency causes the amplitude of the RF wave to decay exponentially, and therefore the average voltage along the metal transmission line will drop below the low frequency value. Secondly, the velocity mismatch between the RF wave and optical wave discussed before will also greatly impact the effective switching voltage seen by the optical wave. Thirdly, the characteristic

Table 15.3 *Optical power bandwidth of 3 dB achievable at different combinations of TC thicknesses assuming $\lambda = 1.55\,\mu m$, $W_{gap} = 2.4$ or $1.6\,\mu m$ (for $d_{cor} = 0.6$ or $1.5\,\mu m$), $\varepsilon_{EO} = 3$, $r_{33} = 150\,pm\,V^{-1}$, $L = 5\,mm$ with push-pull operation*

d_{cor} (μm)	V_π (V)	$\sigma_{TC}d_{TC}$ (S)	R_{TC} (Ω cm)	C_{TC} (pF cm^{-1})	$f_{BW\text{-}TC\text{-}RC}$ (GHz)
0.6	0.22	4.6×10^{-5}	0.13	44	50
1.5	0.42	1.4×10^{-4}	0.03	18	525

impedance mismatch between the device and the RF source will cause reflection of the RF wave from the device back to the RF source and therefore decrease the effective driving voltage coupled into the device. By taking into consideration the RF loss coefficient, the velocity mismatch, the TC RC voltage drop, and the transmission line characteristic impedance simulated before, the effective voltage applied across the EO material along the TC electrodes is:

$$V_{eff}(z,f,V_{appl}) = V_{appl} \cdot \frac{V_{eo}(f)}{V_{metal}(f)} \cdot \frac{T}{1-\Gamma_L \Gamma_S e^{-2\gamma_\mu L}} \cdot \left[e^{-\alpha_{RF} x} e^{j(\frac{\omega}{v_{Opt}} - \beta_{RF})z} + \Gamma_L e^{-\alpha_{RF}(2L-x)} e^{j\omega \frac{z}{v_{Opt}} - j\beta_{RF}(2L-z)} \right] \tag{15.11}$$

Here V_{appl} is the RF source voltage, $\Gamma_L = \frac{Z_L - Z_m}{Z_L + Z_m}$ and $\Gamma_s = \frac{Z_s - Z_m}{Z_s + Z_m}$ are the reflection coefficients of the RF wave (if the line impedance Z_m is different from the source impedance Z_s or the load impedance Z_L; RF reflection will appear both at the entry point and the terminus of the transmission line), L is the length of the device. $V_{eo}(f)/V_{metal}(f)$ defined as $r_{RC}(f)$ is the TC RC cut-off factor. The RF reflection can be eliminated by designing the structure of the device to have $Z_m = Z_s = Z_L$. Then $\Gamma_L = \Gamma_s = 0$ and Equation (15.11) becomes:

$$V_{eff}(x,f) = V_{appl} \cdot \frac{V_{eo}(f)}{V_{metal}(f)} \cdot e^{-\alpha_{RF} x} \cdot e^{j(\frac{\omega}{v_{gopt}} - \beta_{RF})x} \tag{15.12}$$

The velocity mismatch term $e^{j(\frac{\omega}{v_{gopt}} - \beta_{RF})x}$ can also be eliminated if we can achieve perfect velocity matching, or $\beta_{RF} = \omega/v_{RF} = \omega/v_{gopt}$ (which means $n_{RF} = n_{gopt}$). Then, $(\omega/v_{RF}) - \beta_{RF} = 0$ and Equation (15.12) becomes:

$$V_{eff}(x,f) = V_{appl} \cdot \frac{V_{eo}(f)}{V_{metal}(f)} \cdot e^{-\alpha_{RF} x} \tag{15.13}$$

The averaged effective switching voltage $V_{av\text{-}eff}(f)$ in a modulator with interaction length L is found by integrating the effective switching voltage $V_{eff}(x,f)$ seen by the optical wave from 0 to L:

$$V_{av-eff}(f) = \int_0^L V_{eff}(x,f,V_{appl}) = V_{appl} \cdot \frac{V_{eo}(f)}{V_{metal}(f)} \cdot \frac{(1-e^{-\alpha_{RF}(f)L})}{\alpha_{RF}(f)L} \tag{15.14}$$

From Equation (15.14), we can see that if the RF reflections and velocity mismatch can be eliminated, the averaged effective switching voltage $V_{av\text{-}eff}(f)$ is determined mainly by two factors: one is the TC RC voltage drop factor $r_{RC}(f)$, which is a function of frequency f as discussed before, and the other is the RF decay factor $(1-e^{-x})/x$, in which $x = \alpha_{RF}(f)L$. The RF loss coefficient is also a function of frequency f because the RF decay coefficient $\alpha_{RF}(f)$ will increase with frequency f. The electrical bandwidth of an EO modulator can then be found by solving for the frequency f_{BW} at which the $V_{av\text{-}eff}$ drops to $V_{appl}/2$ or

$$r_{RC}(f_{BW}) \cdot \frac{(1-e^{-\alpha_{RF}(f_{BW})L})}{\alpha_{RF}(f_{BW})L} = \frac{V_{av-eff}(f_{BW})}{V_{appl}} = 0.5 \tag{15.15}$$

In other words, if we have a targeted bandwidth f_{BW}, the requirement for $\alpha_{RF}(f)L$ is given by:

$$1-e^{-x} = \frac{0.5}{r_{RC}(f_{BW})}x, \quad \text{here } x = \alpha_{RF}(f_{BW})L \tag{15.16}$$

In short, Equation (15.16) indicates that if most of the frequency cut-off is already determined by the TC RC voltage drop frequency cut-off r_{RC} [e.g. when $r_{RC}(f) \sim 0.5$], there is little room left for the RF loss, and a small $\alpha_{RF}(f)$ will push it to the cut-off (when the voltage drops to half). Thus, Equation (15.16) enables one to design or trade-off between the TC RC voltage drop frequency cut-off and the RF loss frequency cut-off, and it is similar for the frequency cut-offs due to velocity mismatch and frequency-dependant impedance mismatch.

15.3 Full Design: A Detailed Example of High-Frequency Modulator Design

The basic design discussed above does not include the design of the transmission line, which is important for achieving a broad modulation bandwidth. Due to the relatively complex geometry, design of the transmission line requires numerical simulation. In order to simulate the frequency response of the entire modulator, including both the TC bridge electrode structure and the metallic transmission line structure, we use commercial electromagnetic simulation software based on the finite element method [23]. Figure 15.12 shows the

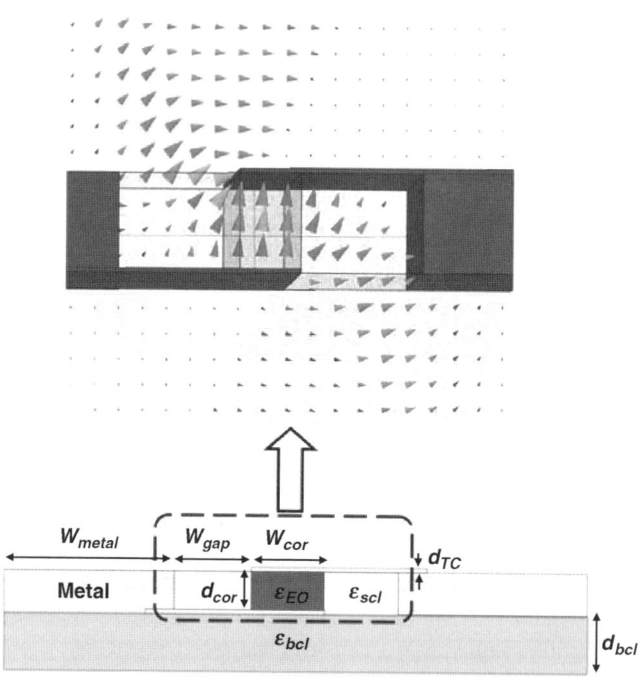

Figure 15.12 *A TC parallel plates, metal coplanar slot-line structure, and the parameters that can be tuned*

structure of interest. In this structure, many parameters can be tuned such as the dimensions of the metal and TC electrodes, the material dielectric constant of the bottom cladding layer, and so on. The ultimate goal is to design a structure that can work at or above 40 GHz with the half wave voltage below 1 V, device length about or less than 1 cm, and RF transmission line impedance of around 50 Ω.

More specifically, in order to have a broad frequency bandwidth structure, there are several design goals to achieve:

1. Transmission line impedance $Z_0 = 50\,\Omega$ in order to work with a standard microwave system.
2. RF-optical propagating refractive index matching. If the optical propagating refractive index is $n_{opt} = 1.6 \sim n_{gopt}$ (this can be obtained by numerical simulation), then the RF propagating refractive index n_{RF} should be targeted to be $n_{RF} = 1.6$, or the RF effective dielectric constant ε_{RF} should be close to $\varepsilon_{RF} = 2.56$.
3. The modulator 3 dB RC cut-off bandwidth for optical intensity modulation should be larger than the targeted bandwidth f_{BW}, preferably about double to leave room for the other frequency cut-off factors. This is because the frequency response is not only cut off by RC voltage drop but also cut off by the RF loss and velocity mismatch at high frequency.

In order to fully understand the high frequency operation of the modulator structure, we must analyze and design the structure step by step, from a metallic-transmission-line-in-air-only structure with the EO material dielectric constant ε_{EO} present (MTLIA-EO) to the full EO modulator (FEOM) structure, including the TC bridge electrodes and the substrate.

15.3.1 MTLIA-EO Structure

The first step is to analyze and design the MTLIA-EO structure. In this case, we set TC conductivity $\sigma_{TC} = 0\ \mathrm{S\,cm^{-1}}$, and substrate bottom cladding dielectric constant $\varepsilon_{sub} = \varepsilon_{bcl} = 1$ in Figure 15.10, while the EO and side claddings are present with $\varepsilon_{EO} = \varepsilon_{scl} = 3$. In the simulation we study three types of metal: gold, copper, and a perfect electric conductor (PEC) as the material for the MTLIA-EO structure.

From Figure 15.13(a) we can see that the RF loss coefficient α_{RF} with gold or copper is small, around $\alpha_{RF} = 0.1\ \mathrm{cm^{-1}}$ at 40 GHz. Also copper electrodes have lower RF loss than gold electrodes because of the higher conductivity. We will see later that with the addition of the TC, the RF loss will become somewhat larger but can be managed. Figure 15.13(b) shows the transmission line propagating RF dielectric constant ε_{RF} $(=n_{RF}^2)$. We see that ε_{RF} is much smaller than the square of refractive index of the EO waveguide core (n_{opt}^2 = 2.56 from the numerical simulation the optical waveguide effective refractive index n_{opt} is 1.6) because a large part of the RF mode is in air. The characteristic impedance of the transmission line is around 85 Ω. As we will see later, with the addition of the TC, the line impedance drops to ~50 Ω due to TC's resistive-capacitive loading effect.

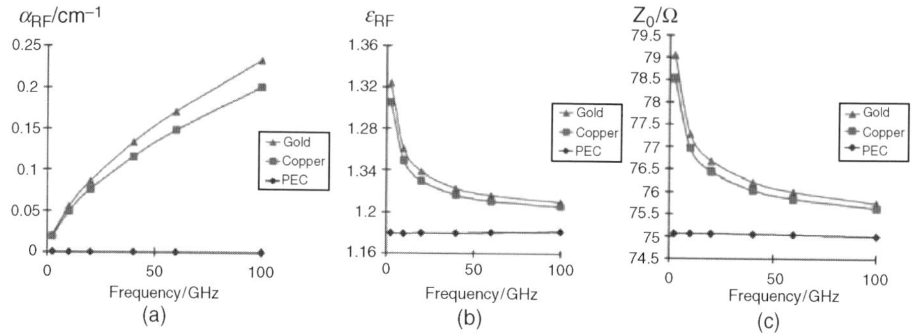

Figure 15.13 *Computational results for the MTLIA-EO structure using gold, copper and a perfect electric conductor (PEC) as the metallic material: (a) RF loss coefficient* α_{RF}*; (b) transmission line propagating dielectric constant* ε_{RF}*; and (c) transmission line characteristic impedance* Z_0*. We assumed the structure of Figure 15.10 with* $\sigma_{TC}=0\,S\,cm^{-1}$, $\sigma_{Gold}=4.1^{*}10^{5}\,S\,cm^{-1}$, $\sigma_{Copper}=5.8^{*}10^{5}\,S\,cm^{-1}$, $\varepsilon_{TC}=3$, $\varepsilon_{EO}=3$, $\varepsilon_{scl}=3$, $\varepsilon_{bcl}=1$, $W_{metal}=250\,\mu m$, $d_{metal}=d_{cor}=1.5\,\mu m$, $d_{TC}=130\,nm$, $W_{cor}=1\,\mu m$, $W_{gap}=1.6\,\mu m$ *and* $T_{bcl}=10\,\mu m$

15.3.2 FEOM Structure Including the TC Bridge Electrodes in the Vacuum

After studying the MTLIA-EO structure described above, we then study the FEOM structure by adding the TC-based bridge electrodes. We first explore different TC conductivities σ_{TC} from 5 to 20 S cm^{-1}.

From Figure 15.14(a), we see thatthe first impact of the TC layer is that the RF loss α_{RF} becomes higher than that of the MTLIA-EO structure. When $\sigma_{TC}=10\,S\,cm^{-1}$, α_{RF} is $\sim$1.1 cm^{-1} at 40 GHz, which is higher than that for the case with only gold electrodes ($\sigma_{TC}=0\,S\,cm^{-1}$) for which $\alpha_{RF}=0.1\,cm^{-1}$ at 40 GHz (see Figure 15.13). Here, we see that a frequency limitation is the RF propagation loss when TC is present and α_{RF} is $\sim$1.1 cm^{-1} at 40 GHz. We can call this the TC RF loss factor and label its bandwidth limitation as $f_{BW\text{-}TC\text{-}RFL}$.

From Figure 15.14(b), we see that the RF transmission line propagating dielectric constant ε_{RF} also changes from that of the MTLIA-EO only structure (from $\varepsilon_{RF}=1.2$ to 2.2). This is because the TC-based bridge electrodes move more RF energy into the EO core region, which has a high dielectric constant of $\varepsilon_{EO}=3$. However, since part of the RF mode is still in the air due to the metallic coplanar structure, ε_{RF} is still less than ε_{EO}. From Figure 15.14(c), we see that the characteristic impedance also drops from the MTLIA-EO structure value of 76 to 57 Ω because the resistive TC bridge electrodes now play a part in determining the impedance of the hybrid structure due to its mainly resistive-capacitive loading. To investigate the TC RC cut-off frequency discussed above, we compute the ratio of V_{eo}/V_{metal} versus frequency for different TC conductivies in Figure 15.14(d). Here V_{eo} is the voltage across the two TC electrodes at the EO core, and V_{metal} is the voltage across the two metallic electrodes. Besides the TC RC cut-off bandwidth $f_{BW\text{-}TC\text{-}RC}$, the full modulator bandwidth is also affected by $f_{BW\text{-}TC\text{-}RFL}$, the metal RF loss bandwidth $f_{BW\text{-}Metal\text{-}RFL}$, and the velocity matching bandwidth $f_{BW\text{-}Vel}$. From this section, it looks like the main frequency limitation for this particular example is from $f_{BW\text{-}TC\text{-}RFL}$ (it is lower than $f_{BW\text{-}TC\text{-}RC}$). The full

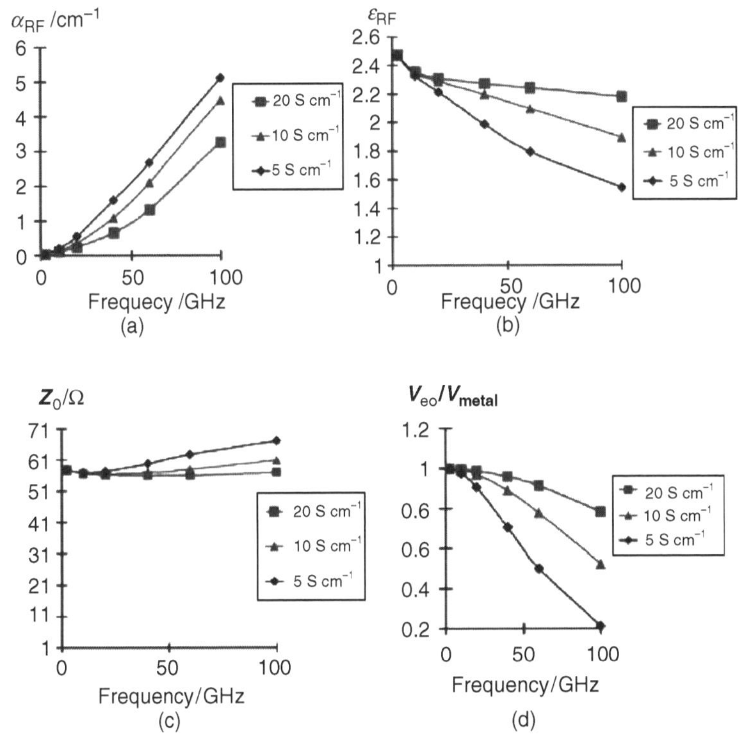

Figure 15.14 Computational results for the FEOM structure with gold transmission line: (a) RF loss coefficient α_{RF}; (b) transmission line propagating dielectric constant ε_{RF}; (c) transmission line characteristic impedance Z_0; and (d) V_{eo}/V_{metal} plot showing the TC RC frequency cut-off factor. We assume the structure of Figure 15.10 with $\sigma_{TC} = 5$, 10 and 20 S cm^{-1}, $\sigma_{Metal} = \sigma_{Gold}$ $4.1{*}10^5$ S cm^{-1}, $\varepsilon_{TC} = 3$, $\varepsilon_{EO} = 3$, $\varepsilon_{bcl} = 1$, $\varepsilon_{scl} = 3$, $d_{metal} = d_{cor} = 1.5\,\mu m$, $d_{TC} = 130$ nm, $W_{gap} = 1.6\,\mu m$, $W_{cor} = 1\,\mu m$, $W_{metal} = 250\,\mu m$ and $d_{bcl} = 10\,\mu m$

modulator bandwidth is extracted by taking into consideration all the factors as discussed in Section 15.3.5.

15.3.3 The Effect of Substrate Dielectric Constant

The transmission line effective dielectric constant, which is the key to achieving velocity matching, is also dependent on the material dielectric constant of the bottom cladding forming the substrate. This is because a significant portion of the RF field will be in the substrate.

The effect of the dielectric constant of the bottom cladding for the FEOM structure is shown by the family of curves in Figure 15.15. From the curves we see that the main effect of the dielectric constant of the bottom cladding material ε_{bcl} is that it can significantly change the effective propagating dielectric constant ε_{RF} of the transmission

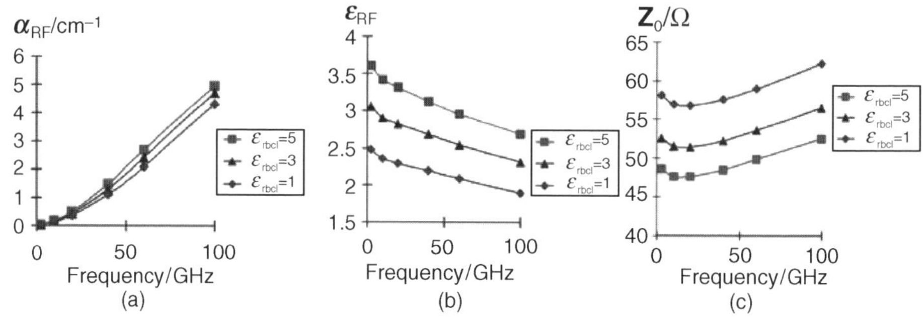

Figure 15.15 *Computational results for the FEOM modulator structure with gold transmission line, illustrating the effect of the dielectric constant of the bottom cladding layer ε_{bcl} on: (a) RF loss coefficient α_{RF}; (b) transmission line propagating dielectric constant ε_{RF}; and (c) transmission line characteristic impedance Z_0. We assumed the geometry of Figure 15.10 with $d_{TC} = 130\,nm$, $\sigma_{TC} = 10\,S\,cm^{-1}$, $\sigma_{Metal} = \sigma_{Gold} = 4.1^* 10^5\,S\,cm^{-1}$, $\varepsilon_{TC} = 3$, $\varepsilon_{EO} = 3$, $W_{metal} = 250\,\mu m$, $d_{metal} = d_{cor} = 1.5\,\mu m$, $W_{gap} = 1.2\,\mu m$, $W_{cor} = 1\,\mu m$ and $d_{bcl} = 10\,\mu m$*

line on top of it. When ε_{bcl} increases from 1 to 5, ε_{RF} increases from 2.2 to 3.12. In terms of transmission line effective refractive index n_{RF}, which is the square root of ε_{RF}, it increases from 1.48 to 1.77. The RF loss coefficient α_{RF} and transmission line characteristic impedance Z_0 also change a small amount but the changes are not as significant as that of the transmission line effective dielectric constant ε_{RF}. Hence, tuning the substrate dielectric constant ε_{bcl} is a good way to tune ε_{RF}, which is important for tuning the RF optical velocity matching.

15.3.4 Width of the Metal Electrodes

Compared with the TC material, the substrate material and the EO material, the dimensions of the metallic electrodes are easier to engineer in order to achieve the final design goals.

The effect of the width of the metallic electrodes W_{metal} for the FEOM structure is shown by the family of curves in Figure 15.16. From Figure 15.16, we see that by adjusting the width of the metallic electrodes, the transmission line effective dielectric constant ε_{RF} can be tuned towards matching the effective optical index $n_{opt} = 1.6$ (which is obtained from the numerical simulation of the optical waveguide). Furthermore, the characteristic impedance can be tuned to be $\sim 50\,\Omega$. At the same time, the RF voltage decay coefficient α_{RF} at 40 GHz is $\sim 1.2\,cm^{-1}$. Using Figure 15.4, $\alpha_{RF} = 1.2\,cm^{-1}$ translates to a voltage drop coefficient r_{RC} of 0.78.

15.3.5 Overall Frequency Response of the Effective Switching Voltage

Figure 15.17(a), (b), (c) and (d) shows RF loss coefficient α_{RF}, transmission line effective dielectric constant ε_{RF}, transmission line characteristic impedance Z_0 of a specific structure versus frequency, and the voltage drop coeffcient V_{eo}/V_{metal}, respectively. From the curves

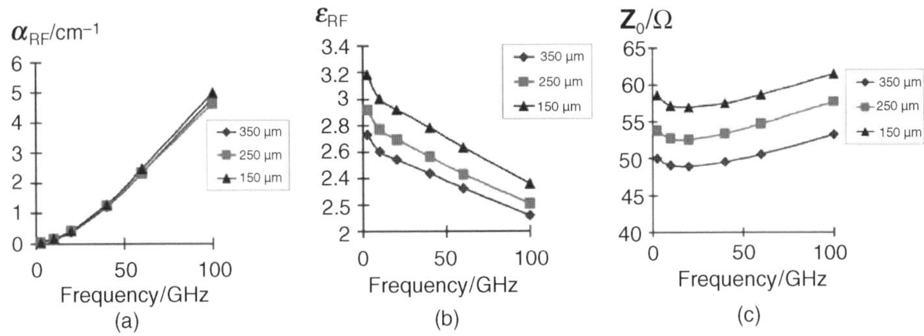

Figure 15.16 *Computational results for the FEOM modulator structure illustrating the width of the gold metallic electrode on: (a) RF loss coefficient* α_{RF}*; (b) transmission line effective dielectric constant* ε_{RF}*; and (c) transmission line characteristic impedance* Z_0*. We assume the geometry of Figure 15.10 with* $d_{TC} = 130\,nm$, $\sigma_{TC} = 10\,S\,cm^{-1}$, $\sigma_{Metal} = \sigma_{Gold} = 4.1^{*}10^{5}\,S\,cm^{-1}$, $\varepsilon_{TC} = 3$, $\varepsilon_{EO} = 3$, $\varepsilon_{bcl} = 2.5$, $\varepsilon_{scl} = 3$, $W_{metal} = 150$, *250 and* $350\,\mu m$, $d_{metal} = d_{cor} = 1.5\,\mu m$, $W_{gap} = 1.6\,\mu m$, $W_{cor} = 1\,\mu m$, $d_{bcl} = 10\,\mu m$

we can see that the RF loss increases with frequency and lower TC conductivity leads to a higher RF loss, which is expected. The effective RF dielectric constant is tuned to the square of effective optical index ($1.6^2 = 2.56$). The characteristic impedance is designed to be 50 Ω. Figure 15.17(e) shows the overall frequency response taking into consideration the effects shown in Figure 15.17(a)–(d). Note that when the TC conductivity σ_{TC} is about $10\,S\,cm^{-1}$ and the RF source voltage $V_{appl} = 1$ V, the effective voltage V_{eff} (which is the overall effective switching voltage on the EO layer) drops to ~0.5 V at 56 GHz. This means that the final effective modulator bandwidth is above 40 GHz for this particular case. Thus, this detailed numerical simulation gives the bandwidth of the EO modulator.

From the previous discussion of the relationship between $V_{\pi}L$ and r, it can be seen that if the material EO coefficient is above $30\,pm\,V^{-1}$, a 1-cm-long TC-based EO modulator with push-pull MZI configuration only requires a half-wave voltage $V_{\pi}/2 = 1$ V. Therefore we can see that, the TC-based structure can help the organic EO modulator to achieve ultra-low voltage (sub 1 V), ultra compact size (sub 1 cm), and a broad bandwidth (40 GHz).

15.4 Experimental Realization of a TC-Based Organic EO Modulator and Measurement Result

We have reported initial fabrication and characterization results for a TC-based organic EO modulator [24]. This initial work was not designed for high-speed operation, but demonstrated the low voltage achieved. To fabricate the modulator, a 2.8 μm thick SiO_2 layer was first deposited on a GaAs substrate wafer as the bottom cladding by plasma enhanced chemical vapor deposition (PECVD). The bottom TC electrode was then grown by MOCVD at 450 °C. The nonlinear organic EO material AJL8/APC was then spin-coated onto the bottom ZnO electrode to form a 1.5 μm thick layer as the waveguide core. A 1.5-μm-thick sacrificial poly(4-hydroxystyrene) (PVP) poling protective layer was next spin-coated onto

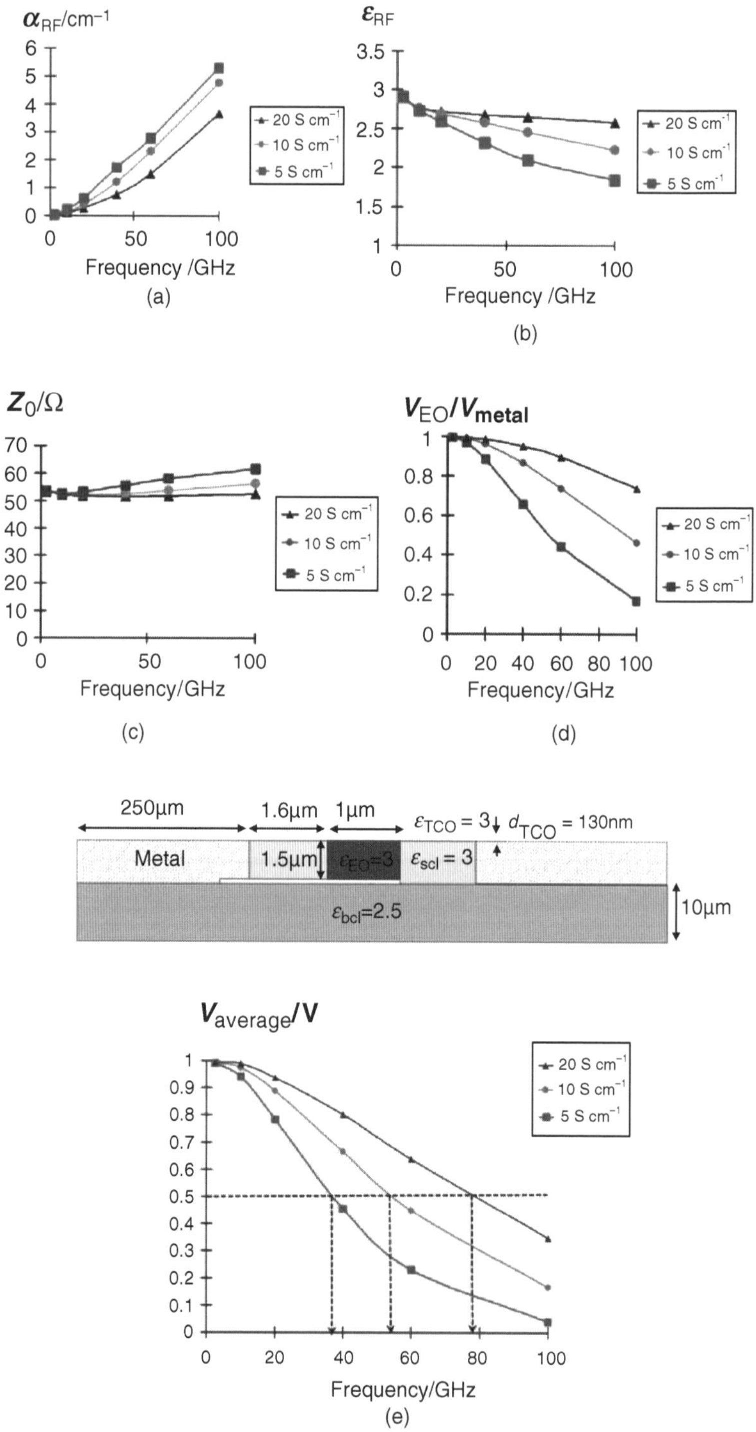

αRF/cm−1
6
5
4
3
2
1
0
0
50
100
Frequency /GHz
20 S cm−1
10 S cm−1
5 S cm−1
(a)
εRF
3.5
3
2.5
2
1.5
1
0
50
100
Frequency /GHz
20 S cm−1
10 S cm−1
5 S cm−1
(b)
Z0/Ω
70
60
50
40
30
20
10
0
0
50
100
Frequency/GHz
20 S cm−1
10 S cm−1
5 S cm−1
(c)
VEO/Vmetal
1
0.8
0.6
0.4
0.2
0
0 20 40 60 80 100
Frequency/GHz
20 S cm−1
10 S cm−1
5 S cm−1
(d)
250μm
1.6μm
1μm
εTCO = 3
dTCO = 130nm
Metal
1.5μm
εEO=3
εscl = 3
εbcl=2.5
10μm
Vaverage/V
1
0.9
0.8
0.7
0.6
0.5
0.4
0.3
0.2
0.1
0
0
20
40
60
80
100
Frequency/GHz
20 S cm−1
10 S cm−1
5 S cm−1
(e)

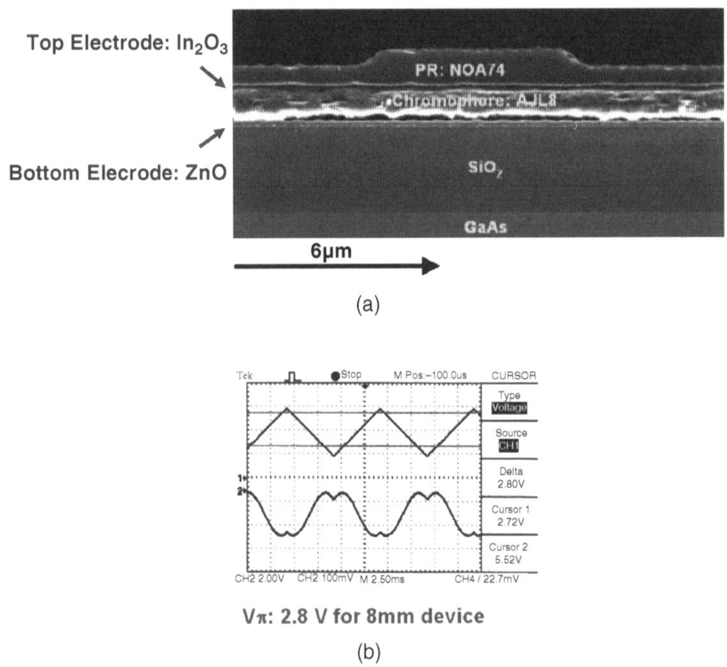

Figure 15.18 *(a) The cross-section view of a TC-based organic EO modulator using AJL8 from the University of Washington; (b) measurement result showing a 8-mm-long device with a non push-pull configuration that has half wave voltage as low as* $V_\pi = 2.8\,V$ *which corresponds to 1.4 V in the push-pull configuration*

the EO layer before deposition of the gold electrode for contact poling. The film was next poled at 140 °C with a poling voltage of 300 V for 5 min, and then cooled down to room temperature while maintaining the electrical field applied to the sample. After the poling process, the poling gold electrode was etched away using gold etchant and the protective PVP layer was removed by methanol. Then, a 20 nm thick In_2O_3 top TC electrode was deposited using an IAD system at room temperature (IAD enables the deposition of film with reasonable quality even at room temperature, which helps to avoid the organic EO material deterioration at higher temperatures), and the pattern was defined by using a shadow mask. To confine the optical mode laterally, a top rib waveguide formed by UV-cured polymer Norland Optical Adhesive 74 (NOA74) with a thickness of 0.8 µm and width of 3 µm was fabricated by reactive

Figure 15.17 *Computational results of the final TC-based modulator structure: (a) RF loss coefficient* α_{RF}*; (b) transmission line effective dielectric constant* ε_{RF} *of the active region; (c) transmission line characteristic impedance* Z_0 *of the active region; (d) the voltage drop from the metal transmission line to the EO waveguide core due to the RC cut-off effect; and (e) the full frequency response of the effective applied voltage. We assume* $V_{appl} = 1\,V$, $L = 5\,mm$, $\sigma_{TC} = 5$, *10 and 20 S cm*$^{-1}$, $\sigma_{Metal} = \sigma_{Gold} = 4.1 * 10^5\,S\,cm^{-1}$, $\varepsilon_{TC} = 3$, $\varepsilon_{EO} = 3$, $\varepsilon_{bcl} = 2.5$, $\varepsilon_{scl} = 3$, W_{metal} *250 µm,* $d_{metal} = d_{cor} = 1.5\,\mu m$, $d_{TC} = 130\,nm$, $W_{gap} = 1.6\,\mu m$, $W_{cor} = 1\,\mu m$ *and* $d_{bcl} = 10\,\mu m$

ion etching (RIE) using photoresist AZ1505 as the etching mask. The measurement results for the device are shown in Figure 15.18(b). An 8-mm-long device with non push-pull configuration has V_{π} as low as 2.8 V, which corresponds to 1.4 V in a push-pull configuration.

Acknowledgements

The authors acknowledge support by DARPA/ONR (SP01P7001R-A1/N00014-00-C) and the NSF MRSEC Program (DMR-0520513). EO materials for this work were provided by Professor Alex Jen's group in the Department of Materials Science and Engineering, University of Washington.

References

[1] A. Yariv, *Optical Electronics*, CBS College Publishing, New York, 1985.
[2] T. L. Paoli and J. E. Ripper, Direct modulation of semiconductor lasers, *Proc. IEEE*, **58**, 1457–1465 (1970).
[3] A. Faniuolo, G. Tartarini, P. Bassi, P. Faccin and A. Casini, Effects of directly modulated laser chirp on the performances of radio over fiber systems, *International Topical Meeting on Microwave Photonics*, 283–286 (2003).
[4] D. N. Nikogosyan, *Nonlinear Optical Crystals: A Complete Survey*, Springer, Berlin, 2005.
[5] B. Kuhlow. Modulators, in *Laser Fundamentals. Part 2*, Springer, Berlin, 2006.
[6] W. S. C. Chang, *RF Photonic Technology in Optical Fiber Links*, Cambridge University Press, Cambridge, 2002.
[7] D. M. Pozar, *Microwave Engineering*, 2nd edition, John Wiley & Sons, Ltd, Chichester, 1998.
[8] L. R. Dalton, W. H. Steier, B. H. Robinson, C. Zhang, A. Ren, S. Garner, A. Chen, T. Londergan, L. Irwin, B. Carlson, L. Fifield, G. Phelan, C. Kincaid, J. Amend and A. Jen, From molecules to opto-chips: organic electro-optic materials, *J. Mater. Chem.* **9**, 1905–1920 (1999).
[9] Y. Shi, W. Lin, D. J. Olson, J. H. Bechtel, H. Zhang, W. H. Steier, C. Zhang and L. R. Dalton, Electro-optic polymer modulators with 0.8 V half-wave voltage. *Appl. Phys. Lett.* **77**, 1–3 (2000).
[10] Y. Enami, C. T. DeRose, C. Loychik, D. Mathine, R. A. Norwood, J. Luo, A. K.-Y. Jen, and N. Peyghambarian, Low half-wave voltage and high electro-optic effect in hybrid polymer/sol-gel waveguide modulators, *Appl. Phys. Lett*, **89**, 143506 (2006).
[11] Y. Enami, C. T. Derose, D. Mathine, C. Loychik, C. Greenlee, R. A. Norwood, T. D. Kim, J. Luo, Y. Tian, A. K.-Y. Jen and N. Peyghambarian, Hybrid polymer/sol–gel waveguide modulators with exceptionally large electro–optic coefficients, *Nature Photonics*, **1**, 180–185 (2007).
[12] R. Soref and J. Larenzo, All-silicon active and passive guided-wave components for $\lambda = 1.3$ and 1.6 μm, *IEEE J. Quantum Electron.* **22**, 873–879 (1986).
[13] J. K. Luo and H. Thomas, Transport properties of indium tin oxide/p-InP structures, *Appl. Phys. Lett.* **62**, 705 (1993).
[14] M. J. Tsai, A. L. Fahrenbruch and R. H. Bube, Sputtered oxide/indium phosphide junctions and indium phosphide surfaces, *J. Appl.Phys.* **51**, 2696 (1980).
[15] M. Yan, M. Lane, C. R. Kannewurf and R. P. H. Chang, Highly conductive epitaxial CdO thin films prepared by pulsed laser depostion, *Appl. Phys. Lett.*, **78**, 2342 (2001).
[16] C. H. Liu and R. P. H. Chang, Theoretical and experimental study of impact of electric field on the atomic layer epitaxy of ZnO on alpha-Al_2O_3 surface, *J. Chem. Phys.* **116**, 8139–8143 (2002).

[17] N. L. Edelman, A. C. Wang, J. A. Belot, A. W. Metz, J. R. Babcock, A. M. Kawaoka, J. Ni, M. V. Metz, C. J. Flaschenriem, C. L. Stern, L. M. Liable-Sands, A. L. Rheingold, P. R. Markwork, R. P. H. Chang, M. P. Chudzik, C. R. Kannewurf and T. J. Marks, Synthesis and characterization of volatile, fluorine-free beta-ketoiminate lanthanide MOCVD precursors and their implementation in low-temperature growth of epitaxial CeO_2 buffer layers for superconducting electronics, *Inorg. Chem.* **41**, 5005–5023 (2002).
[18] X. Wang, S. Yang, J. Wang, M. Li, X. Jiang, G. Du, X. Liu and R. P. H. Chang, Structural and optical properties of ZnO film by plasma-assisted MOCVD, *Opt. Quant. Electron.* **34**, 883–891 (2002).
[19] M. Yan, Y. Koide, J. R. Babcock, P. R. Markworth, J. A. Belot, T. J. Marks and R. P. H. Chang, Selective area atomic layer epitaxy growth of ZnO features on microcontact patterned substrates, *Appl. Phys. Lett*, **79**, 1709–1711 (2001).
[20] R. Asahi, A. Wang, J. R. Babcock, N. L. Edleman, A.W. Metz, M. A. Lane, V. P. Dravid, C. R. Kannewurf, A. J. Freeman and T. J. Marks, First-principles calculations for understanding high conductivity and optical transparency in InxCd1-xO films, *Thin Solid Films* **411**, 101–105 (2002).
[21] A. J. Freeman, K. R. Poeppelmeier, T. O. Manson, R. P. H. Chang and T.J. Marks, Chemical and thin-film strategies for new transparent conducting oxides, *MRS Bull.* **25**, 45–51 (2000).
[22] A. Wang, N. L. Edleman, J. R. Babcock, T. J. Marks, M. A. Lane, P. W. Brazis and C. R. Kannewurf, Metal-organic chemical vapor deposition of In-Zn-Sn-O and In-Ga-Sn-O transparent conducting oxide thin films, *MRS Symp. Ser.* **607**, 345–352 (2000).
[23] Ansoft Corporation, High Frequency Structure Simulator.
[24] G. Xu, Z. Liu, J. Ma, B. Liu, S. T. Ho, L. Wang, P. Zhu, T. J. Marks, J. Luo and A. Jen, Organic EO modulator using transparent conducting oxides as electrodes, *Optics Express* **13**, 7380–7385 (2005).

16

Naphthalenetetracarboxylic Diimides as Transparent Organic Semiconductors

Kevin Cua See and Howard E. Katz
Department of Materials Science and Engineering, Johns Hopkins University, USA

16.1 Introduction

In the mid-1990s, there was only a hint of the variety of organic transistor semiconductors that would emerge in the years to follow. A perusal of review articles from that time [1, 2] reveals about a dozen well-characterized molecular solids and the beginnings of semiconducting polymers. Thiophenes were constituents of the overwhelming majority of these materials, with some mention of pentacene, C_{60}, and phthalocyanines (Figure 16.1). The primary goal was to reach charge carrier mobility above $0.1\,cm^2\,V^{-1}\,s^{-1}$ in field effect transistors (FETs), to establish competitiveness with amorphous silicon. Reaching this goal with pentacene had been considered a major achievement [3–6]. Most of these early experiments were done with hole transporters because hole energy levels in organics were much more environmentally accessible and compatible with metal work functions than electron energies, in contrast to the case for most inorganic semiconductors.

The importance of electron-carrying semiconductors for organic transistors was recognized because of the desire for complementary logic, but there was no apparent route to high-mobility, air-stable versions of such compounds. Results had been published on devices using C_{60}, [7, 8], rare-earth bis(phthalocyanines) [9], diphenylperylenetetracarboxylic diimide [10] and intentionally doped tetracyanoquinodimethane [11]; the first one being quite sensitive to oxygen and the others with low mobility.

Transparent Electronics: From Synthesis to Applications Edited by Antonio Facchetti and Tobin J. Marks

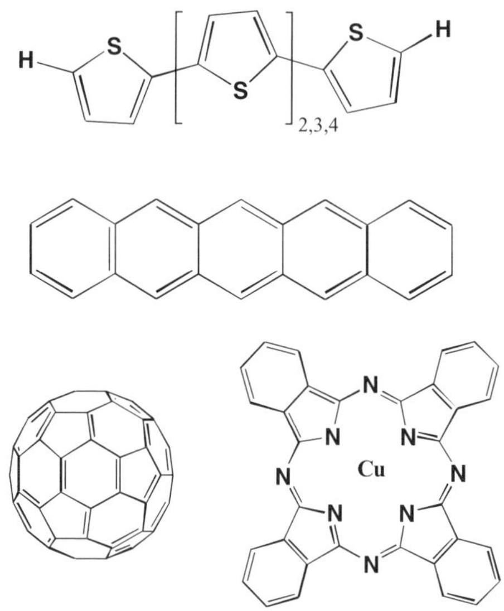

Figure 16.1 *Molecular structures (from top) of thiophene oligomers, pentacene, C-60 and copper phthalocyanine*

Transparency would have been considered a secondary advantage to organic semiconductors of that era, and in fact, none of these original semiconductors were transparent in the visible spectral region. On the contrary, some, including electron transporters, were being developed simultaneously for organic light emitting diodes, and were thus designed with visible-region transitions in mind. While it was understood that electronics for display control and windshield mounting would desirably be transparent, the necessity for mobility combined with a modicum of environmental stability was paramount.

Tetracarboxylic diimide and dianhydride semiconductors (Figure 16.2) had been studied for their fundamental properties and some assorted device applications in the early 1990s. Naphthalenetetracarboxylic dianhydride (NTCDA) was used in an organic semiconductor superlattice with perylenetetracarboxylic dianhydride (PTCDA) in an attempt to create an organic version of a quantum well structure [12, 13]. A substituted naphthalenetetracarboxylic diimide (NTCDI) glass was shown to have a fair mobility using time-of-flight photocurrent methods, and was modeled using a hopping formalism [14, 15]. The mobility was also found to be field-dependent, foreshadowing a detail that became apparent later in organic transistor studies. PTCDA showed signs of ambipolar transport, depending on molecular orientation of films and device architecture [16]. Other investigations focused on optical and dielectric properties, rectifying barriers between the anhydrides and other inorganic and organic semiconductors, photoelectrochemistry of thin films, molecular spectroscopy, and self-assembly of radical anions in solution [17–26].

16.2 Initial Demonstration of NTCDI Semiconductor FETs

In 1996, the Bell Labs group undertook an examination of compounds that had already been established as electron acceptors or electron transporters in other applications to see whether

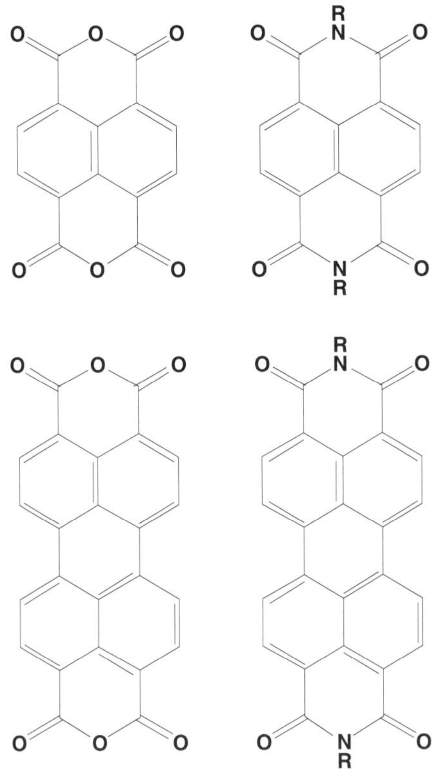

Figure 16.2 *NTCDA, NTCDI, PTCDA and PTCDI structures (clockwise from upper left)*

such compounds would function in FETs. Three prototypical compounds were selected: NTCDA, NTCDI (R = H), and tetracyanonaphtho-2,6-quinodimethane (TCNNQ). These compounds were viewed as having the minimal conjugation and substitution levels to support charge transport, and had low enough "lowest unoccupied molecular orbital" (LUMO) energies to accept electrons from an external circuit. The tetracarboxylic compounds were also transparent to visible light [27], as shown in Figure 16.3. Ironically, the FETs were bottom contact; because the electronics mindset at that time was still rooted in silicon technology, electrodes were prepatterned using photolithography on silicon substrates. The ideas of lithography on an organic surface, or even using shadow masks for top contact devices, were not yet in the mainstream. Mobilities exceeding 0.001 $cm^2 V^{-1} s^{-1}$ were achieved for NTCDA. Although these values were observed when precise substrate temperatures were used during deposition, and only under vacuum, this result was arguably the first demonstration of significant FET mobility using a transparent organic semiconductor, though at the time, it was the *n*-channel aspect of the devices that was emphasized.

From a chemistry point of view, one promising aspect of the NTCDI result is that it offered a straightforward means of optimizing the molecular solid morphology for higher in-plane mobility through substitution of the imide nitrogens with side chains that could promote two-dimensional self-assembly. A classic reaction had been developed to prepare

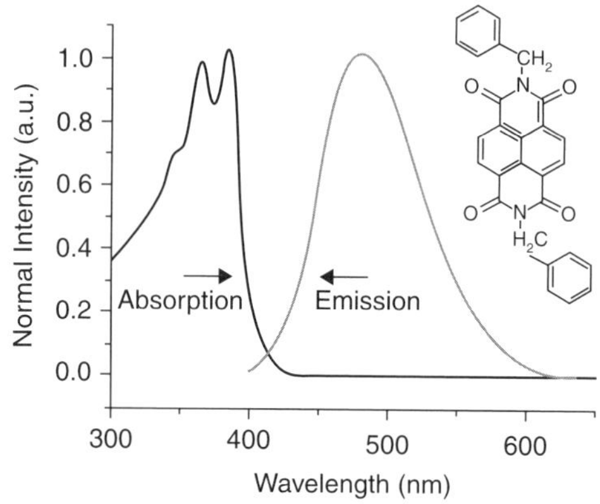

Figure 16.3 *Absorption and emission spectra of dibenzyl NTCDI. (Reprinted with permission from [27]. Copyright (2008) American Chemical Society)*

N-alkyl and N-aryl NTCDIs starting from NTCDA and primary amines, requiring inexpensive quinoline as solvent and zinc acetate as catalyst [28]. Products often precipitated directly from the reaction mixture, and could be purified by washing, recrystallization, and vacuum sublimation. The reaction was tolerant to any functional group that was not especially prone to reaction with nucleophiles or Lewis acids.

Utilizing this chemistry to expand the lengths and self-assembly properties of NTCDI semiconductor molecules, a device breakthrough was achieved. In 2000, the Bell Labs group showed that thermally evaporated thin films of various N,N′-substituted NTCDIs could show field effect electron mobilities up to $10^{-1}\,cm^2\,V^{-1}\,s^{-1}$, an order of magnitudeimprovement over most previous *n*-channel materials, in FETs with top-contact gold electrodes [29, 30]. More importantly, the authors showed that by incorporating perfluoroalkyl chains at the N, N′ positions, these high mobilities could be achieved in air. The key N,N′ substituents and resulting device properties are shown in Figure 16.4.

Two dramatic details point out the effects of the side chains. First, the longer, octyl chains of **1** and **2** greatly increase the thin film ordering compared with the parent NTCDI, and even compared with the *N,N′*-bis(heptafluorobutyl) derivative **4** (Figure 16.5), which also showed moderately high transistor mobility in air. The octyl derivatives show at least six orders of diffraction with narrow and intense principal peaks. Diffraction peaks from shorter-chain NTCDIs are much less discernible. Thus, the crystallization motif of the octyl NTCDIs is dominated by a propensity for the naphthalene cores and the alkyl chains to self-segregate, forming layers of aromatic rings that are close enough for π-overlap, among which electrons can easily hop and thus display high mobility.

Secondly, even slight fluoroalkyl substitution makes a major difference in the magnitude of the thin film mobility and environmental stability. The heptafluorobutyl NTCDI **4** shows much higher mobility than the parent, and considerable air stability as well. The bis (trifluoromethylbenzyl) derivative **3** (Figure 16.6) is also an excellent *n*-channel semiconductor in air with a mobility $>0.1\,cm^2\,V^{-1}\,s^{-1}$, while a methylbenzyl derivative with a very closely related crystal structure showed no *n*-channel activity in a thin film [32]. Recently, additional experiments verifying the effectiveness of trifluoromethylbenzyl and trifluoromethoxybenzyl groups have been reported [33].

1 R: 

2 R: 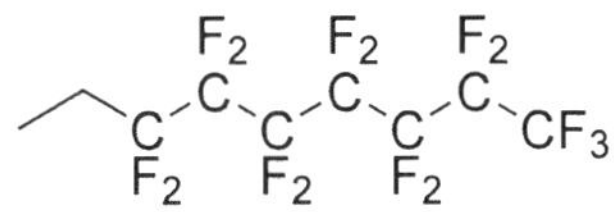

3 R: 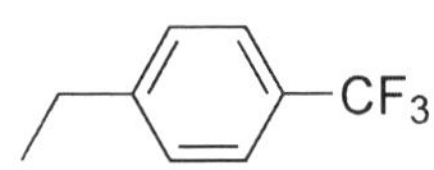

4 R: 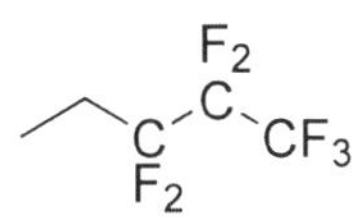

Compound	T_{sub}	μ_{air} cm^2 V^{-1} s^{-1}	μ_{vacum} cm^2 V^{-1} s^{-1}
1	70	10^{-6}	0.16
2	70	0.05	Slight decrease
3	98	0.12	Small increase
4	Room temperature	0.01	Not measured

Figure 16.4 *Summary of NTCDIs (R groups shown) and device properties (T_{sub} = substrate temperature during thermal evaporation). (Reprinted with permission from [31]. Copyright (2000) American Chemical Society)*

The perfluoroalkyl chains are clearly crucial to air stability of the compounds. The explanation proposed for this was that the larger atomic radius of the fluorine atoms as compared with the aliphatic analog created a dense protective layer which prohibited the penetration of electron quenching agents like oxygen and water. This hypothesis was bolstered by electrochemical measurements which indicated negligible shifts in reduction potential due to varied functionality at the N,N′ positions, and similar mobilities of **2** and **3** in vacuum. Further supporting the theory of the densely packed barrier to quenching is the fact that compound **1** exhibited substantially reduced mobilities on exposure to air.

The hypothesis that this is a kinetic stabilization is based on an analysis by deLeeuw that is summarized here. In order to engineer air stable *n*-channel small molecule semiconductors, oxidation of the redical anions by oxygen and water must be avoided. In other words, it must be thermodynamically favorable for electrons to remain on the organic semiconductor molecules instead of being transferred to other molecules in the environment.

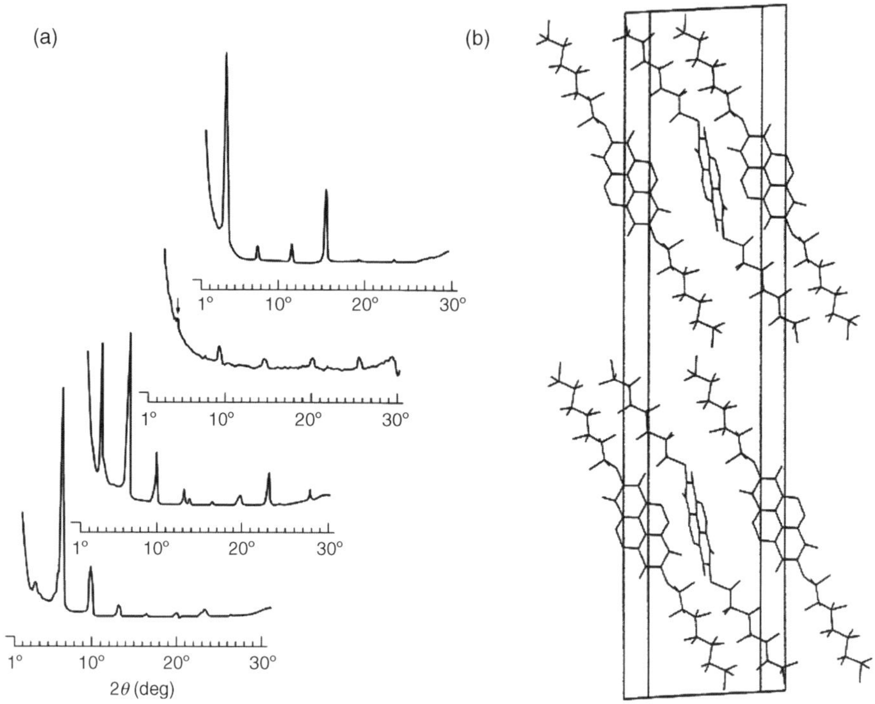

Figure 16.5 *(a) X-ray diffractograms (from top) of films of* ***1***, ***4***, ***2*** *cast from solution, and* ***2*** *vapor deposited. (b) X-ray single crystal structure of* ***2***

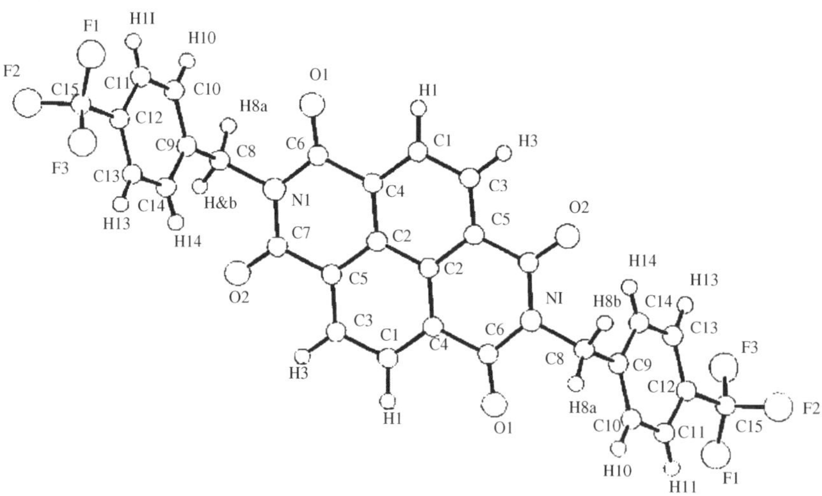

Figure 16.6 *Single crystal X-ray structure of bis(trifluoromethylbenzyl) NTCDI* ***3***. *(Reprinted with permission from [32]. Copyright (2001) Wiley-VCH)*

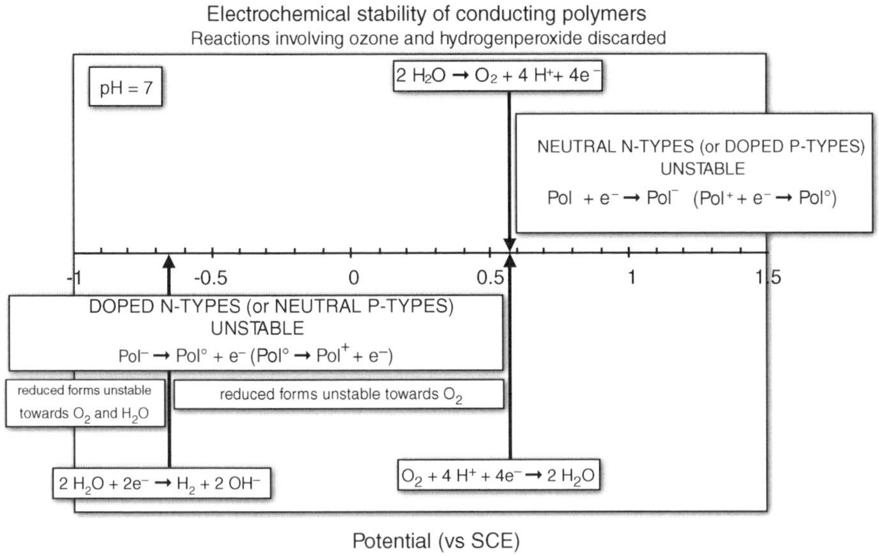

Figure 16.7 *Conditions for electrochemical stability of n-channel organic semiconductors. (Reprinted with permission from [34]. Copyright (1997) Elsevier Ltd)*

The free energy must then be less when the electron resides on the molecule than when it is transferred for reduction of oxygen or water. deLeeuw *et al.* have worked out the conditions for electron stability in terms of electrochemical redox potentials as displayed in Figure 16.7 [34].

From the strict application of Figure 16.7, the reduction potential required of a molecule so that its radical anion would be stable to exposure to both oxygen and water is 0.57 V or greater. As a practical matter, such a high reduction potential is very difficult to achieve, and there are examples of air stable *n*-channel materials with much lower reduction potentials. deLeeuw *et al.* presume that there is likely an overpotential (energy barrier to oxygen reduction) on the order of 0.5–1 V. In this case then, the threshold for electron stability shifts to a reduction potential around 0 V vs. SCE not accounting for acid-base interactions. This is a more manageable value; however, it is known that the materials studied (N,N′-substituted naphthalenetetracarboxylic diimides) have even lower reduction potentials around –0.4 V [35]. Furthermore, a very recent study suggests that the mechanism (or even the importance) of the fluoroalkyl stabilization of electrons and/or promotion of electron mobility may not be the same for tetracarboxylic diimides that do or do not lie within the electrochemical stability window for their radical anions [36]. Thus, it is possible that the fluorinated chains in more stabilized compounds, as opposed to core-unsubstituted NTCDIs, are influencing interfacial or intergrain properties of the semiconductor films, instead of or in addition to bulk packing, vapor diffusion and chemical reactivity.

Some other details emerged from the initial study that would have implications for future work. For example, a fair mobility of 0.001 $cm^2\,V^{-1}\,s^{-1}$ was observed in air from the *N*-hydroxyhexyl derivative, pointing out that utterly nonpolar conditions are not an absolute requirement for *n*-channel operation, and also that it is possible to have a chemically interactive functionality on the surface of an *n*-channel film. A similar mobility was more recently reported for dibutyl-NTCDI in air [37] and dibenzyl NTCDI in air [27]. The hydroxyhexyl and benzyl derivatives are somewhat amenable to solution processing. Also, thiol modifications of gold bottom contacts were useful in enabling charge injection from high work function metals that did not interpenetrate the semiconductors, as vapor-deposited top contacts tend to do.

16.3 Further Structural Elaboration of NTCDI Molecular Semiconductors

Interest in the side chain chemistry of NTCDIs has been recently renewed. Two new fluoroalkylated benzyl naphthalenetetracarboxylic diimides **5** and **6** (Figure 16.8) were synthesized using the same anhydride condensation discussed above, from commercially available amines [38]. Compound **5**, having the extended perfluoroalkyl chain, displayed a maximum mobility of 0.57 $cm^2\,V^{-1}\,s^{-1}$ and excellent on/off ratios ranging from 10^6 to 10^8. Typical electron mobilities for this compound were 0.2–0.4 $cm^2\,V^{-1}\,s^{-1}$ in FETs, for example, as shown in Figure 16.8. The mobilities were surprisingly insensitive to substrate surface treatment, and films showed extraordinarily precise layer morphology judging by both the orders of reflection and additional fringing in the X-ray diffractogram. These results compare favorably with the above mentioned compounds and require only a one step synthesis. The combination of simplicity of synthesis, air stability and electronic

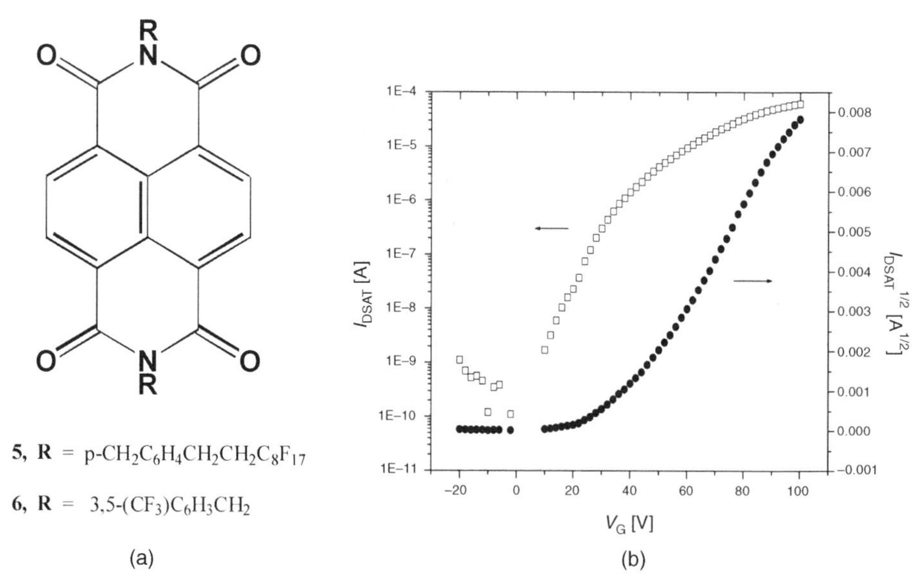

Figure 16.8 (a) Recent NTCDI compound structures and (b) transfer characteristics for **5** in a FET on silicon/silicon oxide. The semiconductor film was deposited on a 140 °C substrate. (Reprinted with permission from [38]. Copyright (2008) American Chemical Society)

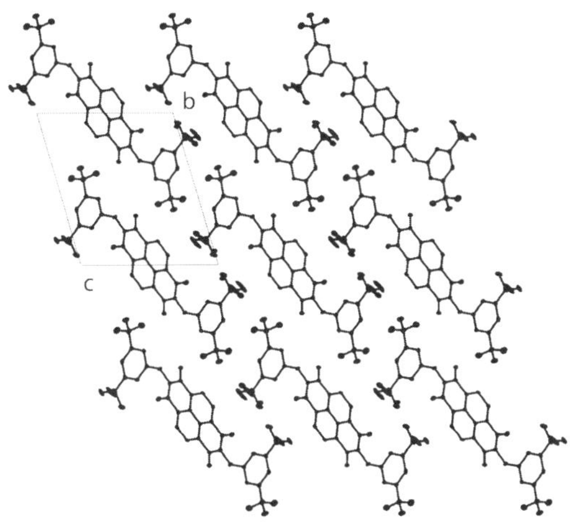

Figure 16.9 X-ray crystal structure of **6**. One layer is shown, the bc plane. π stacking occurs along the a axis, perpendicular to the plane of the diagram. (Reprinted with permission from [38]. Copyright (2008) American Chemical Society)

performance puts these among the leading *n*-channel organic FET (OFET) semiconductors explored to date. Note that the stabilizing effect of the fluoroalkyl side chain is manifest even though the chain is several atoms from the conjugated core. Compound **6** had mobilities of 0.1–0.2 $cm^2 V^{-1} s^{-1}$ depending on the surface treatment, and was notable for a stacked bulk crystal structure that is an exception to the general trend for herringbone packing in this class of organic solids (Figure 16.9).

A recent patent application from a Kodak-based group [39] strongly suggests that the cyclohexyl substituent would be particularly effective at promoting charge carrier mobility in NTCDIs. Mobility values $\gg 1\, cm^2 V^{-1} s^{-1}$ are claimed under inert atmosphere. The applicants infer that various other cycloalkyl substituents would be similarly useful, though no further details are given. Other NTCDI derivatives, included those with various N-aryl substituents, are claimed in related applications [40, 41].

An alternative approach to improving transport characteristics and air stability is to functionalize the conjugated core, in order to alter the electronic structure of the molecule and thereby enhance the thermodynamic stability towards oxidation by water and air, as discussed above [42]. This has been successfully demonstrated by Jones *et al.*, who reported air stable, transparent transistors with high mobility without the need for perfluorinated chains [43]. The halogenation and substitution chemistry developed for the synthesis is a significant contribution to the preparative methodology for NTCDI compounds, and condensed aromatic rings in general [43].

Figure 16.10 shows the synthetic route for core-cyano mono- and di-substituted NTCDI derivatives. The addition of the cyano groups shifts the reduction potential to higher values, making electrons more stable versus the oxidizing agents air and water. The dicyano product, designated NDI-8CN_2, shows a mobility in the range of 0.1–0.2 $cm^2 V^{-1} s^{-1}$, considerable air stability, on/off ratio of 100–1000, and operation on substrates with a variety of functionalizations. The compounds are very transparent with a band gap of $\sim$3.0 eV (Figure 16.11). The film morphology consists of elongated, interconnected grains,

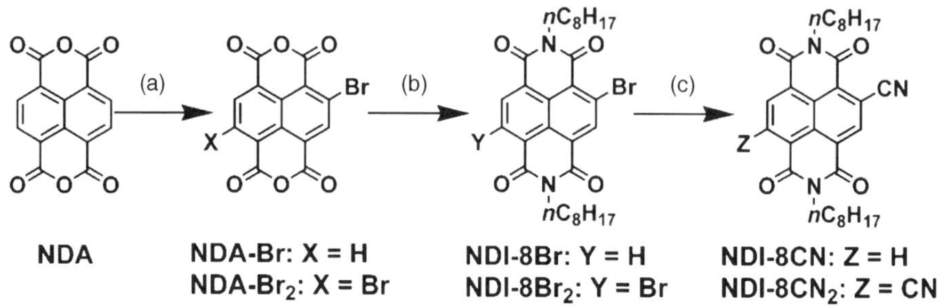

Figure 16.10 *Synthetic route to NDI-8CN and NDI-8CN$_2$ where conditions are: (a) Br$_2$/I$_2$, oleum; (b) n-octyl amine, HOAc; (c) CuCN, DMF. (Reprinted with permission from [43]. Copyright (2007) American Chemical Society)*

rather than the more tile-like structure seen with nonfunctionalized cores. However, this morphology is clearly among the favorable ones for high mobility. The monocyano compound has a platelike morphology, but the plates may be somewhat separated, resulting in lower mobility of 0.005 $cm^2 V^{-1} s^{-1}$. The cyano groups stabilize electrons through the conventional resonance model, thus bringing the threshold voltage to less positive values but also promoting conductivity in the absence of the gate voltage. The transparency associated with NTCDI compounds is retained. No other NTCDI shows mobility $>0.02\ cm^2 V^{-1} s^{-1}$ in air with only alkyl substitution. Comparisons of these compounds with related PTCDIs are extensively discussed in a recent primary publication [35]. Both the fluorinated side chain designs and the core halogenations/cyanation chemistry have been widely applied to PTCDI skeletons, but because of their highly absorbing chromophores, they are outside the scope of this chapter.

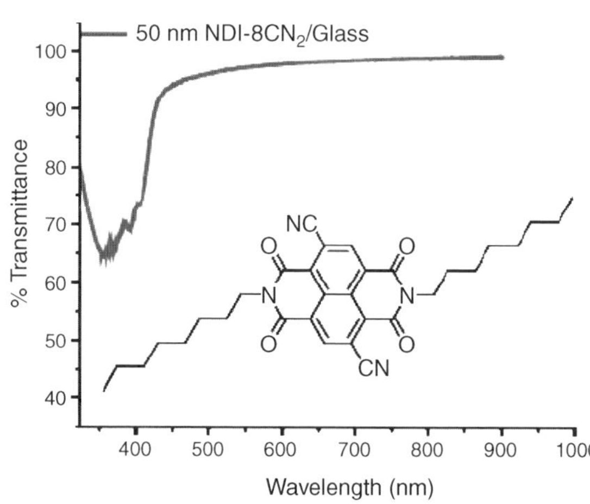

Figure 16.11 *Transmission optical spectrum of a 50 nm vapor-deposited thin film of NDI-8CN$_2$ on glass demonstrating the impressive transparency of this material between 400 nm and 800 nm. (Reprinted with permission from [43]. Copyright (2007) American Chemical Society)*

16.4 Use of NTCDI Semiconductors in Multifunctional Transistors

NTCDI-based semiconductors were used in two other devices that are related to OFETs. Both add value because of capabilities that go beyond simple on/off switching. First, the perfluorinated octyl compound **2** was used in nonvolatile 'memory' FETs. Devices using either silica or poly(4-methylstyrene) gate dielectrics were held at depletion gate voltages for several minutes and threshold voltage shifts were observed [44] (Figure 16.12). Thus, the function of such FETs could be tuned in the field, or even during manufacture, by the intentional application of this form of gate bias stress. Different degrees of circuit amplification, pixel current, or oscillator frequency could be obtained this way. However, for fast switching applications, this kind of bias stress effect is undesirable and considerable effort is being devoted to understanding and avoiding it. There are multiple possible mechanisms for this nonvolatility, and they are highly dependent on the nature of both the semiconductor and dielectric.

Secondly, an NTCDI heterostructure was used as the basis of an OFET that was responsive to polar vapors, especially to a phosphonate model of nerve agents [45]. Compound **2** was used as a base layer, and bis(3-hydroxybenzyl) NTCDI was deposited on top. On account of the very low surface energy of the base layer, the hydroxyl compound grew in an island morphology [45] (Figure 16.13). The resulting heterostructure showed a much more reproducible response to dimethyl methylphosphonate than did films of **2**

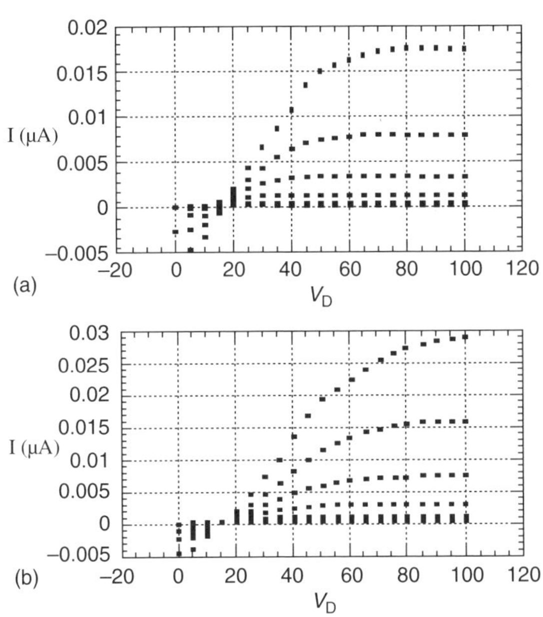

Figure 16.12 *FET current–voltage characteristics of* ***2*** *on poly(4-methylstyrene) before (a) and after (b) application of –200 V (depletion) to the gate. Gate voltages are zero to +100, bottom to top in each figure, increasing in 20 V increments. (Reprinted with permission from [44]. Copyright (2002) American Institute of Physics)*

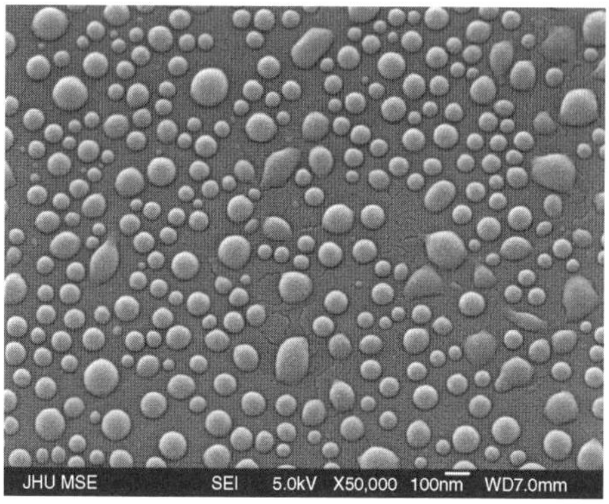

Figure 16.13 *Island morphology of bis(3-hydroxybenzyl) NTCDI grown on a base layer of* ***2***. *(Reprinted with permission from [45]. Copyright (2007) Wiley-VCH)*

alone, probably because of the hydrogen bonding ability of the phenolic surface groups on the islands to the phosphonate analyte molecules.

16.5 Conclusion

NTCDI molecular solids remain among the few choices available for transistor activity in transparent organic films, especially for electron transport. Mobilities are in the 0.1–1 $cm^2 V^{-1} s^{-1}$ range that is associated with amorphous silicon, and the chemical stability is excellent. The small molecular size allows moderate temperature processability, and in some cases, the resistance to doping results in on/off ratios well over one million. Control of device hysteresis and establishment of fast, reproducible deposition processes are two main challenges for the further utilization of these semiconductors in flexible transparent circuits.

Acknowledgements

Results from Johns Hopkins University reported in this manuscript were obtained with the support of the JHU Applied Physics Laboratory and the Air Force Office of Scientific Research (contract FA9550-06-1-0076).

References

[1] F. Garnier, *Current Opinion in Solid State and Materials Science* **2**, 455 (1997).
[2] A. J. Lovinger and L. J. Rothberg, *Journal of Materials Research* **11**, 1581 (1996).
[3] D. J. Gundlach, Y. Y. Lin, T. N. Jackson, *et al.*, *IEEE Electron Device Letters* **18**, 87 (1997).
[4] Y. Y. Lin, D. J. Gundlach, S. F. Nelson, *et al.*, *IEEE Electron Device Letters* **18**, 606 (1997).
[5] Y. Y. Lin, D. J. Gundlach, S. F. Nelson, *et al.*, *IEEE Transactions on Electron Devices* **44**, 1325 (1997).
[6] J. G. Laquindanum, H. E. Katz, A. J. Lovinger, *et al.*, *Chemistry of Materials* **8**, 2542 (1996).

[7] R. C. Haddon, A. S. Perel, R. C. Morris, *et al.*, *Applied Physics Letters* **67**, 121 (1995).
[8] K. Hoshimono, S. Fujimori, S. Fujita, *et al.*, *Japanese Journal of Applied Physics Part 2* **32**, L1070 (1993).
[9] G. Guillaud, M. Alsadoun, M. Maitrot, *et al.*, *Chemical Physics Letters* **167**, 503 (1990).
[10] G. Horowitz, F. Kouki, P. Spearman, *et al.*, *Advanced Materials* **8**, 242 (1996).
[11] A. R. Brown, D. M. Deleeuw, E. J. Lous, *et al.*, *Synthetic Metals* **66**, 257 (1994).
[12] F. F. So and S. R. Forrest, *Physical Review Letters* **66**, 2649 (1991).
[13] F. F. So, S. R. Forrest, Y. Q. Shi, *et al.*, *Applied Physics Letters* **56**, 674 (1990).
[14] P. M. Borsenberger, W. T. Gruenbaum, E. H. Magin, *et al.*, *Japanese Journal of Applied Physics Part 1* **35**, 6135 (1996).
[15] P. M. Borsenberger, W. T. Gruenbaum and E. H. Magin, *Physica Status Solidi B* **190**, 555 (1995).
[16] J. R. Ostrick, A. Dodabalapur, L. Torsi, *et al.*, *Journal of Applied Physics* **81**, 6804 (1997).
[17] I. H. Campbell, P. S. Davids, J. P. Ferraris, *et al.*, *Synthetic Metals* **80**, 105 (1996).
[18] M. Adachi, Y. Murata and S. Nakamura, *Journal of Physical Chemistry* **99**, 14240 (1995).
[19] M. Sadrai, L. Hadel, R. R. Sauers, *et al.*, *Journal of Physical Chemistry* **96**, 7988 (1992).
[20] C. J. Zhong, W. S. V. Kwan and L. L. Miller, *Chemistry of Materials* **4**, 1423 (1992).
[21] J. Danziger, J. P. Dodelet and N. R. Armstrong, *Chemistry of Materials* **3**, 812 (1991).
[22] J. F. Penneau, B. J. Stallman, P. H. Kasai, *et al.*, *Chemistry of Materials* **3**, 791 (1991).
[23] G. Tamizhmani, J. P. Dodelet, R. Cote, *et al.*, *Chemistry of Materials* **3**, 1046 (1991).
[24] Y. Hirose, A. Kahn, V. Aristov, *et al.*, *Physical Review B* **54**, 13748 (1996).
[25] D. Y. Zang, F. F. So and S. R. Forrest, *Applied Physics Letters* **59**, 823 (1991).
[26] P. H. Schmidt, S. R. Forrest and M. L. Kaplan, *Journal of the Electrochemical Society* **133**, 769 (1986).
[27] Y. L. Lee, H. L. Hsu, S. Y. Chen, *et al.*, *Journal of Physical Chemistry C* **112**, 1694 (2008).
[28] A. Rademacher, S. Maerkle and H. Langhals, *Chemische Berichte* **115**, 2927 (1982).
[29] H. E. Katz, A. J. Lovinger, J. Johnson, *et al.*, *Nature* **404**, 478 (2000).
[30] J. G. Laquindanum, H. E. Katz, A. Dodabalapur, *et al.*, *Journal of the American Chemical Society* **118**, 11331 (1996).
[31] H. E. Katz, J. Johnson, A. J. Lovinger, *et al.*, *Journal of the American Chemical Society* **122**, 7787 (2000).
[32] H. E. Katz, T. Siegrist, J. H. Schon, *et al.*, *Chemphyschem* **2**, 167 (2001). Note that single crystal device data in that paper are not reliable.
[33] C. C. Kao, P. Lin, C. C. Lee, *et al.*, *Applied Physics Letters* **90**, 212101 (2007).
[34] D. M. deLeeuw, M. M. J. Simenon, A. R. Brown, *et al.*, *Synthetic Metals* **87**, 53 (1997).
[35] B. A. Jones, A. Facchetti, M. R. Wasielewski, *et al.*, *Journal of the American Chemical Society* **129**, 15259 (2007).
[36] R. T. Weitz, K. Amsharov, U. Zschieschang, *et al.*, *Journal of the American Chemical Society* **130**, 4637 (2008).
[37] T. B. Singh, S. Erten, S. Gunes, *et al.*, *Organic Electronics* **7**, 480 (2006).
[38] K. See, C. Landis, A. Sarjeant, *et al.*, *Chemistry of Materials* **20**, 7032614 (2008).
[39] D. Shukla, D. C. Freeman, S. F. Nelson, *et al.*, Eastman Kodak Co.
[40] D. Shukla, S. F. Nelson, D. C. Freeman, US application number 20060237712.
[41] D. Shukla, D. C. Freeman, S. F. Nelson, *et al.*, US application number 20070096084.
[42] D. M. de Leeuw, M. M. J. Simenon, A. R. Brown, *et al.*, *Synthetic Metals* **87**, 53 (1997).
[43] B. A. Jones, A. Facchetti, T. J. Marks, *et al.*, *Chemistry of Materials* **19**, 2703 (2007).
[44] H. E. Katz, X. M. Hong, A. Dodabalapur, *et al.*, *Journal of Applied Physics* **91**, 1572 (2002).
[45] K. C. See, A. Becknell, J. Miragliotta, *et al.*, *Advanced Materials* **19**, 3322 (2007).

17

Transparent Metal Oxide Semiconductors as Gas Sensors

Camilla Baratto, Elisabetta Comini, Guido Faglia, Matteo Ferroni, Andrea Ponzoni, Alberto Vomiero and Giorgio Sberveglieri

University of Brescia and CNR-INFM SENSOR Laboratory, Brescia, Italy

17.1 Introduction

The evolution of nanotechnology has recently introduced the synthesis of crystalline nanostructures, usually referred to as nanowires, nanorods or nanobelts, which share the feature of very small lateral dimension and extremely high aspect ratio. Metal oxide nanowires and their great functional potential are attracting a great interest as they exhibit physical properties which are significantly different from the polycrystalline counterpart owing to the confinement effects arising from the nanosized lateral dimensions.

This chapter summarizes the established descriptive model of gas sensing with polycrystalline nanostructures of transparent conducting oxides (TCOs) such as tin oxide and indium oxide and highlights the impact of the introduction of single crystalline nanowires on sensor technology.

The chapter also deals with the methodology of fabrication and the technological issues concerning control of dimension, homogeneity and uniformity of the nanowires. The basic mechanism driving the nucleation and growth is discussed starting from the different strategies presently adopted for synthesis. In general, TCO nanowires are single crystalline structures with well-defined and uniform chemical composition, surface terminations, free from dislocation and other extended defects.

Transparent Electronics: From Synthesis to Applications Edited by Antonio Facchetti and Tobin J. Marks

The acknowledged model of chemical sensing with polycrystalline nanostructures is extended and adapted to take into account the peculiar characteristics of nanowires. High crystalline quality makes nanowires very promising for the development of a new generation of gas sensors reducing instabilities, typical in polycrystalline systems, associated with grain coalescence and drift in electrical properties. In addition, surface effects are enhanced because of the increased specific surface, resulting in a promotion enhancement of catalytic activity and surface adsorption: key properties for superior chemical sensor performance.

It will be shown that nanowires can be used as conventional resistors and may be used as optical-based sensors and even integrated in silicon-based micromachined devices, opening up new opportunities for application. The emerging field of optical sensing and the recent achievement in the detection of traces of dangerous gases are reported to convince the reader about such opportunities.

17.2 Sensing with Nanostructures

For chemical sensing metal oxide (MOX) gas sensors are generally operated in air in the temperature range between 500 K and 800 K, as conduction is electronic and oxygen vacancies are doubly ionized and fixed.

Due to the reactivity of MOX surface atoms, which lack binding partners, molecules from the gas phase are adsorbed at the surface. The first step of association of gas species with a solid surface is physisorption, afterwards the physisorbed species can be chemisorbed – often named ionosorption when the adsorbate acts as a surface state – if they exchange electrons with the semiconductor surface [1]. The adsorption isobar, that is the volume adsorbed as a function of temperature at a constant pressure, is characterized at low temperature by physisorption and at high temperature by equilibrium chemisorptions.

In semiconductor statistics the physi- and chemisorbed atoms and molecules are represented by surface localized states in the semiconductor energy gap, whose occupation is given by the same Fermi- Dirac distribution, physisorption corresponding to unoccupied and chemisorption to occupied states. The appearance of surface-localized acceptor states in MOX induces a charge transfer between bulk and surface in order to establish thermal equilibrium between the two. The charge transfer results in a non-neutral region (with a non-zero electric field) within the semiconductor bulk, usually referred to as the surface space charge region (SCR) [2]. In *n*-type MOX with surface states of acceptor type, a depletion SCR is established and the height of the surface barrier is:

$$V_s = \frac{qN_d}{2\varepsilon} x_0^2 = \frac{qN_{ta}^{-2}}{2\varepsilon N_d} \tag{17.1}$$

where x_0 is the width of the depletion layer and N_{ta} is the density of occupied surface states.

The overall resistance of the sensor element is determined by the charge transfer process produced by surface reactions and by the transport mechanism from one electrode to the other through the sensing layer. In polycrystalline devices the carriers must overcome by thermionic emission the energy barrier created at the intergrain contacts in order to cross from one grain to the neighbor [3]. Since the film has many barriers, the external applied voltage drop across any one is small and the energy height is equal to qV_s, as

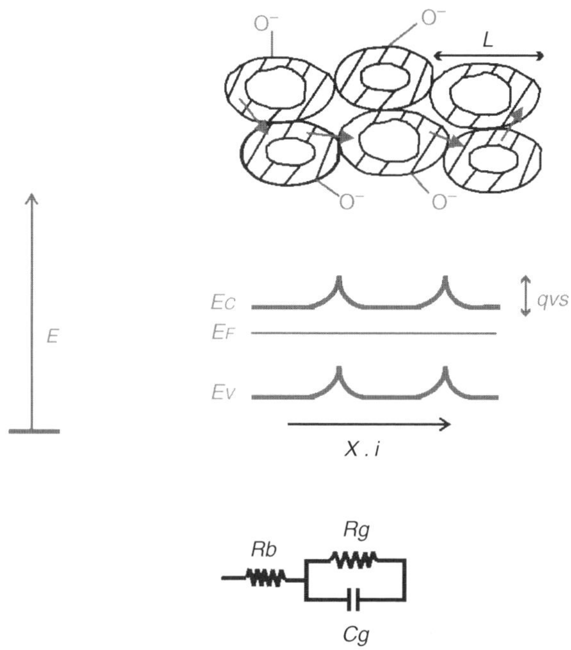

Figure 17.1 *Conduction mechanisms, bands and equivalent circuit for a polycrystalline layer*

shown in Figure 17.1. The conductance of a polycrystalline gas sensor can therefore be expressed as:

$$G = G_0 e^{\frac{-qV_s}{kT}} = G_0 e^{\frac{-q^2 N_{ta}^{-2}}{2\varepsilon N_d kT}} \qquad (17.2)$$

where G_0 is a pre-exponential term independent to a first approximation from the surface adsorption and temperature.

The process of gas detection is intimately related to the reactions between the species to be detected and ionosorbed surface oxygen. In the 400–800 K temperature range oxygen ionosorbs over SnO_2 and other oxides in a molecular (O_2^-) and atomic form (O^-) [4]; when a reducing gas like CO comes into contact with the surface, it oxidizes to CO_2 by reacting with ionosorbed oxygen, releasing electrons from surface states to the conduction band. The overall effect at equilibrium is shrinking of the density of ionosorbed oxygen, i.e. occupied surface acceptor states and consequently the height of the surface barrier. Direct adsorption is also proposed for the gaseous species – like strongly electronegative NO_2 – whose effect is to decrease sensor conductance ($e^- + NO_{2,ads} \leftrightarrow NO_{2,ads}^-$). The occupation of surface states, which are much deeper in the band gap than oxygen's, increases the surface potential and reduces the overall sensor conductance. Another important ubiquitous species that ionosorbs over MOX surfaces is water [4]. The chemisorption of water onto oxide from air can be very strong, forming an 'hydroxylated surface', where the OH^- ion is bound to the cation and the H^+ ion to the oxide anion. The overall effect of water vapor is to increase the surface conductance.

Addition of a small amount of noble metals over the MOX surface, such as Au, Pd, Pt and Ag particles, can speed up surface reactions and improve selectivity towards target gas species. The activation of species involved in a surface reaction may be the dissociation of a molecule, the ionisation of the species or some other intermediate reaction. From the energetic point of view, the effect of catalysis is to provide a more favorable reaction path.

As for electrical transduction, the easiest measurable physical quantity that can be measured is the sensor DC conductance. The standard extracted figure is the sensor response towards a target gas concentration, which is usually defined as the (relative) change of conductance, with the obvious relationship:

$$S_G = \frac{G_S}{G_0} = \frac{G_S - G_0}{G_0} + 1 = S_G^* + 1 \tag{17.3}$$

where G_0 is the steady state conductance in air and G_S is the steady state conductance in the target gas concentration.

In the case of an oxidizing target species, resistance increases following gas introduction and the sensor response is defined as the (relative) change of resistance:

$$S_R = \frac{R_S}{R_0} = \frac{R_S - R_0}{R_0} + 1 = S_R^* + 1 \tag{17.4}$$

Starting from the sensor response it is possible to derive the sensor response curve, that is the representation of the steady state output as a function of the input concentration [5]. The sensor response curve is frequently called, erroneously, a sensitivity curve. Instead, sensitivity is the derivative of the sensor response curve.

17.3 Synthesis of Nanostructures for Sensing

17.3.1 Nanowires of SnO_2

Tin dioxide nanowires integrated into gas sensing devices were grown by physical vapor deposition (PVD), chemical vapor deposition (CVD), electrospinning or template-based synthesis.

PVD consists of a phase transformation from a solid phase (source material), an intermediate vapor phase and a final solid phase (nanowires) deposited onto the substrate. Evaporation is the oldest PVD process, the source material is heated to the point of vaporization and the vapor phase condenses on the substrate. In order to control the composition of the deposited material and to allow a lower operation temperature, the process is generally performed in vacuum.

In the particular case of nanowire preparation, evaporation is one of the most explored techniques in recent papers on account of its cheap experimental set-up and the good crystalline quality of the produced structures. The conventional experimental set-up for the oxide deposition consists of an alumina furnace capable of achieving high temperatures, in order to activate oxide decomposition and promote evaporation. The metal oxide powder is placed in the higher temperature region and a gas carrier transports the evaporated oxide towards the substrates. The controlled pressure of the inert atmosphere and the temperature gradient within the furnace allow condensation and nucleation downstream of the gas flow. The growth chamber has to be designed in order to obtain the proper temperature gradient

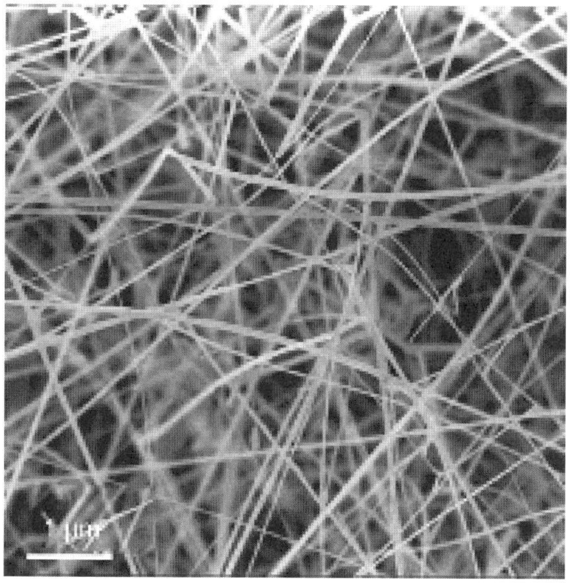

Figure 17.2 *Vapor phase growth of SnO_2 nanowires over polycrystalline alumina. (Reprinted with permission from ref. [7]. Copyright 2006 Springer Science + Business Media)*

for one-dimensional nanostructure formation. Pressure, temperature gradient and carrier flux have to be precisely controlled in order to guarantee the reproducibility for deposition.

The evaporation source can be either metallic tin or tin oxide: in the first case the evaporation can be obtained at lower temperatures with a faster rate, while in the second case higher temperatures are necessary and the evaporation process is slower but more reproducible. Tin oxide was used by Comini *et al.* [6, 7]. Each nanowire was a single crystal without the presence of dislocations; its morphology and structure, such as growth direction and surface planes, were well defined, and the surfaces clean and atomically flat. The nanowires had a rectangular cross-section with an average width of 200 nm, width- to-thickness ratio of 5–10 and length of up to a few millimeters [6]. Figure 17.2 shows a scanning electron microscopy (SEM) image of tin dioxide nanowires prepared from vapor phase deposited on alumina substrates.

Metallic tin was used by Ying *et al.* [8]. The SnO_2 nanowires had a rectangular cross-section, a diameter in the range of 50–200 nm and a length of up to tens of micrometers. High-resolution transmission electron microscopy (TEM) revealed that the SnO_2 nanowires were single crystals with longitudinal directions along the (001) direction of the tetragonal rutile SnO_2.

Another deposition technique is CVD. It consists of the transformation of gaseous molecules (precursors) into a solid material in the form of nanowires due to a chemical reaction. CVD is a widely used method for depositing a large variety of materials. In a typical CVD process reactant gases (in general diluted in carrier gases) enter the reaction chamber at room temperature. The gas is then heated while approaching the deposition surface. The reactant gases may undergo homogeneous chemical reactions in the vapor phase before striking the surface, depending on the operation conditions. Near the surface the gas reactants slow down due to viscous drag and chemical composition changes. Heterogeneous reactions of the source gases or reactive intermediate species (formed from homogeneous pyrolysis) occur at the deposition surface forming the deposited material. Gaseous reaction by-products are then transported out of the reaction chamber.

The fundamental steps during the CVD deposition process are summarized as follows: the first step is vaporization and transport of the precursor molecules into the reactor; the second step is diffusion of the precursor molecules towards the surface. Then, if we do not have a homogeneous chemical reaction before hitting the substrate, the third step is the adsorption on the surface. Finally, there will be the decomposition and incorporation into solid films with desorption into the gas phase of by-products. A generic CVD system includes gas and vapor delivery lines to a reactor main chamber, a substrate loading and unloading assembly, an energy source, vacuum systems, an exhaust system (by-product removal), process control, measurement gauges and safety equipment. Hazardous vapor phase reactants and products are frequently used and produced by chemical reactions, so ventilation, sensors, and alarms are needed for safety reasons.

Tin dioxide nanowires were prepared by CVD [9]. They were grown on gold-coated substrates by the vapor deposition of $[Sn(OtBu)_4]$, which acts as a single delivery source for the Sn and O necessary for the formation of the SnO_2 phase. Under constant precursor flux, chemical vapor growth of one-dimensional structures was achieved in the 550–750 °C temperature range. The gaseous species liberated during the chemical vapor growth of SnO_2 nanowires (from the decomposition of $[Sn(OtBu)_4]$) were tert-butyl alcohol and isobutene. High resolution (HR) TEM measurements confirmed the high structural uniformity and defect-free nature of the single nanowires. As shown in Figure 17.3, the nanowires turned out to be single crystals, orientated along the [100] direction.

Electrospinning is a template free synthesis method. It uses an electrical field to form fibers. The high voltage induces a liquid jet formation. The electrostatic repulsions between the surface charges and the evaporation of solvent stretches the continuously electrified jet and a solid fiber is formed. The elongation by bending instability results in the fabrication of uniform fibers with nanometric diameter. In 1934 the first patent was filed describing the operation of electrospinning, when Formalas proposed an apparatus for producing polymer fibers by taking advantage of the electrostatic repulsions between surface charges [10]. The experimental set-up for electrospinning consists of a substrate, a metallic needle and a high voltage power supply. The metallic needle is connected to a syringe with the solution for electrospinning. A high voltage is applied to the metallic needle.

Tin dioxide nanowires can be also prepared by electrospinning [11]: the solution used for electrospinning was a poly(vinyl alcohol) (PVA)/$SnCl_4{\cdot}5H_2O$ composite with PVA concentrations ranging from 5 to 8 wt%. The electrode-to-collector distance varied from 0.5 to 2.6 cm and the applied voltage was changed accordingly between 5 kV and 10 kV. The angle of the syringe was settled as 60°. All electrospun nanowires were dried for 24 h at 100 °C for complete evaporation of the solvents. The SnO_2 nanowires were several hundred micrometers in length and randomly distributed on the substrate. TEM analyses demonstrated their polycrystalline nature, with a rutile phase (see Figure 17.4).

Nanowire growth can be obtained also by means of templates. The template confines the growth of the nanostructure in one dimension. Particular care must be taken in the preparation of the template and also in the technique used for filling the template pores. Moreover, at the end of the process the template must be dissolved. The template can be

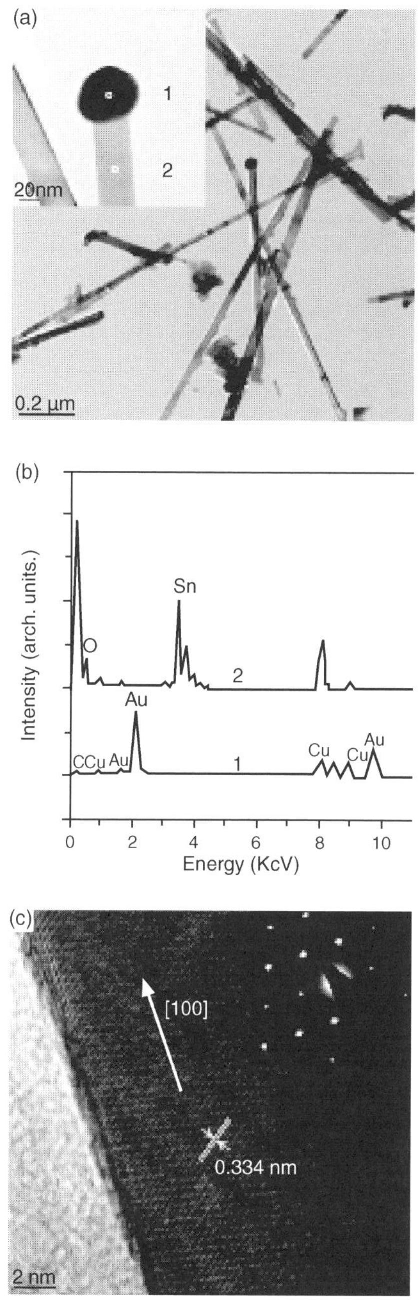

Figure 17.3 *(a) TEM micrograph of SnO_2 nanowires deposited at 700 °C. (b) Nanoprobe (5 nm) energy dispersive X-ray (EDX) spectra of an SnO_2 nanowire recorded on the tip (1) and middle of the wire (2). (c) HR TEM image of a single crystalline SnO_2 nanowire grown at 700 °C with the corresponding selected area electron diffraction (SAED) pattern (inset). (Reprinted with permission from ref. [9]. Copyright 2005 Wiley-VCH)*

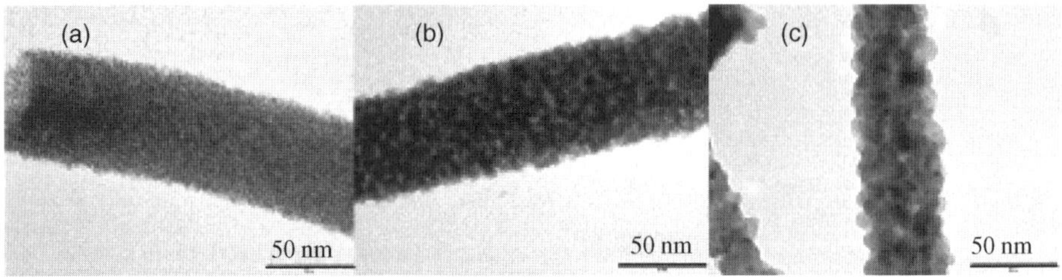

Figure 17.4 TEM image of nanowires prepared by electrospinning and annealed at (a) 300, (b) 500 and (c) 700 °C. (Reprinted with permission from ref. [11]. Copyright 2008 Elsevier Ltd)

prepared by anodization, a well-established process for the preparation of hexagonal close packed, highly ordered alumina membranes and other porous templates. The most common template is alumina. There are advantages with respect to the other techniques such as low costs, repeatability and potential compatibility with silicon technology. The nanowire section and nanowire-to-nanowire distance can be controlled by the pore diameter and pore-to-pore distance that are easily adjusted by the anodization voltage. Furthermore the array of nanowires is well ordered due to the template constraints. Particular care must be taken for template dissolution that can destroy nanowire alignment. The main disadvantage, common to solution-based techniques, is the low crystalline quality of the prepared nanowires.

The most used procedures for deposition inside the template pores are electrochemical growth and sol-gel methods. Despite its simplicity template assisted growth produces polycrystalline nanowires and this can limit their potentialities for fundamental studies and applications.

Zhang *et al.* presented SnO_2 nanowires fabricated by an electrochemical method [12]: TEM and X-ray diffraction analysis confirmed the nanowire integrity after oxidation and its complete oxidation to rutile polycrystalline SnO_2.

17.3.2 Nanowires of In_2O_3

The thermodynamically stable crystallographic structure of bulk In_2O_3 is the $Ia\bar{3}$ body-centered cubic lattice (spatial group 206). Such a configuration does not exhibit preferential orientations with respect to growth during formation of bulk crystals, and in fact In_2O_3 typically includes faces normal to the directions [100], [010], [001]. For this reason anisotropic growth has to be forced to obtain quasi one-dimensional nanostructure. A series of methodologies has been developed for the purpose, mainly based on three concepts: (1) heterocatalysis by application of foreign material on the substrate [13, 14]; (2) homocatalysis by application of In_2O_3 catalyst on the substrate [15–17]; (3) heteroepitaxial growth of oriented nanowires on single crystalline substrate (as shown in Figure 17.5) [18]. In addition to these principal mechanisms, which act at the level of condensation of volatile species at the condensation site, foreign catalyst can be added to the precursors, to induce a carbothermal reaction and lower the evaporation temperature [16].

In heterocatalysis using gold droplets, In_2O_3 nanowires grow according to the vapor-liquid-solid (VLS) mechanism, and the preferential direction of growth is governed by precipitation of In_2O_3 at the base of the droplet of eutectic composition.

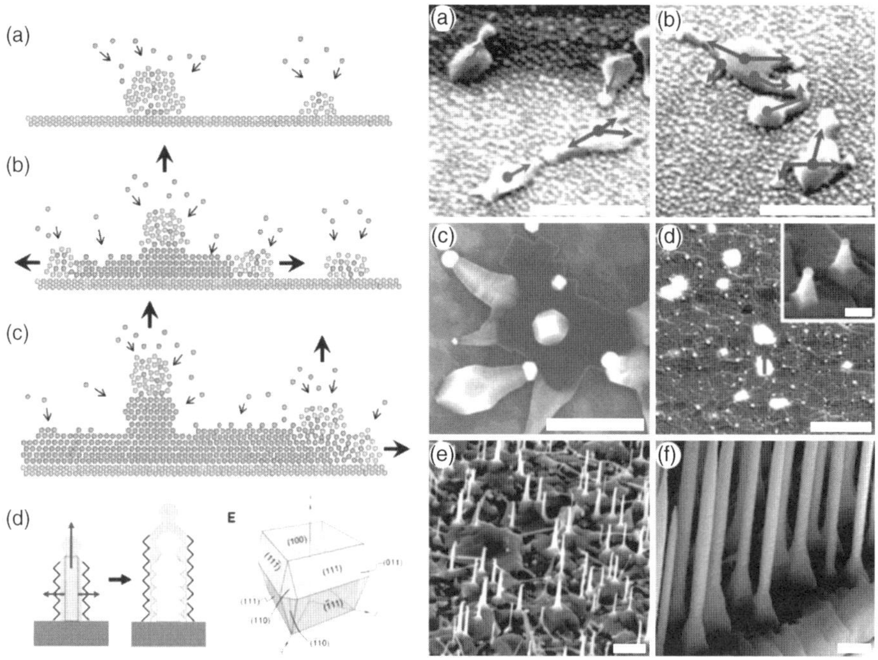

Figure 17.5 *Dynamic nucleation and epitaxial In_2O_3 nanowire and thin film growth. (A–C). Schematic diagrams illustrating the competitive growth modes during the growth of the nanowires and underlying thin film. (D) An illustration depicting simultaneous catalyst-directed vertical nanowire growth and sidewall epitaxial deposition. (E) A three-dimensional view showing the crystallographic relationships between the nanofacets and the growth plane of the nanowire. (A′–F′) A series of SEM images revealing various stages of the controlled time-sequenced growth process. (A′–B′) SEM images revealing generation of daughter alloyed catalytic heads (indicated by arrows) from respective parent alloyed heads at heating duration: $T_{1000^\circ C} = 0$ min. (C′) A top view SEM image obtained at $T_{1000^\circ C} = 2$ min, showing two-dimensional epitaxial thin film formation and lateral expansion assisted by Au alloyed catalytic heads. (D′) An SEM image showing almost complete coverage of the sapphire substrate with an In_2O_3 thin film at $T_{1000^\circ C} = 5$ min. The arrows indicate two exposed regions (bright spots) of the substrate. The inset shows initial formation of In_2O_3 nanowires with their pyramidal bases. (E′) An SEM image obtained at $T_{1000^\circ C} = 20$ min showing evolution of a regular array of In_2O_3 nanowires with an underlying In_2O_3 thin film. (F′) An SEM image showing the regular pyramidal bases and bodies of the In_2O_3 nanowires at $T_{1000^\circ C} = 60$ min. The underlying thin film has a thickness ~800 nm. All images except (C′) are 45° perspective views. Scale bars: 500 nm for (A′), (B′), (C′), and (F′); 2 μm and 1 μm for (D′) and (E′), respectively, and 150 nm for the inset of (D′). (Reprinted with permission from ref. [18]. Copyright 2004 American Chemical Society)*

Homocatalysis has been obtained either by applying a very thin In layer on the substrate [15] or by the carbothermal reaction method, starting from In_2O_3 powders [16]. In the first case, nanowires nucleate from indium-based clusters, the lateral dimension of the nanowire is smaller than the size of the cluster, and seeding reduces the lateral dimensions with respect to unseeded substrate. The nanowires exhibit pyramidal termination at the apex, which is shown in Figure 17.6. The facets of the pyramid are crystal planes with higher

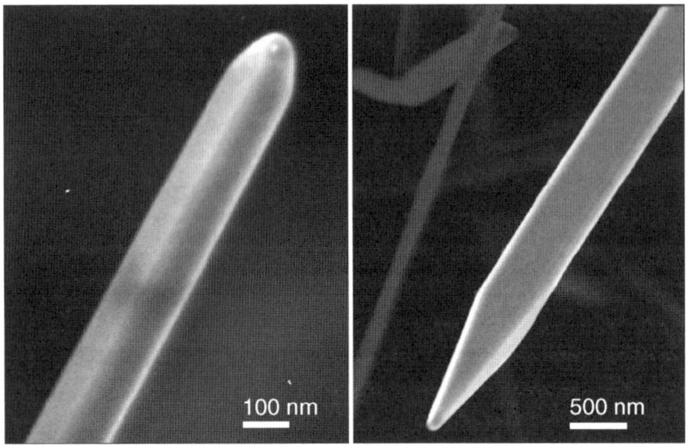

Figure 17.6 *SEM images of the crystal nature of the In_2O_3 nanowires grown on seeded substrate. The section appears to be squared, and the apex of the wires is tapered. No metallic droplet is present over the tapered tips, according to a VS growth mechanism. (Reprinted with permission from ref. [15]. Copyright 2007 American Chemical Society)*

index with respect to the lateral sides and indicate that the vapor-solid (VS) mechanism was predominant. In the second case, the carbothermal reaction leads to formation of metal In a droplet at the substrate that is not completely oxidized at the substrate, and gives rise to a self-catalytic VLS condensation (see Figure 17.7). The main advantage of homocatalysis is the lack of heterogeneous structures like gold nanoparticles at the apex of the nanowires. In fact, for the purpose of functional characterization, the presence of catalytically active metal particles is expected to strongly affect the interaction with adsorbed gases [19, 20].

Simultaneous application of single crystalline substrate [α-oriented sapphire or yttrium stabilized zirconia (YSZ)] and heterogeneous catalyst allows selection of the direction of growth owing to the heteroepitaxial mechanism induced by the substrate, and the formation of a one-dimensional structure due to the limitation of lateral dimensions induced by the catalytic cluster. Both α-sapphire and YSZ are ideal for heteroepitaxial growth of ITO films

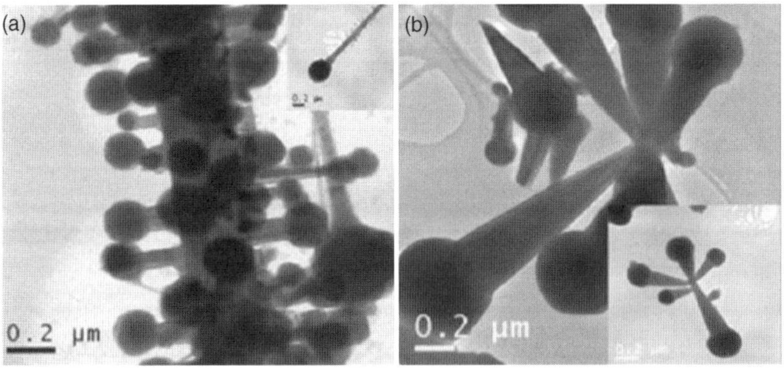

Figure 17.7 *TEM images of In_2O_3 nanostructures: (a) nanotrees; (b) nanobouquets. (Reprinted with permission from ref. [16]. Copyright 2004 Elsevier Ltd)*

with a bixbyite crystal structure (cubic, lattice constant $a = 1.01$ nm) because of their structure: hexagonal for α-sapphire, fluorite crystal structure (cubic, lattice constant $a = 0.514$ nm) for YSZ, which perfectly matches the lattice of ITOs. Under these conditions the main mechanism responsible for one-dimensional confinement of the growth is the VLS mechanism acting at the apex of the nanowire; however, simultaneous VS direct absorption on the lateral sides of the nanowires is detected, which determines thickening of the wires as the condensation proceeds.

One crucial parameter determining the growth direction of the nanowires is the condensation temperature. While [100]-oriented nanowires were typically obtained on α-sapphire substrate at high temperature (between 800 °C and 1000 °C) [18, 21], [111]-oriented nanowires were recently obtained at lower temperature (about 620 °C). Such results demonstrate that the driving force of the epitaxial relationship between the lattice of the substrate and of the nanowires can be overcome in suitable conditions, and that other mechanisms of total energy reduction lead the growth direction and the final crystallographic nature of In_2O_3 nanowires.

The fabraction of nanostructures has been greatly improved and controlled doping or creation of heterostructures for enhancement of the transport properties of In_2O_3 one-dimensional nanostructures has been also exploited [22–24]. The growth of branched nanowire structures in which semiconducting In_2O_3 nanowire arrays with variable densities were epitaxially grown on metallic ITO nanowire backbones was obtained by sequential seeding of the backbone and repeated condensation. The electronic conductivity of the novel heterostructured SnO_2-In_2O_3 nanowires was two orders of magnitude better than that of the pure SnO_2 nanowires due to the formation of Sn-doped In_2O_3 caused by the incorporation of Sn into the In_2O_3 lattice.

17.4 Gas Sensing with Nanowires

17.4.1 The Sensing Mechanism of Nanowires

Tin dioxide has been studied for many years as a polycrystalline thin or thick film gas sensor. To prevent grain growth, thus increasing stability of the device, single crystalline nanosized wires have been used since the first paper in 2002, which reported that SnO_2 nanowire performance is comparable with that of thin films [6].

In nanowire-based sensors the current flows parallel to the surface as schematically shown in Figure 17.8. If the lateral size d is small, almost all the carriers are trapped in surface states and only a few thermal activated carriers are available for conduction. The shape of the band is parabolic or flat based on the lateral size being greater or smaller than the semiconductor Debye length ($L_D = \sqrt{\frac{\varepsilon kT}{q^2 N_d}}$, typical values of L_D for tin oxide ranges from 130 to 10 nm when temperature changes from 400 to 700 K) [3]. In both cases carriers thermally activated from surface states are responsible for conduction. In this configuration the transition from activated to strongly not activated carrier density, produced by target gases species, greatly affects sensor conductance. Indeed when considering nanowire bundles, the conduction mechanism is dominated by the inherent intercrystalline boundaries at nanowire connections – like in polycrystalline samples – rather than by the intracrystalline characteristics.

The metal semiconductor junction that forms at the interface between the layer and the contacts can play a role in gas detection, enhanced by the fact that the metal used for the

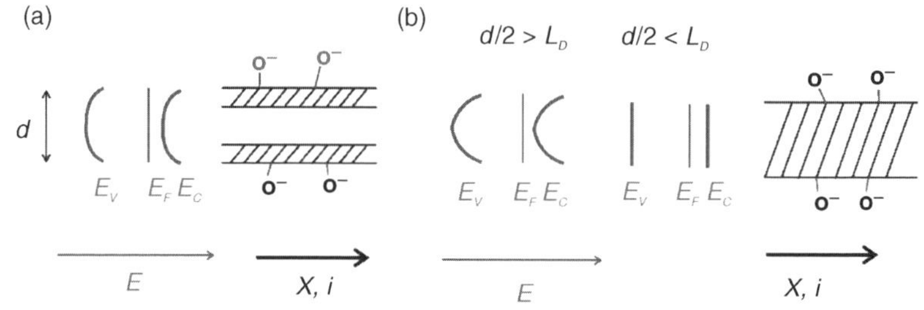

Figure 17.8 *Conduction mechanisms and bands for a nanowire when grains are (a) partially or (b) fully depleted*

contact acts also as a catalyst. The contact resistance is more important for single nanowires (since it is in series to the semiconductor resistance) than for bundles (where it is connected to a large number of resistances).

17.4.2 Chemoresistive Sensing Properties of SnO_2 Nanowires

From the great number of articles presented in the literature on SnO_2 nanowires for gas sensing, only a few will be presented here, distinguishing between single nanowire and multiple nanowire devices.

In a single nanowire device the current flows along the direction of the wire from one contact to the other. Gas absorption creates a space charge region that gives rise to a channel; carrier depletion depends on the ratio between the diameter of the wire and L_D: if it is close to one, carriers thermally activated from surface states are responsible for conduction and switching between conducting and nonconducting behavior is expected as a consequence of gas interaction.

The technological challenge of realizing a single nanowire device is contact deposition. Usually nanowires are dispersed in a solvent and drop coated on prepatterned contacts; sometimes a nanomanipulation step is required to align the nanowire between them. Further contacts, usually made of Pt or Au, are then deposited by focused ion beam (FIB) (Figure 17.9) or electron-beam lithography (EBL) to improve existing electrical contacts or to extend prepatterned contacts.

The papers of Hernández-Ramírez *et al.* [26, 27] deal with the challenge of performing reliable electrical contacts on one individual nanostructure. They pointed out that the contact resistance contribution is much more important than the nanowire resistance; the main part of the measured contact resistance is believed to originate in the Pt-SnO_2 nanowire junction. This is often true also in other studies even if it is not always analyzed and recognized. Sensor response to CO of an individual nanowire depends on its radius as expected: for a radius smaller than 25 nm (when the radius approaches L_D) an increase in sensor response and in the sensitivity is observed.

Hernández-Ramírez *et al.* [28] studied the cross-sensitivity effects between H_2O and CO gases in a single nanowire device: they demonstrated that the resistance variation towards a CO/H_2O mixed pulse did not correspond to the sum recorded with separate injections. Figure 17.10 shows the reversible resistance drops observed for injection of water pulses.

In some cases single nanowires can be implemented in a micromachined substrate, to decrease power consumption during high temperature operation [29]. Similar to MOX thin

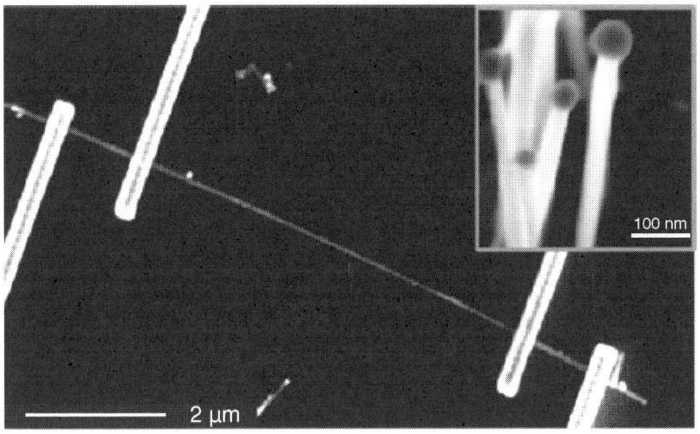

Figure 17.9 Four Pt contacts on a SnO_2 nanowire deposited by FIB nanolithography. (Reprinted with permission from ref. [25]. Copyright 2008 Wiley-VCH)

film sensors, the response signals and response times for NH_3 and CO (100, 50 and 25 ppm) improve with increasing temperature.

An alternative way to activate gas adsorption and desorption is the use of light activation: an explosive environment is a typical application where the use of heated sensors is not favorable. Law *et al.* [30] used light with wavelength higher or slightly smaller than the band gap: in this latter case (365 nm radiation) the NO_2 (5–100 ppm) detection is better than that obtained with 254 nm radiation. Photodesorption of surface species is responsible for recovery of baseline, not observed in dark.

Functionalization of nanowires is used to enhance sensing capabilities towards a specific gas, thus increasing selectivity. Usually the nanowire is coated with transition metal catalysts (such as Au, Pt and Pd) or other metal oxide sensing materials (such as CuO, NiO and ZnO). For example, good results were reported for deposition of NiO particles (*p*-type material) on a SnO_2 nanowire, with enhancement of sensitivity to CO both on a single nanowire [31] and

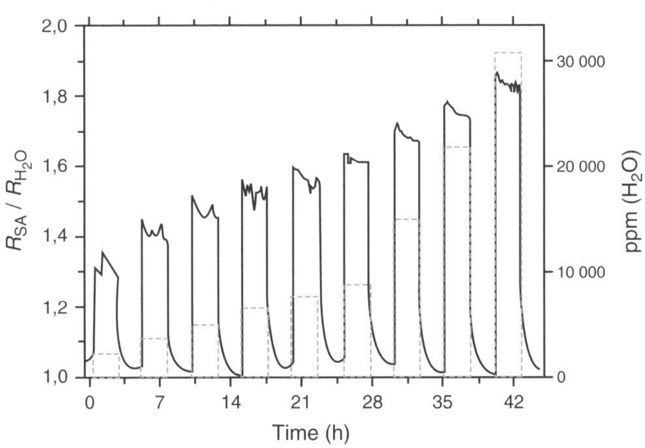

Figure 17.10 Electrical response of a SnO_2 nanowire with a diameter of 25 nm to water vapor: the gas carrier was dry synthetic air, and working temperature 295 °C. (Reprinted with permission from ref. [28]. Copyright 2007 Institute of Physics Publishing)

on multiple nanowires [32]. Ni or Pd deposition improved the response and kinetics of response of single SnO_2 nanowires to O_2, H_2 and CO [31]. There is an optimum amount of catalyst for which the effect is enhanced, and a saturation effect for excessive quantities.

In the case where a multiple nanowire device is realized, the conduction mechanism is dominated by the intercrystalline boundaries at nanowire connections – as in polycrystalline sensitive layers – rather than by the intracrystalline characteristics. The natural distribution of the dimension of the wire inside the bundles will also affect the switch between conducting and nonconducting behavior. For bundles of nanowires the contact deposition is not as difficult, indeed it is easier than in a single nanowire. In the case of low-temperature fabrication, nanowires can be directly grown over the contacts. If the deposition technique involves higher temperatures, nanowires are grown in a selected area of the substrate, leaving free regions for further contact deposition. As a third way, mats of nanowires can be dispersed in a solvent and placed over pre-atterned contacts.

Comini *et al.* [6] demostrated for the first time in 2002 the use of bundles of SnO_2 nanobelts with a rectangular cross-section (200 nm long–20/40 nm thick) as a conductometric gas sensor. The nanobelts were sensitive to gaseous polluting species like CO and NO_2 as well as to ethanol and ozone [33]. Similar results in ethanol were obtained by Ying *et al.* [34] with SnO_2 nanowhiskers (nanostructure diameter in the 50–200 nm range).

Jung *et al.* [35] reported significant results with multiple SnO_2 nanowires bridged across trenched electrodes. Ohmicity of the contacts was previously controlled by a current–voltage curve in air and after that tests in NO_2 were performed. Figure 17.11 shows the response to NO_2 pulses in the range of 1–5 ppm at a working temperature of 200 °C.

17.4.3 Chemical Warfare Agents Detected by SnO_2

This section is dedicated to the contribution of tin oxide nanowires to the detection of chemical warfare agents (CWAs) emphasizing the innovative features they could introduce respect to thin or thick films traditionally developed as gas sensors.

In gas sensing application, CWA detection is one of the most critical and challenging tasks, due to the required high sensitivity, selectivity and short response time.

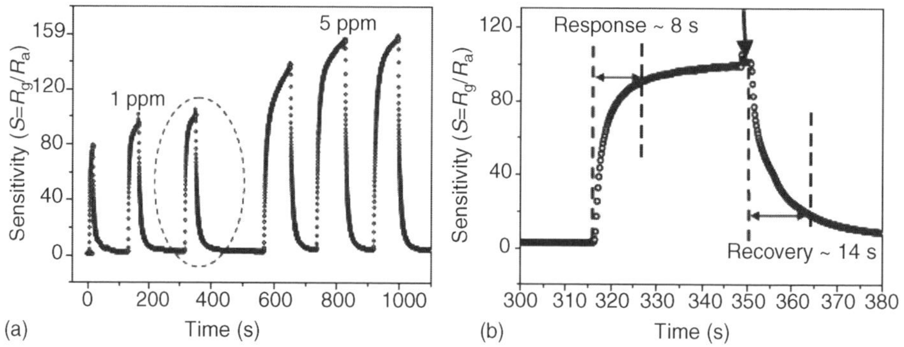

Figure 17.11 *A multiple nanowire device with SnO_2 nanowires bridged across contacts. Sensing performances with NO_2 at a working temperature of 200 °C. (a) Sensitivity versus time for three pulses of NO_2 (1 ppm and 5 ppm). (b) Detail of the dynamic response to 1 ppm. (Reprinted with permission from ref. [35]. Copyright 2008 Springer Science + Business Media)*

CWAs are often classified according to their intended uses and effects on the human body as nerve, blister, choking and blood agents (limited to the best known compounds). Among these, nerve gases are considered the most dangerous, due to their high toxicity. The Immediately Dangerous for Life and Health[1] (IDLH) value of these compounds, such as Sarin ($C_4H_{10}FO_2P$) or Soman ($C_7H_{16}FO_2P$), range from a few to a few tens of ppb [36]. Consequently, most studies related to CWA detection focus on nerve agents, using simulant compounds that mimic the behavior of the corresponding CWA but featuring reduced toxicity. Dimethyl methylphosphonate ($C_3H_9O_3P$), abbreviated to DMMP, is one of the widest used compounds as a Sarin nerve agent simulant [36, 37].

The interaction between DMMP and metal oxides is quite complex and involves DMMP decomposition. As a consequence, different phosphorous compounds are developed and remain adsorbed over the oxide surface at temperatures as high as 500–600 °C [38, 39]. To reduce poisoning effects arising from such compounds, metal oxides are usually heated at temperatures above 350 °C (the sublimation temperature of P_2O_5).

Concerning the low detection limit of tin oxide nanowires, their capability to respond to DMMP at concentrations close to the Sarin IDLH value (30 ppb) has been achieved both using sensors based on tin oxide nanowires mesh [40] and a tin oxide single nanowire [41].

As an example, the variation of the tin oxide nanowire mesh conductance upon exposure to 50 and 70 ppb of DMMP is reported in Figure 17.12. The response of tin oxide prepared by the rheotaxial growth and thermal oxidation (RGTO) method [42], chosen as reference thin film technology, is also shown for comparison.

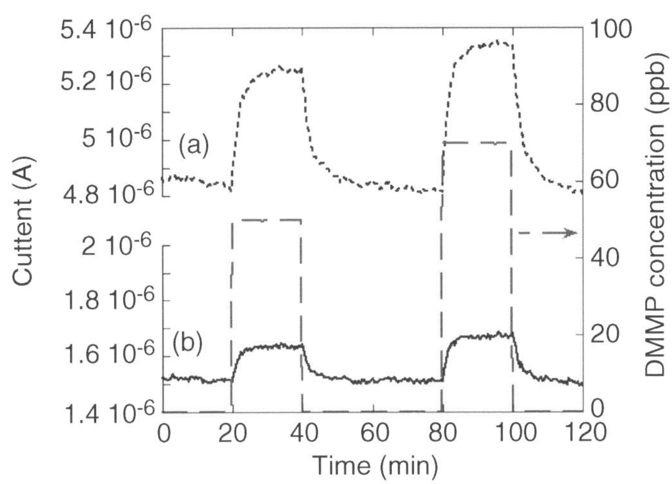

Figure 17.12 *Response of tin oxide nanowire mesh (a) and tin oxide thin film (b) to 0.05 and 0.07 ppm of DMMP in air. The working temperature is 500 °C for both sensors. The long recovery time required by the nanowire signal to recover its baseline value is ascribed to poisoning effects induced by DMMP exposure*

[1] The IDLH value corresponds to the gas concentration in air that would cause permanent adverse health effects after 30 min of unprotect exposure. A person exposed to a chemical at its IDLH concentration has 30 min to leave the contaminated environment without having irreversible health effects or dying (depending on the gas effects on the human body).

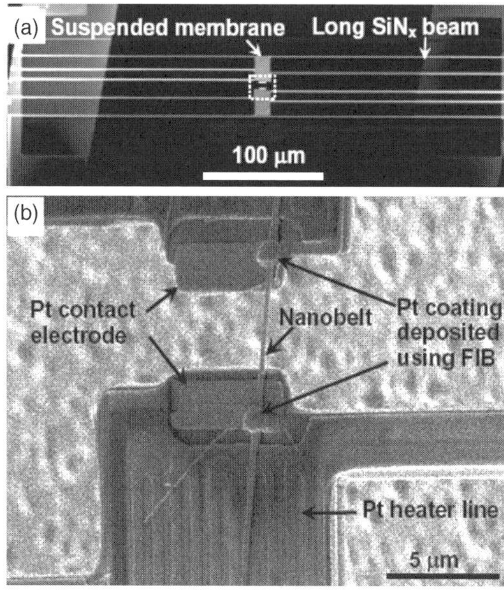

Figure 17.13 *Scanning electron micrograph of the microheater sensor (a) based on a single tin oxide nanobelt (b). (Reprinted with permission from ref. [41]. Copyright 2005 American Institute of Physics)*

It is evident that both layers exhibit large recovery times, sensor conductance taking about 20 min to recover its baseline value. Increasing the DMMP concentration up to a few hundreds of ppb increases the recovery time too and about 90 min are needed to complete the baseline-conductance recovery. These phenomena are usually ascribed to the above DMMP poisoning effects [37, 38, 40].

According to Yu *et al.* [41], the use of single nanowires may strongly reduce such undesired effects. They used drop coating to disperse tin oxide nanobelts over a micro-hotplate and dielectrophoresis to align nanobelts between the Pt electrodes. The structure of a sensor based on a single nanobelt is reported in Figure 17.13.

The principal difference with respect to nanowire (nanobelts) mesh is the absence of nanowire–nanowire interfaces, which are expected to behave similarly to grain boundary interfaces featured by thin and thick films. The absence of such interfaces and a robust Pt–nanobelt contact (obtained by Pt coating both electrode–nanobelt contacts with the FIB technique) are proposed as the key features to obtain the short recovery time observed with this device. The conductance variation of the single nanobelt device heated at the working temperature of 500 °C upon exposure to 78 and 53 ppb of DMMP is reported in Figure 17.14. It is evident that the signal recovers its baseline in less than a minute.

According to Ponzoni *et al.* [40], the detection of acetonitrile and dipropylene glycol methyl ether (DPGME) turned out to be less difficult than the detection of DMMP simulants for cyanide and blistering agent, respectively. Tin oxide nanowire mesh and RGTO thin films can detect such compounds at concentrations lower than the IDLH value of the respective CWAs without featuring poisoning effects typical of DMMP (and nerve agents).

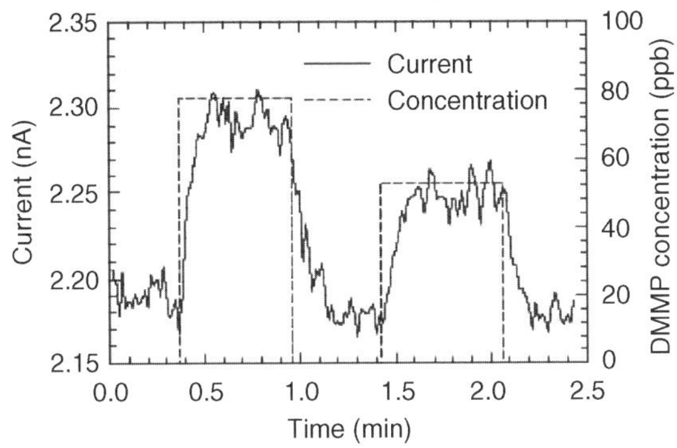

Figure 17.14 *Response of the single nanowire device to 78 and 53 ppb of DMMP in air. Sensor temperature is 500°C. The short recovery time is evident, suggesting that DMMP poisoning effects are reduced by the use of similar devices, see text for details. (Reprinted with permission from ref. [41]. Copyright 2005 American Institute of Physics)*

17.4.4 Transistor Devices Based on a Single SnO_2 Nanowire

A transistor device based on a single nanowire is shown schematically in Figure 17.15. Bundles of nanowires are removed from substrates and dispersed in solvents; contacts are prepared, usually by electron beam lithography, after drying the solution over Si heavily doped substrates endowed with a SiO_2 thermally grown insulating layer. The gate back contact is deposited over the Si substrate.

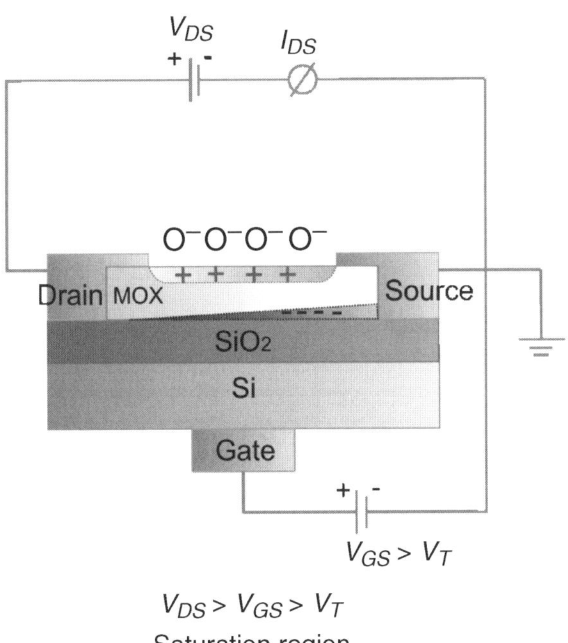

Figure 17.15 *Schematic cross-section of a single nanowire n-type transistor device in saturation*

As for the electrical behavior a majority carrier channel in accumulation mode is created upon application of a positive gate bias as in *n*-channel thin film transistors (TFTs) [43, 44].

Many independent studies showed how the additional gate voltage tool, which is introduced by configuring a single nanowire in a field effect configuration, allows tuning the availability of electrons for surface adsorption and desorption reactions and the sensing and recovering performance of the devices. By applying a gate potential the well known electroadsorbitive effect [45] can be exploited by modulating the position of the electrochemical potential and the availability of electrons for surface reactions.

Zhang *et al.* [46–48] carried out a detailed study of the effects of the gate voltage in a SnO_2 based single nanowire transistor (SNT) following O_2 and CO introduction in N_2 at 553 K. They showed that the rates and extent of oxidation and reduction reactions taking place on a SnO_2 SNT can be modified by changing through the gate voltage the electron density in the wire.

As shown in Figure 17.16, when operated in the active zone ($V_{GS} > V_T = -6.2$ V) the conductance decreases abruptly following O_2 introduction, the amount of this reduction being lower at smaller V_{GS}. When CO is introduced the conductance increases achieving a new steady state value; the observed conductance increase on admitting CO is not monotonic with V_{GS} but achieves a maximum value in the range $V_{GS} = -2$ to 0 V. Interestingly, even at $V_{GS} = -6$ V, when oxygen ionosorption is not expected to take place any more, CO increases conductance.

Zhang *et al.* were able to estimate surface coverage with ionosorbed oxygen θ – that is the fraction of nanowire surface covered with ionosorbed oxygen – as a function of V_{GS}. The rate and extent of oxygen ionosorption and the resulting rate and extent of catalytic CO oxidation reaction on the nanowire surface could be controlled and even entirely halted by applying a sufficiently negative gate potential.

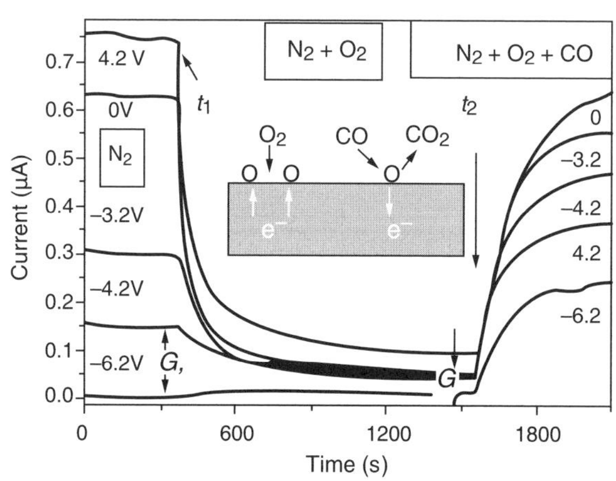

Figure 17.16 *Current kinetic response of a SnO_2 based SNT to the addition of 10 sccm of O_2 to 100 sccm of flowing N_2 gas at time t_1 and the addition of 5 sccm of CO at time t_2 as a function of gate voltage ($V_{DS} = 0.2$ V, $T = 553$ K). (Reprinted with permission from ref. [46]. Copyright 2004 American Chemical Society)*

17.4.5 Optical Sensing with SnO_2 Nanowires

Single crystalline nanowires of SnO_2 show a photoluminescence (PL) spectrum at room temperature with a broad emission band in the visible range from 400 to 600 nm. The origin of this emission is still debated in the literature [49–51]: it cannot be assigned to band-edge or to exciton recombination once one considers the wide band gap energy of SnO_2. Thus, transitions involving defective states within the band gap have to be invoked. The role of radiative states for the visible PL emission is commonly attributed to states arising from oxygen vacancies [51, 52].

In 2005 Faglia *et al.* [53] confirmed that SnO_2 nanowires exhibit interesting gas-sensitive PL properties, since exposure to low concentrations of NO_2 (few ppm concentrations) reversibly quenched the visible PL emission of nanowires at 120 °C. The interest in an all-optical gas sensor is great because electrical contacts are not needed. NO_2 adsorbs over the surface creating competitive nonradiative recombination paths: this quenches PL emission without peak shift. The kinetics of the involved processes was analyzed by taking the PL spectrum every 5 s and calculating the area under the broad PL band peak for every sampling. The response time needed to reach 90% of the steady state step response is equal to about 30 s. The adsorption process is reversible: 90% of PL baseline is recovered after restoring air in about 600 s. Moreover, the amount of quenching is not influenced by the relative humidity, which is maintained constant at 0, 30 and 70% during measurements. Therefore, a very important feature is that NO_2 can be detected at low concentration with almost no interference from water vapor. Besides, the authors investigated the effect on PL of NH_3 (50 ppm) and CO (1000 ppm) without observing any appreciable quenching effect.

Lettieri *et al.* [54] investigated the same PL quenching as a function of temperature, demonstrating that bigger quenching is observed at room temperature. They studied in detail the mechanism of PL quenching by NO_2 adsorption with continuous wave (CW) time resolved photoluminescence (TRPL) [55]. The ratio of PL area in gas over PL area in air and the relative variation of area under the PL peak (Figure 17.17) can be plotted versus NO_2 concentration. While the decrease in PL intensity is linearly proportional to surface density of adsorbed molecules (assuming a Langmuir-like adsorption of gas molecule on the semiconductor surface), the recombination rates are not significantly affected by interaction

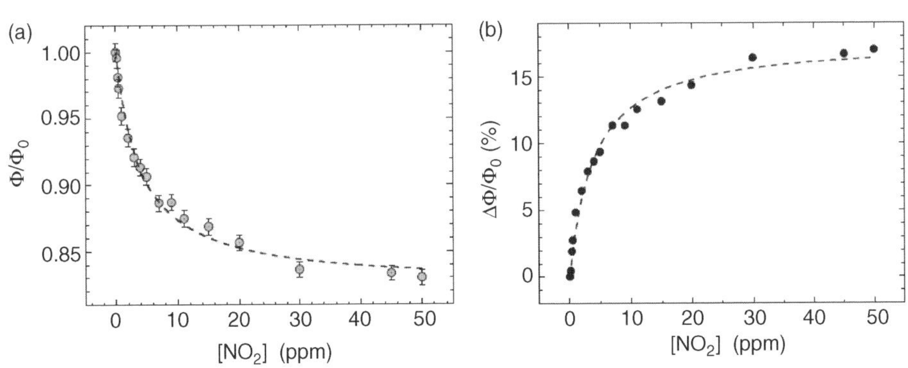

Figure 17.17 *(a) Quenching ratio and (b) relative variation of the PL yield as a function of the NO_2 concentration. (Reprinted with permission from ref. [55]. Copyright 2008 Institute of Physics)*

with NO_2. This suggests that NO_2 acts as a static quencher, determining a change in the number of states acting as radiative recombination centers.

17.5 Chemoresistive Sensing Properties of In_2O_3 Nanowires

Typically In_2O_3 was used as polycrystalline thin or thick film for detection of oxidizing gases. As a nanowire it is quite difficult to obtain due to its cubic cell. Probably this is the reason why there are less articles describing its gas sensing application with respect to those on SnO_2 [15, 56–60]. Single nanowires and bundles of In_2O_3 nanowires were used as gas sensors. As in the case of SnO_2, the diameter of the wires is the parameter that controls the sensor's performance. For smaller wires an increase in surface to volume ratio is observed and this explains the increased sensor response for devices composed of a bundle of nanowires.

Kaur *et al.* [59] reported the use of single crystalline whiskers of very large size (100–300 μm in diameter) for H_2 and NO gas sensing at room temperature. Even if the dimension of the wire is huge, which also means high conductance of the wire (current was of the order of 10 ohm), space charge induced by gas adsorption at the surface of the whisker is sufficient to give a response to 200 ppb of H_2S at room temperature. As seen in Figure 17.18, response to H_2S saturates for a concentration higher than 10 ppm. The authors pointed out also that different wires gave different responses, a well-known problem that can be ascribed to differences in dimension of the wires and to differences in the contact of the devices. As the sensors were operated at room temperature, recovery of the initial conductance is a major concern.

Ryu *et al.* [57] realized a device with many single In_2O_3 whiskers deposited over a hotplate with interdigitated contacts for detection of H_2, ethanol and CO down to

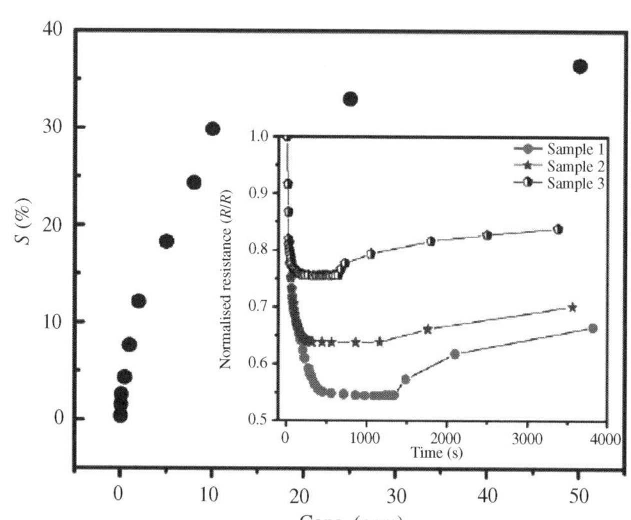

Figure 17.18 Response to H_2S of a single crystal whisker and (inset) response–recovery curves of three whiskers to 10 ppm of H_2S. (Reprinted with permission from ref. [59]. Copyright 2008 American Chemical Society)

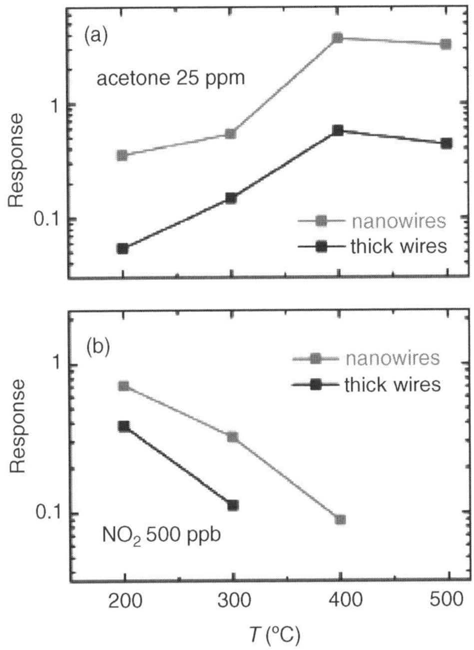

Figure 17.19 Response of In_2O_3 nanowires and thick wires to (a) 25 ppm of acetone and (b) 500 ppb of NO_2. (Reprinted with permission from ref. [15]. Copyright 2007 American Chemical Society)

concentrations of 1, 10 and 50 ppm, respectively. The use of a micromachined hotplate gave many advantages in terms of power consumption of the device, as already known for conventional sensors based on thin film metal oxides. Response to ethanol, CO and H_2 diluted in dry air peaked at 275 °C.

Vomiero *et al.* [15] tested sensing capabilities of devices with different average size of the wires in the bundle, grown by the evaporation-condensation method. The electrical conductance was lower for thinner wires (about 100 nm). The results on acetone and NO_2 sensing showed that the thin nanowires feature a response higher than that of thick wires (about 500 nm), as reported in Figure 17.19. Measurements were performed in humid air, and the lowest concentration used for NO_2 (500 ppb) was very low, comparable with that used with a conventional thick or thin film gas sensor.

Other authors [56, 60] reported the use of indium oxide nanorods mixed with adhesive paste to detect ethanol. The authors also dealt with different gases (NH_3, benzene, toluene, methane and gasoline) with lower response.

17.5.1 Transistor Devices Based on a Single In_2O_3 Nanowire

The first observed effect of the gaseous environment on a single In_2O_3 nanowire transistor has been the shift of V_{GS} necessary to induce the conducting channel inside the nanowires: Zhou and colleagues [61, 62] showed that at room temperature V_T of a SNT increases following the introduction of NO_2, as shown in Figure 17.20.

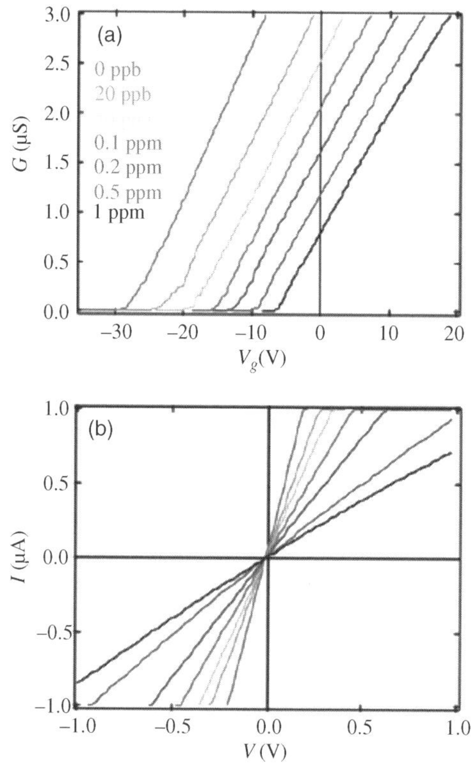

Figure 17.20 (a) Conductivity of a In_2O_3 SNT as a function of gate voltage at growing concentration (left to right) of NO_2. (b) Current–voltage curves at $V_{GS} = 0$. (Reprinted with permission from ref. [62]. Copyright 2004 American Chemical Society)

Results are in agreement with NO_2 ionosorption over In_2O_3 capturing electrons from the conduction band. As a consequence, a greater threshold gate voltage V_T is required in order to build up the majority carrier accumulation conductive (see Figure 17.20) channel.

In addition, Zhou and colleagues observed a relationship between the surface treatment, nanowire doping concentration and electrical response in the presence of NH_3 [63, 64], a species which is known to inject electrons in the conduction band when adsorbed. For In_2O_3 nanowires whose surfaces have been cleaned via UV illumination in vacuum, V_T has been found to decrease for lightly doped nanowires and to increase for heavily doped nanowires upon exposure. In contrast, in air V_T has been consistently observed to decrease upon NH_3 exposure, regardless of the doping concentration.

17.5.2 Chemical Warfare Agents Detected by Indium Oxide

Similarly to Section 17.4.3, this section is dedicated to reviewing the use of indium oxide nanowires to detect CWAs.

In Ponzoni *et al.* [40], indium oxide nanowire mesh as well as indium oxide thin films were used to detect different chemicals, including CWA simulants. DMMP, acetonitrile and DPGME were used as simulants for Sarin nerve gas, cyanide compounds and blistering

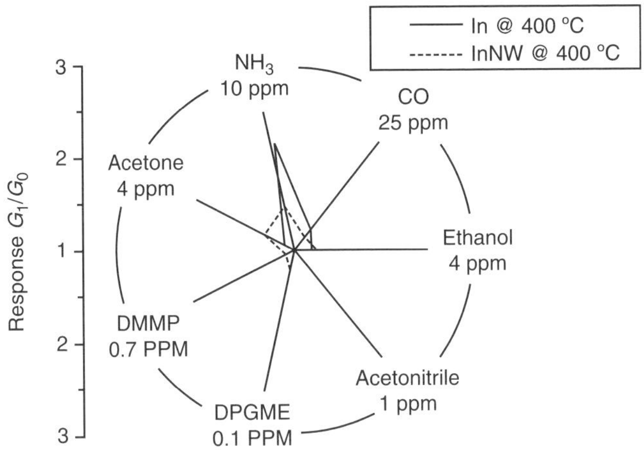

Figure 17.21 *Response to different chemicals exhibited by indium oxide thin film (In, continuous line) and nanowire mesh (InNW, dashed line) heated at 400 °C. (Reprinted with permission from ref. [40]. Copyright 2008 IEEE)*

gases, respectively. Ethanol, CO. NH_3 and acetone were used as false-alarm gases that could arise by everyday activities.

Results indicated that both indium oxide nanowires and thin films are able to sense DPGME and acetonitrile at concentrations lower than the IDLH value of the respective CWAs, corresponding to a few tens of ppm and a few ppm, respectively [65]. The detection of DMMP at concentrations close to the Sarin IDLH value (30 ppb) turned out to be more difficult: indium oxide nanowires exhibit appreciable response down to 200 ppb, while the response of indium oxide thin films vanishes at a DMMP concentration as low as 500 ppb. The comparison between the sensing performances of the two materials is reported in Figure 17.21. It is evident that these materials feature different specific sensitivity: nanowires were more sensitive to gases such as DMMP, DPGME and acetone, while thin films showed higher sensitivity to NH_3 and CO. These results led authors to argue that the different performance of these materials cannot simply be ascribed to the low lateral dimensions of nanowires, otherwise they would be more sensitive than thin films to each compound. A variety of phenomena have been invoked, including the different adsorption mechanisms that may occur over the nanowire and thin film surfaces. The former feature a high degree of crystallinity, corresponding to well defined crystalline planes and featuring atomic sharp termination while the latter are composed of rounded grains, thus exposing different crystallographic planes to the gaseous environment.

References

[1] S. R. Morrison, *The Chemical Physics of Surfaces*, Plenum Press, New York (1978).
[2] T. Wolkenstein, *Electronic Processes on Semiconductor Surfaces during Chemisorption*, Plenum Press, New York (1991).

[3] J. W. Orton and M. J. Powell, The Hall effect in polycrystalline and powdered semiconductors, *Rep. Progr. Phys.*, **43**, 1265 (1980).
[4] N. Barsan and U. Weimar, Conduction model of metal oxide gas sensors, *J. Electroceram.*, **7**, 143 (2001).
[5] A. D'Amico and C. Di Natale, A contribution on some basic definitions of sensor properties, *IEEE Sensors J.*, **1**, 183 (2001).
[6] E. Comini, G. Faglia, G. Sberveglieri, Z. Pan and Z. Wang, Stable and highly sensitive gas sensors based on semiconducting oxide nanobelts, *Appl. Phys. Lett.*, **81**, 1869–1871 (2002).
[7] E. Comini, S. Bianchi, G. Faglia, M. Ferroni, A. Vomiero and G. Sberveglieri, Functional nanowires of tin oxide, *Appl. Phys. A*, **89**, 73–76 (2007).
[8] Z. Ying, Q. Wan, Z. T. Song and S. L. Feng, SnO_2 nanowhiskers and their ethanol sensing characteristics, *Nanotech.*, **15**, 1682–1684 (2004).
[9] S. Mathur, S. Barth, H. Shen, J. Pyun and U. Werner, Size-dependent photoconductance in SnO_2 nanowires, *Small*, **1**, 713–717 (2005).
[10] A. Formalas,US patent 1 975 504 (1934).
[11] Y. Zhang, X. He, J. Li, Z. Miao and F. Huang, Fabrication and ethanol-sensing properties of micro gas sensor based on electrospun SnO_2 nanofibers, *Sens. Actuators, B*, **132**, 67–73 (2008).
[12] Y. Zhang, A. Kolmakov, Y. Lilach and M. Moskovits, Electronic control of chemistry and catalysis at the surface of an individual tin oxide nanowire, *J. Phys. Chem. B*, **109**, 1923–1929 (2005).
[13] J. Lao, J. Huang, D. Wang and Z. Ren, Self-assembled In_2O_3 nanocrystal chains and nanowire networks, *Adv. Mater.*, **16**, 65–69 (2004).
[14] J. Zhang, X. Qing, F. Jiang and Z. Dai, A route to Ag-catalyzed growth of the semiconducting In_2O_3 nanowires, *Chem. Phys. Lett.*, **371**, 311–316 (2003).
[15] A. Vomiero, S. Bianchi, E. Comini, G. Faglia, M. Ferroni and G. Sberveglieri, Controlled growth and sensing properties of In_2O_3 nanowires, *Cryst. Growth, Des.*, **7**, 2500–2504 (2007).
[16] K. C. Kam, F. L. Deepak, A. K. Cheetham and C. N. R. Rao, In_2O_3 nanowires, nanobouquets and nanotrees, *Chem. Phys. Lett.*, **397**, 329–334 (2004).
[17] X. S. Peng, G. W. Meng, J. Zhang, X. F. Wang, Y. W. Wang, C. Z. Wang and L. D. Zhang, Synthesis and photoluminescence of single-crystalline In_2O_3 nanowires, *J. Mater. Chem.*, **12**, 1602–1605 (2002).
[18] P. Nguyen, H. T. Ng, T. Yamada, M. K. Smith, J. Li, J. Han and M. Meyyappan, Direct integration of metal oxide nanowire in vertical field-effect transistor, *Nano Lett.*, **4**, 651–657 (2004).
[19] M. S. Chen and D. W. Goodman, The structure of catalytically active gold on titania, *Science*, **306**, 252 (2004).
[20] J. M. C. Soares, P. Morrall, A. Crossley, P. Harris and M. Bowker, Catalytic and noncatalytic CO oxidation on Au/TiO_2 catalysts, *J. Catal.*, **219**, 17 (2003).
[21] A. Vomiero, M. Ferroni, E. Comini, G. Faglia and G. Sberveglieri, Preparation of radial and longitudinal nanosized heterostructures of In_2O_3 and SnO_2, *Nano Lett.*, **7**, 3553–3558 (2007).
[22] Q. Wan, E. N. Dattoli, W. Y. Fung, W. Guo, Y. Chen, X. Pan and W. Lu, High-performance transparent conducting oxide nanowires, *Nano Lett.*, **12**, 2909–2915 (2006).
[23] D.-W. Kim, I.-S. Hwang, S. J. Kwon, H.-Y. Kang, K.-S. Park, Y.-J. Choi, K.-J. Choi and J.-G. Park, Highly conductive coaxial SnO_2-In_2O_3 heterostructured nanowires for Li ion battery electrodes, *Nano Lett.*, **7**, 3041–3045 (2007).
[24] P. Nguyen, S. Vaddiraju and M. Meyyappan, Indium and tin oxide nanowires by vapor-liquid-solid growth technique, *J. Electron. Mater.*, **35**, 200–206 (2006).
[25] F. Hernández-Ramírez, D. Prades, A. Tarancón, S. Barth, O. Casals, R. Jimenez-Diaz, E. Pellicer, J. Rodriguez, J. R. Morante, M. A. Juli, S. Mathur and A. Romano-Rodriguez, Insight into the role of oxygen diffusion in the sensing mechanisms of SnO_2 nanowires, *Adv. Funct. Mater.*, **18**, 1 (2008).
[26] F. Hernández-Ramírez, A. Tarancón, A. Romano-Rodríguez and J. R. Morante, Electrical contacts and gas sensing analysis of individual metal oxide nanowires and 3-D nanocrystal networks, *IEEE J. Trans. Sens. Micromach.*, **126**, 537–547 (2006).

[27] F. Hernández-Ramírez, A. Tarancón, A. Romano-Rodríguez, O. Casals, J. Arbiol and J. R. Morante, High response and stability in CO and humidity measures using a single SnO_2 nanowire, *Sens. Actuators, B*, **121**, 3–17 (2007).
[28] F. Hernandez-Ramirez, S. Barth, A. Tarancon, O. Casals, E. Pellicer, J. Rodriguez, R. Romano-Rodriguez, J. R. Morante, S. Barth and S. Mathur, Water vapor detection with individual tin oxide nanowires, *Nanotech.*, **18**, 424016 (2007).
[29] D. C. Meier, S. Semancik, B. Button, E. Strelcov and A. Kolmakov, Coupling nanowire chemiresistors with MEMS microhotplate gas sensing platforms, *Appl. Phys. Lett.*, **91**, 063118 (2007).
[30] M. Law, H. Kind, B. Messer, F. Kim and P. Yang, Photochemical sensing of NO_2 with SnO_2 nanoribbon nanosensors at room temperature, *Angew. Chem. Int. Ed.*, **41**, 2405–2408 (2002).
[31] A. Kolmakov, The effect of morphology and surface doping on sensitization of quasi-1D metal oxide nanowire gas sensors, *Proc. SPIE* 6370:63700X1-8 (2006).
[32] Q. Kuang, C. S. Lao, Z. Li, Y. Z. Liu, Z. X. Xie, L. S. Zheng and Z. L. Wang, *J. Phys. Chem.*, **112**, 11539–11544 (2008).
[33] G. Sberveglieri, C. Baratto, E. Comini, G. Faglia, M. Ferroni, A. Ponzoni and A. Vomiero, Synthesis and characterization of semiconducting nanowires for gas sensing, *Sens. Actuators, B*, **121**, 208–213 (2007).
[34] Z. Ying, Q. Wan, Z. T. Song and S. L. Feng, SnO_2 nanowhiskers and their ethanol sensing characteristics, *Nanotech.*, **15**, 1682–1684 (2004).
[35] T. H. Jung, S. I. Kwon, J. H. Park, D. G. Lim, Y. J. Choi and J. G. Park, SnO_2 nanowires bridged across trenched electrodes and their gas sensing characteristics, *Appl. Phys. A*, **91**, 707–710 (2008).
[36] Y. Sun and K. Y. Ong, *Detection Technologies for Chemical Warfare Agents and Toxic Vapors*, CRC Press, Boca Raton (2005).
[37] A. A. Tomchenko, G. P. Harmer and B. T. Marquis, Detection of chemical warfare agents using nanostructured metal oxide sensors, *Sens. Actuators, B*, **108**, 41 (2005).
[38] E. Brunol, F. Berger, M. Fromm and R. Planade, Detection of dimethyl methylphosphonate (DMMP) by tin dioxide-based gas sensor: response curve and understanding of the reactional mechanism, *Sens. Actuators, B*, **120**, 35–41 (2006).
[39] L. Cao, S. R. Segal, S. L. Suib, X. Tang and S. Satyapal, Thermocatalytic oxidation of dimethyl methylphosphonate on supported metal oxides, *J. Catal.*, **194**, 61–70 (2000).
[40] A. Ponzoni, C. Baratto, S. Bianchi, E. Comini, M. Ferroni, M. Pardo, M. Vezzoli, A. Vomiero, G. Faglia and G. Sberveglieri, Metal oxide nanowire and thin-film based gas sensors for chemical warfare simulants detection, *IEEE Sensors J.*, **8**, 735–742 (2008).
[41] C. Yu, Q. Hao, S. Saha, L. Shi, X. Y. Kong and Z. L. Wang, Integration of metal oxide nanobelts with microsystems for nerve agent detection, *Appl. Phys. Lett.*, **86**, 063101 (2005).
[42] G. Sberveglieri, Recent developments in semiconducting film gas sensors, *Sens. Actuators, B*, **23**, 103–109 (1995).
[43] C. D. Dimitrakopoulos and P. R. L. Malenfant, Organic thin film transistors for large area electronics, *Adv. Mater.*, **14**, 99 (2002).
[44] D. B. Mitzi, L. L. Kosbar, C. E. Murray, M. Copel and A. Afzali, High mobility ultrathin semiconducting films prepared by spin coating, *Nature*, **428**, 299 (2004).
[45] M. Bogner, A. Fuchs, K. Scharnagl, R. Winter, T. Doll and I. Eisele, Electrical field impact on the gas adsorptivity of thin metal oxide films, *Appl. Phys. Lett.*, **17**, 2524 (1998).
[46] Y. Zhang, A. Kolmakov, S. Chretien, H. Metiu and M. Moskovits, Control of catalytic reactions at the surface of a metal oxide nanowire by manipulating electron density inside it, *Nano Lett.*, **4**, 403 (2004).
[47] A. Kolmakov and M. Moskovits, Chemical sensing and catalysis by one-dimensional metal-oxide nanostructures, *Ann. Rev. Mater. Res.*, **34**, 151 (2004).
[48] Y. Zhang, A. Kolmakov, Y. Lilac and M. Moskovits, Electronic control of chemistry and catalysis at the surface of an individual tin oxide nanowire, *J. Phys. Chem. B*, **109**, 1923 (2005).
[49] J. Hu, Y. Bando, Q. Liu and D. Goldberg, Laser-ablation growth and optical properties of wide and long single-crystal SnO_2 ribbons, *Adv. Funct. Mater.*, **13**, 493–496 (2003).

[50] J. H. He, T. H. Wu, C. L. Sin, K. M. Li, L. J. Chen, Y. L. Chueh, L. J. Chou and Z. L. Wang, Beaklike SnO_2 nanorods with strong photoluminescent and field-emission properties, *Small*, **2**, 116–120 (2006).
[51] S. Luo, P. K. Chu, W. Liu, M. Zhang and C. Lin, Origin of low-temperature photoluminescence from SnO_2 nanowires fabricated by thermal evaporation and annealed in different ambients, *Appl. Phys. Lett.*, **88**, 183112 (2006).
[52] J. X. Zhou, M. S. Zhang, J. M. Hong and Z. Yin, Raman spectroscopic and photoluminescence study of single-crystalline SnO_2 nanowires, *Solid State Commun.*, **138**, 242–246 (2006).
[53] G. Faglia, C. Baratto, G. Sberveglieri, M. Zha and A. Zappettini, Adsorption effects of NO_2 at ppm level on visible photoluminescence response of SnO_2 nanobelts, *Appl. Phys. Lett.*, **86**, 011923 (2005).
[54] S. Lettieri, A. Bismuto, P. Maddalena, C. Baratto, E. Comini, G. Faglia, G. Sberveglieri and L. Zanotti, Gas sensitive light emission properties of tin oxide and zinc oxide nanobelts, *J. Non-Cryst. Solids*, **352**, 1457–1460 (2006).
[55] S. Lettieri, A. Setaro, C. Baratto, E. Comini, G. Faglia, G. Sberveglieri and P. Maddalena, On the mechanism of photoluminescence quenching in tin dioxide nanowires by NO_2 adsorption, *New J. Phys.*, **10**, 043013 (2008).
[56] C. Xianfeng, W. Caihong, J. Dongli and Z. Chenmou, Ethanol sensor based on indium oxide nanowires prepared by carbothermal reduction reaction, *Chem. Phys. Lett.*, **399**, 461–464 (2004).
[57] K. Ryu, D. Zhang and C. Zhou, High-performance metal oxide nanowire chemical sensors with integrated micromachined hotplates, *Appl. Phys. Lett.*, **92**, 093111 (2008).
[58] A. Ponzoni, C. Baratto, S. Bianchi, E. Comini, M. Ferroni, M. Pardo, M. Vezzoli, A. Vomiero, G. Faglia and G. Sberveglieri, Metal oxide nanowire and thin-film-based gas sensors for chemical warfare simulants detection, *IEEE Sensors J.*, **8**, 735 (2008).
[59] M. Kaur, N. Jain, K. Sharma, S. Bhattacharya, M. Roy, A. K. Yagi, S. K. Gupta and J. V. Yakhmi, Room-temperature H_2S gas sensing at ppb level by single crystal In_2O_3 whiskers, *Sens. Actuators, B*, **133**, 456–461 (2008).
[60] J. Xu, Y. Chen and J. Shen, Ethanol sensor based on hexagonal indium oxide nanorods prepared by solvothermal methods, *Mater. Lett.*, **62**, 1363–1365 (2008).
[61] C. Li, D. H. Zhang, X. L. Liu, S. Han, T. Tang, J. Han and C. W. Zhou, In_2O_3 nanowires as chemical sensors, *Appl. Phys. Lett.*, **82**, 1613 (2003).
[62] D. H. Zhang, Z. Q. Liu, C. Li, T. Tang, X. L. Liu, S. Han, B. Lei and C. W. Zhou, Detection of NO_2 down to ppb levels using individual and multiple In_2O_3 nanowire devices, *Nano Lett.*, **4**, 1919 (2004).
[63] D. J. Zhang, C. Li, X. L. Liu, S. Han, T. Tang and C. W. Zhou, Doping dependent NH_3 sensing of indium oxide nanowires, *Appl. Phys. Lett.*, **83**, 1845 (2003).
[64] C. Li, D. H. Zhang, B. Lei, S. Han, X. L. Liu and C. W. Zhou, Surface treatment and doping dependence of In_2O_3 nanowires as ammonia sensors, *J. Phys. Chem. B*, **107**, 12451 (2003).
[65] M. Utriainen, E. Karpanoja and H. Paakkanen, Combining miniaturized ion mobility spectrometer and metal oxide gas sensor for the fast detection of toxic chemical vapors, *Sens. Actuators, B*, **93**, 17–24 (2003).

Index

Note: page numbers in *italics* refer to figures and tables.

Transparent Electronics: From Synthesis to Applications Edited by Antonio Facchetti and Tobin J. Marks